Detection
of
Signals
in
Noise

SECOND EDITION

Detection

of

Signals

in

Noise

SECOND EDITION

Robert N. McDonough
The Johns Hopkins University
Applied Physics Laboratory
Laurel, Maryland

Anthony D. Whalen
AT&T Bell Laboratories
Whippany, New Jersey

Academic Press
An Imprint of Elsevier
San Diego   New York   Boston   London   Sydney   Tokyo   Toronto

Academic Press and the authors have expended great effort to ensure that the material presented in this volume is correct. However, neither Academic Press nor the authors guarantee accuracy and completeness of the material included herein, and neither Academic Press nor the authors shall be responsible for any errors, omissions, or damages arising from the use of this volume.

Academic Press, Inc.
An Imprint of Elsevier
525 B Street, Suite 1900, San Diego, California 92101-4495

United Kingdom Edition published by
Academic Press Limited
24-28 Oval Road, London NW1 7DX

Library of Congress Cataloging-in-Publication Data

McDonough, Robert N.
      Detection of signals in noise / by Robert N. McDonough, A. D.
Whalen. -- 2nd ed.
         p.   cm.
      Includes index.
      ISBN-13: 978-0-12-744852-7
      ISBN-10: 0-12-744852-7
         1. Signal detection.   2. Signal processing--Statistical methods.
      3. Random noise theory.   4. Signal processing--Digital techniques.
      I. Whalen, A. D.   II. Title.
      TK5102.9.M39   1995
      621.382'24--dc20   94-24910
      ISBN-13: 978-0-12-744852-7   CIP
      ISBN-10: 0-12-744852-7

Transferred to Digital Printing 2007

R.N.M. dedicates his contributions to this work to his wife, Natalia, in appreciation of her patience, understanding, and encouragement, and to their children Paul and Stefan.

A.D.W. dedicates his contribution to this work to all of the people who, over a period of 20 years, kept the first edition alive by using it in their academic, intellectual, and professional life. An author could not have asked for higher satisfaction.

Contents

6 *Detection of Known Signals*

7 *Detection of Signals with Random Parameters*

8 Multiple Pulse Detection of Signals

9 Detection of Signals in Colored Gaussian Noise

10 Estimation of Signal Parameters

11 *Extensions*

Preface

The cordial reception given to the first edition of this work encouraged the publisher and the original author (A.D.W.) to plan a second edition. The extent of developments in the practice of signal detection since the appearance of the first edition dictated a thorough revision. A new author (R.N.M.) was therefore invited to participate in the project, and the present volume is the result. Much new material has been added, although the scope of the topics considered and the philosophy of their treatment have not changed. The additions are intended to add background and, in some cases, generality to the treatments of the first edition.

Our aim is unchanged and can be summarized in a quote from the Preface to the first edition:

> This book is intended to serve as an introduction to the principles and applications of the statistical theory of signal detection. Emphasis is placed on those principles which have been found to be particularly useful in practice. These principles are applied to detection problems encountered in digital communications, radar, and sonar. In large part the book has drawn upon the open literature: texts, technical journals, and industrial and university reports. The author[s'] intention was to digest this material and present a readable and pedagogically useful treatment of the subject.

The topics covered are delineated in the Table of Contents. In comparison with the first edition, they are treated in more 'cases using discrete formulations, in addition to the formulations in terms of continuous time signals that characterized the first edition. Because of the wide application of the fast Fourier transform as a system element, a significant part of the development considers Fourier coefficients as the observables to be processed. Because of this, and because of the convenient calculation of the complex envelope of a signal that is now possible, much of the present development is phrased specifically for the processing of complex observables. However, the treatment of real discrete observables, such as samples of real time functions, is not neglected.

In our treatments of hypothesis testing in Chapter 5 and of estimation theory in Chapter 10, we have made the developments somewhat more general than in the first edition. This is intended to make clear the limitations of the methods discussed, as well as to point out extensions.

Since the publication of the first edition, essentially all of the material that we discuss has migrated from the original journal articles and reports into texts and monographs. We have therefore made fewer citations to the original literature than earlier. Rather, we have appended a bibliography of standard works and have occasionally made specific citations where needed in the text.

The first edition grew out of a Bell Telephone Laboratories in-house course presented by A.D.W. Thus, the book should be suitable for a course in signal detection, at either the graduate or the advanced undergraduate level. Beyond that, we have strongly aimed the book at practicing engineers who need a text for either self-study or reference. Throughout the text, equations of particular importance have been indicated with an asterisk before the equation number.

The preparation of this second edition was supported in part by The Johns Hopkins University Applied Physics Laboratory, through the granting of a Stuart S. Janney Fellowship to R.N.M. In addition, both authors are indebted to their respective organizations for sustained opportunities over the years to work and learn in ways that made it possible for us to produce this book. The friendly help of Jacob Elbaz in preparing the illustrations is gratefully acknowledged.

1 Probability

*E*ngineering is often regarded as a discipline which deals
in precision. A well-ordered nature evolves in ways
which are in principle perfectly measurable and completely predictable.
On the contrary, in the problems that will concern us in this book, nature
presents us with situations which are only partially knowable through
our measurements, which evolve in ways that are not predictable with
certainty, and which require us to take actions based on an imperfect
perception of their consequences. The branch of applied mathematics
called probability theory provides a framework for the analysis of such
problems. In this chapter we will give a brief summary of the main ideas
of probability theory which will be of use throughout the remainder of
the book.

1.1 PROBABILITY IN BRIEF

Probability can be discussed at the intuitive level that deals with the events observed in everyday life. If a coin is tossed many times, it is noted that heads and tails appear about equally. We assign a "probability" of 0.5 to the appearance of a head on any particular toss and proceed to calculate the probability of, say, 3 heads in 5 tosses. Generality and clarity of thought follow if we introduce a more mathematical framework for the subject of probability. This was developed in the 1930s. We will summarize that view here, in order to introduce the main elements of the theory and some terminology.

The subject of probability is concerned with abstract experiments, each defined by a triple of elements $\{\mathscr{S}, \sigma, P\}$. The first is the set $\mathscr{S}$ of the *elementary outcomes* of the experiment. These are defined such that exactly one of them occurs each time the experiment is performed. The second element is a set σ of *events,* each of which can be said to have occurred or not to have occurred depending on which elementary outcome occurs in a performance of the experiment. The third element, P, is a rule of assignment of a number (its *probability*) to each event of the set σ.

The construction of an *abstract experiment* begins by specifying all its elementary outcomes. For example, the rolling of one of a pair of idealized perfect dice (a die) is an abstract experiment. The elementary outcomes could be that the die lands with one of the numbers 1 through 6 on top. Regardless of the experiment, in a case such as this, in which the elementary outcomes are countable, we will label them by the symbols ζ_i, $i = 1, 2, \ldots$. On the other hand, if the abstract experiment is such that its elementary outcomes are not countable, we will label them by a continuous variable ω, as $\zeta(\omega)$. Such an experiment might be, for example, the measurement (with infinite precision) of a voltage which can have any real number for its value. The definition of the elementary outcomes of an experiment is entirely arbitrary, except that they must be complete and mutually exclusive; that is, exactly one of them must occur at each performance of the experiment.

The elementary outcomes $\zeta \in \mathscr{S}$ (where $\in$ means "is an element of the set") of the experiment $\{\mathscr{S}, \sigma, P\}$ can be considered as a collection of points in a point set space, the *sample space* $\mathscr{S}$ of the experiment. Although the space $\mathscr{S}$ can be abstractly defined without reference to points, the language of point sets provides a convenient way to visualize the outcomes ζ as points in a geometric space. In the case of a countable set of outcomes ζ_i, the points are discrete, whereas with a continuum of outcomes $\zeta(\omega)$ the points are visualized as being in completely filled regions of the space.

From the elementary outcomes $\zeta \in \mathscr{S}$ of an experiment we can then construct events, which are defined as sets of elementary outcomes.

Events are generally denoted by script letters. For example, in the idealized die-throwing experiment, some events are:

$\mathcal{A}$: Either the number 2 or the number 5 shows uppermost.
$\mathcal{B}$: The uppermost number is even.
$\mathcal{C}$: The number 5 shows uppermost.
$\mathcal{D}$: A number less than 5 shows uppermost.
$\mathcal{S}$: Some number shows uppermost.

These events can also be described by the set of index numbers of the elementary outcomes they comprise:

$$
\begin{aligned}
\mathcal{A} &= \{2, 5\}, \\
\mathcal{B} &= \{2, 4, 6\}, \\
\mathcal{C} &= \{5\}, \\
\mathcal{D} &= \{1, 2, 3, 4\}, \\
\mathcal{S} &= \{1, 2, 3, 4, 5, 6\}.
\end{aligned}
\tag{1.1}
$$

Note that we formally distinguish between the single elementary outcome (point), say ζ_5, and the event (set) $\{\zeta_5\}$ comprising that single outcome.

An event is said to occur if any one of the elementary outcomes constituting the event occurs. The set $\mathcal{S}$ of all outcomes, the sample space, is also called the *certain event,* because by definition it must occur in any trial of the experiment.

New events can be defined from old events by introducing rules of combination. These are exactly the rules of point set theory. The sum (union, logical OR) and product (intersection, logical AND) of two events are introduced:

$$
\begin{aligned}
\mathcal{C} &= \mathcal{A} + \mathcal{B}, \\
\mathcal{C} &= \mathcal{A}\mathcal{B}.
\end{aligned}
$$

These are defined respectively as the set of the elementary outcomes (points) which belong to either $\mathcal{A}$ or $\mathcal{B}$ or to both (Fig. 1.1a) and the set of elementary outcomes which belong to both $\mathcal{A}$ and $\mathcal{B}$ (Fig. 1.1b). We also introduce the difference of two events, $\mathcal{A} - \mathcal{B}$, in that order, defined as all the outcomes contained in the event $\mathcal{A}$ which are not also contained in the event $\mathcal{B}$ (Fig. 1.1c). We will agree always to consider the certain event $\mathcal{S}$ as a member of any set of events we discuss. Then we can introduce the *complement* (negation) of an event as

$$
\overline{\mathcal{A}} = \mathcal{S} - \mathcal{A},
$$

which is the set of all the elementary outcomes not contained in the event $\mathcal{A}$ (Fig. 1.1d). It is then necessary also to include the *null event* $\Phi = \mathcal{S} - \mathcal{S}$, which is the event consisting of no elementary outcomes. Finally, we will mention the *symmetric difference* of two sets (Fig. 1.1e), defined as

$$
\mathcal{A} \, \Delta \, \mathcal{B} = (\mathcal{A} - \mathcal{B}) + (\mathcal{B} - \mathcal{A}).
$$

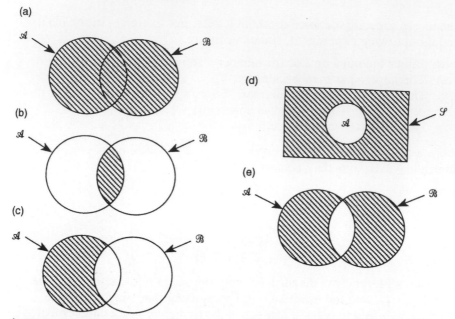

FIGURE 1.1   Combinations of events (a) Union $\mathcal{A} + \mathcal{B}$; (b) intersection $\mathcal{A}\mathcal{B}$; (c) difference $\mathcal{A} - \mathcal{B}$; (d) negation $\overline{\mathcal{A}}$; (e) symmetric difference $\mathcal{A} \triangle \mathcal{B}$.

The second element σ of an abstract experiment $\{\mathcal{S}, \sigma, P\}$ can now be defined. We define the set σ of events as a collection of events which is closed under (possibly infinite) summation and under negation. That is, for any event $\mathcal{A} \in \sigma$ we require that $\overline{\mathcal{A}} \in \sigma$, and for any collection of $\mathcal{A}_i \in \sigma$, we require that

$$\sum_{i=1}^{\infty} \mathcal{A}_i \in \sigma,$$

where the summation is in the sense of the union operation. These two postulates about σ are sufficient to guarantee that σ also contains all the (possibly infinite) products (intersections) of its members and all symmetric differences of its members.

In algebra, any set of elements with the properties assumed above for the set σ of events is called a σ-algebra. In an older terminology, specific to probability theory, the term *field of probability* or *σ-field* is used. That invites confusion, however, because a "field" is a collection whose elements have properties different from those of an algebra. The term is still in use, however.

A σ-algebra of events thus defined on elements of the sample space $\mathcal{S}$ is such that any events constructed by set operations on events of σ will also be members of σ. The specification of such a set σ is a technical matter, however, because in applications the events of interest are always clear and can always be imbedded in a σ-algebra. It is enough to know that the possibility exists, because we are interested only in the events themselves and usually do not need to characterize the σ-algebra explicitly.

The third element of the abstract experiment $\{\mathscr{S}, \sigma, P\}$ is a real-valued function $P(\mathscr{A})$ defined over the events $\mathscr{A}$ of the set σ. It assigns a *probability* to each event $\mathscr{A}$ of a σ-algebra σ. The number $P(\mathscr{A})$ reflects the extent of our belief that, if the abstract experiment were performed, event $\mathscr{A}$ would occur. For example, for the events Eq. (1.1) of the die-rolling experiment, we might assign:

$$P(\mathscr{A}) = 1/3,$$
$$P(\mathscr{B}) = 1/2,$$
$$P(\mathscr{C}) = 1/6,$$
$$P(\mathscr{D}) = 2/3,$$
$$P(\mathscr{S}) = 1.$$

These assignments would indicate we believe the die to be "fair."

Such an assignment of probabilities cannot be made arbitrarily. For example, we feel that "something is wrong" if we assign probability 1/6 to each face of a fair die and probability 2/3 to rolling an even number. In general, the assignment of probabilities P to the sets $\mathscr{A}$ of σ must be such that the following axioms are fulfilled:

$$P(\mathscr{A}) \geq 0, \qquad \mathscr{A} \in \sigma;$$
$$P(\mathscr{S}) = 1;$$
$$P(\textstyle\sum \mathscr{A}_i) = \sum P(\mathscr{A}_i) \quad \text{for } \mathscr{A}_i \in \sigma \text{ and } \mathscr{A}_i \mathscr{A}_j = \Phi, i \neq j. \qquad *(1.2)$$

(In this last, the summation of events is the union, and infinite sums are allowed. The events are to be disjoint, with no elementary outcomes in common.)

From the axioms Eq. (1.2) there follow all the usual properties of probability. For example, because any event $\mathscr{A}$ and the event $\overline{\mathscr{A}}$ (that $\mathscr{A}$ does not occur) satisfy

$$\mathscr{A} + \overline{\mathscr{A}} = \mathscr{S}, \qquad \mathscr{A}\overline{\mathscr{A}} = \Phi,$$

we have from Eq. (1.2) that

$$P(\mathscr{A} + \overline{\mathscr{A}}) = P(\mathscr{A}) + P(\overline{\mathscr{A}}) = P(\mathscr{S}) = 1,$$

so that

$$P(\overline{\mathscr{A}}) = 1 - P(\mathscr{A}).$$

In the special case of this last with $\mathscr{A} = \mathscr{S}$, the certain event, we have $\overline{\mathscr{A}} = \mathscr{S} - \mathscr{S} = \Phi$, and the probability of the null event is

$$P(\Phi) = 1 - P(\mathcal{S}) = 0.$$

The axioms Eq. (1.2) of the probability function $P(\mathcal{A})$ suffice to build up the probabilities of any events $\mathcal{A} \in \sigma$ from those assigned to the elementary outcomes $\zeta \in \mathcal{S}$. That latter assignment, however, can be done in many ways. The specifics depend on the aspects of the physical situation being modeled. For example, in the case of a fair die, we take $P(\{\zeta_i\}) = 1/6$, $i = 1, 6$. There then follow, for example,

$$P(\{\text{even}\}) = P(\{2\}) + P(\{4\}) + P(\{6\}) = 1/2,$$
$$P(\{\geq 3\}) = P(\{3\}) + \cdots + P(\{6\}) = 2/3,$$
$$P(\{\text{even}\} \text{ AND } \{\geq 3\}) = P(\{4\}) + P(\{6\}) = 1/3,$$
$$P(\{\text{even}\} \text{ OR } \{\geq 3\}) = P(\{2\}) + \cdots + P(\{6\}) = 5/6.$$

All the various manipulations of set algebra are useful. For example, from the identities

$$\mathcal{A} + \mathcal{B} = \mathcal{A} + (\overline{\mathcal{A}}\mathcal{B}),$$
$$\mathcal{B} = \mathcal{S}\mathcal{B} = (\mathcal{A} + \overline{\mathcal{A}})\mathcal{B} = (\mathcal{A}\mathcal{B}) + (\overline{\mathcal{A}}\mathcal{B}),$$

and from the axioms of the probability function, it follows that for any events $\mathcal{A}$, $\mathcal{B}$

$$P(\mathcal{A} + \mathcal{B}) = P(\mathcal{A}) + P(\mathcal{B}) - P(\mathcal{A}\mathcal{B}). \qquad *(1.3)$$

In particular, in the fair die experiment,

$$P(\{\text{even}\} \text{ OR } \{\geq 3\})$$
$$= P(\{\text{even}\}) + P(\{\geq 3\}) - P(\{\text{even}\} \text{ AND } \{\geq 3\})$$
$$= 1/2 + 2/3 - 1/3 = 5/6.$$

With the final axioms Eq. (1.2), the experiment $\{\mathcal{S}, \sigma, P\}$ is well defined and amenable to analysis. The extent to which such an abstract experiment $\{\mathcal{S}, \sigma, P\}$ relates to a specific engineering question at hand is not part of the study of probability theory. Rather, it is a crucial aspect of the application of probability theory to real problems. In this chapter, and indeed in much of the book, we will be concerned only with idealized experiments and with the use of probability theory to make statements about them.

1.2 CONDITIONAL PROBABILITY AND STATISTICAL INDEPENDENCE

We will often have need to calculate the probability of some event $\mathcal{A}$ assuming the knowledge that another event $\mathcal{B}$ has occurred. This is called the *conditional probability* of the event $\mathcal{A}$ given the event $\mathcal{B}$ and is defined as

$$P(\mathcal{A} \mid \mathcal{B}) = P(\mathcal{A}\mathcal{B})/P(\mathcal{B}), \qquad P(\mathcal{B}) \neq 0. \qquad *(1.4)$$

If $P(\mathcal{B}) = 0$, the quantity indicated is not defined. The events $\mathcal{A}$, $\mathcal{B}$ must belong to the same space σ of events. A typical application is to the case that $\mathcal{A}$ is the selection of a particular message for transmission over a channel, while $\mathcal{B}$ is the reception of data resulting from operation of the channel on the message. We will want to compute the probabilities of various messages $\mathcal{A}$ having been sent in the case that we have received a particular set $\mathcal{B}$ of data.

EXAMPLE 1.1 Suppose that we roll a fair die. Let the event $\mathcal{A} = \{1, 3, 5\}$ be that an odd number comes up, and let $\mathcal{B} = \{1, 2, 3\}$ be the event that a number less than 4 comes up. The combined event $\mathcal{A}\mathcal{B} = \{1, 3\}$ has probability 1/3. Also, $P(\mathcal{B}) = 1/2$. Then, from Eq. (1.4), $P(\mathcal{A} \mid \mathcal{B}) = (1/3)/(1/2) = 2/3$, which is the probability of the event $\mathcal{A} = \{\text{odd}\}$, given that the event $\mathcal{B} = \{1, 2, 3\}$ has occurred. The conditioning reduces the sample space $\mathcal{S} = \{1, \ldots, 6\}$ to the space $\mathcal{S}_{\mathcal{B}} = \{1, 2, 3\}$.

Two events $\mathcal{A}$, $\mathcal{B}$ in the same set σ of events are called *independent* provided

$$P(\mathcal{A}\mathcal{B}) = P(\mathcal{A})P(\mathcal{B}). \qquad (1.5)$$

The probability on the left is the probability that both events $\mathcal{A}$ and $\mathcal{B}$ occur and is often written $P(\mathcal{A}, \mathcal{B})$. It is the sum of the probabilities of the elementary outcomes of the experiment which belong to both the event $\mathcal{A}$ and the event $\mathcal{B}$. For example, letting $\mathcal{A}$, $\mathcal{B}$ be respectively the events that a fair die comes up odd and that the die comes up less than 4, we have

$$P(\mathcal{A}\mathcal{B}) = P(\{1, 3\}) = 1/3 \neq P(\mathcal{A})P(\mathcal{B}) = (1/2)(1/2) = 1/4.$$

We conclude that the events are not independent, in the sense of the definition. The word also corresponds to our usual usage, in that knowing the die shows a number less than 4 changes our degree of belief that it has come up odd.

An equivalent definition of independence follows from introducing the conditional probability Eq. (1.4) of the two events into the definition Eq. (1.5). From Eq. (1.5), two events are independent provided

$$P(\mathcal{A}\mathcal{B}) = P(\mathcal{A} \mid \mathcal{B})P(\mathcal{B}) = P(\mathcal{A})P(\mathcal{B}), \qquad P(\mathcal{B}) \neq 0,$$

so that

$$P(\mathcal{A} \mid \mathcal{B}) = P(\mathcal{A}), \qquad P(\mathcal{B}) \neq 0. \qquad (1.6)$$

This last indicates that our degree of belief about the occurrence of the event $\mathcal{A}$ is unaffected by knowledge that the event $\mathcal{B}$ has occurred. This is in accord with the everyday meaning of the term "independence."

Combined Experiments

That two events $\mathcal{A}$, $\mathcal{B}$ which we want to consider together must belong to the same set σ of events is no real restriction. We can always expand the set σ to encompass all events of interest by expanding the definition of the abstract experiment and the set $\mathcal{S}$ of its elementary outcomes. That is, suppose that $\mathcal{A} \in \sigma_1$, $\mathcal{B} \in \sigma_2$ are events in two different sets of events. Let $\mathcal{S}_1$, $\mathcal{S}_2$ be the corresponding sample spaces of elementary outcomes ζ_{1i}, ζ_{2j}. Aggregate together the two sets of outcomes by defining new elementary outcomes $\zeta_{ij} = (\zeta_{1i}, \zeta_{2j})$ consisting of all ordered pairs of outcomes from the two sets of elementary outcomes. These points ζ_{ij} make up a new sample space $\mathcal{S}$, called the *Cartesian product* of the constitutent spaces $\mathcal{S}_1$, $\mathcal{S}_2$, and written $\mathcal{S} = \mathcal{S}_1 \times \mathcal{S}_2$. Now a σ-algebra of combined events can be built on the space $\mathcal{S}$.

The probabilities of events based on a Cartesian product space $\mathcal{S} = \mathcal{S}_1 \times \mathcal{S}_2$ are related to the probabilities of the events in the constituent spaces. Suppose as above that the elementary outcomes in $\mathcal{S}_1$ and $\mathcal{S}_2$ are ζ_{1i} and ζ_{2j}. Then $\mathcal{S}_2 = \Sigma \, \zeta_{2j}$, where the sum is over all j and is in the sense of the union of events. By the definition of Cartesian product we have

$$\zeta_{1i} \times \mathcal{S}_2 = \sum_j \zeta_{1i} \times \zeta_{2j}.$$

Because the events $\zeta_{1i} \times \zeta_{2j} \in \sigma$ are disjoint (their pairwise intersections are the null event Φ), the probabilities of the events in σ are properly defined provided we have

$$P(\zeta_{1i} \times \mathcal{S}_2) = \sum_j P(\zeta_{1i} \times \zeta_{2j}).$$

Because the joint space probability on the left is the probability that ζ_{1i} occurs in the first experiment and anything occurs in the second, we reasonably assign to it the value $P_1(\zeta_{1i})$. The result is the formula for computing the probabilities in the first experiment (the *marginal probabilities*) from those in the joint experiment:

$$P_1(\zeta_{1i}) = \sum_j P(\zeta_{1i}, \zeta_{2j}). \tag{1.7}$$

EXAMPLE 1.2   Suppose we toss a coin and, separately, roll a die. The first experiment has elementary outcomes $\mathcal{S}_1 = \{H, T\}$, and the second has $\mathcal{S}_2 = \{1, \ldots, 6\}$. If for some reason we want to consider together the tossing of a coin and the rolling of a die, we need only consider a new abstract experiment with elementary outcomes $\mathcal{S} = \{H1, H2, \ldots, T6\}$, which we label as ζ_i, $i = 1, 12$. Suppose that these elementary outcomes have probabilities respectively of (1/12, 1/12, 1/12, 1/12, 1/12, 1/12, 1/24, 1/8, 1/24, 1/8, 1/24, 1/8). (The only restriction in assigning these is that they be nonnegative and add to unity.) From Eq. (1.7) the marginal probability of heads is

$$P_1(H) = \sum_{i=1}^{6} P(\zeta_i) = 1/2,$$

and that of an even number on the die is

$$P_2(\text{even}) = \sum_I P(\zeta_i) = 3/12 + 3/8 = 5/8,$$

where the index set is $I = (2, 4, 6, 8, 10, 12)$. Clearly, this is not a fair die; our assignment of probabilities models a different kind of die.

For this die and coin, the conditional probabilities of rolling an even number, given that we know the coin has come up heads or tails, from Eq. (1.4) are

$$P(\text{even} \mid H) = P(\text{even}, H)/P_1(H) = (3/12)/(1/2) = 1/2,$$
$$P(\text{even} \mid T) = P(\text{even}, T)/P_1(T) = (3/8)/(1/2) = 3/4.$$

Since then, for example,

$$P(\text{even} \mid H) = 1/2 \neq P_2(\text{even}) = 5/8,$$

we conclude from Eq. (1.6) that the two constitutent experiments of tossing the coin and rolling the die are not independent. That is, the outcome of the coin toss affects the properties of the die that is rolled. That might come about, for example, if we were to choose one of two dies, depending on the outcome of the coin toss.

The definition of independence given in Eq. (1.5) for two events $\mathcal{A}$, $\mathcal{B}$ extends to the independence of any number of events $\mathcal{A}_i$. A collection of events $\mathcal{A}_i$ is said to be (mutually) independent provided the events are pairwise independent:

$$P(\mathcal{A}_i \mathcal{A}_j) = P(\mathcal{A}_i)P(\mathcal{A}_j), \quad i \neq j,$$

and in addition if, for trios of events,

$$P(\mathcal{A}_i, \mathcal{A}_j, \mathcal{A}_k) = P(\mathcal{A}_i)P(\mathcal{A}_j)P(\mathcal{A}_k), \quad i \neq j \neq k,$$

with similar relations for the events taken by fours, by fives, etc.

1.3 PROBABILITY DISTRIBUTION FUNCTIONS

In the previous two sections, we introduced the idea of an abstract experiment as a triple of elements $\{\mathcal{S}, \sigma, P\}$, with $\mathcal{S}$ (the sample space) a collection of mutually exclusive and exhaustive outcomes ζ of the experiment, which we can think of as elements of a point set. The σ-algebra σ is the set of all events $\mathcal{A}$ formed from elements ζ taken from $\mathcal{S}$, constructed according to certain rules. In particular, σ contains all the elementary outcomes from $\mathcal{S}$ and all unions, intersections, differences, and negations of its elements. The probability $0 \le P(\mathcal{A}) \le P(\mathcal{S}) = 1$ is a real-valued function whose domain is the events in the σ-algebra. Now we need to introduce the idea of a random variable.

A *random variable* is any real finite-valued function $X(\zeta)$ whose domain is the elementary outcomes ζ from the sample space $\mathcal{S}$ corresponding to an abstract experiment. In any particular performance of the experiment underlying a random variable, some elementary outcome ζ_0 will occur. This will in turn indicate a specific value $X(\zeta_0)$ for the random variable.

Specifying some property that the value of a random variable is to have induces an event in the σ-algebra of events of the abstract experiment. The event is defined as the union of the elementary outcomes $\zeta \in \mathcal{S}$ for which the random variable $X(\zeta)$ has the requested property. For example, in the die-rolling experiment, we can take as elementary outcomes the 6 numbers which could appear uppermost. A random variable $X(\zeta)$ is then defined by specifying the 6 values $X_i = X(\zeta_i)$ it should take for the 6 possible elementary outcomes of the experiment. These might be specified by a relation such as $X_i = 2i - 5$, $i = 1, 6$. Then the requirement $1.5 \le X \le 8$ induces the event

$$\mathcal{A} = \{\zeta: 1.5 \le X \le 8\} = \{\zeta_i: 3.25 \le i \le 6.5\}$$
$$= \{4\} + \{5\} + \{6\}, \tag{1.8}$$

where the notation $\{\zeta: 1.5 \le X \le 8\}$ means "the set of points ζ for which the value $X(\zeta)$ satisfies $1.5 \le X \le 8$."

With such associations of requirements on a random variable $X(\zeta)$ with events in the σ-algebra of an experiment, we can then speak of the event that the random variable satisfies the requirement. That is, for example, in the particular case of the assignment $X = 2i - 5$ relative to the fair die experiment, we speak of the event that $1.5 \le X \le 8$. Furthermore, we can speak of the probability of the event $1.5 \le X \le 8$ as the probability of the corresponding event Eq. (1.8) in the σ-algebra based on the experiment. That is,

$$P(1.5 \le X \le 8) = P(\{4\} + \{5\} + \{6\}) = 1/2.$$

In engineering, it is the values of random variables which are central to applications. That is because our measurements are real numbers produced by an instrument observing some system. The state of operation of the system has been determined by the particular outcome of a performance of some abstract experiment by some entity (nature, for example, or the sender of a message). In the study of the values of a random variable corresponding to an experiment, the *distribution function* of the random variable is central. By definition, that is a function

$$P_X(x) = P(X \le x). \qquad \qquad *(1.9)$$

It is worth reiterating carefully the meaning of the symbols in this definition. On the right, X is a real-valued function $X(\zeta)$ defined over the elementary outcomes $\zeta \in \mathcal{S}$ of some abstract experiment $\{\mathcal{S}, \sigma, P\}$. The inequality is in the usual sense, and x is some real number. The requirement $X \le x$ induces an event $\mathcal{A} = \{\zeta: X(\zeta) \le x\} \in \sigma$, and the number $P_X(x)$ is the probability assigned to that event by the function $P: P_X(x) = P(\mathcal{A})$.

In the fair die example above, the random variable $X = 2i - 5$ has values $X = (-3, -1, 1, 3, 5, 7)$. With the particular assignment $P(\zeta_i) = 1/6$ the distribution function $P_X(x)$ is as shown in Fig. 1.2a. A different

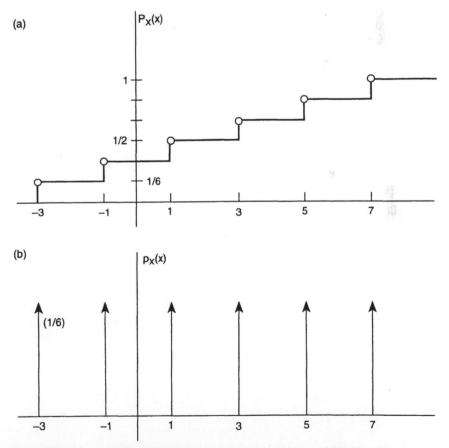

FIGURE 1.2 (a) Distribution function $P_X(x)$ of the random variable $X = 2i - 5$ in the experiment of casting a fair die. (b) Density function as a sequence of Dirac impulses.

assignment of values X to the outcomes i, or different probabilities $P(\zeta_i)$, would result in a different distribution function for X.

Because the values of a distribution function are the probabilities of various events, from the axioms Eq. (1.2) satisfied by any probability function P we must have

$$P_X(-\infty) = P(\{\zeta: X \le -\infty\}) = P(\Phi) = 0,$$
$$P_X(+\infty) = P(\{\zeta: X \le \infty\}) = P(\mathcal{S}) = 1,$$
$$P_X(x + dx) = P(\{\zeta: X \le x + dx\})$$
$$= P(\{\zeta: X \le x\} \text{ OR } \{\zeta: x < X \le x + dx\})$$
$$= P(\{\zeta: X \le x\}) + P(\{\zeta: x < X \le x + dx\})$$
$$\ge P(\{\zeta: X \le x\}) = P_X(x). \qquad (1.10)$$

The last follows because the events $\{\zeta: X \le x\}$ and $\{\zeta: x < X \le x + dx\}$ have no elementary outcomes in common. We conclude from Eq. (1.10) that any distribution function $P_X(x)$ is monotonically nondecreasing between the values 0 at $x = -\infty$ and 1 at $x = +\infty$. Rewriting the last equation of Eq. (1.10) yields another important property of a distribution function:

$$P(x_1 < X \le x_2) = P_X(x_2) - P_X(x_1). \qquad *(1.11)$$

From this, letting $x_2 = x$ and $x_1 = x - \varepsilon$, and letting $\varepsilon \Rightarrow 0$, there follows

$$P_X(x) - P_X(x^-) = P(\{\zeta: X = x\}). \qquad (1.12)$$

This is illustrated for example in Fig. 1.2a.

It is often useful to consider two different sets of measurements relative to the same experiment. We then define random variables, say $X(\zeta)$, $Y(\zeta)$, and consider the *bivariate (joint) distribution function*

$$P_{XY}(x, y) = P(\{\zeta: X \le x\} \text{ AND } \{\zeta: Y \le y\}) = P(X \le x, Y \le y),$$

where the last form is an abbreviation of the second form. From the probabilities of the various events involved it is easy to see that

$$P_{XY}(x, -\infty) = P_{XY}(-\infty, y) = 0,$$
$$P_{XY}(+\infty, +\infty) = 1,$$
$$P_{XY}(x, +\infty) = P_X(x),$$
$$P_{XY}(+\infty, y) = P_Y(y), \qquad (1.13)$$

with the obvious generalizations for multivariate distribution functions $P_{XY\ldots}(x, y, \ldots)$.

1.4 CONTINUOUS RANDOM VARIABLES

A *continuous* random variable is defined as one with a continuous distribution function, such as in Fig. 1.3a. A *discrete* random variable is one with a piecewise constant distribution function, as in Fig. 1.2a. A

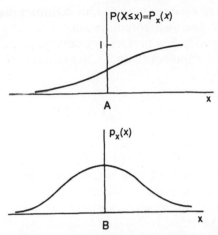

FIGURE 1.3 Probability functions of a continuous random variable. (a) The cumulative distribution function. (b) The density function.

mixed random variable combines the two types and has a distribution function as in Fig. 1.4a. Because a continuous random variable has a continuous distribution function $P_X(x)$, the derivative dP_X/dx exists everywhere. It is called the (probability) *density function* of the random variable X:

$$p_X(x) = dP_X(x)/dx = dP(X \le x)/dx. \qquad *(1.14)$$

An example is shown in Fig. 1.3b. As any $P_X(x)$ is monotone nondecreasing, $p_X(x) \ge 0$ everywhere.

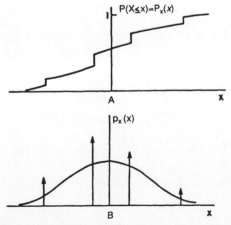

FIGURE 1.4 Probability functions of a mixed random variable. (a) The cumulative distribution function. (b) The density function.

A discrete or mixed random variable by definition has a distribution function which is not everywhere continuous. Formally, we still define the corresponding density function as the derivative of the distribution function, but we allow impulses (Dirac δ functions) in the density. Thus, we write

$$p_X(x) = \frac{dP_X^c}{dx} + \sum_i P_i \delta(x - x_i),$$

where P_X^c indicates the continuous part of the distribution, and at (isolated) points x_i of discontinuity $P_i = P_X(x_i) - P_X(x_i^-)$. Figure 1.2b shows an example of a density function for a discrete random variable, and Fig. 1.4b shows the same for a mixed random variable.

In most of our work, the random variables of interest will be of the continuous type. By the definition Eq. (1.14), the density function is related to the distribution function as

$$P_X(x) = P(X \le x) = \int_{-\infty}^x p_X(u) \, du,$$

so that

$$P(a < X \le b) = P_X(b) - P_X(a) = \int_a^b p_X(x) \, dx. \qquad *(1.15)$$

That is,

$$p_X(x) \, dx = P(\{\zeta : x < X \le x + dx\}). \qquad (1.16)$$

In the case of a multivariate distribution function, we define a multivariate density, for example,

$$p_{XY}(x, y) = \partial^2 P_{XY}(x, y)/\partial x \, \partial y. \qquad (1.17)$$

Then

$$P(a < X \le b, c < Y \le d) = \int_a^b \int_c^d p_{XY}(x, y) \, dy \, dx. \qquad (1.18)$$

In particular,

$$P_X(x) = P_{XY}(x, +\infty) = P(-\infty < X \le x, -\infty < Y \le +\infty)$$
$$= \int_{-\infty}^x \int_{-\infty}^\infty p_{XY}(u, v) \, dv \, du,$$
$$P_Y(y) = P_{XY}(+\infty, y) = \int_{-\infty}^\infty \int_{-\infty}^y p_{XY}(u, v) \, dv \, du.$$

From these follow the *marginal densities*

$$p_X(x) = \frac{dP_X(x)}{dx} = \int_{-\infty}^\infty p_{XY}(x, v) \, dv, \qquad *(1.19)$$

$$p_Y(y) = \frac{dP_Y(y)}{dy} = \int_{-\infty}^{\infty} p_{XY}(u, y)\, du. \qquad *(1.20)$$

Equation (1.4) defines the conditional probability of one event $\mathcal{A}$, conditioned by another event $\mathcal{B}$ in the same σ-algebra. We now introduce the conditional probability distribution function of a random variable $X(\zeta)$. This is

$$\begin{aligned}
P_{X|\mathcal{B}}(x \mid \mathcal{B}) &= P(\{\zeta: X \le x\} \mid \mathcal{B}) \\
&= P(X \le x, \mathcal{B})/P(\mathcal{B}), \qquad P(\mathcal{B}) \ne 0.
\end{aligned} \qquad (1.21)$$

Often we will be interested in $\mathcal{B} = \{\zeta: a < Y \le b\}$, in which case we have

$$P_{X|Y}(x \mid a < Y \le b) = [P_{XY}(x, b) - P_{XY}(x, a)]/[P_Y(b) - P_Y(a)].$$

Letting $a = y$, $b = y + dy$, as $dy \Rightarrow 0$ this becomes

$$P_{X|Y}(x \mid Y = y) = [\partial P_{XY}(x, y)/\partial y]/p_Y(y), \qquad p_Y(y) \ne 0.$$

The *conditional probability density* corresponding to this conditional distribution function is

$$\begin{aligned}
p_{X|Y}(x \mid Y = y) &= \partial P_{X|Y}(x \mid Y = y)/\partial x = [\partial^2 P_{XY}(x, y)/\partial y \partial x]/p_Y(y) \\
&= p_{XY}(x, y)/p_Y(y),
\end{aligned} \qquad *(1.22)$$

or, in a common notation,

$$p(x \mid y) = p(x, y)/p(y), \qquad p(y) \ne 0.$$

This last causes the same symbol (p) to do heavy duty as three different functions in the same equation.

Two random variables X, Y, are said to be *independent* provided the events $x < X \le x + dx$, $y < Y \le y + dy$ are independent for all values x, y. That is,

$$\begin{aligned}
P(\{x < X \le x + dx\} &\text{ AND } \{y < Y \le y + dy\}) \\
&= \int_x^{x+dx} \int_y^{y+dy} p_{XY}(u, v)\, dv\, du = P_{XY}(x, y)\, dx\, dy \\
&= P(x < X \le x + dx)P(y < Y \le y + dy) \\
&= P_X(x)\, dx\, P_Y(y)\, dy.
\end{aligned}$$

That is,

$$\begin{aligned}
P_{XY}(x, y) &= P_X(x)P_Y(y), \\
p_{XY}(x, y) &= \partial^2 [P_X(x)P_Y(y)]/\partial x\, \partial y = p_X(x)p_Y(y),
\end{aligned} \qquad (1.23)$$

so that

$$p_{X|Y}(x \mid y) = p_{XY}(x, y)/p_Y(y) = p_X(x). \qquad (1.24)$$

As we will often do hereafter, we can write a multidimensional random variable as a (column) vector $\mathbf{Z}$ of random variables Z_i. The corresponding multidimensional distribution function is $P_Z(z) = P(Z_1 \leq z_1, \ldots, Z_n \leq z_n)$, with density $p_Z(z) = \partial^n P_Z(z)/\partial z_1 \ldots \partial z_n$. [This latter, scalar, nth order partial derivative is different from the vector $\partial P_Z(z)/\partial z$, which is defined as the row vector of the n first-order partial derivatives $\partial P_Z(z)/\partial z_i$.] If the variables Z are divided into two groups, X and Y, it is easy to see that

$$p_X(x) = \int_{-\infty}^{\infty} p_{XY}(x, y) \, dy,$$

where the integral is multidimensional. The two sets of variables are said to be independent provided $p_{XY}(x, y) = p_X(x)p_Y(y)$. Furthermore, the conditional density is defined as

$$p_{X|Y}(x \mid y) = p_{XY}(x, y)/p_Y(y). \tag{1.25}$$

The vector random variables x, y are found to be independent (of one another) if $p_{X|Y}(x \mid y) = p_X(x)$. [This does not imply that the random variables $x_1, \ldots, x_n$ are independent; that would require their independence by pairs, by threes, etc., in the sense that $p_X(x_i, \ldots, x_j) = p_X(x_i) \cdots p_X(x_j)$.]

1.5 FUNCTIONS OF RANDOM VARIABLES

In engineering, we are often interested in families of random variables $X(t_i)$, $i = 1, n$, which are labeled by time instants. Typically, a family $X(t_i)$ is the input to some system, and the system generates another family $Y(t_i)$ as its output. The system defines the output samples as some function of the input samples: $Y(t_i) = h_i[X(t_1), \ldots, X(t_n)]$. Analysis of the characteristics of the output samples as random variables, given the characteristics of the input samples, requires calculation of the distribution functions of the one set, Y, from those of the other set, X.

To illustrate the process we will use, consider first a single random variable (RV) Y whose values are a function of the values of a single RV X: $y = h(x)$. Suppose also that h is one-to-one; that is, not only does a unique value y correspond to each x but also each y is produced by only one value of x. That is, there exists a well-defined function h^{-1} such that $x = h^{-1}(y)$ assigns a unique value x to each value y. Then the distribution function of Y is easily found from that of X as follows.

Suppose first that $h(x)$ is monotonically increasing. Then

$$\begin{aligned} P_Y(y) &= P(Y \leq y) = P[h(X) \leq y] \\ &= P[X \leq h^{-1}(y)] = P_X[h^{-1}(y)]. \end{aligned}$$

On the other hand, for a monotonically decreasing $h(x)$ we have

$$P_Y(y) = P[h(X) \le y] = P[X \ge h^{-1}(y)] = 1 - P[X < h^{-1}(y)]$$
$$= 1 - P[X \le h^{-1}(y)] = 1 - P_X[h^{-1}(y)].$$

In this last, we have assumed that X is a continuous random variable, so that $P_X(x)$ is continuous, and hence $P_X(x^-) = P_X(x)$. Together with the assumption of strict monotonicity of the *system function* $h(x)$, this is sufficient to ensure also that Y is a continuous RV, having a continuous distribution $P_Y(y)$.

From the continuous distribution function $P_Y(y)$, the probability density function of the RV Y follows as

$$p_Y(y) = dP_Y(y)/dy = \pm [dP_X(x)/dx|_{h^{-1}(y)}] [dh^{-1}(y)/dy],$$

where the choice of sign is for $h(x)$ respectively monotonically increasing or decreasing. This can be written more simply as

$$p_Y(y) = p_X[h^{-1}(y)]|dx(y)/dy| = p_X[x(y)]/|dy/dx|. \qquad *(1.26)$$

Some examples below will illustrate the use of Eq. (1.26). Figure 1.5 illustrates the procedure. A region dy of the y domain is shown at location y_0. Via the function $y = h(x)$, this region came from the region dx at x_0. The probability to be assigned to the y region is just the probability assigned to the x region:

$$p_Y(y_0)|dy| = p_X(x_0)|dx|.$$

The inverse function $x = h^{-1}(y)$ and the relation

$$dx = (dx/dy)_0 \, dy = dy/(dy/dx)_0$$

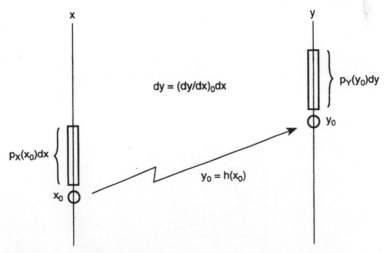

FIGURE 1.5 Mapping of the density $p_X(x)$ to $p_Y(y)$, under a transformation $y = h(x)$.

move us from the x domain, where we know $p_X(x)$, to the y domain:

$$p_Y(y_0)|dy| = p_X(x_0)|dx| = p_x[x(y_0)]|dy/(dy/dx)_0|.$$

Canceling $|dy|$ and letting (x_0, y_0) be a general (x, y) gives Eq. (1.26).

EXAMPLE 1.3   Suppose that the system function is $y = h(x) = ax + b$. Then from Eq. (1.26):

$$p_Y(y) = (1/|a|)\, p_X[(y - b)/a].$$

In particular, if X has the normalized *exponential density* $p_X(x) = \exp(-x)$, $x \geq 0$, then

$$p_Y(y) = (1/|a|)\, \exp[-(y - b)/a], \qquad (y - b)/a \geq 0.$$

Figure 1.6 shows the result of this calculation for $a = 2$, $b = 0$, and Fig. 1.7 illustrates the case $a = 1$, $b \geq 0$.

In case the continuous system function $y = h(x)$ is not strictly monotonic, the inverse function $x = h^{-1}(y)$ is not defined. (An example is $y = x^2$, in which case any y corresponds to two values of x: $x = \pm\sqrt{y}$.) We then consider separately the intervals of x over which $h(x)$ is strictly monotonic. If we assume that the critical points of $h(x)$ (the points at which $dh/dx = 0$) are isolated, then the strictly monotonic sections of $h(x)$ are separated by those isolated points. As $h(x)$ is strictly monotonic on each interval I_i between critical points (Fig. 1.8), for any specified value of y there will be at most one solution x_i of $y = h(x)$ on each interval I_i. That is, on each I_i the function $h(x)$ has a well-defined inverse $x = h^{-1}(y)$. Then

$$P_Y(y) = P(Y \leq y) = \sum_i P(X \leq x_i) + \sum_{i'} P(X \geq x_{i'}),$$

where the first sum is over all points lying in intervals I_i for which $y = h(x)$ has a solution for x and for which $h(x)$ is monotonically increasing, and the second sum is over the corresponding points of intervals for which $h(x)$ is monotonically decreasing.

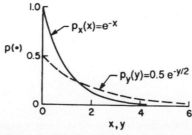

FIGURE 1.6   Density functions for Example 1.3.

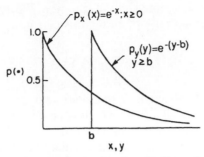

FIGURE 1.7 Density functions for Example 1.3.

Manipulating this last expression term by term, in accordance with the above procedure for a system function which is strictly monotonic over $(-\infty, \infty)$, yields finally

$$p_Y(y) = \sum_i p_X[x_i(y)]/|dy/dx|_{x_i(y)}, \qquad *(1.27)$$

where the sum is over all the solutions of $y = h(x)$ for the particular value y of interest.

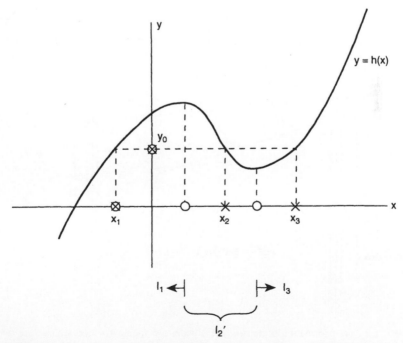

FIGURE 1.8 Three points x_i where the distribution $P_X(x_i)$ relates to the distribution $P_Y(y_0)$, showing critical points separating regions of monotonicity of $y(x)$.

Figure 1.9 illustrates this more general situation. We assign to the region dy around $y = y_0$ the probability of all regions of x from which that y_0 can arise. That is,

$$p_Y(y_0)|dy| = \sum_i p_X(x_i) |dx_i|,$$

where the absolute value sign is a reminder that probability is always nonnegative. The values x_i of x from which we could reach the value y_0 of y are the multiple solutions x_i of the equation $y_0 = h(x)$. The corresponding regions dx_i are of size $dx_i = (dx/dy)_i \, dy$. Substituting in the above and cancelling dy leads to Eq. (1.27).

EXAMPLE 1.4   Let $y = x^2$. For $y < 0$ there are no solutions for x, and for $y > 0$ there are two: $x = \pm\sqrt{y}$. Since $dy/dx = 2x$, from Eq. (1.27) we have

$$p_Y(y) = p_X(+\sqrt{y})/|2\sqrt{y}| + p_X(-\sqrt{y})/|-2\sqrt{y}|, \qquad y > 0. \quad (1.28)$$

Also $p_y(y) = 0$, $y < 0$, because $P_Y(y) = P[Y \le (y < 0)] = 0$, so $dP_Y/dy = 0$ for $y < 0$. Proceeding further, suppose in particular that

$$p_X(x) = (2\pi)^{-1/2} \exp(-x^2/2), \qquad -\infty < x < \infty \qquad *(1.29)$$

is the normalized *Gaussian density*. Then, with $y = x^2$, we have

$$p_Y(y) = (2\pi y)^{-1/2} \exp(-y/2), \qquad y > 0, \cdot \qquad (1.30)$$

which is the *chi-squared density* with one degree of freedom.

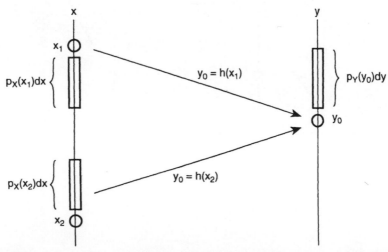

FIGURE 1.9   Transformation of density $p_X(x)$ under a mapping $y = h(x)$ which does not have a single-valued inverse. The interval dy induces multiple intervals dx from which it might arise.

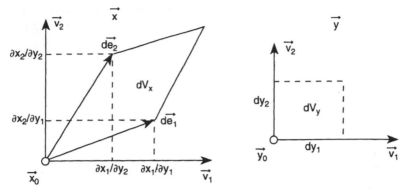

FIGURE 1.10   Edges de_1, de_2 of volume dV_x in X-space induced by volume dV_y in Y-space.

Most transformations of interest are multidimensional, rather than scalar as we have considered so far. That is because typical systems have memory, so that output samples y_i depend on more than one input sample x_i. Accordingly, we need to consider a system function $\mathbf{y} = h(\mathbf{x})$, where the vectors $\mathbf{x}$, $\mathbf{y}$ might or might not have the same dimension. Let us suppose first that they do, and for the moment suppose further that the transformation is such that a unique inverse exists: $\mathbf{x} = h^{-1}(\mathbf{y})$.

As sketched in Fig. 1.10, let dV_y represent the volume of a rectangular parallelepiped with edges of lengths $dy_1, \ldots, dy_n$. These can be regarded as lengths dy_i along the unit coordinate vectors $\mathbf{v}_i$ of a rectangular coordinate system, so that $dV_y = dy_1 \cdots dy_n$. We want to find the density $p_Y(\mathbf{y})$ of the RV $\mathbf{Y}$ at some point $\mathbf{y}_0$. As in the one-dimensional case, we equate the probability of $\mathbf{y}$ being in a volume dV_y at $\mathbf{y}_0$ to the probability of $\mathbf{x}$ being in the volume dV_x at $\mathbf{x}_0$ from which the dV_y in question arose. That is,

$$p_Y(\mathbf{y}_0)\, dV_y = p_X(\mathbf{x}_0)\, dV_x. \tag{1.31}$$

We know $p_X(\mathbf{x}_0)$ and want $p_Y(\mathbf{y}_0)$. To that end, we need to express the volume dV_x in terms of dV_y.

Each edge $(dy_i)\mathbf{v}_i$ corresponds to an edge de_i with coordinates $(\partial x_j/\partial y_i)_0\, dy_i$, that is,

$$de_i = \sum_{j=1}^{n} \left(\frac{\partial x_j}{\partial y_i}\right)_0 dy_i\, \mathbf{v}_j.$$

We can arrange the coordinates of the vectors de_i as columns of a matrix:

$$A = \{a_{ji}\} = \{(\partial x_j/\partial y_i)_0\, dy_i\},$$

where a_{ji} is the element in row j and column i of the matrix A.

The volume of a parallelepiped, such as dV_x in Fig. 1.10, with certain column vectors as its edges, is the unsigned value of the determinant of the matrix formed with the edge vectors as columns. (The signed value of the determinant is the Jacobian of the transformation.) That is, the volume dV_x is just

$$dV_x = |\det(A)| = |\det(\partial\mathbf{x}/\partial\mathbf{y})_0|\, dV_y,$$

factoring out the dy_i from the columns of the determinant. Hence, finally, using this in Eq. (1.31) and canceling dV_y, the density of the transformed variables is

$$p_Y(\mathbf{y}) = p_X[\mathbf{x}(\mathbf{y})]|\det(\partial\mathbf{x}/\partial\mathbf{y})|. \qquad (1.32)$$

Since

$$\det(\partial\mathbf{x}/\partial\mathbf{y}) = \det[(\partial\mathbf{y}/\partial\mathbf{x})^{-1}] = 1/\det(\partial\mathbf{y}/\partial\mathbf{x}),$$

this can be written alternatively as

$$p_Y(\mathbf{y}) = p_X[\mathbf{x}(\mathbf{y})]/|\det(\partial\mathbf{y}/\partial\mathbf{x})|. \qquad *(1.33)$$

Here $\partial\mathbf{y}/\partial\mathbf{x}$ is a matrix A with elements $a_{ij} = \partial y_i/\partial x_j$.

EXAMPLE 1.5 Suppose that x_1, x_2 are independent normalized Gaussian variables with identical densities as in Eq. (1.29). As they are independent, their joint density is

$$p_X(x_1, x_2) = p(x_1)p(x_2) = (1/2\pi)\exp[-(x_1^2 + x_2^2)/2].$$

Suppose we are interested in new variables $y_1 = x_1 + x_2$, $y_2 = x_1 - x_2$. The Jacobian is

$$\det\left(\frac{\partial\mathbf{y}}{\partial\mathbf{x}}\right) = \begin{vmatrix} \partial y_1/\partial x_1 & \partial y_1/\partial x_2 \\ \partial y_2/\partial x_1 & \partial y_2/\partial x_2 \end{vmatrix} = \begin{vmatrix} 1 & 1 \\ 1 & -1 \end{vmatrix} = -2,$$

and the inverse transformation is $x_1 = (y_1 + y_2)/2$, $x_2 = (y_1 - y_2)/2$. Then

$$p_Y(y_1, y_2) = (1/4\pi)\exp[-(y_1^2 + y_2^2)/4].$$

It is interesting to note that this factors in such a way as to indicate that Y_1, Y_2 are independent RVs, each with density

$$p(y) = (1/2\sqrt{\pi})\exp(-y^2/4).$$

EXAMPLE 1.6 In later chapters we will be interested in the transformation

$$r = (x^2 + y^2)^{1/2} > 0, \qquad \phi = \mathrm{Tan}^{-1}(y/x),$$

where by the notation Tan^{-1} we mean the principal value: $-\pi < \phi \le \pi$. The Jacobian is

$$\frac{\partial(r, \phi)}{\partial(x, y)} = \begin{vmatrix} x/r & y/r \\ -y/r^2 & x/r^2 \end{vmatrix} = 1/r,$$

and the inverse transformation is $x = r \cos \phi$, $y = r \sin \phi$. Again assuming X, Y to be independent normalized Gaussians, we have

$$p_{R\Phi}(r, \phi) = (r/2\pi) \exp(-r^2/2) = (1/2\pi)[r \exp(-r^2/2)].$$

The marginal densities are

$$p_R(r) = \int_{-\pi}^{\pi} p_{R\Phi}(r, \phi) \, d\phi = r \exp(-r^2/2),$$

$$p_\Phi(\phi) = \int_0^\infty p_{R\Phi}(r, \phi) \, dr = 1/2\pi.$$

That is, R is *Rayleigh distributed* and Φ has the *uniform density*. Since

$$p_{R\Phi}(r, \phi) = p_R(r)p_\Phi(\phi),$$

the random variables R, Φ are independent.

It may well happen that the multidimensional transformation $\mathbf{y} = h(\mathbf{x})$ does not have a unique inverse. That is, there may be multiple solutions $\mathbf{x}_i$ of the equation $\mathbf{y}_0 = h(\mathbf{x})$ for some or all values $\mathbf{y}_0$. For any particular value $\mathbf{y}_0$ of interest, let $\mathbf{x}_{0i}$ be the set of solutions of this equation. Then in the above development the region dV_x induced in the $\mathbf{X}$ space by a differential excursion around $\mathbf{y}_0$ consists of multiple subregions of the type in Fig. 1.10, each located at one of the solutions $\mathbf{x}_{0i}$. The volume of each such region is the unsigned Jacobian of the transformation evaluated at $\mathbf{x}_{0i}$, and the volume of the region dV_x is their sum. Consequently, as in Eq. (1.27),

$$p_Y(\mathbf{y}) = \sum_i \frac{p_X[\mathbf{x}_i(\mathbf{y})]}{|\det[(\partial\mathbf{y}/\partial\mathbf{x})|_{\mathbf{x}_i}]|}, \tag{1.34}$$

where the sum is over all solutions $\mathbf{x}_i$ of the equation $\mathbf{y} = h(\mathbf{x})$ for the particular value $\mathbf{y}$ on the left.

It may be that, in the transformation $\mathbf{y} = h(\mathbf{x})$, the dimension of $\mathbf{y}$ is less than that of $\mathbf{x}$. A useful technique then consists of supplementing $\mathbf{y}$ with additional variables $\mathbf{z}$, chosen for convenience and such that the dimension of the combined set $(\mathbf{y}, \mathbf{z})$ is the dimension of $\mathbf{x}$. The above techniques can then be applied to find $p_{YZ}(\mathbf{y}, \mathbf{z})$, from which follows

$$p_Y(\mathbf{y}) = \int_{-\infty}^{\infty} p_{YZ}(\mathbf{y}, \mathbf{z}) \, d\mathbf{z}. \tag{1.35}$$

EXAMPLE 1.7 Suppose X_1, X_2 are independent normalized Gaussian RVs, as in Eq. (1.29), and that we are interested in $y = x_1 + x_2$. We introduce $z = x_2$ and easily compute

$$p_{YZ}(y, z) = |\partial(y, z)/\partial(x_1, x_2)|^{-1} (1/2\pi) \exp\{-[(y - z)^2 + z^2]/2\}$$
$$= (1/2\pi) \exp\{-[y^2 - 2yz + 2z^2]/2\}.$$

Then

$$p_Y(y) = \int_{-\infty}^{\infty} p_{YZ}(y, z) \, dz = \left(\frac{1}{2\sqrt{\pi}}\right) \exp\left(\frac{-y^2}{4}\right),$$

which is in accord with the result of Example 1.5.

EXAMPLE 1.8   As a final example, consider again that X_1, X_2 are normalized independent Gaussian variables, as in Eq. (1.29), and let us find the density of the ratio $y = x_1/x_2$. As an auxiliary variable we introduce $z = x_2$ and seek the density $p_{YZ}(y, z)$. The Jacobian $\det[\partial(y, z)/\partial(x_1, x_2)] = 1/x_2 = 1/z$. The inverse transformation is $x_1 = yz$, $x_2 = z$. Then

$$p_{YZ}(y, z) = (|z|/2\pi) \exp[-(1 + y^2)z^2/2],$$

and

$$p_Y(y) = \int_{-\infty}^{\infty} p_{YZ}(y, z) \, dz$$

$$= \frac{1}{\pi} \int_{0}^{\infty} z \exp\left[\frac{-(1 + y^2)z^2}{2}\right] dz = \frac{1/\pi}{1 + y^2}, \qquad (1.36)$$

which is the normalized *Cauchy density*.

1.6 CHARACTERISTIC FUNCTIONS

The *characteristic function* corresponding to a probability density function is useful in discussing the properties of a random variable. By definition, the characteristic function of the probability density $p_X(x)$ of a random variable X is the conjugate of its Fourier transform:

$$\Phi_X(j\omega) = \int_{-\infty}^{\infty} p_X(x) \exp(j\omega x) \, dx. \qquad *(1.37)$$

The *moment generating function* of the RV X is also defined and is

$$M_X(s) = \int_{-\infty}^{\infty} p_X(x) \exp(sx) \, dx. \qquad (1.38)$$

The moment generating function is defined for a complex variable s. The characteristic function is just the moment generating function evaluated on the j axis of the complex s plane, in the same way that the Fourier transform of a function is the (two-sided) Laplace transform evaluated on the j axis. Notice, however, that the moment generating function is defined as the mirror image in the origin of the Laplace transform of the probability density. That is,

$$M_X(s) = \mathscr{L}[p_X(x)]|_{s \to -s}. \qquad (1.39)$$

Correspondingly, the characteristic function $\Phi(j\omega)$ is the complex conjugate of the Fourier transform of the density $p_X(x)$.

Because the characteristic function $\Phi_X(j\omega)$ and the density $p_X(x)$ form a Fourier transform pair in the variable $-\omega$, the inverse Fourier relationship yields at once

$$p_X(x) = (1/2\pi) \int_{-\infty}^{\infty} \Phi_X(j\omega) \exp(-j\omega) \, d\omega. \qquad *(1.40)$$

The characteristic function always exists, since

$$|\Phi_X(j\omega)| \leq \int_{-\infty}^{\infty} p_X(x) \, dx = 1.$$

However, the moment generating function may fail to exist for certain densities. The Cauchy density of Example 1.8 is an example. We have

$$M_X(s) = \frac{1}{\pi} \int_{-\infty}^{\infty} \left[\frac{\exp(sx)}{(1 + x^2)} \right] \, dx.$$

This fails to converge at either $+\infty$ or $-\infty$ for any value of s other than $s = j\omega$, for which case we recover the characteristic function.

EXAMPLE 1.9 By completing the square in the exponent, it is easy to see that the characteristic function of the normalized Gaussian density, Eq. (1.29), is

$$\Phi_X(\omega) = \int_{-\infty}^{\infty} (2\pi)^{-1/2} \exp\left(\frac{-x^2}{2}\right) \exp(j\omega x) \, dx = \exp\left(\frac{-\omega^2}{2}\right). \qquad *(1.41)$$

That is, the characteristic function of a Gaussian density is again a Gaussian.

In the case of a multivariate density function $p_X(\mathbf{x})$ of dimension n, the characteristic function is the (conjugate of the) multivariate Fourier transform:

$$\Phi_x(j\omega) = \int_{-\infty}^{\infty} p_X(\mathbf{x}) \exp(j\omega^T \mathbf{x}) \, d\mathbf{x}, \qquad *(1.42)$$

where the integral is n-dimensional and the symbol T indicates the transpose, so that $\omega^T \mathbf{x}$ is the dot product: $\omega^T \mathbf{x} = \Sigma \, \omega_i x_i$. The corresponding inversion formula is

$$p_X(\mathbf{x}) = \int_{-\infty}^{\infty} \Phi_X(j\omega) \exp(-j\omega^T \mathbf{x}) \, d\mathbf{f}, \qquad *(1.43)$$

where as usual $\omega = 2\pi f$.

An important use of the characteristic function is in finding the density of the sum of two random variables given the joint density of the two RVs. Let X, Y be the given RVs, with known joint density $p_{XY}(x, y)$. We seek the density of the sum $z = x + y$. Introducing the auxiliary variable $w = x$, we consider the transformation from (x, y) to (w, z). The Jacobian is $\det[\partial(w, z)/\partial(x, y)] = 1$, and the inverse transformation is $x = w$, $y = z - w$. Therefore, the joint density of the new variables w, z is

$$p_{WZ}(w, z) = p_{XY}(w, z - w).$$

Finally, the desired density of $z = x + y$ follows as

$$p_Z(z) = \int_{-\infty}^{\infty} p_{WZ}(w, z) \, dw = \int_{-\infty}^{\infty} p_{XY}(w, z - w) \, dw. \qquad *(1.44)$$

This last expression is particularly useful in the case that X and Y are independent RVs. Then

$$p_{XY}(x, y) = p_X(x)p_Y(y),$$

so that, with $Z = X + Y$,

$$p_Z(z) = \int_{-\infty}^{\infty} p_X(w)p_Y(z - w) \, dw = p_X(z) * p_Y(z), \qquad (1.45)$$

where the symbol * indicates the convolution defined by the integral. Because the Fourier transform of a convolution is the product of the Fourier transforms of the functions convolved, in terms of characteristic functions we then have

$$\begin{aligned} \Phi_Z(j\omega) &= [\mathcal{F}\{p_Z(j\omega)\}]^* \\ &= [\mathcal{F}\{p_X(j\omega)\}]^* [\mathcal{F}\{p_Y(j\omega)\}]^* = \Phi_X(j\omega)\Phi_Y(j\omega), \qquad (1.46) \end{aligned}$$

with the corresponding expression for the moment generating functions:

$$M_Z(s) = M_X(s)M_Y(s). \qquad (1.47)$$

These results can be iterated to determine the characteristic function of the sum of any number of random variables, from which the probability density of the sum follows as the inverse Fourier transform.

EXAMPLE 1.10   Suppose that X and Y are independent normalized Gaussian RVs, as in Eq. (1.29). Then their characteristic functions are identical and, from Example 1.9, equal to

$$\Phi(j\omega) = \exp(-\omega^2/2).$$

Accordingly, the characteristic function of their sum Z is

$$\Phi_Z(j\omega) = \Phi_X(j\omega)\Phi_Y(j\omega) = \exp(-\omega^2).$$

The density function of the sum follows as the inverse transform:

$$p_Z(j\omega) = \mathcal{F}^{-1}\{\exp(-\omega^2)\} = (4\pi)^{-1/2} \exp(-z^2/4).$$

This last is an (unnormalized) Gaussian density, and the result of this example is a special case of the general fact, that we will prove below, that any linear combination of Gaussian RVs, independent or not, is another Gaussian RV.

1.7 EXPECTATION AND MOMENTS

In analyzing and designing systems involving random phenomena, either inputs or system characteristics, we are not usually interested in any particular realization of the random situation. That is, we are not concerned only with the result of one particular choice nature has made for the outcome of the underlying abstract experiment. Rather, we are usually more concerned with the performance of the system "on the average," taken over all the elementary outcomes of the abstract experiment which nature might bring to pass during future operation of the system. The concept of *expectation,* or *expected value,* of a random variable gives us information about that.

We are used to the idea of the "average" of n measurements x_i as the number

$$\text{aver}(x_i) = \frac{1}{n} \sum_{i=1}^{n} x_i. \tag{1.48}$$

We now view this as the sum of all n possible values $x_i = x(\zeta_i)$ which might arise as the result of an experiment carried out by nature, each weighted by the particular choice of probability $P(\zeta_i) = 1/n$ that each of the n values x_i might arise. It is specifically implied that we believe any value has the same chance of being picked by nature as any other.

The generalization of this idea is the expected value, or expectation, or *ensemble average,* or *mean* of a random variable X, defined as

$$\mathscr{E}(X) = \int_{-\infty}^{\infty} x p_X(x) \, dx. \qquad\qquad *(1.49)$$

Here we consider all values of X such that $x < X \le x + dx$ and add them up weighted by their probabilities $P(\{\zeta: x < X \le x + dx\}) = p_X(x) \, dx$. In case the RV X is of discrete or mixed type, the distribution function $P_X(x)$ has discontinuities, as in Fig. 1.2a: $P_X(x_0^+) = P(x_0 < X \le x_0 + dx) = P_X(x_0^-) + \Delta P$, where $\Delta P = P(\{\zeta: X = x_0\})$. The density function $p_X(x)$ then has an impulse of strength ΔP: $p_X(x_0) = (\Delta P)\delta(x - x_0)$, as in Fig. 1.2b. Then the expected value expression Eq. (1.49) becomes

$$\mathscr{E}(X) = \int_{-\infty}^{\infty} x p_X^c(x) \, dx + \sum_i x_i P_i. \tag{1.50}$$

Here $p_X^c(x) = dP_X^c(x)/dx$ is the derivative of the continuous part of $P_X(x)$ and does not contain the impulses of $p_X(x)$, and x_i are the points of discontinuity of $P_X(x)$ and P_i are the nonzero probabilities that X takes on the values x_i. In the case of a discrete RV, the continuous part of $p_X(x)$ is not present, and the expectation expression reduces to a simple sum like Eq. (1.48).

In the common case that we are interested in an RV Y which is a function of another RV X: $Y = h(X)$, the expected value of Y is, by definition,

$$\mathscr{E}(Y) = \int_{-\infty}^{\infty} y p_Y(y) \, dy.$$

However, we often know the density $p_X(x)$ but not directly the density $p_Y(y)$. We could certainly calculate the latter, by the methods of the previous section. However, if we are interested only in $\mathscr{E}(Y)$, it is unnecessary labor to determine $p_Y(y)$. Rather, we can use the computation

$$\mathscr{E}(Y) = \mathscr{E}[h(X)] = \int_{-\infty}^{\infty} h(x) p_X(x) \, dx. \qquad *(1.51)$$

That is, instead of integrating over all the values y of Y, weighted by the probability $p_Y(y) \, dy = P(y < Y \le y + dy)$ that Y has each of those values, we integrate over the values x of X which cause Y to have those values, weighted by the probability that X has the values x. It is not necessary that $h(x)$ have an inverse; that is, there may be more than one value x for which $h(x)$ has a particular value y. The density $p_X(x)$ accounts for the probabilities of all those values. The same relationship holds for multidimensional RVs and

$$\mathscr{E}(\mathbf{Y}) = \mathscr{E}[h(\mathbf{X})] = \int_{-\infty}^{\infty} h(\mathbf{x}) p_X(\mathbf{x}) \, d\mathbf{x}. \qquad (1.52)$$

It is not necessary that $\mathbf{X}$ and $\mathbf{Y}$ have the same dimension.

EXAMPLE 1.11   As in Example 1.6, suppose that X, Y are independent normalized Gaussian RVs and that we are interested in $\mathscr{E}[(X^2 + Y^2)^{1/2}]$. In the earlier example, we determined that $r = (x^2 + y^2)^{1/2}$ had the Rayleigh density

$$p(r) = r \exp(-r^2/2), \qquad r \ge 0. \qquad (1.53)$$

Therefore, by definition,

$$\mathscr{E}(R) = \int_0^{\infty} r^2 \exp\left(\frac{-r^2}{2}\right) dr = \left(\frac{\pi}{2}\right)^{1/2}.$$

However, knowing $p_{XY}(x, y) = p_X(x) p_Y(y)$, we can compute directly (converting to polar coordinates)

$$\mathscr{E}(R) = \int_{-\infty}^{\infty} (x^2 + y^2)^{1/2} \left(\frac{1}{2\pi}\right) \exp\left[\frac{-(x^2 + y^2)}{2}\right] dx \, dy$$

$$= \frac{1}{2\pi} \int_{-\pi}^{\pi} \int_0^{\infty} r \exp\left(\frac{-r^2}{2}\right) r \, dr \, d\phi = \int_0^{\infty} r^2 \exp\left(\frac{-r^2}{2}\right) dr,$$

just as above.

The *moments* of the probability density of an RV are the expected values of the various powers of the RV. That is, the nth moment of the RV X is defined as

$$m_n = \mathscr{E}(X^n) = \int_{-\infty}^{\infty} x^n p_X(x)\, dx. \tag{1.54}$$

The first moment m_1 is the mean of the RV X, and the second moment m_2 is the mean square. Also defined are the *central moments* of the RV X:

$$\mu_n = \mathscr{E}[(X - \mathscr{E}X)^n] = \int_{-\infty}^{\infty} (x - m_1)^n p_X(x)\, dx. \tag{1.55}$$

The first central moment is obviously $\mu_1 = 0$. The second central moment is the *variance*,

$$\mu_2 = \sigma^2 = \mathscr{E}[(X - \mathscr{E}X)^2], \qquad \text{*(1.56)}$$

and its square root σ is the *standard deviation*. The standard deviation is a measure of the spread of the density of an RV around its mean value. The variance can also be written

$$\sigma^2 = \mathscr{E}(X^2) - (\mathscr{E}X)^2, \tag{1.57}$$

which can be useful in calculations.

EXAMPLE 1.12 As shown in Fig. 1.11a, consider the uniform density: $p_X(x) = 1/a, \; x_0 < x \le x_0 + a$. We have

$$\mathscr{E}(X) = \int_{x_0}^{x_0+a} x\left(\frac{1}{a}\right) dx = x_0 + \frac{a}{2},$$

$$\sigma^2 = \int_{x_0}^{x_0+a} \left[x - \left(x_0 + \frac{a}{2}\right)\right]^2 \left(\frac{1}{a}\right) dx = \int_{-a/2}^{a/2} \left(\frac{u^2}{a}\right) du = \frac{a^2}{12}.$$

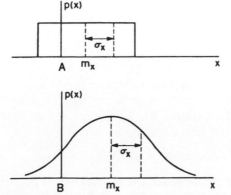

FIGURE 1.11 Examples to show mean and standard deviation for (a) uniform distribution and (b) Gaussian distribution.

In a common application, this indicates that a quantizer (at level a), which introduces a uniform random error $\pm a/2$ in converting an analog voltage into a digital value, has a random output with standard deviation $\sigma = a/\sqrt{12}$.

EXAMPLE 1.13   Consider the (unnormalized) Gaussian density (Fig. 1.11b):

$$p_X(x) = (2\pi\sigma^2)^{-1/2} \exp[-(x - m)^2/2\sigma^2].   \tag{1.58}$$

Using the change of variable $u = (x - m)/\sigma$, we have

$$\mathscr{E}(X) = (2\pi\sigma^2)^{-1/2} \int_{-\infty}^{\infty} x \exp\left[\frac{-(x - m)^2}{2\sigma^2}\right] dx = m,$$

$$\mathscr{E}[(X - m)^2] = (2\pi\sigma^2)^{-1/2} \int_{-\infty}^{\infty} (x - m)^2 \exp\left[\frac{-(x - m)^2}{2\sigma^2}\right] dx = \sigma^2.$$

These identify the parameters m, σ in the density as its mean and standard deviation.

Similar moments are introduced for the case of two or more RVs. For example, for two RVs, X, Y, with joint density $p_{XY}(x, y)$, the joint moment of order (n, k) is introduced as

$$\mathscr{E}(X^n Y^k) = m_{n,k} = \int_{-\infty}^{\infty} x^n y^k p_{XY}(x, y) \, dx \, dy,   \tag{1.59}$$

with the corresponding central moment

$$\mu_{n,k} = \mathscr{E}[(X - m_x)^n (Y - m_y)^k],   \tag{1.60}$$

where m_x, m_y are the means of X and Y. The moment μ_{11} is of enough importance to merit a special name, the *covariance* of X and Y:

$$C_{xy} = \mu_{11} = \mathscr{E}[(X - m_x)(Y - m_y)].   \text{*(1.61)}$$

The *correlation coefficient* between two RVs X and Y is the normalized covariance:

$$\rho_{xy} = C_{xy}/(\sigma_x \sigma_y),   \tag{1.62}$$

where σ_x, σ_y are the standard deviations of x and y.

The second-order moments of RVs are of particular importance. In the case of a multidimensional RV $\mathbf{X}$, the second moments $\mathscr{E}(X_i X_j)$ are conveniently organized into a matrix, as

$$R_x = \mathscr{E}(\mathbf{X}\mathbf{X}^\mathrm{T}),$$

where the superscript T indicates the transpose, so $\mathbf{X}^\mathrm{T}$ is the row vector whose elements are the same as those of the column vector $\mathbf{X}$. Thus R_x

has element in row i and column j which is

$$(R_x)_{ij} = \mathcal{E}(X_i X_j).$$

Correspondingly, we can arrange the covariances of the elements X_i as

$$C_x = \mathcal{E}(\mathbf{X} - \mathbf{m}_x)(\mathbf{X} - \mathbf{m}_x)^T,$$

where $\mathbf{m}_x = \mathcal{E}(\mathbf{X})$. For two vectors $\mathbf{X}$, $\mathbf{Y}$ of RVs X_i, Y_i, we arrange the covariances between the two sets as

$$C_{xy} = \mathcal{E}(\mathbf{X} - \mathbf{m}_x)(\mathbf{Y} - \mathbf{m}_y)^T.$$

For a collection of more than two RVs, moments are defined in the obvious way. For example, with four RVs we could define

$$m_{ijkl} = \mathcal{E}(W^i X^j Y^k Z^l).$$

Two RVs X, Y for which $\mathcal{E}(XY) = \mathcal{E}(X)\mathcal{E}(Y)$ are called *uncorrelated*. If that is the case, then their covariance C_{xy} and correlation coefficient ρ_{xy} are both zero. If $\mathcal{E}(XY) = 0$, then the RVs X, Y are called *orthogonal*, for reasons which we will discuss later.

There is a strong link between the characteristic function (or moment generating function) of an RV and its moments. In fact, from the definition of the characteristic function:

$$\Phi_X(j\omega) = \int_{-\infty}^{\infty} \exp(j\omega x) p_X(x) \, dx = \mathcal{E}[\exp(j\omega X)],$$

by differentiating under the integral sign we obtain

$$d^n\Phi_X(j\omega)/d\omega^n\big|_{\omega=0} = j^n \int_{-\infty}^{\infty} x^n p_X(x) \, dx = j^n m_n. \qquad (1.63)$$

That is, the moments of the RV are obtained from the derivatives of its characteristic function evaluated at the origin. It is interesting to note that we therefore have the Taylor series (provided all moments exist)

$$\Phi_X(j\omega) = \sum_{n=0}^{\infty} [j^n m_n/n!]\omega^n, \qquad (1.64)$$

so that the moments of the probability density of an RV determine its characteristic function, in principle, and therefore, from the inverse Fourier transform relation, the probability density itself.

Similarly, for the moment generating function of Eq. (1.38) we have

$$M_X(s) = \sum_{n=0}^{\infty} \frac{m_n s^n}{n!}.$$

This indicates that the moment generating function will fail to exist if any of the moments m_n of the RV fail to exist. Taking into account Eq. (1.63), that means that the moment generating function will not exist if any derivative of the characteristic function fails to exist at the origin. The Cauchy density is an example.

1.8 COMPLEX RANDOM VARIABLES

Some important quantities are considered to be complex numbers, for example, the complex form of the Fourier series coefficients of a periodic time function. As we will see later, this leads to the need for a notation to deal formally with random variables $Z(\zeta)$ whose values are complex numbers. Accordingly, we define a complex random variable

$$Z(\zeta) = X(\zeta) + jY(\zeta) \tag{1.65}$$

as the indicated sum of two real RVs X and Y, where $j = \sqrt{-1}$.

The probability density $p_Z(z)$ of such a complex RV is defined as the joint density:

$$p_Z(z) = p_{XY}(x, y).$$

Thereby $p_Z(z)$ is a real number.

The mean of a complex RV is

$$m_z = \mathscr{E}(Z) = \mathscr{E}(X) + j\mathscr{E}(Y) = \int_{-\infty}^{\infty} (x + jy)p_{XY}(x, y) \, dx \, dy, \tag{1.66}$$

or, defining $dz = dx \, dy$ as the infinitesimal area in the complex z plane,

$$m_z = \mathscr{E}(Z) = \int_{-\infty}^{\infty} zp_Z(z) \, dz. \tag{1.67}$$

Furthermore,

$$\mathscr{E}(|Z|^2) = \int_{-\infty}^{\infty} |x + jy|^2 p_{XY}(x, y) \, dx \, dy$$

$$= \int_{-\infty}^{\infty} |z|^2 p_Z(z) \, dz = \mathscr{E}(X^2) + \mathscr{E}(Y^2). \tag{1.68}$$

The variance is, by definition,

$$\sigma_z^2 = \mathscr{E} \, | \, z - m_z|^2 = \sigma_x^2 + \sigma_y^2 = \mathscr{E}|z|^2 - m_z^2. \tag{1.69}$$

For two complex RVs $W = U + jV$, $Z = X + jY$, by definition

$$p_{WZ}(w, z) = p_{UVXY}(u, v, x, y).$$

The covariance is defined as

$$C_{wz} = \mathscr{E}(w - m_w)(z - m_z)^*. \tag{1.70}$$

The correlation coefficient is again

$$\rho_{wz} = C_{wz}/\sigma_w\sigma_z \tag{1.71}$$

and in general is complex.

For complex vector RVs, we generalize the definitions in the previous sections as

$$C_x = \mathscr{E}(\mathbf{X} - \mathbf{m}_x)(\mathbf{X} - \mathbf{m}_x)', \tag{1.72}$$
$$C_{xy} = \mathscr{E}(\mathbf{X} - \mathbf{m}_x)(\mathbf{Y} - \mathbf{m}_y)', \tag{1.73}$$

where the prime means the conjugate of the transpose. We will make use of these definitions in Chapter 11.

Exercises

1.1 Consider a coin toss experiment where the probability of a head is p, and that of a tail is $q = 1 - p$.

(a) Show that the probability of exactly i heads in N independent trials is given by the binomial distribution

$$\binom{N}{i}p^i q^{N-i}, \qquad 0 \le i \le N$$

(b) Show that the mean and variance of i are Np and Npq respectively.

(c) Show that the characteristic function, $E\{e^{j\omega i}\}$, is equal to $(pe^{j\omega} + q)^N$.

(d) Find the first two moments of i by using the characteristic function.

1.2 The Poisson distribution is given by

$$P(k) = \frac{e^{-\mu}\mu^k}{k!}, \qquad k \text{ integer}, \quad k \ge 0$$

(a) Show that the mean and variance are equal to μ.

(b) Show that the characteristic function is $\exp[\mu(e^{j\omega} - 1)]$.

1.3 The exponential probability density function is

$$p(x) = \frac{1}{\sigma}\exp\left[-\left(\frac{x-a}{\sigma}\right)\right], \qquad x \ge \alpha, \quad \sigma > 0$$

(a) Show that the mean and variance of x are $\alpha + \sigma$ and σ^2 respectively.

(b) Show that the characteristic function is

$$C(j\omega) = \frac{e^{j\omega\alpha}}{1 - j\omega\sigma}$$

1.4   The uniform probability density function with zero mean value may be expressed

$$p(x) = \begin{cases} 1/2a, & -a \leq x \leq a \\ 0, & \text{otherwise} \end{cases}$$

Show that the characteristic function is

$$C(j\omega) = \frac{1}{a\omega} \sin \omega a$$

1.5   The Gaussian (also called normal) probability density function with mean and variance denoted by μ and σ^2 respectively is

$$p(x) = \frac{1}{(2\pi)^{1/2}\sigma} \exp \frac{(x - \mu)^2}{-2\sigma^2}, \qquad -\infty < x < \infty$$

(a)   Show that the characteristic function is equal to

$$C(j\omega) = \exp\left(j\mu\omega - \frac{\omega^2\sigma^2}{2} \right)$$

(b)   The odd central moments are obviously zero. By successive differentiation of $\int_{-\infty}^{\infty} e^{-\alpha x^2} \, dx$ with respect to α, show that the even order central moments of the Gaussian variable are

$$\mathcal{E}\{X^m\} = 1 \cdot 3 \cdot 5 \dots (m - 1)\sigma^m, \qquad m \text{ even}$$

1.6   Assume Y is a Gaussian random variable with mean and variance μ and σ^2 respectively. We say that the random variable X is lognormally distributed if $Y = \ln X$, and Y is Gaussian.

(a)   Show that the log normal density may be written

$$p(x) = \frac{1}{(2\pi)^{1/2}\sigma x} \exp \frac{(\ln x - \mu)^2}{-2\sigma^2}, \qquad x \geq 0$$

(b)   Show that the first and second moments of x are

$$\mathcal{E}\{X\} = \exp\left(\mu + \frac{\sigma^2}{2} \right)$$

$$\mathcal{E}\{X^2\} = \exp(2\mu + 2\sigma^2)$$

1.7   A quantizer characteristic or analog-to-digital encoder is shown in Fig. 1.12. The input, denoted X, is usually a continuous variable. The output Y is a discrete variable.

(a)   Write an expression for the probability density of the output in terms of the density function $p(x)$ of the input.

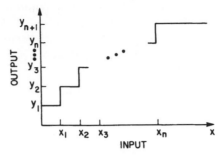

FIGURE 1.12 Quantizer characteristic (analog-to-digital converter).

(b) Assume $p(x) = e^{-x}$, $x \geq 0$. What is the probability density function of the sum of two statistically independent samples of Y.

1.8 Assume the random variables $X_1, X_2, \ldots, X_n$ are statistically independent with means μ_i and variances σ_i^2.

(a) Show that the mean and variance of the sample mean, defined as $\bar{x} = (1/n) \sum_{i=1}^{n} x_i$, are

$$\frac{1}{n} \sum_{i=1}^{n} \mu_i \quad \text{and} \quad \frac{1}{n^2} \sum_{i=1}^{n} \sigma_i^2$$

respectively.

(b) Suppose the X_i's are identically distributed Gaussian variables with zero-mean values, what is the probability density function of $\bar{x}$. Show that $\bar{x}$ is a Gaussian random variable.

(c) Suppose the X_i's are identically distributed exponential variables, that is,

$$p(x) = \frac{1}{\sigma} \exp\left[-\left(\frac{x - \alpha}{\sigma} \right) \right], \qquad x \geq \alpha, \sigma > 0$$

Is $\bar{x}$ exponentially distributed?

1.9 In the preceding problem assume the random variables X_i are statistically dependent. Define

$$\sigma_{|i-j|}^2 \triangleq E\{(X_i - \mu_i)(X_j - \mu_j)\}$$

Show that the variance of the sample mean may be expressed as

$$\frac{\sigma_0^2}{n} + \frac{2}{n} \sum_{i=1}^{n-1} \left(1 - \frac{i}{n}\right) \sigma_i^2$$

1.10 Derive an expression for the density function of a product of two continuous variables such as $Z = XY$ where $x, y \geq \varepsilon > 0$.

1.11 Denote n statistically independent identically distributed random variables as $X_1, X_2, \ldots, X_n$ and their density and distribution functions by $f(x)$ and $F(x)$ respectively. Define a random variable which is the maximum of the X_i's; that is $Y = \max\{X_1, X_2, \ldots, X_n\}$. Show that the probability density function of Y is

$$p(y) = nF^{n-1}(y)f(y)$$

What is the result if $f(x) = e^{-x}$, $x \geq 0$?

1.12 Assume for random variables W and X that

$$E\{W\} = E\{X\} = 0, \qquad E\{WX\} = \rho$$

and the variance of each is σ^2. Consider the transformation

$$Y = aW, \qquad Z = bW + cX$$

where a, b, and c are constants. Find a, b, and c such that $E\{Y^2\} = E\{Z^2\} = 1$ and $E\{YZ\} = 0$.

1.13 Assume we have the discrete random variables S, N, and R with the possible values

$$s_i, \quad i = 1, \ldots, U$$
$$n_i, \quad i = 1, \ldots, V$$
$$r_i, \quad i = 1, \ldots, W$$

(a)   show that

$$P(s_i \mid r_j) = \frac{P(r_j \mid s_i)P(s_i)}{\sum_{i=1}^{U} P(r_j \mid s_i)P(s_i)}$$

This is called Bayes' theorem.

(b)   Suppose further that $R = S + N$, and assume $P(s_1 = 1) = P(s_2 = -1) = \frac{1}{2}$ and $P(n_1 = 1) = P(n_2 = -1) = \frac{1}{2}$. Determine $P(s_i \mid r_j)$ for all i and j. (This problem can be cast as a detection or estimation problem. We want to guess the value of S given that we know only R. We would therefore find useful the probability that S takes on each value given that R is known.)

1.14 For continuous random variables W and X, show that

$$p(w \mid x) = \frac{p(x \mid w)p(w)}{\int p(x \mid w)p(w)\,dw}$$

This is Bayes' theorem expressed in terms of probability density functions (see Exercise 1.13).

1.15 For the continuous random variable Y, show that

$$\int_{-\infty}^{\infty} p(y \mid x) \, dy = 1$$

1.16 For continuous random variables W, X, and Y show that

$$p(w \mid x, y)p(x \mid y) = p(w, x \mid y)$$

1.17 Suppose the random variables W, X, Y, and Z are such that $p(w, x \mid y, z) = p(w \mid y)p(x \mid z)$. Show that

   (a) $p(w \mid y, z) = p(w \mid y)$ or $p(x \mid y, z) = p(x \mid z)$

   (b) $p(w, z \mid y) = p(w \mid y)p(z \mid y)$ or $p(x, z \mid y) = p(x \mid z)p(z \mid y)$

   (c) $p(z \mid y, w) = p(z \mid y)$ or $p(y \mid z, x) = p(y \mid z)$

2 *Random Processes*

*I*n the engineering systems we will be concerned with, something is ongoing in time. Systems have input and output time functions, or the noise in a radio channel continues as time goes on. In this chapter, we extend the ideas of probability to the treatment of waveforms which progress in time in a fashion such that the future is more or less unknowable from observations of the past and present.

2.1 INTRODUCTION

In the last chapter, a random variable X was defined as a real-valued function $X(\zeta)$ over the points ζ representing the outcomes of an abstract experiment $\{\mathcal{S}, \sigma, P\}$. At each trial of the experiment, nature chooses exactly one point $\zeta \in \mathcal{S}$ as elementary outcome, and the experiment in turn presents us with the value $X(\zeta)$ of the random variable. We extended this idea to a multidimensional random variable $\mathbf{X}(\zeta)$, in which each elementary outcome is associated with an *n*-dimensional (column) vector of random variables. In particular, we could interpret the elements $X_i(\zeta)$ of that vector as the samples of some waveform at certain specified instants of time. To each elementary outcome ζ selected by nature there corresponds one set of samples $\{X_i(\zeta)\}$. In the language of this chapter, that set is called a *realization* of a discrete time *random* (or *stochastic*) *process*. The set $\{X_i(\zeta)\}$ considered for all values ζ is the random process itself.

In computations, and in digital system analysis and design, we always deal with a finite number of time samples of whatever waveforms are of interest. For such a case, the random process $\mathbf{X}(\zeta) = \{X_i(\zeta)\}$ is the appropriate concept to apply. However, in the analog world, the time samples t_i become infinitesimally close together, and we can think in terms of a continuous variable t which can take on any real value. In that case, rather than the discrete time random process $X_i(\zeta) = X(t_i, \zeta)$, we think of a continuous time process $X(t, \zeta)$. For each elementary outcome ζ of the experiment, nature selects a time function $X(t, \zeta)$, which represents a particular waveform that would be observed in the system of interest for each particular choice of point $\zeta \in \mathcal{S}$ in the sample space of the abstract experiment. Such a collection of random variables $X(t, \zeta)$ indexed by a continuous variable t is called a continuous time random (or stochastic) process.

In Chapter 1, we defined a discrete random variable (RV) X as one with a piecewise constant distribution function $P_X(x)$. A continuous RV is one with $P_X(x)$ continuous. In terms of random processes, a continuous time random process $X(t, \zeta)$ might have a distribution $P_X(x)$ with discontinuities. Then we have a mixed random process in continuous time. On the other hand, $X(t, \zeta)$ might have any value and therefore represent a continuous RV. Similarly, a discrete time random process $X(t_i, \zeta)$ can be either discrete, with quantized amplitude values, or continuous, with any value X_i allowed. In practice, we always compute with discrete random processes with values $x_j(t_i)$ from a quantized time-sampled signal.

Examples of a (continuous time) random process $X(t, \zeta)$ would be the temperatures indicated on the strip chart of an analog recording thermometer or the noise voltage observed at the output of a communications or radar receiver. Figure 2.1 shows some possible waveforms. There $x^{(i)}(t) = X(t, \zeta_i)$ is the particular waveform we would observe if nature chose the point $\zeta_i \in \mathcal{S}$ as the elementary outcome of the experiment at the *i*th trial. Each such waveform is called a *realization*, or *sample function*, of the random process, and the collection of all such waveforms for a given abstract experiment is called the *ensemble* of sample functions.

As indicated in Fig. 2.1, the process $X(t, \zeta)$ can be considered in either of its two dimensions. If we restrict attention to a specific time instant t_j, we deal with an ensemble of random variables $\{X(t_j, \zeta)\}$, considered as a

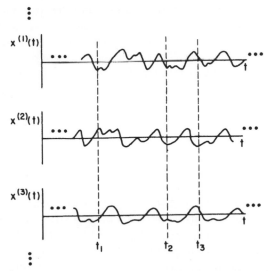

FIGURE 2.1 Members (sample functions) of an ensemble.

function of ζ. On the other hand, if we restrict attention to a single sample function $X(t, \zeta_i)$ of the random process, we deal with an ordinary function of time, that one selected by nature if the trial of the elementary experiment yielded outcome ζ_i. Depending on the purpose at hand, one or the other viewpoint would be appropriate.

Henceforth we will usually use a shorthand notation. As we have emphasized, a random process $X(t, \zeta)$ is a function which has values $x(t, \zeta)$. Unless we need to be explicit, henceforth we will let context distinguish between $X(t, \zeta)$ and $x(t, \zeta)$ and simply use the latter notation. Furthermore, we will assume the reader is now familiar with the idea that a point space underlies all our discussions and that we can drop the explicit dependence on ζ and write just $x(t)$ for $X(t, \zeta)$. Finally, because we will not have much occasion henceforth to refer to $P(\mathscr{A} \in \sigma)$, the probability function defined over the events of the σ-algebra of sets on the sample space $\mathscr{S}$, we will use the notations $P(x)$ and $p(x)$ for the probability distribution and density of the RV X.

2.2 RELATION TO PROBABILITY

In our applications, we will observe some data, say the values $x(t)$ of some sample function of a random process as time evolves. From that, we want to make some conclusions about the situation being observed. For example, we might observe the voltage output of a communications receiver and from that attempt to determine what message has been sent to us by nature (in reality, the sender of the message, whose behavior is presumably not completely known to us, else we would not need the communication channel). As another example, we might have available the power output of a radar receiver for each pulse at a time after transmission corresponding to a particular range, from which we want to decide whether or not a target is present at that range. Again, we might want to predict the future course of a waveform from observations of its past.

In all these cases, an element of randomness is present. We proceed in the face of randomness by ascribing a certain degree of regularity to the situation. Specifically, we model the phenomena as random processes, in the technical sense of the word we defined above, and assume we know the probability structure of the processes involved. That is, we assume we know the values of the probability function $P(\mathcal{A} \in \sigma)$ and how our observables $x(t)$ relate to the point set $\mathcal{S}$ of the underlying abstract experiment. We are able to do that productively because we have available models for the physical processes which introduce randomness (noise), models of the way message sources behave, knowledge of the way in which our instruments convert observables to observations, etc.

All the analytical machinery introduced in the previous chapter, having to do with random variables, carries over without change to random processes. For example, we can speak of the mean value $m(t)$ of a random process $x(t)$ as

$$m(t) = \mathcal{E}[x(t)] = \int_{-\infty}^{\infty} x(t)p[x(t), t] \, dx(t), \tag{2.1}$$

in which t plays the role of a parameter in the integral. Note especially here the notation $p[x(t), t]$ for the density, or just $p(x, t)$ for brevity. The density depends, first, on the values x of the random variable X but also on the time to which those values are indexed.

We will often want to deal with values of the random process $x(t)$ at more than one time instant. Then we consider together the two random variables $x(t_1)$, $x(t_2)$, and their corresponding joint probability density $p[x_1(t_1), x_2(t_2); t_1, t_2]$, or simply $p(x_1 \, x_2; t_1, t_2)$. Extending this to n time samples of a continuous random process, all the properties discussed earlier (Sect. 1.4) for a multidimensional probability density hold for the time samples, for example:

$$\int_{-\infty}^{\infty} p(x_1, \ldots, x_n; t_1, \ldots, t_n) \, dx_1 \cdots dx_n = 1,$$

$$\int_{-\infty}^{\infty} p(x_1, \ldots, x_k, x_{k+1}, \ldots, x_n) \, dx_{k+1} \cdots dx_n = p(x_1, \ldots, x_k),$$

$$p(x_1, \ldots, x_k \mid x_{k+1}, \ldots, x_n) = p(x_1, \ldots, x_n)/p(x_{k+1}, \ldots, x_n), \tag{2.2}$$

where in the last two expressions, for brevity, we drop the time indexes.

Furthermore, we say that the set of samples $\{x_1, \ldots, x_k\}$ is independent of the set $\{x_{k+1}, \ldots, x_n\}$ if

$$p(x_1, \ldots, x_n) = p(x_1, \ldots, x_k)p(x_{k+1}, \ldots, x_n), \qquad (2.3)$$

again dropping the implied time parameters. This latter is equivalent to the requirement that

$$p(x_1, \ldots, x_k \mid x_{k+1}, \ldots x_n) = p(x_1, \ldots, x_k). \qquad (2.4)$$

In this last, we have in mind a fixed set of time instants t_i, $i = 1, n$. If we have two random processes $x(t)$, $y(t)$, in the same way we can speak of the independence of the set of samples $x(t_i)$, $i = 1, n$, from the set $y(t_i)$, $i = 1, n$, for each specific choice of times $\{t_i\}$. If these two sets of samples are independent of each other for every choice we might make of time instants, then the two random processes $x(t)$, $y(t)$ are said to be independent. That is, the processes $x(t)$, $y(t)$ are independent if, for any choice of m, n and for any choices t_i, $i = 1, m$; t_i', $i = 1, n$, we have

$$p(x_1, \ldots, x_m, y_1, \ldots, y_n) = p(x_1, \ldots, x_m)p(y_1, \ldots, y_n).$$

2.3 ENSEMBLE CORRELATION FUNCTIONS

As we will be dealing extensively with random processes, observed over intervals of time, the probability relations of the process from one time instant to another are very important. That is, for a process $x(t)$ we are interested in the density $p(x_1, \ldots, x_n; t_1, \ldots, t_n)$ for all choices that we might make of n and of the particular instants t_i, $i = 1, n$. It is useful to distinguish two kinds of random processes, those whose properties do not depend on the absolute origin of the time axis and those whose properties do depend on absolute time. The former are called (strictly) *stationary* processes, and the latter are called *nonstationary*. The terminology is in analogy with that of a stationary system, whose input–output properties do not depend on time, in contrast to a nonstationary system, such as a system with switches thrown at specific time instants, whose properties do change with time.

That is, a random process is called (strictly) stationary provided

$$p(x_1, \ldots, x_n; t_1, \ldots, t_n) = p(x_1, \ldots, x_n; t_1 - \tau, \ldots, t_n - \tau), \quad (2.5)$$

for all choices of n, t_i ($i = 1, n$), and τ. If this expression holds true, it indicates that the time indexes can be changed arbitrarily (by varying τ), as long as all time differences remain unchanged. (That follows because we can change τ without changing anything on the left of the equality, hence without changing the value of the quantity on the right.) Hence, the absolute time of the origin is irrelevant, and the density depends only on the differences among the time instants of the samples. On the other hand, if Eq. (2.5) does not hold, absolute time does matter, and the process is not stationary.

In our applications, it turns out that we often deal with real Gaussian random processes $x(t)$, which are processes such that any collection of time samples $x(t_i)$ has a multivariate Gaussian density (Chapter 4), and with linear dynamic systems. As we will see, that gives particular prominence to the first joint moment of pairs of time samples of the process. That is, we will often deal with the joint moment

$$R_x(t_1, t_2) = \mathscr{E}[x(t_1) x(t_2)] = \int_{-\infty}^{\infty} x_1 x_2 p(x_1, x_2; t_1, t_2) \, dx_1 \, dx_2. \quad (2.6)$$

This is defined as the (ensemble) *autocorrelation function* of the process $x(t)$. For a real random process $x(t)$, we then have that the variance is

$$\sigma_x^2(t) = \mathscr{E}\{[x(t) - m_x(t)]^2\}$$
$$= \mathscr{E}x^2(t) - m_x^2(t) = R_x(t, t) - m_x^2(t), \quad (2.7)$$

where $m_x(t) = \mathscr{E}x(t)$ is the mean of the process $x(t)$.

In the common case of a stationary random process, we can introduce the lag variable τ in Eq. (2.6) and consider time instants t_1, $t_2 = t_1 - \tau$. Since the process is stationary, we know that the density depends on time only through the differences of the sample times. That is, in this bivariate case,

$$p(x_1, x_2; t_1, t_2) = p(x_1, x_2; t_1 - t_2) = p(x_1, x_2; \tau).$$

Then the autocorrelation function Eq. (2.6) becomes

$$R_x(t_1, t_1 - \tau) = \mathscr{E}x(t_1)x(t_1 - \tau)$$
$$= \int_{-\infty}^{\infty} x(t_1)x(t_1 - \tau)p[x(t_1), x(t_1 - \tau); \tau] \, dx_{t_1} \, dx_{t_1 - \tau} = R_x(\tau).$$
$$*(2.8)$$

Furthermore, since absolute time is irrelevant for a stationary process, the first-order density $p(x; t)$ is independent of time, and the mean

$$m(t) = \mathscr{E}[x(t)] = \int_{-\infty}^{\infty} x_t p(x_t) \, dx_t = m$$

is constant, as is the variance

$$\sigma_x^2 = R_x(0) - m_x^2.$$

A random process $x(t)$ such that its mean is constant: $\mathscr{E}[x(t)] = m$, and its correlation function depends only on time difference: $R_x(t_1, t_2) = R_x(t_1 - t_2)$, is called *wide-sense stationary*. A stationary process is also wide-sense stationary, as we just showed, but the converse need not be true. It turns out in practice that wide-sense stationarity is often enough to lead to considerable simplifications, whereas strict stationarity is very difficult to verify for a particular situation.

In addition to the autocorrelation function $R_x(t_1, t_2)$ of a random process $x(t)$, it is convenient to define the *autocovariance function*,

$$C_x(t_1, t_2) = \mathscr{E}[x(t_1) - m_x(t_1)][x(t_2) - m_x(t_2)]$$
$$= R_x(t_1, t_2) - m_x(t_1)m_x(t_2), \qquad *(2.9)$$

and the normalized autocovariance function, which is the *correlation coefficient*:

$$\rho_x(t_1, t_2) = C_x(t_1, t_2)/[\sigma_x(t_1)\sigma_x(t_2)]. \qquad (2.10)$$

For a wide-sense stationary process, as we have seen the mean and variance are constant, and the autocorrelation function Eq. (2.8) depends only on the time difference $t_1 - t_2$. It follows that the autocovariance function Eq. (2.9) and the normalized autocovariance Eq. (2.10) also depend only on the difference of the two sample times in question.

So far, we have spoken only of real random variables. It will be convenient for later discussions to deal with complex random variables, as introduced in Sect. 1.8. These are just RVs defined as $Z = X + jY$, where X, Y are real. Using the definition

$$p_Z(z) = p_{X,Y}(x, y), \qquad (2.11)$$

the autocorrelation function of a complex random process $Z(t)$ is defined as

$$R_Z(t_1, t_2) = \mathscr{E}[z(t_1)\, z(t_2)^*]$$
$$= \int_{-\infty}^{\infty} (x_1 + jy_1)(x_2 - jy_2)p(x_1, y_1, x_2, y_2; t_1, t_2)\, dx_1\, dy_1\, dx_2\, dy_2,$$
$$(2.12)$$

and the mean is

$$m_z(t) = \mathscr{E}z(t) = m_x(t) + jm_y(t).$$

The variance of a complex RV is defined as

$$\sigma_Z^2(t) = \mathscr{E}|z - m_z|^2 = \mathscr{E}|z(t)|^2 - |m_z(t)|^2$$
$$= R_Z(t, t) - |m_z(t)|^2 = \mathscr{E}[(x - m_x)^2 + (y - m_y)^2] = \sigma_x^2(t) + \sigma_y^2(t).$$
$$(2.13)$$

We are often concerned with two data streams $x(t)$, $y(t)$ and their relationships. If these are both considered as random processes (assumed complex for generality), much of that information is carried by the *cross-correlation function*:

$$R_{xy}(t_1, t_2) = \mathscr{E}[x(t_1)y(t_2)^*]$$
$$= \mathscr{E}[y(t_2)^*x(t_1)] = R_{yx}^*(t_2, t_1). \qquad (2.14)$$

If the two processes $x(t)$, $y(t)$ are wide-sense stationary, then, as in Eq. (2.8),

$$R_{xy}(t, t - \tau) = R_{xy}(\tau). \qquad (2.15)$$

From Eq. (2.14) it then follows, for wide-sense stationary processes, that

$$R_{xy}(\tau) = R_{yx}^*(-\tau), \qquad *(2.16)$$

which has the special case, for a real process, that

$$R_{xy}(\tau) = R_{yx}(-\tau). \qquad (2.17)$$

In the case of an autocorrelation function, Eq. (2.16) shows that

$$R_x(-\tau) = R_x^*(\tau) \qquad *(2.18)$$

if the process is complex, or, for a real process,

$$R_x(-\tau) = R_x(\tau). \qquad (2.19)$$

That is, the autocorrelation function of a real wide-sense stationary random process is an even function of the lag time τ.

Some other properties are of interest. For a wide-sense stationary, possibly complex, process $x(t)$, from the definition Eq. (2.12) there follows

$$R_x(0) = \mathscr{E}|x(t)|^2 \geq 0,$$

with equality if and only if all values of $x(t)$ are zero with probability 1. Furthermore, it can be shown that

$$|R_{xy}(\tau)| \leq [R_x(0) R_y(0)]^{1/2}, \qquad (2.20)$$

or, in particular for an autocorrelation function,

$$|R_x(\tau)| \leq R_x(0). \qquad (2.21)$$

EXAMPLE 2.1   Let us find the autocorrelation function Eq. (2.8) of the random process $x(t) = \cos(\omega_0 t + \theta)$, where the phase angle θ is an RV with the uniform probability density over $[0, 2\pi)$: $p(\theta) = 1/2\pi$, $0 \leq \theta < 2\pi$. This is an example of a periodic random process, in that every sample function is periodic in t. We have

$$R_x(t, t - \tau) = \mathscr{E}\{\cos[\omega_0 t + \theta] \cos[\omega_0(t - \tau) + \theta]\}$$
$$= \tfrac{1}{2}\mathscr{E}\{\cos(\omega_0\tau) + \cos[2(\omega_0 t + \theta) - \omega_0\tau]\}.$$

Since, for uniform θ and any a, we have

$$\mathscr{E} \cos(a + \theta) = \int_0^{2\pi} \left(\frac{1}{2\pi}\right) \cos(a + \theta) \, d\theta = 0,$$

and since the expected value of any constant (not an RV), such as $\cos(\omega_0\tau)$, is just the constant, we have

$$R_x(t, t - \tau) = (1/2) \cos(\omega_0\tau) = R_x(\tau). \qquad (2.22)$$

That is, the autocorrelation function depends only on the lag variable τ and not on absolute time. Noting further that the mean of the process is

$$\mathcal{E}\{\cos(\omega_0 t + \theta)\} = \int_0^{2\pi} \left(\frac{1}{2\pi}\right) \cos(\omega_0 t + \theta) \, d\theta = 0,$$

which is independent of time, we conclude that the process is wide-sense stationary.

EXAMPLE 2.2 A simple model of a binary-coded communication signal is that shown in Fig. 2.2. The waveform can take the values ± 1 in each interval of length T and might or might not switch to the other value at the end of the interval, independent of the current value. We assume the signal started in the infinite past and will continue into the infinite future. Since we don't know what structure, if any, the messages will have, we will assume the two possible values of the wave in any interval are equally likely: $P(\pm 1) = 1/2$ at every value of t. We seek the mean and autocorrelation function of the random process represented by this waveform $x(t)$. This is the simplest case of a class of signals called random telegraph waves.

The mean is the easy part. Since $x(t)$ can take on only the isolated values ± 1, it is a discrete random process with a probability density composed only of delta functions. We have

$$\mathcal{E}[x(t)] = \int_{-\infty}^{\infty} x(t)p[x(t)] \, dx(t)$$

$$= \int_{-\infty}^{\infty} x[\tfrac{1}{2} \delta(x - 1) + \tfrac{1}{2} \delta(x + 1)] \, dx = \tfrac{1}{2}(1 - 1) = 0.$$

To determine the (auto)correlation function, we need the joint density of the RVs $x(t)$ and $x(t - \tau)$. That is, we need the probabilities $P(x_t = \pm 1, x_{t-\tau} = \pm 1)$ corresponding to the four possible situations. We first consider that $|\tau| > T$. Then x_t and $x_{t-\tau}$ are independent, as we assume the value in one interval has no influence on the value in any other interval. That is, for each of the four cases we have

$$P(x_t \mid x_{t-\tau}) = P(x_t) = \tfrac{1}{2},$$

so that

$$P(x_t, x_{t-\tau}) = P(x_t \mid x_{t-\tau}) P(x_{t-\tau}) = (\tfrac{1}{2})(\tfrac{1}{2}) = \tfrac{1}{4}, \qquad |\tau| > T.$$

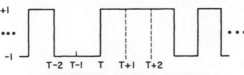

FIGURE 2.2 A binary waveform.

Then

$$R_x(\tau) = \tfrac{1}{4}[(+1)(+1) + (+1)(-1) + (-1)(+1) + (-1)(-1)] = 0, |\tau| > T.$$

If $|\tau| < T$, and if t and $t - \tau$ are such that a potential switching point $t = kT$ is between them, and if a switch does occur, then x_t and $x_{t-\tau}$ have opposite sign. If there is no switch point between them, or if there is a point and no switch actually occurs, then they have the same sign. The two points in question are separated by a distance $|\tau| < T$, and the switch points are separated by T. Then there will be a potential switch point between t and $t - \tau$ in a proportion $|\tau|/T$ of the cases. Since on every interval each value ± 1 of the waveform has equal probability, the probability that a switch actually occurs at any potential switch point is 1/2.

For $|\tau| \leq T$ we can now calculate, for example,

$$\begin{aligned}
P(x_t = 1 \mid x_{t-\tau} = 1) &= \text{Prob}(\{\text{No switch point intervenes}\}) \\
&\quad + \text{Prob}(\{\text{A switch point intervenes}\} \text{ AND} \\
&\qquad \{\text{No switch occurs}\}) \\
&= (1 - |\tau|/T) + (|\tau|/T)(\tfrac{1}{2}) = 1 - |\tau|/2T, \\
P(x_t = 1 \mid x_{t-\tau} = -1) &= \text{Prob}(\{\text{A switch point intervenes}\} \text{ AND} \\
&\qquad \{\text{A switch occurs}\}) \\
&= (|\tau|/T)(\tfrac{1}{2}) = |\tau|/2T, \qquad |\tau| < T.
\end{aligned}$$

The other two cases are similar, and we find

$$\begin{aligned}
P(x_t = 1, x_{t-\tau} = 1) &= P(x_t = -1, x_{t-\tau} = -1) \\
&= P(x_t = -1 | x_{t-\tau} = -1) P(x_{t-\tau} = -1) \\
&= (1 - |\tau|/2T)(\tfrac{1}{2}), \\
P(x_t = 1, x_{t-\tau} = -1) &= P(x_t = -1, x_{t-\tau} = 1) \\
&= |\tau|/4T, \qquad |\tau| < T.
\end{aligned}$$

Then finally we have (Fig. 2.3):

$$\begin{aligned}
R_x(t, t - \tau) &= \sum_{4 \text{ cases}} x_t x_{t-\tau} P(x_t, x_{t-\tau}) \\
&= 1 - |\tau|/T, \qquad |\tau| < T.
\end{aligned}$$

Again we have a wide-sense stationary process.

For a random process which is the product of two other processes, the two being independent of one another and wide-sense stationary, there is a useful relation among the autocorrelation functions. For $z(t) = x(t)y(t)$, we have

$$\begin{aligned}
R_z(t, t - \tau) &= \mathscr{E}[z(t)z^*(t - \tau)] \\
&= \mathscr{E}[x(t)x^*(t - \tau)y(t)y^*(t - \tau)] \\
&= \mathscr{E}[x(t)x^*(t - \tau)\, \mathscr{E}]y(t)y^*(t - \tau)] = R_x(\tau)R_y(\tau). \qquad (2.23)
\end{aligned}$$

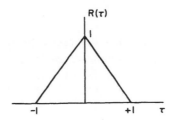

FIGURE 2.3   Correlation function for binary waveform of Fig. 2.2.

The factoring follows because, for two independent processes, the mean of the product is the product of the means:

$$\mathcal{E}(xy) = \int_{-\infty}^{\infty} xyp(x, y) \, dx \, dy = \int_{-\infty}^{\infty} xyp(x)p(y) \, dx \, dy$$

$$= \int_{-\infty}^{\infty} xp(x) \, dx \int_{-\infty}^{\infty} yp(y) \, dy = \mathcal{E}(x)\mathcal{E}(y). \tag{2.24}$$

EXAMPLE 2.3   Suppose $x(t)$ is a real wide-sense stationary random process which amplitude modulates a carrier $y(t) = \cos(\omega_0 t + \theta)$, where, as in Example 2.1, the carrier phase θ is a uniform random variable and is independent of the modulating wave. Then from Eq. (2.23) the autocorrelation of the modulated waveform

$$z(t) = x(t) \cos(\omega_0 t + \theta)$$

is

$$R_z(\tau) = R_x(\tau)R_y(\tau) = \tfrac{1}{2}R_x(\tau) \cos(\omega_0\tau),$$

where we use Eq. (2.22). This is sketched in Fig. 2.4 for a slowly changing $R_x(\tau)$.

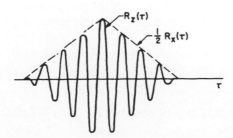

FIGURE 2.4   Autocorrelation function for Example 2.3.

2.4 TIME AVERAGES

In the theory of probability, time averages are not of much interest. All our discussions above have dealt with ensemble averages, and we have needed to know the values of the probability functions $P(\mathcal{A} \in \sigma)$ for the various events $\mathcal{A}$ in the σ-algebra of the underlying abstract experiment. We used those in the form of probability densities of the various RVs of interest. In practice, however, time averages are often all we can calculate, as a data stream is just one sample function from the ensemble. In order to obtain multiple sample functions, we in principle would need to operate, say, multiple radar sets looking at the same surveillance region. The alternative we have assumed so far is to postulate the probability functions from whatever knowledge we have of the physical processes producing the data set we are observing.

Alternatively, time averages can often stand in for ensemble averages. That is, we can consider the approximation

$$\mathcal{E}[x(t)] \cong \langle x(t) \rangle = \lim_{T \Rightarrow \infty} \left(\frac{1}{T} \right) \int_{-T/2}^{T/2} x(t) \, dt, \tag{2.25}$$

where the symbol $\langle \ \rangle$ is defined by the integral. It is necessary to discuss whether or not such a procedure is reasonable.

For each particular sample function $x(t)$, the time average $\langle x(t) \rangle$ will in general have a different constant for its value. It might be expected to be unusual that the number $\langle x(t) \rangle$ would be the same, no matter what sample function $x(t)$ nature happened to choose as the result of a particular trial of the abstract experiment underlying the random process $x(t)$. However, it is often the case that the approximation Eq. (2.25) is an equality, at least in probability—that is, in the sense that $\mathcal{E}[x(t)] = \langle x(t) \rangle$ except for sample functions which have zero probability of occurring. If that is true, the process $x(t)$ is said to be *mean ergodic*. For that to be the case, roughly speaking, in the evolution of any particular sample function $x(t)$, any value that can occur will, if we wait long enough, and in fact in the same proportion as indicated by the probability density $p(x)$.

More generally, a process is said to be *ergodic* if we can compute the ensemble expectation of any functional of the process as the limit of a time average, with equality of the two in the sense of probability. That is, for any function $h(x)$,

$$\mathcal{E}h(x) = \langle h(x) \rangle = \lim_{T \Rightarrow \infty} \left(\frac{1}{T} \right) \int_{-T/2}^{T/2} h[x(t)] \, dt, \tag{2.26}$$

where the equality holds except possibly for sample functions $x(t)$ which have probability zero.

To test whether or not a particular process is ergodic is difficult. However, it is often clear from physical reasoning whether or not the assumption is reasonable. For example, a process such that each time segment is independent of all the others, and which is not tied to an origin in absolute time, might be expected to be ergodic. In such a case, an infinite number of independent time segments might substitute for the infinite number of performances of the underlying experiment.

Fortunately, in the important case of a Gaussian random process, there is a simple criterion for ergodicity. A wide-sense stationary Gaussian process $x(t)$ is ergodic provided its autocorrelation function is such that

$$\int_{-\infty}^{\infty} R_x(\tau)\, d\tau < \infty.$$

On the other hand, it is easy to write down processes which are not ergodic.

EXAMPLE 2.4   Consider the random process

$$x(t) = A\,\cos(\omega_0 t + \theta),$$

where A is an RV with Rayleigh density $p(A) = A\,\exp(-A^2/2)$, $A \geq 0$, and θ has the uniform density on $[0, 2\pi)$, with A and θ being independent. Using Eq. (2.24), the ensemble mean is

$$\mathscr{E}(x) = \mathscr{E}(A)\mathscr{E}[\cos(\omega_0 t + \theta)] = 0,$$

since, as in Example 2.1, the mean of the second factor is 0. Also, applying Eq. (2.25),

$$\langle x(t)\rangle = A\,\cos(\theta)\,\lim_{T \Rightarrow \infty}[\sin(\omega_0 T/2)]/(\omega_0 T/2) = 0.$$

Thus the process is mean ergodic. However,

$$\mathscr{E}(x^2) = \mathscr{E}(A^2)\mathscr{E}[1 + \cos 2(\omega_0 t + \theta)]/2$$

$$= \tfrac{1}{2}\mathscr{E}(A^2) = \tfrac{1}{2}\int_0^{\infty} A^3 \exp(-A^2/2)\, dA = 1,$$

while

$$\langle x^2(t)\rangle = A^2/2 \neq 1.$$

The process is therefore not ergodic.

In the analysis of data, an important operation is computation of the time (auto)correlation function of a sample function of a process. For a complex RV, this is

$$\mathscr{R}_x(\tau) = \lim_{T \Rightarrow \infty}\left(\frac{1}{T}\right)\int_{-T/2}^{T/2} x(t)x^*(t - \tau)\, dt. \qquad (2.27)$$

The time cross-correlation of sample functions of two complex processes is

$$\mathcal{R}_{xy}(\tau) = \lim_{T \to \infty} \left(\frac{1}{T}\right) \int_{-T/2}^{T/2} x(t)y^*(t - \tau) \, dt. \tag{2.28}$$

In the case of ergodic processes, these are equal in probability to respectively the ensemble auto- and cross-correlation functions.

2.5 POWER SPECTRAL DENSITY

In engineering analysis information is often gained by decomposition of signals into their sinusoidal components through some version of Fourier analysis. Partly this is because many phenomena of interest are nearly periodic, and a Fourier analysis concentrates information about their time behavior into a few scalar Fourier coefficients. Furthermore, uniquely among all waveforms, complex exponentials $\exp(-\sigma t) \cos(\omega t + \theta)$, including the special case of the sinusoids $\cos(\omega t + \theta)$, pass through linear systems unchanged in waveshape. Therefore, Fourier decompositions lead to easy analysis of linear system response in terms of transfer functions. In this section we will introduce briefly the special version of Fourier analysis appropriate to random processes.

Any periodic function $x(t) = x(t + T)$ has a Fourier series decomposition into sinusoids:

$$x(t) = \frac{1}{T} \sum_{k=-\infty}^{\infty} X_k \exp\left(\frac{j2\pi kt}{T}\right), \tag{2.29}$$

where

$$X_k = \int_{-T/2}^{T/2} x(t) \exp\left(\frac{-j2\pi kt}{T}\right) dt, \tag{2.30}$$

provided the function is absolutely integrable over a period:

$$\int_{-T/2}^{T/2} |x(t)| \, dt < \infty.$$

In passing to the case of nonperiodic functions (periodic functions with infinite period: $T \Rightarrow \infty$), it is noted in Eq. (2.29) that the frequency spacing $1/T$ of the Fourier components becomes infinitesimal: $1/T \Rightarrow df$, so that the discrete set of frequencies k/T passes over to a continuous variable: $k/T \Rightarrow f$. The sum Eq. (2.29) over the discrete k components passes to an integral, and it is argued heuristically that, from Eqs. (2.29), (2.30), we obtain the Fourier transform pair:

$$x(t) = \int_{-\infty}^{\infty} X(j\omega) \exp(j\omega t) \, df, \tag{2.31}$$

$$X(j\omega) = \int_{-\infty}^{\infty} x(t) \exp(-j\omega t) \, dt, \tag{2.32}$$

(where $\omega = 2\pi f$), provided

$$\int_{-\infty}^{\infty} |x(t)| \, dt < \infty. \tag{2.33}$$

Then $X(j\omega)$ is well defined, because

$$|X(j\omega)| = \left| \int_{-\infty}^{\infty} x(t) \exp(-j\omega t) \, dt \right| \leq \int_{-\infty}^{\infty} |x(t)| \, dt < \infty. \tag{2.34}$$

The difficulty is that Eq. (2.33) is not satisfied for the usual kind of random process we want to deal with. Amplifier noise, for example, goes on forever (unless we turn off the amplifier), and there is no rationale for assuming a decay of the process toward zero as some origin recedes into the past.

On the other hand, it is reasonable to assume that any process likely to be of interest to us has finite average power. Therefore we are willing to let the integral Eq. (2.33) diverge, but we will require

$$\lim_{T \to \infty} \frac{1}{T} \int_{-T/2}^{T/2} |x(t)|^2 \, dt < \infty. \tag{2.35}$$

Then the function $x(t)/\sqrt{T}$ has a Fourier transform in the limit. We then redefine the constant in the Fourier series, and write Eqs. (2.29), (2.30) as the pair:

$$x(t) = \frac{1}{\sqrt{T}} \sum_{k=-\infty}^{\infty} X_k' \exp\left(\frac{j2\pi kt}{T}\right), \tag{2.36}$$

$$X_k' = \frac{1}{\sqrt{T}} \int_{-T/2}^{T/2} x(t) \exp\left(\frac{-j2\pi kt}{T}\right) \, dt. \tag{2.37}$$

In the limit $T \Rightarrow \infty$, the restriction to finite power in Eq. (2.35) guarantees that X_k' exists, because the limit is the Fourier transform of $x(t)/\sqrt{T}$.

However, in the inversion formula Eq. (2.36) we now have $1/\sqrt{T} \Rightarrow \sqrt{(df)}$, which is not such that we obtain an ordinary integral representation for $x(t)$, as we did in Eq. (2.31).

Nonetheless, we proceed. The quantity $X_k'/\sqrt{T}$ is an infinitesimal of order $\sqrt{(df)}$ which depends on frequency and on the function $x(t)$. We denote its limit, as $T \Rightarrow \infty$, as $X(f, df)$:

$$\lim_{T \Rightarrow \infty} (X_k'/\sqrt{T}) = X(f, df). \tag{2.38}$$

Then we can write the pair Eqs. (2.36), (2.37) in the limit as

$$x(t) = \int_{-\infty}^{\infty} X(f, df) \exp(j\omega t), \qquad (2.39)$$

$$X(f, df) = \lim_{T \Rightarrow \infty} \left(\frac{1}{T}\right) \int_{-T/2}^{T/2} x(t) \exp(-j\omega t) \, dt. \qquad (2.40)$$

This identifies $x(t)$ as a linear combination of phasors $\exp(j\omega t)$ with complex amplitudes $X(f, df)$. Just as for any phasor, the average power in each such component is $|X(f, df)|^2$.

Now suppose that $x(t)$ is a sample function of some random process. That is, in Eqs. (2.39), (2.40) we really have $x(t, \zeta)$. Then we are not likely to be interested in the power of the phasor components $X(f, df, \zeta)$ of some particular sample function. However, we are interested in the (ensemble) average power, over the entire ensemble of sample functions $x(t, \zeta)$, in each phasor component. That is just

$$\mathscr{P}_x(f) = \mathscr{E}|X(f, df)|^2. \qquad (2.41)$$

Since $X(f, df)$ is an infinitesimal of order $\sqrt{(df)}$, $\mathscr{P}_x$ is an infinitesimal of order df. If the random process $x(t)$ is stationary, which we henceforth assume, the expected value Eq. (2.41) is independent of time and we can write

$$\mathscr{P}_x(f) = S_x(f) \, df, \qquad (2.42)$$

where $S_x(f)$ is called the *power spectral density* (or *power spectrum*) of the random process $x(t)$.

The power spectrum $S_x(f)$ in Eq. (2.42) has a remarkable relation to the (ensemble) autocorrelation function $R_x(\tau)$ of the process in Eq. (2.8). Once we choose to regard $x(t)$ as a stationary random process $x(t, \zeta)$, the phasor amplitudes $X(f, df)$ become random variables $X(f, df, \zeta)$. It is a fact that they are uncorrelated random variables over frequency. That is, $X(f_1 < f \leq f_1 + df, \zeta)$ and $X(f_2 < f \leq f_2 + df, \zeta)$ are uncorrelated, provided the intervals $(f_1, f_1 + df]$ and $(f_2, f_2 + df]$ do not overlap. Then we can compute the autocovariance function (for a complex process, in general) as

$$R_x(\tau) = \mathscr{E}x(t)x^*(t - \tau)$$

$$= \int_{-\infty}^{\infty} \int_{-\infty}^{\infty} \mathscr{E}[X(f, df)X^*(f', df')] \exp\{j[\omega t - \omega'(t - \tau)]\}$$

$$= \int_{-\infty}^{\infty} \mathscr{E}|X(f, df)|^2 \exp(j\omega\tau) = \int_{-\infty}^{\infty} S_x(f) \exp(j\omega\tau) \, df. \qquad *(2.43)$$

Here the third equality follows because, from the uncorrelated property,

$$\mathscr{E}X(f, df)X^*(f', df') = \mathscr{E}|X(f, df)|^2\delta(f - f'). \qquad (2.44)$$

The last equality in Eq. (2.43) is from Eqs. (2.41), (2.42).

Equation (2.43) shows that the autocorrelation function $R_x(\tau)$ is the inverse Fourier transform of the power spectral density $S_x(f)$ of the process $x(t)$. Therefore the power spectral density is the Fourier transform of the autocorrelation function:

$$S_x(f) = \int_{-\infty}^{\infty} R_x(\tau) \exp(-j\omega\tau) \, d\tau. \qquad *(2.45)$$

This result is the Wiener–Khintchin theorem, for which we hasten to add we have only given a heuristic argument.

It follows from Eq. (2.43) that

$$R_x(0) = \mathscr{E}|x(t)|^2 = \sigma_x^2 + |m_x|^2 = \int_{-\infty}^{\infty} S_x(f) \, df. \qquad *(2.46)$$

Since $\mathscr{E}|x|^2$ is the ensemble average instantaneous power of the sample functions, Eq. (2.46) identifies $S_x(f) \, df$ as the power attributable to phasors $X(f, df)$ with frequencies in the band $(f, f + df]$.

Estimation of the Power Spectrum

In the analysis of data, in seeking the ensemble power spectrum $S_x(f)$ we sometimes compute the time autocorrelation function Eq. (2.27) of an available sample function of a random process:

$$\mathscr{R}_x(\tau) = \lim_{T \Rightarrow \infty} \left(\frac{1}{T}\right) \int_{-T/2}^{T/2} x(t)x^*(t - \tau) \, dt. \qquad (2.47)$$

If the process $x(t)$ is ergodic, $\mathscr{R}_x(\tau)$ is the same as the ensemble correlation function $R_x(\tau)$. Then the power spectral density of the sample function, defined as

$$\mathscr{S}_x(f) = \int_{-\infty}^{\infty} \mathscr{R}_x(\tau) \exp(-j\omega\tau) \, d\tau,$$

is the same as the ensemble power spectral density $S_x(f)$. In practice, however, an infinite span of data $x(t)$ is not available for use in Eq. (2.47). In that case, it is necessary to consider carefully in what sense the density $\mathscr{S}_x(f)$ is an approximation to the power spectrum $S_x(f)$.

There is an enormous literature on this question of power spectrum estimation. One good way to proceed is to break the available data $x(t)$ into some number n of segments $x_i(t)$ of length T. Then (using the fast Fourier transform to compute $X_i(f_k)/\Delta t$) we find Fourier coefficients $X_i(f)$ over each segment of length T, with the definition Eq. (2.30):

$$X_i(f_k) = \int_{t_i-T/2}^{t_i+T/2} x_i(t) \exp(-j\omega_k t) \, dt,$$

where t_i is the center of the segment $x_i(t)$. Then, following Eqs. (2.41), (2.42), and using Eq. (2.40), one uses some form of approximation to the mean, such as

$$S_x(f_k) = \mathscr{E}|X(f, df)|^2/\Delta f \cong \left(\frac{1}{n}\right) \sum_i \frac{[|X_i(f_k)|/T]^2}{\Delta f}$$

$$= \left(\frac{1}{nT}\right) \sum_i |X_i(f_k)|^2, \tag{2.48}$$

where the last step follows because the frequency spacing of the Fourier coefficients is $\Delta f = 1/T$.

In case the random process $x(t)$ contains any pure sinusoidal components, including a zero-frequency (DC) component, the power spectral density $S_x(f)$ will contain impulse functions. Just as in the case of the Fourier transform of a pure sinusoid, this simply indicates that noninfinitesimal power is contained in a differential frequency interval around the frequency of the sinusoid.

EXAMPLE 2.5 As shown in Example 2.1, the process $x(t) = \cos(\omega_0 t + \theta)$, with θ a uniform RV, has autocorrelation function

$$R_x(\tau) = (1/2) \cos(\omega_0 \tau) = (1/4)[\exp(j\omega_0\tau) + \exp(-j\omega_0\tau)].$$

Since

$$\mathscr{F}^{-1}\{\delta(f - f_0)\} = \int_{-\infty}^{\infty} \delta(f - f_0) \exp(j2\pi ft) \, df = \exp(j\omega_0 t),$$

from Eq. (2.45) we have

$$S_x(f) = \mathscr{F}\{R_x(\tau)\} = (1/4)[\delta(f - f_0) + \delta(f + f_0)].$$

Any power spectral density has the following properties. First, from Eqs. (2.41), (2.42),

$$S_x(f) = \mathscr{E}|X(f, df)|^2/df \ge 0, \tag{2.49}$$

so that the spectrum of any random process is real and non-negative. Furthermore, if $x(t)$ is real, so that $R_x(\tau)$ is real, from Eqs. (2.19), (2.45) we have

$$S_x(-f) = \int_{-\infty}^{\infty} R_x(\tau) \exp(j\omega\tau) \, d\tau$$

$$= \int_{-\infty}^{\infty} R_x(-\tau) \exp(j\omega\tau) \, d\tau = S_x(f), \tag{2.50}$$

so that the power spectrum of a real process is an even function of frequency. Going the other way, it can be shown that any non-negative function is the power spectrum of some process and that any even non-negative function is the power spectrum of a real process.

EXAMPLE 2.6 The power spectrum of the binary wave of Example 2.2 is

$$S_x(f) = \int_{-T}^{T} \left(1 - \frac{|\tau|}{T}\right) \exp(-j\omega\tau) \, d\tau$$

$$= 2 \int_{0}^{T} \left(1 - \frac{\tau}{T}\right) \cos(\omega\tau) \, d\tau = \left(\frac{2}{\omega^2 T}\right)[1 - \cos(\omega T)]$$

$$= T\left\{\left[\sin\left(\frac{\omega T}{2}\right)\right] \Big/ \left(\frac{\omega T}{2}\right)\right\}^2. \tag{2.51}$$

EXAMPLE 2.7 The modulated waveform of Example 2.3, $y(t) = x(t) \cos(\omega_0 t + \theta)$, with carrier phase θ a uniform RV, has autocorrelation function

$$R_y(\tau) = (1/2)R_x(\tau) \cos(\omega_0 \tau) = (1/4)R_x(\tau)[\exp(j\omega_0 \tau) + \exp(-j\omega_0 \tau)].$$

The power spectrum is therefore

$$S_y(f) = (1/4)[S_x(f - f_0) + S_x(f + f_0)].$$

This is sketched in Fig. 2.5, for the particular case that $x(t)$ is the random binary waveform of Example 2.2, with power spectrum Eq. (2.51).

Cross-Spectral Densities

We are often interested in the relation between two jointly wide-sense stationary processes $x(t)$, $y(t)$. In particular, we might want to know the correlation between the powers of phasors at the same frequency in the Fourier representations of the two waveforms:

$$\mathcal{P}_{xy}(f) = \mathcal{E}[X(f, df) Y^*(f, df)]. \tag{2.52}$$

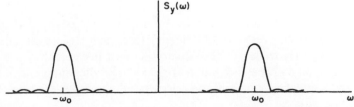

FIGURE 2.5 Power spectral density of sine wave amplitude modulated by a binary waveform.

In the same way as discussed for a single process, this leads to a (cross) spectral desity which is the Fourier transform of the cross-correlation function Eq. (2.14):

$$S_{xy}(f) = \frac{\mathscr{P}_{xy}(f)}{df} = \int_{-\infty}^{\infty} R_{xy}(\tau) \exp(-j\omega\tau) \, d\tau, \qquad *(2.53)$$

so that

$$R_{xy}(\tau) = \int_{-\infty}^{\infty} S_{xy}(f) \exp(j\omega\tau) \, df. \qquad *(2.54)$$

In particular, from this last we have

$$\mathscr{E}[x(t)y^*(t)] = R_{xy}(0) = \int_{-\infty}^{\infty} S_{xy}(f) \, df.$$

Using Eq. (2.16), the cross-spectral density Eq. (2.53) has the property that

$$S_{xy}^*(f) = \int_{-\infty}^{\infty} R_{xy}^*(\tau) \exp(j\omega\tau) \, d\tau = \int_{-\infty}^{\infty} R_{yx}(-\tau) \exp(j\omega\tau) \, d\tau$$

$$= \int_{-\infty}^{\infty} R_{yx}(\tau) \exp(-j\omega\tau) \, d\tau = S_{yx}(f). \qquad (2.55)$$

For real processes, $R_{xy}(\tau)$ is real, and Eq. (2.55) shows that

$$S_{xy}^*(f) = S_{xy}(-f) = S_{yx}(f). \qquad (2.56)$$

The cross-spectral density $S_{xy}(f)$ is estimated in practice from finite data spans of single sample functions $x(t)$, $y(t)$ in the same way as done for the spectral density of a single process. For example, where $X_i(f_k)$, $Y_i(f_k)$ are the Fourier series coefficients Eq. (2.30) over segments of length T, we have in some sense that

$$S_{xy}(f_k) \cong \left(\frac{1}{nT}\right) \sum_i X_i(f_k) Y_i^*(f_k).$$

2.6 RESPONSE OF LINEAR FILTERS

Linear systems are omnipresent in the applications of interest to us. First, some systems are naturally linear, to a high degree of fidelity. Second, we often try to design systems to be linear because they are easy to analyze and have properties which are easy to predict and control. The input $x(t)$ to a linear system and the corresponding output are related by a convolution integral

$$y(t) = \int_{-\infty}^{\infty} h(t, t')x(t') \, dt', \qquad (2.57)$$

where $h(t, \tau)$ is the impulse response, or Green's function, of the system. The impulse response $h(t, \tau)$ is the output of the system if the input is a Dirac delta function applied at time $t = \tau$: $x(t) = \delta(t - \tau)$. That is, from Eq. (2.57),

$$\int_{-\infty}^{\infty} h(t, t')\delta(t' - \tau) \, dt' = h(t, \tau).$$

As we have written the convolution Eq. (2.57), it applies to a system such that the waveform $h(t, \tau)$ in time t of the response to an impulse at time τ depends on the time τ at which the impulse occurs. That means that the system response characteristics depend on absolute time. Such a system is called *nonstationary* (not to be confused with a nonstationary random process, for which the governing probabilities depend on absolute time). In the other case, that the system properties do not depend on absolute time, the response $h(t, \tau)$ at time t depends only on how long ago the input was applied. That is, the response at t to an impulse at τ depends only on the difference $t - \tau$: $h(t, \tau) = h(t - \tau)$. [In this, we use a common abuse of notation by naming two different functions with the same letter in the same equation. What we mean is that the value $h(t, \tau)$ of a function in two variables is equal to the value $h(t - \tau)$ of a function of one variable.]

In the case of a stationary linear system, the convolution Eq. (2.57) then becomes

$$y(t) = \int_{-\infty}^{\infty} h(t - t')x(t') \, dt' = \int_{-\infty}^{\infty} h(t')x(t - t') \, dt'. \qquad (2.58)$$

It is a simple mathematical result that then the Fourier transform $Y(j\omega)$ of the output $y(t)$ is the product

$$Y(j\omega) = H(j\omega)X(j\omega), \qquad (2.59)$$

where $H(j\omega)$, the Fourier transform of the impulse response $h(t)$, is the *transfer function* of the system. A stationary linear system is *stable* if its impulse response $h(t)$, the response at time t to an impulse that occurs at time zero: $x(t) = \delta(t)$, is bounded. We will speak only of stable systems. That is, we assume $|h(t)| < \infty$.

The intimate connection between stationarity and linearity, as expressed by the convolution integral Eq. (2.58), and the Fourier transform relation Eq. (2.59), is easy to understand. Suppose we succeed in expressing the input as a superposition of exponentials:

$$x(t) = \sum_k X_k \exp(j\omega_k t),$$

where the sum might actually be an integral and the coefficients X_k would be written in the notation $X(j\omega_k)$ if the function $x(t)$ were not periodic.

Then from Eq. (2.58) the output is

$$y(t) = \sum_k X_k \int_{-\infty}^{\infty} h(t') \exp[j\omega_k(t - t')] \, dt'$$

$$= \sum_k H(j\omega_k)X_k \exp(j\omega_k t), \tag{2.60}$$

where

$$H(j\omega) = \int_{-\infty}^{\infty} h(\tau) \exp(-j\omega\tau) \, d\tau \tag{2.61}$$

is the Fourier transform of the impulse response. That is, the coefficients in the output representation Eq. (2.60) are $Y_k = H(j\omega_k)X_k$, which is just the Fourier transform relation Eq. (2.59) in different notation.

Random Inputs

All of the foregoing discussion holds whether we regard $x(t)$ as a deterministic input or as a sample function $x(t, \zeta)$ of a random process. In the latter case, the output is also a sample function $y(t, \zeta)$ of a new random process, and we might inquire as to its probabilistic properties, in terms of those of the input process.

First, consider the means $\mathscr{E}[x(t)] = m_x(t)$ and $\mathscr{E}[y(t)] = m_y(t)$. From Eq. (2.57) we have

$$m_y(t) = \mathscr{E} \int_{-\infty}^{\infty} h(t, t')x(t') \, dt' = \int_{-\infty}^{\infty} h(t, t')\mathscr{E}[x(t')] \, dt'$$

$$= \int_{-\infty}^{\infty} h(t, t')m_x(t') \, dt'. \tag{2.62}$$

In the special case of a stationary process, this becomes

$$m_y = m_x \int_{-\infty}^{\infty} h(t, t') \, dt'.$$

If further the system is stationary, we have

$$m_y = m_x \int_{-\infty}^{\infty} h(\tau) \, d\tau = H(0)m_x, \qquad *(2.63)$$

where $H(0)$ is the transfer function evaluated at zero frequency (DC).

The power spectral densities of the input and output stationary processes are related very simply for a stationary system. As in Eq. (2.39), the input process has the expansion

$$x(t) = \int_{-\infty}^{\infty} X(f, df) \exp(j\omega t),$$

where $X(f, df)$ is of order $\sqrt{(df)}$ and $\mathscr{E}|X(f, df)|^2 = S_x(f)\, df$. Because the system is linear and the response to $\exp(j\omega t)$ is just $H(j\omega) \exp(j\omega t)$, the output representation is

$$y(t) = \int_{-\infty}^{\infty} H(j\omega)X(f, df) \exp(j\omega t).$$

This identifies

$$Y(f, df) = H(j\omega)X(f, df), \tag{2.64}$$

so that

$$S_y(f)\, df = \mathscr{E}|Y(f, df)|^2 = |H(j\omega)|^2 \mathscr{E}|X(f, df)|^2 = |H(j\omega)|^2 S_x(f)\, df,$$

and hence the output power spectrum is

$$S_y(f) = |H(j\omega)|^2 S_x(f). \tag{*2.65}$$

EXAMPLE 2.8 Suppose that the input random process is white noise (a random process having the same power at every frequency). That is,

$$S_x = N_0/2, \qquad -\infty < f < \infty. \tag{2.66}$$

(Here and elsewhere we will use $N_0/2$, rather than N_0, in order that N_0 would be the power density if only positive frequencies were considered.) Clearly this is an idealization, since from Eq. (2.46) the ensemble average instantaneous power in the sample functions of the process is

$$\mathscr{E}|x(t)|^2 = \int_{-\infty}^{\infty} \left(\frac{N_0}{2}\right) df = \infty.$$

However, as we shall see, it is a useful idealization. Suppose further that the system is an RC circuit, as in Fig. 2.6, with voltage transfer function

$$H(j\omega) = (1/j\omega C)/(R + 1/j\omega C) = 1/(1 + j\omega RC).$$

It follows at once from Eq. (2.65) that the power spectrum of the output is

$$S_y(f) = S_x(f)/[1 + (\omega RC)^2] = (N_0/2)/[1 + (\omega RC)^2].$$

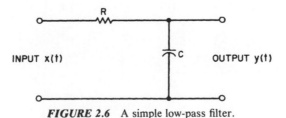

FIGURE 2.6 A simple low-pass filter.

We can note in passing that the output autocorrelation function is then

$$R_y(\tau) = \mathcal{F}^{-1}[S_y(f)] = (N_0/4RC)\exp(-|\tau|/RC).$$

This is an even function of τ, as it must be for $R_y(\tau)$ to be the autocorrelation function of a real random process.

Although it is seldom convenient to do so for a stationary system and process, the autocorrelation of the output can also be obtained directly from the input autocorrelation and the system impulse response. (For a nonstationary system or input process we must proceed in that way however.) For stationary system and input we have

$$R_y(\tau) = \mathcal{E}y_t y_{t-\tau}^* = \int_{-\infty}^{\infty}\int_{-\infty}^{\infty} h(u)h^*(v)\mathcal{E}[x(t-u)x^*(t-\tau-v)]\,du\,dv$$

$$= \int_{-\infty}^{\infty}\int_{-\infty}^{\infty} h(u)h^*(v)R_x(\tau-u+v)\,du\,dv. \tag{2.67}$$

This again yields the input–output expression for the power spectrum, as follows:

$$S_y(f) = \int_{-\infty}^{\infty} R_y(\tau)\exp(-j\omega\tau)\,d\tau$$

$$= \int_{-\infty}^{\infty}\int_{-\infty}^{\infty} h(u)\,h^*(v)\int_{-\infty}^{\infty} R_x(\tau-u+v)\exp(-j\omega\tau)\,d\tau\,dv\,du$$

$$= \int_{-\infty}^{\infty}\int_{-\infty}^{\infty} h(u)h^*(v)\exp[-j\omega(u-v)]\,du\,dv\,S_x(f)$$

$$= |H(j\omega)|^2 S_x(f),$$

after making the change of variable $w = \tau - u + v$ in the inner integral.

EXAMPLE 2.9   Consider the same situation as in Example 2.8. Since $S_x(f) = N_0/2,\ -\infty < f < \infty$, we have the input autocorrelation as

$$R_x(\tau) = \mathcal{F}^{-1}[S_x(f)] = (N_0/2)\delta(\tau). \tag{2.68}$$

The system impulse response is

$$h(t) = \mathcal{F}^{-1}\{1/(1+j\omega RC)\} = (1/RC)\exp(-t/RC)\,U(t),$$

where $U(t)$ is the unit step function: $U(t) = 1,\ t \geq 0$. Then from Eq. (2.67):

$$R_y(\tau) = \int_{-\infty}^{\infty}\int_{-\infty}^{\infty} \left(\frac{1}{RC}\right)^2 \exp\left[\frac{-(u+v)}{RC}\right]$$

$$\times\, U(u)U(v)\left(\frac{N_0}{2}\right)\delta(\tau-u+v)\,du\,dv$$

$$= \left(\frac{1}{RC}\right)^2 \left(\frac{N_0}{2}\right) \int_0^\infty \int_0^\infty \exp\left[\frac{-(u + v)}{RC}\right] \delta(\tau - u + v) \, du \, dv$$

$$= \left(\frac{1}{RC}\right)^2 \left(\frac{N_0}{2}\right) \int_0^\infty \exp\left[\frac{-(\tau + 2v)}{RC}\right] U(\tau + v) \, dv$$

$$= \left(\frac{1}{RC}\right)^2 \left(\frac{N_0}{2}\right) \exp\left(\frac{-\tau}{RC}\right) \int_{\max(0,-\tau)}^\infty \exp\left(\frac{-2v}{RC}\right) \, dv$$

$$= \left(\frac{N_0}{4RC}\right) \exp\left\{\frac{-[\tau + 2\max(0, -\tau)]}{RC}\right\} = \left(\frac{N_0}{4RC}\right) \exp\left(\frac{-|\tau|}{RC}\right).$$

This example illustrates the careful way in which the limits of integration must be handled when working in the time domain rather than the frequency domain.

EXAMPLE 2.10   The running integrator is a linear system with output

$$y(t) = \int_{t-T}^t x(u) \, du. \tag{2.69}$$

The impulse response of the system is by definition

$$h(t) = \int_{t-T}^t \delta(u) \, du = 1, \qquad t - T \le 0 \le t, 0 \le t \le T,$$

so that the transfer function is

$$H(j\omega) = \int_0^T \exp(-j\omega t) \, dt = \frac{1 - \exp(-j\omega T)}{j\omega}$$

$$= T \exp\left(\frac{-j\omega T}{2}\right) \frac{[\sin(\omega T/2)]}{(\omega T/2)}.$$

Hence the output power spectrum in terms of the input power spectrum is

$$S_y(f) = T^2\{[\sin(\omega T/2)]/(\omega T/2)\}^2 S_x(f).$$

It is often useful to know the cross-correlation between the input and output random processes of a linear stationary system. Using the representation Eq. (2.39) of the input as a superposition of phasors, we have

$$x(t) = \int_{-\infty}^\infty X(f, df) \exp(j\omega t),$$

and similarly for the output. The cross-spectral density is

$$S_{yx}(f) = \mathcal{E}[Y(f, df)X^*(f, df)]/df$$
$$= H(j\omega)\mathcal{E}[X(f, df)X^*(f, df)]/df = H(j\omega)S_x(f), \tag{2.70}$$

using Eq. (2.64). Hence also, from Eq. (2.56),

$$S_{xy}(f) = S_{yx}*(f) = H*(j\omega)S_x*(f) = H*(j\omega)S_x(f).$$

Applying the inverse Fourier transform to Eq. (2.70) results in

$$R_{yx}(\tau) = h(\tau)*R_x(\tau) = \int_{-\infty}^{\infty} h(u)R_x(\tau - u) \, du, \qquad (2.71)$$

so that, from Eq. (2.16),

$$R_{xy}(\tau) = R_{yx}*(-\tau) = \int_{-\infty}^{\infty} h*(u)R_x(\tau + u) \, du, \qquad (2.72)$$

using the fact that $R_x^*(-\tau) = R_x(\tau)$ [Eq. (2.18)]. For a real system and a real input process, Eq. (2.72) becomes

$$R_{xy}(\tau) = \int_{-\infty}^{\infty} h(u)R_x(\tau + u) \, du.$$

As a final relation, consider the case of two systems with the same input, illustrated in Fig. 2.7. Again using the decomposition Eq. (2.39) of the input and output processes into phasors, we have at once

$$S_{y_1,y_2}(f) = \mathscr{E}[H_1(j\omega)X(f, df)H_2^*(j\omega)X*(f, df)]/df$$
$$= H_1(j\omega)H_2^*(j\omega)S_x(f). \qquad (2.73)$$

If the systems have Fourier spectra over frequency bands which do not overlap, it follows that $S_{y_1,y_2}(f) \equiv 0$, so that $R_{y_1,y_2}(\tau) \equiv 0$ and the two output processes are uncorrelated.

By cascading two convolutions and being careful to note that, in Eq. (2.73), $H_2^*(j\omega)$ means $[H_2(j\omega)]^*$, which is the transform of the time function $h_2^*(-t)$, it follows from Eq. (2.73) that

$$R_{y_1,y_2}(\tau) = \int_{-\infty}^{\infty}\int_{-\infty}^{\infty} h_1(u)h_2^*(-v)R_x(\tau - u - v) \, du \, dv$$

$$= \int_{-\infty}^{\infty}\int_{-\infty}^{\infty} h_1(u)h_2^*(v)R_x(\tau - u + v) \, du \, dv. \qquad (2.74)$$

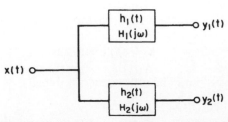

FIGURE 2.7   A single-input, multiple-output linear system.

Exercises

2.1 Assume the random processes $x(t)$ and $y(t)$ are individually and jointly stationary.
    (a)   Find the autocovariance function of $z(t) = x(t) + y(t)$.
    (b)   Repeat for the case when $x(t)$ and $y(t)$ are uncorrelated.
    (c)   Repeat for uncorrelated signals with zero means.

2.2 Assume the stationary random process $x(t)$ is periodic with period T, that is, $x(t) = x(t - T)$. Show that $R_x(\tau) = R_x(\tau + T)$.

2.3 For the complex function $z(t) = e^{j(\omega t + \theta)}$, assume θ is uniformly distributed $(0, 2\pi)$. Find

$$E\{z(t)z^*(t - \tau)\} \qquad \text{and} \qquad E\{z(t)z(t - \tau)\}$$

2.4 A first-order Markov process has the property that

$$p(x_n | x_{n-1}, x_{n-2}, \ldots, x_1) = p(x_n | x_{n-1})$$

For such a process show that

$$p(x_n, x_{n-1}, \ldots, x_1) = p(x_n | x_{n-1})p(x_{n-1} | x_{n-2}) \cdots p(x_2 | x_1)p(x_1)$$

2.5 Assume that $x(t)$ is a stationary input to a time-invariant linear filter with transfer function $H(j\omega)$. Denote its output by $y(t)$. Show that

$$S_y(\omega) = H^*(j\omega)S_{yx}(\omega)$$

2.6 White noise with autocorrelation function $(N_0/2)\delta(\tau)$ is inserted into a filter with $|H(j\omega)|^2$ as shown in Fig. 2.8. What is the total noise power measured at the filter output?

2.7 For white noise with $R(\tau) = (N_0/2)\delta(\tau)$ as an input to the low-pass RC filter (Example 2.8), show that the autocorrelation function of the output can be expressed as

$$R(t_3 - t_1) = \frac{R(t_3 - t_2)R(t_2 - t_1)}{R(0)}$$

for $t_3 > t_2 > t_1$.

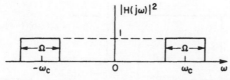

FIGURE 2.8 Absolute value (squared) of transfer function.

2.8   Consider a zero-mean stationary process $x(t)$ inserted into a linear filter which is a truncated exponential

$$h(t) = \begin{cases} ae^{-at}, & 0 \leq t \leq T \\ 0, & \text{otherwise} \end{cases}$$

Show that the power spectral density of the output is

$$\frac{a^2}{a^2 + \omega^2}(1 - 2e^{-aT}\cos \omega T + e^{-2aT})S_x(\omega)$$

2.9   For the system of Fig. 2.9, assume $x(t)$ is stationary. Show that the power spectral density of $y(t)$ is

$$S_y(\omega) = 2S_x(\omega)(1 + \cos \omega T)$$

2.10   Let $x(t)$ be a zero-mean stationary input to a filter $h(t)$, and denote its output by $y(t)$. The output correlation function is denoted by $R_y(\tau)$. Let x_1, x_2, x_3, and x_4 be samples of the process $x(t)$ and assume that

$$E\{x_1 x_2 x_3 x_4\} = E\{x_1 x_2\}E\{x_3 x_4\} + E\{x_1 x_3\}E\{x_2 x_4\} + E\{x_1 x_4\}E\{x_2 x_3\}$$

Show for the output process that

$$E\{y(t_1)y(t_2)y(t_3)y(t_4)\} = R_y(\Delta_{12})R_y(\Delta_{34}) + R_y(\Delta_{13})R_y(\Delta_{24})$$
$$+ R_y(\Delta_{14})R_y(\Delta_{23})$$

where $\Delta_{ij} = t_i - t_j$.

2.11   Assume a signal

$$f(t) = a\cos(t + \phi)$$

where a may or may not be a random variable. Assume ϕ is uniformly distributed in the range $(0, 2\pi)$. Find both the time autocorrelation function and ensemble autocorrelation function. Under what conditions on the amplitude a are the autocorrelation functions equal?

2.12   Consider the random process

$$x(t) = a\cos(\omega t + \theta)$$

where a is a constant, θ is uniformly distributed $(0, 2\pi)$, and ω is a random variable with a probability density which is an even function of

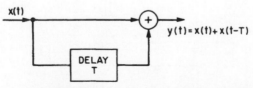

FIGURE 2.9   A linear system.

its argument. That is $p(\omega) = p(-\omega)$. Show that the power spectral density of $x(t)$ is $a^2 \pi p(\omega)$.

2.13 An amplitude-modulated signal is given by

$$y(t) = a(t) \cos(\omega_c t + \theta)$$

where ω_c is a constant, $a(t)$ is a random process, and θ is uniformly distributed $(0, 2\pi)$. Determine the autocorrelation function and power spectral density of $y(t)$.

2.14 For a signal

$$z(t) = x(t) \cos \omega t - y(t) \sin \omega t$$

assume $x(t)$ and $y(t)$ are stationary.
  (a) Find the ensemble correlation function of $z(t)$.
  (b) Repeat the above but with $R_x(\tau) = R_y(\tau)$, and $R_{xy}(\tau) = 0$, thus showing that $R_z(\tau) = R_x(\tau) \cos \omega \tau$.

2.15 For the stationary random process $x(t)$, show that

$$\lim_{\varepsilon \to 0} E\{[x(t + \varepsilon) - x(t)]^2\} = 0$$

if $R_x(t, s)$ is continuous at $t = s$.

2.16 Assume that the derivative of a random process $x(t)$ exists; that is,

$$\frac{dx(t)}{dt} \triangleq \lim_{\varepsilon \to 0} \frac{x(t + \varepsilon) - x(t)}{\varepsilon}$$

exists. In the following problems, assume that the indicated operations exist.
  (a) Find $E\{dx(t)/dt\}$.
  (b) Show that the autocorrelation function of the derivative is $-d^2 R_x(\tau)/d\tau^2$.
  (c) Show $E\{x(t)\, dx(t)/dt\} = dR_x(\tau)/d\tau|_{\tau=0}$.
  (d) Show $E\{x(t)\, dx(t - \tau)/dt\} = -dR_x(\tau)/d\tau$.

2.17 Consider the process $dx(t)/dt$, if it exists, to be generated by passing the process $x(t)$ through a filter with transfer function $H(j\omega) = j\omega$. What are the power cross-spectral density of $x(t)$ and $dx(t)/dt$ and the power spectral density of $dx(t)/dt$?

3 Narrowband Signals

Many of the signals met in practice are narrowband. That is, their spectra have a relatively narrow width B in frequency, concentrated around a relatively high carrier frequency f_c (Fig. 3.1). In the time domain, such signals appear as a slowly varying envelope function $f(t)$ confining the rapid oscillations of the carrier at f_c Hz (Fig. 3.2). Such a situation might be arranged because the propagation medium will not support radiation at the low frequency B directly but will propagate waves at frequency f_c. In radar, for example, the bandwidth of the pulse to be transmitted [the bandwidth of the envelope $f(t)$ in Fig. 3.2] might be 10 MHz, while a carrier frequency of 1 GHz might be required to obtain propagation and antenna directivity with a structure of reasonable size. As another example, the audio speech signal has a bandwidth of about 3 kHz. In communications, thousands of such signals are typically stacked together in frequency by using them to modulate subcarriers of higher frequencies f_c spaced at intervals on the order of the signal bandwidth.

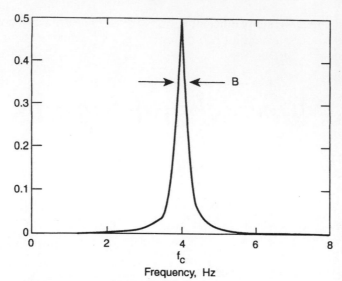

FIGURE 3.1   Magnitude of spectrum of $t \exp(-t) \cos(8\pi t)$.

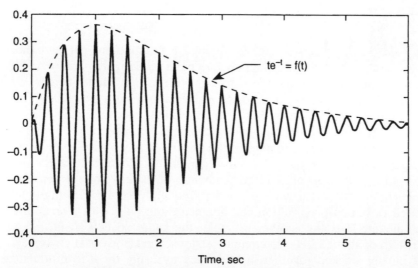

FIGURE 3.2   A 4-Hz carrier modulated by $t \exp(-t)$.

In this chapter, we study efficient ways to treat the information content of such signals, contained in the envelope, while suppressing the largely irrelevant carrier oscillations.

3.1 THE ANALYTIC SIGNAL

Let $f(t)$ be any real signal. Its Fourier spectrum is

$$F(\omega) = \int_{-\infty}^{\infty} f(t) \exp(-j\omega t) \, dt. \qquad (3.1)$$

[Here and in the following we use the notation $F(\omega)$ rather than the customary $F(j\omega)$ for later clarity.] It is evident that the spectrum is conjugate symmetric, in that

$$F(-\omega) = \int_{-\infty}^{\infty} f(t) \exp(j\omega t) \, dt = F^*(\omega). \qquad (3.2)$$

Because of the symmetry in Eq. (3.2), the negative frequencies in the spectrum of a real signal carry redundant information. As so much signal processing is carried out in the frequency domain, it is often convenient to discard that redundant information and deal only with the positive frequency part of the spectrum. The real utility of doing that will be clear only when we consider narrowband signals, as in Fig. 3.1. However, the procedure we discuss here is completely general.

Consider then a real signal $f(t)$ with spectrum $F(\omega)$. From $f(t)$ we want to construct a signal $f_p(t)$ which will retain all the information in $F(\omega)$ for $f = \omega/2\pi \geq 0$, but which will have none of the redundant information for $f < 0$. The simplest such signal is one with spectrum

$$F_p(\omega) = F(\omega)[1 + \text{sgn}(f)], \qquad (3.3)$$

where

$$\text{sgn}(f) = \begin{cases} 1, & f > 0, \\ 0, & f = 0, \\ -1, & f < 0. \end{cases} \qquad (3.4)$$

That is,

$$F_p(\omega) = \begin{cases} 2F(\omega), & f > 0, \\ F(\omega), & f = 0, \\ 0, & f < 0. \end{cases} \qquad *(3.5)$$

The signal $f_p(t)$ with the spectrum Eq. (3.5) is called the *analytic signal* corresponding to the real signal $f(t)$. For reasons we will discuss below, it is sometimes also called the *preenvelope* of the signal $f(t)$.

Since $f_p(t)$ has a spectrum Eq. (3.5) which does not satisfy the symmetry condition Eq. (3.2) required of any real signal, the analytic signal $f_p(t)$ is complex. In fact, although it is a property we will not have need to invoke, $f_p(t)$ is an analytic function of the variable t, if t were taken to be complex: $t = t' + jt''$. This is seen by writing

$$f_p(t) = 2 \int_0^\infty [F_1(\omega) + jF_2(\omega)] \exp(j\omega t) \, df = f_{p1}(t) + jf_{p2}(t),$$

where $F(\omega)$ and $f_p(t)$ are separated into their real and imaginary parts. Using $\exp(j\omega t) = \cos(\omega t) + j \sin(\omega t)$, it is then easy to verify that the Cauchy–Riemann equations are satisfied. That is,

$$\partial f_{p1}/\partial t' = \partial f_{p2}/\partial t'', \qquad \partial f_{p1}/\partial t'' = -\partial f_{p2}/\partial t'.$$

Therefore $f_p(t)$ is an analytic function of the complex variable t.

From the definition Eq. (3.3) follows an important property of the analytic signal, namely

$$f(t) = \text{Re}\{f_p(t)\}. \tag{3.6}$$

That is, the original signal is recovered from its companion (complex) analytic signal by taking the real part. To see that Eq. (3.6) is true, we write

$$g(t) = \text{Re}\{f_p(t)\} = (1/2)[f_p(t) + f_p^*(t)].$$

Recalling that $f_p(t)$ is a complex signal, we have

$$\mathcal{F}\{f_p^*(t)\} = \int_{-\infty}^\infty f_p^*(t) \exp(-j\omega t) \, dt = [F_p(-\omega)]^*.$$

From this and Eq. (3.3) we have

$$G(\omega) = \tfrac{1}{2}\{F_p(\omega) + [F_p(-\omega)]^*\}$$

$$= \tfrac{1}{2} \begin{Bmatrix} 2F(\omega) + 0, & f > 0 \\ F(0) + F(0), & f = 0 \end{Bmatrix} = F(\omega), \qquad f \ge 0.$$

Since $g(t)$ is real, and since the spectrum $G(\omega)$ equals $F(\omega)$ over the nonnegative frequencies, the symmetry condition Eq. (3.2) guarantees that then $G(\omega) = F(\omega)$, so that $g(t) = f(t)$, which is just Eq. (3.6).

EXAMPLE 3.1   Consider the real signal

$$f(t) = \left(\frac{\sin(\pi t)}{\pi t}\right)^2, \tag{3.7}$$

which has the spectrum

$$F(\omega) = 1 - |f|, \qquad |f| \le 1. \tag{3.8}$$

The spectrum of the corresponding analytic signal is

$$F_p(\omega) = 2(1 - f), \qquad 0 < f \le 1,$$
$$F_p(0) = 1.$$

Then

$$f_p(t) = 2 \int_0^1 (1 - f) \exp(j\omega t) \, df$$
$$= \left(\frac{\sin(\pi t)}{\pi t}\right)^2 + \frac{j}{\pi t}\left(1 - \frac{\sin(2\pi t)}{2\pi t}\right). \tag{3.9}$$

Note that $f(t) = \text{Re}\{f_p(t)\}$, as claimed in Eq. (3.6). In this example $F_p(\omega)$ is discontinuous at $\omega = 0$. The transform of Eq. (3.9) is an integral which converges to the midpoint $[F_p(0+) - F_p(0-)]/2$. The defining equation Eq. (3.4) accords with that value.

3.2 NARROWBAND SIGNALS

The main usefulness of the analytic signal comes in dealing with narrowband signals such as illustrated in Fig. 3.1. A common way in which a narrowband signal arises is through modulation of a carrier with an information signal. We will consider that formulation specifically, so that we consider signals

$$f(t) = a(t) \cos[\omega_c t + \phi(t)]. \tag{3.10}$$

Although any signal $f(t)$ can be put into the form Eq. (3.10), we have in mind that $a(t) > 0$ and changes slowly with respect to the carrier period $1/f_c$.

It is useful to rewrite Eq. (3.10) as

$$f(t) = \text{Re}\{\tilde{f}(t) \exp(j\omega_c t)\}, \tag{3.11}$$

where

$$\tilde{f}(t) = a(t) \exp[j\phi(t)] = x(t) + jy(t) \qquad *(3.12)$$

is defined as the *complex envelope* of the real signal $f(t)$. Its absolute value is the (real) envelope:

$$a(t) = |\tilde{f}(t)| = [x^2(t) + y^2(t)]^{1/2}, \tag{3.13}$$

and its phase is the angle modulation of the waveform Eq. (3.10):

$$\phi(t) = \tan^{-1}[y(t)/x(t)], \tag{3.14}$$

where the inverse tangent is the four-quadrant inverse.

Suppose now that the real envelope $a(t)$, the amplitude modulation function of the waveform Eq. (3.10), has a bandwidth B which is narrow with respect to the carrier frequency f_c: $B/2 < f_c$, as illustrated in Fig. 3.3a. From Eq. (3.10) the modulated signal spectrum Fig. 3.3b is

$$
\begin{aligned}
F(\omega) &= \mathscr{F}\{(a/2) \exp[j(\omega_c t + \phi)] + (a/2) \exp[-j(\omega_c t + \phi)]\} \\
&= \tfrac{1}{2}\mathscr{F}\{\tilde{f} \exp(j\omega_c t) + \tilde{f}^* \exp(-j\omega_c t)\} \\
&= \tfrac{1}{2}\{\tilde{F}(\omega - \omega_c) + \tilde{F}^*[-(\omega + \omega_c)]\}. \tag{3.15}
\end{aligned}
$$

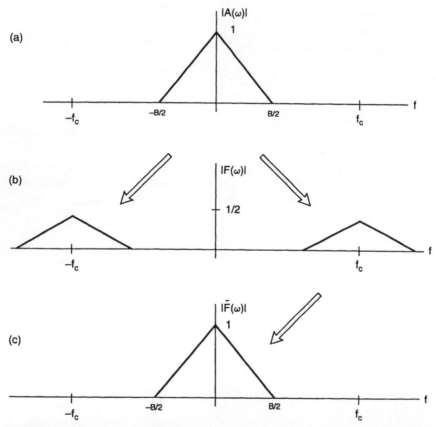

FIGURE 3.3 Spectra of modulated and demodulated waves. (a) Amplitude spectrum of modulating function. (b) Amplitude spectrum of modulated wave. (c) Amplitude spectrum of complex envelope.

From Eq. (3.5) it then follows that

$$F_p(\omega) = \tilde{F}(\omega - \omega_c),$$

since the narrowband nature of $f(t)$ causes the two component spectra in Eq. (3.15) to have no overlap in frequency. It follows that

$$\tilde{F}(\omega) = F_p(\omega + \omega_c), \tag{3.16}$$

as in Fig. 3.3c. Finally,

$$\tilde{f}(t) = f_p(t) \exp(-j\omega_c t). \tag{*(3.17)}$$

The result Eq. (3.17) indicates that the modulations on the signal Eq. (3.10) can be recovered by computing the analytic signal. The real envelope is

$$a(t) = |\tilde{f}(t)| = |f_p(t)|, \tag{3.18}$$

which indicates why the analytic signal $f_p(t)$ is also called the preenvelope of the modulated waveform $f(t)$. Both the amplitude and phase modulations Eqs. (3.13), (3.14) can thus be recovered by computing the spectrum $F(\omega)$ of the modulated signal, separating out and scaling the positive frequency components to obtain the spectrum Eq. (3.5) of the analytic signal, down-shifting in frequency in accord with Eq. (3.16), computing the complex envelope $\tilde{f}(t)$ from $\tilde{F}(\omega)$ and then the modulations $a(t)$, $\phi(t)$ from its amplitude and phase.

EXAMPLE 3.2 Consider the carrier $\cos(\omega_c t)$ amplitude-modulated by the pulse $t \exp(-t)$, $t \geq 0$, as in Fig. 3.2. The spectrum of the modulating signal is

$$A(f) = \int_0^\infty t \exp(-t) \exp(-j\omega t) \, dt = \frac{1}{(j\omega + 1)^2}. \tag{3.19}$$

The modulated signal then has spectrum

$$F(\omega) = \tfrac{1}{2}[A(\omega - \omega_c) + A(\omega + \omega_c)]. \tag{3.20}$$

The amplitude of this is plotted in Fig. 3.1 for the case $f_c = 4$. The bandwidth of the modulating signal Eq. (3.19) is small enough that essentially no part of the shifted spectrum $A(\omega - \omega_c)$ extends into the negative frequency domain. Then, as in Fig. 3.3b and c, in the frequency domain the positive frequency component of Eq. (3.20) can be separated out to recover

$$\tilde{F}(\omega - \omega_c) = 2F(\omega) = F_p(\omega) = A(\omega - \omega_c),$$

from which the modulation spectrum $A(\omega)$ obviously follows.

3.3 HILBERT TRANSFORM

In Eq. (3.3) we defined the analytic signal $f_p(t)$ corresponding to a real signal $f(t)$ by means of its spectrum:

$$F_p(\omega) = F(\omega)[1 + \text{sgn}(f)].$$

We now write this in the form

$$F_p(\omega) = F(\omega) + j\hat{F}(\omega), \qquad (3.21)$$

where

$$\hat{F}(\omega) = H(\omega)F(\omega), \qquad *(3.22)$$

with

$$H(\omega) = -j\,\text{sgn}(f), \qquad (3.23)$$

where $\text{sgn}(f)$ is the sign function defined in Eq. (3.4). The system with transfer function Eq. (3.23) is called a Hilbert transformer, and the signal $\hat{f}(t)$ with spectrum Eq. (3.22) is the *Hilbert transform* of $f(t)$. [Note that $\text{sgn}(0) = 0$, so that $H(\omega)$ has zero response at DC.]

The Hilbert transform $\hat{f}(t)$ of any even signal $f(t)$ is an odd function of time. This follows from Eqs. (3.22), (3.23) as

$$\mathcal{F}\{\hat{f}(-t)\} = \hat{F}(-\omega) = H(-\omega)F(-\omega) = -H(\omega)F(\omega) = -\mathcal{F}\{\hat{f}(t)\},$$

so that

$$\hat{f}(-t) = -\hat{f}(t), \qquad f(t) \text{ even}. \qquad (3.24)$$

Similarly, for an odd $f(t)$,

$$\mathcal{F}\{\hat{f}(-t)\} = H(-\omega)F(-\omega) = [-H(\omega)][-F(\omega)] = H(\omega)F(\omega),$$
$$\hat{f}(-t) = \hat{f}(t), \qquad f(t) \text{ odd}. \qquad (3.25)$$

In the case of a real signal $f(t)$, the spectrum $F(\omega)$ has the symmetry in Eq. (3.2). From Eq. (3.23) we then have

$$\hat{F}(-\omega) = H(-\omega)F(-\omega) = j\,\text{sgn}(f)F^*(\omega)$$
$$= H^*(\omega)F^*(\omega) = \hat{F}^*(\omega).$$

Therefore $\hat{f}(t)$ is real. From Eq. (3.21)

$$f_p(t) = f(t) + j\hat{f}(t), \qquad (3.26)$$

so that then

$$f(t) = \text{Re}\{f_p(t)\}, \qquad \hat{f}(t) = \text{Im}\{f_p(t)\}. \qquad (3.27)$$

Furthermore, combining Eqs. (3.17), (3.26), we have

$$\tilde{f}(t) = [f(t) + j\hat{f}(t)]\exp(-j\omega_c t). \qquad (3.28)$$

Therefore the amplitude modulation (real envelope) $a(t)$ on a signal Eq. (3.10) can be found as

$$a(t) = |\tilde{f}(t)| = [f^2(t) + \hat{f}^2(t)]^{1/2}, \tag{3.29}$$

using Eq. (3.18).

EXAMPLE 3.3   In Example 3.1 we considered the real signal

$$f(t) = \left(\frac{\sin(\pi t)}{\pi t}\right)^2,$$

and found its analytic signal as in Eq. (3.9). We know at once from Eq. (3.27) that the Hilbert transform of this $f(t)$ is

$$\hat{f}(t) = \frac{1}{\pi t}\left(1 - \frac{\sin(2\pi t)}{2\pi t}\right).$$

From Eqs. (3.22), (3.23) the Hilbert transform of $f(t)$ can be written in terms of the convolution

$$\hat{f}(t) = \int_{-\infty}^{\infty} h(t - t')f(t')\, dt'. \tag{3.30}$$

From Eq. (3.23), the spectrum of the Hilbert transformer has infinite bandwidth, so that the impulse response needs to be determined by a limiting process. We also must use the principal value of the improper integral in the inversion. We have

$$h(t) = \lim_{\substack{\alpha \Rightarrow 0 \\ T \Rightarrow \infty}} \int_{-T}^{T} \exp(-\alpha|f|)H(\omega)\exp(j\omega t)\, df$$

$$= \lim_{\alpha \Rightarrow 0} 2\int_{0}^{\infty} \exp(-\alpha f)\sin(\omega t)\, df = 1/(\pi t). \tag{3.31}$$

With this, the Hilbert transform Eq. (3.30) appears as

$$\hat{f}(t) = \left(\frac{1}{\pi}\right)\lim_{T \Rightarrow \infty}\int_{-T}^{T} \frac{f(t')}{t - t'}\, dt'. \tag{3.32}$$

The integral in this is again the principal value of the improper integral. It is necessary because the impulse response Eq. (3.31) exists only in the limiting sense indicated.

The Hilbert transformer Eq. (3.23) has the property

$$1/H(\omega) = j/\text{sgn}(f) = j\,\text{sgn}(f) = -H(\omega), \qquad f \neq 0. \tag{3.33}$$

Equation (3.22) then yields the inverse Hilbert transform:

$$F(\omega) = \hat{F}(\omega)/H(\omega) = -H(\omega)\hat{F}(\omega), \tag{3.34}$$

that is (the integral is the principal value),

$$f(t) = \left(\frac{-1}{\pi t}\right) * \hat{f}(t) = \left(\frac{-1}{\pi}\right) \int_{-\infty}^{\infty} \frac{\hat{f}(t')}{t - t'} \, dt'. \qquad *(3.35)$$

Computation of the Hilbert transform Eq. (3.32) or its inverse Eq. (3.35) in the time domain using convolution is usually not simple, analytically. The procedure can be carried out with sampled waveforms in terms of a discrete convolution, or by a digital filter with the appropriate impulse response, that is, a bandlimited and sampled version of $1/(\pi t)$. Alternatively, the fast Fourier transform (FFT) can be used to compute the spectrum of the signal to be transformed. That is then adjusted by the Hilbert transformer Eq. (3.23) to obtain the spectrum of the Hilbert transform. An inverse FFT then gives the samples of the time function $\hat{f}(t)$.

EXAMPLE 3.4   Consider again the time waveform of Example 3.3:

$$f(t) = \left(\frac{\sin(\pi t)}{\pi t}\right)^2.$$

Its spectrum is strictly bandlimited to $|f| < 1$ Hz. We can therefore sample the signal $f(t)$ at 2 Hz. Using a higher rate for clarity of plotting, we choose 10 Hz and calculate the samples $\{f_i\} = f(i\,\delta t)$, $i = -255,256$, for $\delta t = 0.1$ s. These 512 points span $-25.5 < t < 25.6$, which is adequate to avoid end effects in a circular convolution. The FFT is then taken to produce spectral coefficients $\{F_k\} = \text{FFT}(\{f_i\})$, $k = 0,511$. The samples of the Hilbert transform of $f(t)$ are then computed as the 512-point inverse FFT of the coefficients

$$\begin{aligned}
G_0 &= (0)F_0 = 0, \\
G_k &= -jF_k, &\quad k &= 1,256, \\
G_k &= jF_k, &\quad k &= 257,511.
\end{aligned}$$

The result is plotted in Fig. 3.4, together with the original $f(t)$. The computed transform $\hat{f}(t)$ is indistinguishable at the scale of the plot from the analytic result indicated in Example 3.3.

Properties of the Hilbert Transform

For completeness, and because some of them are useful in discussing the envelopes of modulated waveforms, we will develop here some of the properties of the Hilbert transform.

1. The Hilbert transform of the Hilbert transform $\hat{f}(t)$ of a function $f(t)$ is the negative of the original $f(t)$, with the DC component removed. This follows at once from the definition Eq. (3.23) of the Hilbert transformer:

$$\mathcal{F}[\mathcal{H}\{\hat{f}(t)\}] = H(\omega)\hat{F}(\omega) = H(\omega)H(\omega)F(\omega)$$

$$= H^2(\omega)F(\omega) = -[\text{sgn}(f)]^2 F(\omega) = \begin{cases} -F(\omega), & f \neq 0, \\ 0, & f = 0, \end{cases}$$

$$\mathcal{H}\hat{f}(t) = -f(t) - (\text{DC}). \tag{3.36}$$

2. If a signal $f(t)$ is the convolution of two other signals $u(t)$ and $v(t)$, that is, if $f(t) = u(t) * v(t)$, then the Hilbert transform of $f(t)$ is

$$\hat{f}(t) = u(t) * \hat{v}(t) = \hat{u}(t) * v(t). \tag{3.37}$$

This follows at once from Eq. (3.22):

$$\hat{F}(\omega) = H(\omega)F(\omega) = H(\omega)U(\omega)V(\omega)$$
$$= U(\omega)[H(\omega)V(\omega)] = \mathcal{F}[u(t) * \hat{v}(t)]$$
$$= [H(\omega)U(\omega)]V(\omega) = \mathcal{F}[\hat{u}(t) * v(t)].$$

3. Suppose that a (real or complex) signal $f(t)$ has a spectrum which is bandlimited to $|f| < B/2$. Then

$$\mathcal{H}\{f(t) \cos(\omega_c t)\} = f(t) \sin(\omega_c t),$$
$$\mathcal{H}\{f(t) \sin(\omega_c t)\} = -f(t) \cos(\omega_c t), \tag{3.38}$$

provided that $B/2 < f_c$.

Considering the first of Eq. (3.38), from

$$\cos(\omega_c t) = \tfrac{1}{2}[\exp(j\omega_c t) + \exp(-j\omega_c t)],$$

we have the spectrum

$$x(t) = f(t) \cos(\omega_c t) \Rightarrow \tfrac{1}{2}[F(\omega + \omega_c) + F(\omega - \omega_c)], \tag{3.39}$$

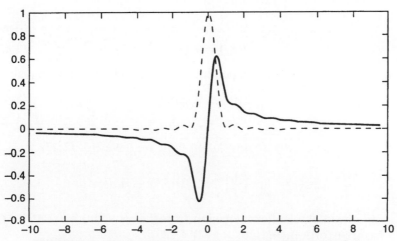

FIGURE 3.4 The function of Example 3.4 and its Hilbert transform.

as sketched in Fig. 3.3b. Applying the Hilbert transformer Eq. (3.23) in the form

$$H(\omega) = \begin{cases} \exp(j\pi/2), & f < 0 \\ 0, & f = 0 \\ \exp(-j\pi/2), & f > 0 \end{cases} \tag{3.40}$$

to the spectrum in Eq. (3.39) yields the spectrum of $\hat{x} = \mathscr{H}\{f \cos(\omega_c t)\}$. Taking account that the range of frequencies over which that spectrum is nonzero is as shown in Fig. 3.3b, the corresponding time function is

$$\begin{aligned}
\hat{x}(t) &= \tfrac{1}{2} \int_{-f_c-B/2}^{-f_c+B/2} F(\omega + \omega_c) \exp\left[j\left(\omega t + \frac{\pi}{2}\right) \right] df \\
&\quad + \tfrac{1}{2} \int_{f_c-B/2}^{f_c+B/2} F(\omega - \omega_c) \exp\left[j\left(\omega t - \frac{\pi}{2}\right) \right] df \\
&= \tfrac{1}{2} \int_{-B/2}^{B/2} F(\omega) \exp(j\omega t) \left\{ \exp\left[j\left(\omega_c t - \frac{\pi}{2}\right) \right] + \exp\left[-j\left(\omega_c t - \frac{\pi}{2}\right) \right] \right\} df \\
&= \cos\left(\omega_c t - \frac{\pi}{2} \right) \int_{-B/2}^{B/2} F(\omega) \exp(j\omega t)\, df = f(t) \sin(\omega_c t).
\end{aligned}$$

The second relation of Eq. (3.38) follows at once by applying Eq. (3.36) to the first relation:

$$\mathscr{H}\{f(t) \sin(\omega_c t)\} = \mathscr{H}[\mathscr{H}\{f(t) \cos(\omega_c t)\}] = -f(t) \cos(\omega_c t).$$

4. The time autocorrelation function Eq. (2.27) of a signal $f(t)$ with zero DC component and that of its Hilbert transform $\hat{f}(t)$ are equal, as are the ensemble autocorrelation functions Eq. (2.6) if the signals are considered as stationary random processes. That is,

$$\mathscr{R}_{\hat{f}}(\tau) = \mathscr{R}_f(\tau), \tag{3.41}$$

$$R_{\hat{f}}(\tau) = R_f(\tau). \tag{3.42}$$

With these relations, the corresponding power spectra are also equal:

$$\mathscr{S}_{\hat{f}}(\omega) = \mathscr{S}_f(\omega), \tag{3.43}$$

$$S_{\hat{f}}(\omega) = S_f(\omega). \tag{3.44}$$

Considering the time autocorrelation first, we have Eq. (2.27) as

$$\mathscr{R}_{\hat{f}}(\tau) = \lim_{T \to \infty} \left(\frac{1}{T} \right) \int_{-T/2}^{T/2} \hat{f}(t) \hat{f}^*(t - \tau)\, dt.$$

For large T, the integral differs negligibly from the convolution $\hat{f}(t) * \hat{f}^*(-t)$, which has transform [using Eqs. (3.22), (3.23)]:

$$\hat{F}(\omega)\hat{F}^*(\omega) = H(\omega)F(\omega)H^*(\omega)F^*(\omega)$$
$$= |H(\omega)|^2 F(\omega)F^*(\omega) = F(\omega)F^*(\omega), \qquad f \neq 0.$$

Therefore in the above integral the Hilbert transforms can be replaced by the corresponding functions themselves, which proves Eq. (3.41).

Equation (3.42) follows more easily as a consequence of Eq. (3.44). Equation (3.22) shows that $\hat{f}(t)$ results from passing the process $f(t)$ through the filter

$$H(j\omega) = -j \, \text{sgn}(f). \tag{3.45}$$

From Eq. (2.65), the filter output has power spectral density

$$S_{\hat{f}}(\omega) = |H(j\omega)|^2 S_f(\omega) = S_f(\omega),$$

which is Eq. (3.44).

Another relation is sometimes useful. From Eq. (3.32) and Eq. (3.52) below we have

$$S_{\hat{f}f}(\omega) = \mathcal{F}\{R_{\hat{f}f}(\tau)\} = \mathcal{F}\{\hat{R}(\tau)\}$$
$$= \mathcal{F}\{(1/\pi t) * R_f(\tau)\} = H(\omega)S_f(\omega). \tag{3.46}$$

Then, from Eqs. (2.70), (3.26), for the analytic signal we have

$$F_p(\omega) = [1 + jH(\omega)]\, F(\omega),$$
$$S_{f_p f}(\omega) = [1 + jH(\omega)]\, S_f(\omega) = S_f(\omega) + jS_{\hat{f}f}(\omega). \tag{3.47}$$

5. The time cross-correlation Eq. (2.28) between a function $f(t)$, having no DC content, and its Hilbert transform $\hat{f}(t)$ is related to the autocorrelation of $f(t)$ by

$$\mathcal{R}_{\hat{f}f}(\tau) = \hat{\mathcal{R}}_f(\tau). \tag{3.48}$$

From Eq. (2.28),

$$\mathcal{R}_{\hat{f}f}(\tau) = \lim_{T \Rightarrow \infty} (1/T) \int_{-T/2}^{T/2} \hat{f}(t)f^*(t - \tau) \, dt. \tag{3.49}$$

For large T, the integral differs only negligibly from the convolution $\hat{f}(t)*f^*(-t)$. Using Eqs. (3.22), (3.23) this function has spectrum

$$\hat{F}(\omega)F^*(-\omega) = H(\omega)[F(\omega)F^*(-\omega)].$$

This is the spectrum of the Hilbert transform of the convolution $f(t)*f^*(-t)$, which for large T differs negligibly from

$$\mathcal{H}\left\{ \int_{-T/2}^{T/2} f(t)f^*(\tau - t) \, dt \right\}.$$

Using this in Eq. (3.49) and reversing the Hilbert transformation and limit operations yields Eq. (3.48).

In the same way, $\mathcal{R}_{f\hat{f}}(\tau)$ involves $f(t)*f^*(-t)$ with transform

$$F(\omega)\hat{F}^*(\omega) = F(\omega)H^*(\omega)F^*(\omega) = -H(\omega)F(\omega)F^*(\omega),$$

where we have used the relation

$$\mathcal{F}\{\hat{f}^*(-t)\} = \int_{-\infty}^{\infty} \hat{f}^*(-t)\exp(-j\omega t)\,dt$$

$$= \int_{-\infty}^{\infty} \hat{f}^*(t)\exp(j\omega t)\,dt$$

$$= \left[\int_{-\infty}^{\infty} \hat{f}(t)\exp(-j\omega t)\,dt\right]^* = \hat{F}^*(\omega). \qquad (3.50)$$

This is the transform of $-\mathcal{H}\{f(t)*f^*(-t)\}$, so that

$$\mathcal{R}_{f\hat{f}}(\tau) = -\hat{\mathcal{R}}_f(\tau). \qquad (3.51)$$

The relation for the ensemble cross-correlation corresponding to Eq. (3.48) follows from Eq. (2.70). Since $y = \hat{f}(t)$ is the output of a filter $H(\omega)$ with input $x(t) = f(t)$, we have

$$S_{\hat{f}f}(\omega) = H(\omega)S_f(\omega).$$

That is,

$$R_{\hat{f}f}(\tau) = \hat{R}_f(\tau). \qquad *(3.52)$$

6. For a real function $f(t)$, the ensemble autocorrelation $R_f(\tau)$ is an even function of its lag variable τ [Eq. (2.19)], as is the time autocorrelation $\mathcal{R}_f(\tau)$. From Eq. (3.24) the corresponding Hilbert transforms $\hat{R}_f(\tau)$ and $\hat{\mathcal{R}}_f(\tau)$ are odd functions of lag τ. Accordingly, from Eqs. (3.48), (3.52), both $\mathcal{R}_{\hat{f}f}(\tau)$ and $R_{\hat{f}f}(\tau)$ are also odd:

$$\mathcal{R}_{\hat{f}f}(-\tau) = \hat{\mathcal{R}}_f(-\tau) = -\hat{\mathcal{R}}_f(\tau) = -\mathcal{R}_{\hat{f}f}(\tau), \qquad (3.53)$$

$$R_{\hat{f}f}(-\tau) = \hat{R}_f(-\tau) = -\hat{R}_f(\tau) = -R_{\hat{f}f}(\tau). \qquad (3.54)$$

7. Collecting together various relations in properties 4 and 5 we have the following properties of the indicated ensemble correlation functions:

$$R_{\hat{f}}(\tau) = R_f(\tau),$$
$$R_{\hat{f}f}(\tau) = \hat{R}_f(\tau),$$
$$R_{f\hat{f}}(\tau) = -\hat{R}_f(\tau),$$
$$R_{\hat{f}f}(-\tau) = -R_{\hat{f}f}(\tau), \qquad f(t)\text{ real},$$
$$R_{f\hat{f}}(-\tau) = -R_{f\hat{f}}(\tau), \qquad f(t)\text{ real}. \qquad *(3.55)$$

It follows further from Eq. (3.55) that, for a real $f(t)$, the signal $f(t)$ and its Hilbert transform $\hat{f}(t)$ are uncorrelated at zero lag:

$$R_{f\hat{f}}(0) = -R_{\hat{f}f}(0) = 0. \qquad (3.56)$$

8. From Eq. (3.38) for the special case $f(t) = 1$ there follows

$$\mathcal{H}\{\cos(\omega_c t + \phi)\} = \cos(\phi)\mathcal{H}\{\cos(\omega_c t)\} - \sin(\phi)\mathcal{H}\{\sin(\omega_c t)\}$$
$$= \cos(\phi) \sin(\omega_c t) + \sin(\phi) \cos(\omega_c t) = \sin(\omega_c t + \phi).$$

$$(3.57)$$

Similarly,

$$\mathcal{H}\{\sin(\omega_c t + \phi)\} = -\cos(\omega_c t + \phi).$$

$$(3.58)$$

9. Equation (3.53) shows that a real function $f(t)$ and its Hilbert transform $\hat{f}(t)$ are orthogonal, in the sense that

$$\lim_{T \Rightarrow \infty} \left(\frac{1}{T}\right) \int_{-T/2}^{T/2} f(t)\hat{f}(t) \, dt = \mathcal{R}_{f\hat{f}}(0) = 0.$$

$$(3.59)$$

3.4 NARROWBAND FILTERS

In Sect. 3.2 we indicated how the redundant information in the negative frequency part of the spectrum $F(\omega)$ of a real signal $f(t)$ could be suppressed by converting from the signal itself to the corresponding analytic signal $f_p(t)$ as in Eq. (3.5). This is especially useful in the case of a narrowband signal which modulates a carrier to produce the modulated signal Eq. (3.10). In that case, the complex envelope $\tilde{f}(t)$ [Eq. (3.12)] carrying the amplitude- and phase-modulating signals is recovered by down-converting the analytic signal in frequency in accordance with Eq. (3.17). In this section, we extend those procedures to deal with computations involving the response of a narrowband filter with a narrowband modulated signal as input.

For any stationary filter with impulse response $g(t)$, the analytic signal corresponding to the output $y(t) = g(t)*x(t)$ in response to an input $x(t)$ is just the response of the filter if the analytic signal $x_p(t)$ were the input rather than $x(t)$ itself. That is,

$$y_p(t) = g(t) * x_p(t).$$

$$(3.60)$$

This follows at once from use of Eq. (3.3):

$$Y_p(\omega) = [1 + \text{sgn}(f)] Y(\omega) = [1 + \text{sgn}(f)]G(\omega)X(\omega)$$
$$= G(\omega)[1 + \text{sgn}(f)]X(\omega) = G(\omega)X_p(\omega).$$

The expression Eq. (3.60) is true whether or not the signals or system are narrowband.

EXAMPLE 3.5 Consider the narrowband signal:

$$x(t) = A \cos(\omega_c t), \qquad |t| \le T/2.$$

Since the modulating pulse is only approximately bandlimited, the spectrum of the modulated signal (Fig. 3.5) has some contribution to the positive frequency part due to the lobe centered at $-f_c$. Nonetheless, if $f_c \gg 1/T$, the portion of the lobe at $-f_c$ which extends into positive frequencies and the portion of the lobe at f_c extending into the negative frequencies will be negligible. Then from Eq. (3.5), approximately,

$$X_p(\omega) = 2X(\omega) = AT \sin[(\pi T(f - f_c))/\pi T(f - f_c)], \qquad f > 0,$$

so that

$$x_p(t) = A \exp(j\omega_c t), \qquad |t| < T/2. \tag{3.61}$$

Now suppose that the filter acting on the original $x(t)$ has an impulse response

$$g(t) = \cos(\omega_c t), \qquad |t| < T/2.$$

The output is (Fig. 3.6):

$$y(t) = g(t) * x(t) = \int_{-\infty}^{\infty} x(t')g(t - t') \, dt'$$

$$= \begin{cases} \int_{-T/2}^{t+T/2} A \cos(\omega_c t') \cos[\omega_c(t - t')] \, dt', & -T < t < 0 \\ \int_{t-T/2}^{T/2} A \cos(\omega_c t') \cos[\omega_c(t - t')] \, dt', & 0 < t < T \end{cases}$$

$$= (AT/2)(1 - |t|/T) \cos(\omega_c t)$$
$$+ (A/2\omega_c) \sin[(\omega_c T)(1 - |t|/T)], \quad |t| < T. \tag{3.62}$$

To proceed without approximation we would need to find the spectrum of $y(t)$ and then the inverse transform of the positive frequency portion. However, we have already assumed that $f_c \gg 1/T$. Then

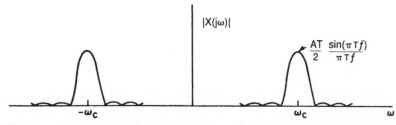

FIGURE 3.5 Magnitude squared of transform of signal $x(t) = A \cos(\omega_c t)$, $|t| \leq T/2$, $f_c \gg 1/T$.

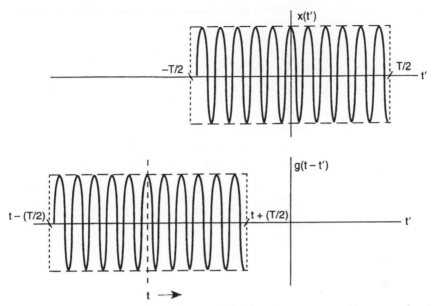

FIGURE 3.6 Convolution of a narrowband filter impulse response with a narrowband signal.

$\omega_c T \gg 1$, and the second term in Eq. (3.62) is very much smaller than the first. That is,

$$y(t) \cong (AT/2)(1 - |t|/T) \cos(\omega_c t). \qquad (3.63)$$

The spectrum of this $y(t)$ is that of $1 - |t|/T$ scaled and shifted up to f_c and down to $-f_c$. As a further consequence of the assumption $\omega_c T \gg 1$, that spectrum has the general appearance of Fig. 3.5. We can then finally use the approximation

$$y_p(t) = (AT/2)(1 - |t|/T) \exp(j\omega_c t), \qquad |t| < T. \qquad (3.64)$$

Proceeding more directly, we can use Eq. (3.60). Then

$$y_p(t) = g(t) * x_p(t)$$

$$= \begin{cases} A \displaystyle\int_{-T/2}^{t+T/2} \cos(\omega_c t') \exp[j\omega_c(t - t')] \, dt', & -T < t < 0 \\[2mm] A \displaystyle\int_{t-T/2}^{T/2} \cos(\omega_c t') \exp[j\omega_c(t - t')] \, dt', & 0 < t < T \end{cases}$$

$$= (AT/2)(1 - |t|/T) \exp(j\omega_c t)$$
$$\quad + (AT/2\omega_c) \sin[\omega_c(1 - |t|/T)], \qquad |t| < T.$$

Again using the assumption $f_c \gg 1/T$, the second term of this last can be dropped, and the result then agrees with Eq. (3.64).

Narrowband Filter Representation

Although Eq. (3.60) is exact, an approximation is useful in application to an input which consists of a relatively narrowband signal modulating a carrier, as in Fig. 3.2 and Example 3.5. It is often the case that the filter applied to such an input has a transfer function which is also of the form shown in Fig. 3.5. The exact relation Eq. (3.60) in the frequency domain is just

$$Y_p(\omega) = G(\omega)X_p(\omega).$$

Since by Eq. (3.5) we have

$$G(\omega) = G_p(\omega)/2, \quad f > 0,$$

this last becomes

$$Y_p(\omega) = \tfrac{1}{2}G_p(\omega)X_p(\omega), \tag{3.65}$$

or, equivalently,

$$y_p(t) = [g_p(t)/2] * x_p(t). \tag{3.66}$$

That is, for preenvelope calculations we can replace a filter with impulse response $g(t)$ having a narrow band around a carrier by a filter with impulse response $(1/2)g_p(t)$.

Usually the complex envelope $\tilde{f}(t)$ is of more interest than the preenvelope. From Eqs. (3.16) and (3.66) we have

$$\tilde{Y}(\omega) = Y_p(\omega + \omega_c) = \tfrac{1}{2}G_p(\omega + \omega_c)X_p(\omega + \omega_c), \tag{3.67}$$

or

$$\tilde{y}(t) = [\tilde{g}(t)/2] * \tilde{x}(t). \tag{3.68}$$

EXAMPLE 3.6   Consider again the input and filter of Example 3.5. The filter

$$g(t) = \cos(\omega_c t), \quad |t| < T/2,$$

has an envelope which is simply

$$\tilde{g}(t) = 1, \quad |t| < T/2.$$

The input preenvelope Eq. (3.61) corresponds to the envelope

$$\tilde{x}(t) = A, \quad |t| < T/2.$$

From Eq. (3.68) the output envelope is

$$
\tilde{y}(t) = \begin{cases}
(A/2) \displaystyle\int_{-T/2}^{t+T/2} dt', & -T < t < 0, \\[3mm]
(A/2) \displaystyle\int_{t-T/2}^{T/2} dt', & 0 < t < T,
\end{cases}
$$

$$
= (AT/2)(1 - |t|/T), \qquad |t| < T,
$$

which agrees with Eq. (3.64).

Equation (3.67) gives the prescription for finding the complex envelope function to use in Eq. (3.68) for a given narrowband carrier-centered filter transfer function $G(\omega)$: Take twice the positive frequency part of $G(\omega)$ to obtain $G_p(\omega)$, then left shift in frequency by ω_c to obtain $\tilde{G}(\omega)$. Carry out the inverse transform if $\tilde{g}(t)$ is needed.

3.5 NARROWBAND PROCESSES

Just as in the case of a real deterministic signal $f(t)$, a random process may also be narrowband. By this is meant that the power spectral density of the process spans a relatively narrow band around a high carrier frequency, as indicated in Fig. 3.1. The material discussed in the preceding sections provides a convenient framework for dealing with the information-bearing modulation signals of such a waveform, while suppressing the extraneous fact that they may have been used to modulate a carrier for purposes of transmission or multiplexing.

In analogy with Eq. (3.10), suppose that a random process has the form

$$
n(t) = a(t) \cos[\omega_c t + \phi(t)], \tag{3.69}
$$

where the carrier $f_c > B/2$, with B the (two-sided) bandwidth containing the spectra of the relatively slowly varying modulations $a(t)$, $\cos[\phi(t)]$. Furthermore, we will assume that the process is zero mean: $\mathscr{E}n(t) = 0$. As in Eq. (3.12), the complex envelope of $n(t)$ can be defined from the amplitude and phase modulation functions as

$$
\tilde{n}(t) = a(t) \exp[j\phi(t)] = x(t) + jy(t). \tag{3.70}
$$

The real and imaginary parts $x(t)$, $y(t)$ of the complex envelope $\tilde{n}(t)$ are called the *in-phase* and *quadrature* components of the signal $n(t)$. From them both the envelope and the phase modulation of $n(t)$ can be obtained as

$$
|a(t)| = [x^2(t) + y^2(t)]^{1/2},
$$
$$
\phi(t) = \tan^{-1}[y(t)/x(t)], \tag{3.71}
$$

where the four-quadrant inverse tangent is required.

Combining Eqs. (3.17) and (3.26), the complex envelope of the wave-form Eq. (3.69) is

$$\tilde{n}(t) = n_p(t) \exp(-j\omega_c t) = [n(t) + j\hat{n}(t)] \exp(-j\omega_c t). \tag{3.72}$$

Equating this to the expression Eq. (3.70), and taking real and imaginary parts, yields

$$x(t) = n(t) \cos(\omega_c t) + \hat{n}(t) \sin(\omega_c t),$$
$$y(t) = \hat{n}(t) \cos(\omega_c t) - n(t) \sin(\omega_c t). \tag{*(3.73)}$$

The expression Eq. (3.72) provides the means of obtaining the quadrature components $x(t)$, $y(t)$ from $n(t)$, and Eq. (3.73) is convenient for use in analysis.

In determining the relations among the correlation functions and power spectra of the quadrature components $x(t)$, $y(t)$ and the real narrowband process $n(t)$, we will need certain of Eqs. (3.55), (3.56), specifically:

$$R_{n\hat{n}}(\tau) = -\hat{R}_n(\tau),$$
$$R_{\hat{n}n}(\tau) = \hat{R}_n(\tau),$$
$$R_{n\hat{n}}(0) = R_{\hat{n}n}(0) = 0,$$
$$R_{\hat{n}}(\tau) = R_n(\tau).$$

We will use these without comment below.

First, we consider the autocorrelation function of the in-phase component $x(t)$ and quadrature component $y(t)$ of the noise process $n(t)$. [Together, $x(t)$ and $y(t)$ are called the quadrature components of $n(t)$.] We have from Eq. (3.73) that

$$R_x(\tau) = \mathscr{E}x(t)x(t - \tau)$$
$$= \mathscr{E}[n(t)n(t - \tau)] \cos(\omega_c t) \cos[\omega_c(t - \tau)]$$
$$+ \mathscr{E}[n(t)\hat{n}(t - \tau)] \cos(\omega_c t) \sin[\omega_c(t - \tau)]$$
$$+ \mathscr{E}[n(t - \tau)\hat{n}(t)] \cos[\omega_c(t - \tau)] \sin(\omega_c t)$$
$$+ \mathscr{E}[\hat{n}(t)\hat{n}(t - \tau)] \sin(\omega_c t) \sin[\omega_c(t - \tau)]$$
$$= [R_n(\tau)/2]\{\cos[\omega_c(2t - \tau)] + \cos(\omega_c \tau)\}$$
$$+ [R_{n\hat{n}}(\tau)/2]\{\sin[\omega_c(2t - \tau)] - \sin(\omega_c \tau)\}$$
$$+ [R_{\hat{n}n}(\tau)/2]\{\sin[\omega_c(2t - \tau)] + \sin(\omega_c \tau)\}$$
$$+ [R_{\hat{n}}(\tau)/2]\{\cos(\omega_c \tau) - \cos[\omega_c(2t - \tau)]\}\}$$
$$= R_n(\tau) \cos(\omega_c \tau) + \hat{R}_n(\tau) \sin(\omega_c \tau). \tag{*(3.74)}$$

In just the same way, it can be shown that

$$R_y(\tau) = R_x(\tau) = R_n(\tau) \cos(\omega_c \tau) + \hat{R}_n(\tau) \sin(\omega_c \tau), \tag{*(3.75)}$$

$$R_{xy}(\tau) = R_n(\tau) \sin(\omega_c \tau) - \hat{R}_n(\tau) \cos(\omega_c \tau), \tag{*(3.76)}$$

$$R_{yx}(\tau) = -R_n(\tau) \sin(\omega_c \tau) + \hat{R}_n(\tau) \cos(\omega_c \tau) = -R_{xy}(\tau). \tag{3.77}$$

In case the processes are not zero mean, the correlation functions are replaced with the corresponding covariance functions. That is,

$$C_x(\tau) = C_y(\tau) = C_n(\tau)\cos(\omega_c\tau) + \hat{C}_n(\tau)\sin(\omega_c\tau), \tag{3.78}$$

$$C_{xy}(\tau) = -C_{yx}(\tau) = C_n(\tau)\sin(\omega_c\tau) - \hat{C}_n(\tau)\cos(\omega_c\tau). \tag{3.79}$$

Some implications of the preceding expressions are worth pointing out explicitly:

1. From Eqs. (3.74), (3.76) it follows that

$$R_n(\tau) = R_x(\tau)\cos(\omega_c\tau) + R_{xy}(\tau)\sin(\omega_c\tau). \tag{3.80}$$

2. From Eq. (3.75) it is clear that, if $n(t)$ is a wide-sense stationary process, so also are $x(t)$ and $y(t)$. Furthermore, Eq. (3.80) shows that a process $n(t)$ constructed from wide-sense stationary processes $x(t)$, $y(t)$ is also wide-sense stationary.

3. Since $n(t)$ has zero mean, so also do $\hat{n}(t)$, $x(t)$, and $y(t)$. Equation (3.75) then shows that the variances of the quadrature components are equal:

$$\sigma_x^2 = R_x(0) = R_y(0) = \sigma_y^2. \tag{3.81}$$

Furthermore, Eq. (3.80) shows that

$$\sigma_n^2 = R_n(0) = R_x(0) = \sigma_x^2 = \sigma_y^2. \tag{3.82}$$

4. Since $x(t)$, $y(t)$ are real random processes, Eq. (2.17) shows that $R_{xy}(\tau) = R_{yx}(-\tau)$. Together with Eq. (3.77) this leads to

$$R_{xy}(-\tau) = -R_{yx}(-\tau) = -R_{xy}(\tau). \tag{3.83}$$

That is, the cross-correlation of the quadrature components is an odd function of the lag variable. Consequently,

$$R_{xy}(0) = -R_{xy}(0) = 0, \tag{3.84}$$

so that samples of $x(t)$ and $y(t)$ taken at the same instant are uncorrelated.

It is a common situation, although not a universal one, that a narrowband signal Eq. (3.69) to be processed has a power spectrum that is symmetric about the carrier frequency. That is,

$$S_n(\omega_c + \Delta\omega) = S_n(\omega_c - \Delta\omega), \qquad 0 \le \Delta\omega < \omega_c. \tag{3.85}$$

With the additional restriction that

$$S_n(\omega) = 0, \qquad \omega \ge 2\omega_c, \tag{3.86}$$

it can be shown that, not only does $R_{xy}(0)$ vanish, as in Eq. (3.84), but also $R_{xy}(\tau)$ vanishes identically. That is, the in-phase and quadrature components $x(t)$, $y(t)$ are uncorrelated, in the sense that $x(t)$, $y(t')$ are uncorrelated random variables for any choice of t, t'.

To see this, consider Eq. (3.76). The spectrum of $R_n(\tau)$ is $S_n(\omega)$, and, from Eqs. (3.46), (3.52), that of $\hat{R}_n(\tau)$ is

$$\hat{S}_n(\omega) = -j \, \text{sgn}(f)S_n(\omega).$$

Then from Eq. (3.76) we have

$$S_{xy}(\omega) = [S_n(\omega - \omega_c) - S_n(\omega + \omega_c)]/2j$$
$$+ j[\text{sgn}(f - f_c)S_n(\omega - \omega_c) + \text{sgn}(f + f_c)S_n(\omega + \omega_c)]/2. \qquad (3.87)$$

The four terms of this sum are shown schematically in Fig. 3.7. The spectral components denoted 1 cancel $(1/j = -j)$, as do those denoted 2. The components denoted 3 are identical, as are those denoted 4. As a result, Eq. (3.87) is

$$S_{xy}(\omega) = j[S_n(\omega + \omega_c) - S_n(\omega - \omega_c)], \qquad |f| < f_c. \qquad (3.88)$$

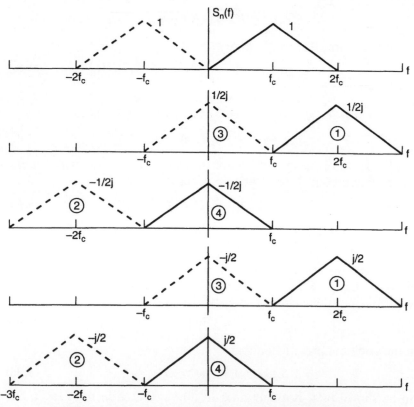

FIGURE 3.7   The spectrum in Eq. (3.87) and its constituent terms.

Furthermore, since $n(t)$ is real, by Eq. (2.50) its spectral density $S_n(\omega)$ is an even function. The special form Eq. (3.85) assumed for $S_n(\omega)$ then guarantees that the components in Fig. 3.7 denoted 3 are equal to those denoted 4. That is, $S_n(\omega - \omega_c) = S_n(\omega + \omega_c)$ over $|f| < f_c$. We conclude from Eq. (3.88) that

$$S_{xy}(\omega) \equiv 0, \qquad R_{xy}(\tau) \equiv 0 \qquad \text{[Special case (Eq. 3.85)].} \qquad (3.89)$$

Furthermore, in this special case of Eq. (3.85), from Eq. (3.80) there follow

$$R_n(\tau) = R_x(\tau) \cos(\omega_c \tau), \qquad (3.90)$$

$$S_n(\omega) = [S_x(\omega - \omega_c) + S_x(\omega + \omega_c)]/2. \qquad (3.91)$$

3.6 DETERMINATION OF THE COMPLEX ENVELOPE

In the case of a signal such as Eq. (3.10), as we have discussed it is often useful to obtain the complex envelope Eq. (3.12). This is especially true if we want to proceed with further calculations using a time-sampled version of the modulated signal. In this section we mention some ways to determine the complex envelope

$$\tilde{f}(t) = a(t) \exp[j\phi(t)] = x(t) + jy(t) \qquad (3.92)$$

from the real signal $f(t)$, where

$$f(t) = \text{Re}[\tilde{f}(t) \exp(j\omega_c t)] = a(t) \cos[\omega_c t + \phi(t)]. \qquad (3.93)$$

As in Fig. 3.3b, we will assume that the real envelope $a(t)$ has a two-sided bandwidth B such that $f_c > B/2$. Often it will be the case that $f_c \gg B/2$.

In principle, one can proceed in terms of the analytic signal (pre-envelope) as prescribed by Eqs. (3.17), (3.26):

$$\begin{aligned}
\tilde{f}(t) &= f_p(t) \exp(-j\omega_c t) = [f(t) + j\hat{f}(t)] \exp(-j\omega_c t) \\
&= [f(t) \cos(\omega_c t) + \hat{f}(t) \sin(\omega_c t)] \\
&\quad + j[\hat{f}(t) \cos(\omega_c t) - f(t) \sin(\omega_c t)].
\end{aligned} \qquad (3.94)$$

The Hilbert transform follows from Eq. (3.32) as $\hat{f}(t) = f(t)*(1/\pi t)$. However, the Hilbert transform is not particularly convenient to implement in the time domain.

Alternatively, the frequency domain implementation of Eq. (3.94) can be used. This is convenient if the original signal $f(t)$ is sampled and Fourier transformed, using the FFT. Then from Eqs. (3.16), (3.3) follows

$$\tilde{F}(\omega) = F_p(\omega + \omega_c) = [1 + \text{sgn}(f + f_c)] F(\omega + \omega_c). \qquad (3.95)$$

That is, the procedure is to sample $f(t)$, compute the FFT, double the positive frequency part of the spectrum, left shift it by the carrier, and compute the inverse FFT. This has the apparent disadvantage that the original waveform $f(t)$ must be sampled at least at the rate $f_s = 2(f_c + B/2)$, which may be quite high if the carrier frequency is high. However, we will see below that this is only apparent.

In a procedure which is often convenient in the analog domain, if the carrier $f_c \gg B/2$, we can proceed directly from Eq. (3.93) as:

$$2f(t) \cos(\omega_c t) = a(t)[\cos(\phi) + \cos(2\omega_c t + \phi)].$$

The first term of this last has a band $|f| < B/2$, while the second term covers a band at $\pm 2f_c$. It is therefore relatively easy to use a low-pass filter to isolate the desired component of Eq. (3.92):

$$x(t) = \mathrm{Re}[\tilde{f}(t)] = a(t) \cos[\phi(t)] = 2[f(t) \cos(\omega_c t)]_{\mathrm{LP}}, \qquad *(3.96)$$

where the LP indicates low-pass filtering of the quantity in brackets. In the same way, it is easy to see that

$$y(t) = a(t) \sin[\phi(t)] = -2[f(t) \sin(\omega_c t)]_{\mathrm{LP}}. \qquad *(3.97)$$

The multiplication of the modulated waveform $f(t)$ by the carrier $\cos(\omega_c t)$ gives rise to the term "in-phase component" for $x(t)$, and multiplication by a waveform $(\sin(\omega_c t))$ 90° out of phase with the carrier leads to the term "quadrature component" for $y(t)$.

An approximate procedure is sometimes useful in order to implement the sampled version of Eqs. (3.96), (3.97). Suppose that the analog waveform Eq. (3.93) is sampled at a rate exactly $f_s = 4f_c$. Then we have consecutive time samples $f_i = f(i/f_s)$ as

$$\begin{aligned} f_i &= a_i \cos(i\pi/2 + \phi_i), \\ f_{i+1} &= a_{i+1} \cos(i\pi/2 + \phi_{i+1} + \pi/2) = -a_{i+1} \sin(i\pi/2 + \phi_{i+1}). \end{aligned}$$

Provided that $f_c \gg B/2$, so that $a(t)$ and $\phi(t)$ change but little over a quarter cycle of the carrier, we have $a_{i+1} \cong a_i$, $\phi_{i+1} \cong \phi_i$, and

$$\begin{aligned} f_{2m} &= (-1)^m a_{2m} \cos(\phi_{2m}) = (-1)^m x_{2m}, \\ f_{2m+1} &\cong (-1)^{m+1} a_{2m} \sin(\phi_{2m}) = (-1)^{m+1} y_{2m}, \end{aligned} \qquad (3.98)$$

where

$$\tilde{f}_m = x_{2m} + jy_{2m} = (-1)^m(f_{2m} - jf_{2m+1})$$

are samples of the complex envelope at rate $f_s' = 2f_c$.

In the case of a high-frequency carrier f_c, sampling of the waveform Eq. (3.93) at a rate slower than $2(f_c + B/2)$ may be feasible, before applying the technique of Eq. (3.98). The effect of sampling any signal $f(t)$ at a rate f_s is to produce a sampled signal $\{f(i/f_s)\}$ whose discrete Fourier transform

$$F_d(\omega) = \sum_{i=-\infty}^{\infty} f_i \exp(-j\omega i/f_s) \qquad (3.99)$$

is periodic in ω with period $2\pi f_s$. Consider a complex envelope $\tilde{f}(t)$ which is bandlimited to $|f| < B/2$, as in Fig. 3.8a. Upon introduction of a carrier with $f_c > B/2$ this yields an analytic signal $f_p(t) = \tilde{f}(t)\exp(j\omega_c t)$ and a corresponding modulated waveform Eq. (3.93),

$$f(t) = \text{Re}[f_p(t)]$$
$$= [(a/2)\exp(j\phi)]\exp(j\omega_c t) + [(a/2)\exp(-j\phi)]\exp(-j\omega_c t). \qquad (3.100)$$

This has the spectrum of Fig. 3.8b.

In accordance with Eq. (3.99), sampling this signal $f(t)$ at a rate f_s yields the spectrum of Fig. 3.8b replicated with period f_s. As indicated in Fig. 3.8c, if we sample appropriately the spectrum region near the carrier can be moved to base band without aliasing. In general, we require $f_c - kf_s - B/2 > 0$ and $-f_c + (k + 1)f_s - B/2 > f_c - kf_s + B/2$. (Figure 3.8c shows the case $k = 2$.) That is, we require

$$(2f_c + B)/(2k + 1) < f_s < (2f_c - B)/2k. \qquad (3.101)$$

The spectrum of Fig. 3.8c is indistinguishable from the sampled spectrum corresponding to the complex envelope of Fig. 3.8a used with a modulating frequency $f_c - kf_s$:

$$f_0(t) = \text{Re}[\tilde{f}(t)\exp[j2\pi(f_c - kf_s)t]].$$

We can then apply any of the above techniques to the samples of $f_0(t)$ to recover its complex envelope $\tilde{f}(t)$, which is also the desired complex

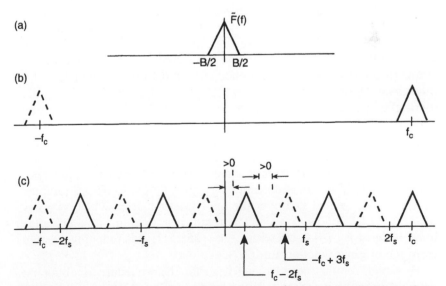

FIGURE 3.8 (a) Spectrum of the complex envelope of a signal. (b) Spectrum of the corresponding modulated signal. (c) Spectrum of modulated signal after subsampling.

envelope of the original signal Eq. (3.100). The technique of Eq. (3.98) can be used in particular, provided the complex envelope $\tilde{f}(t)$ changes only little in the time of a quarter cycle of the new carrier $f_c - kf_s$. The sampling rate must be such that

$$f_s = 4(f_c - kf_s),$$

that is,

$$f_s = 4f_c/(4k + 1). \tag{3.102}$$

Combining this last with Eq. (3.101) yields

$$k < (2f_c - B)/4B. \tag{3.103}$$

The procedure is to choose the largest k for which Eq. (3.103) holds and then to compute f_s from Eq. (3.102).

In case it is not a good approximation that the envelope is nearly constant over a cycle of the new carrier $f_c - kf_s$, a smaller k must be used. The computational savings in this procedure will be the greater the larger is the integer k used in selecting the sampling rate, within the limitation of the bound Eq. (3.103).

EXAMPLE 3.7   Let us consider again the bandlimited signal of Example 3.1, with spectrum as in Fig. 3.3a, with one-sided bandwidth $B/2 = 1$ Hz. If this is used as an envelope $a(t)$ to modulate a carrier with $f_c = 10$ Hz, the resulting modulated wave $f(t)$ in Eq. (3.93) is as shown in Fig. 3.9a. Proceeding directly, we need to sample this at a rate $f_s > 2(f_c + B/2) = 22$ Hz. For purposes of plotting, we have oversampled and used $f_s = 100$. Using 512 points for convenience, we consider $-2.56 \le t \le 2.55$. At the ends of this, the envelope Eq. (3.7) is down to 1/50 of its peak. We then compute the samples $f(i/f_s)$, $i = -256,255$. We treat this as a periodic sequence and shift to the base interval by interchanging the samples f_i, $i = -256,-1$ with f_i, $i = 0,255$. Using the method of Eq. (3.95) we compute the 512-point FFT of the resulting samples, obtaining coefficients F_k, $k = 0,511$. We take the 512 Fourier coefficients of the corresponding analytic signal $f_p(t)$ to be $F_{p0} = F_0$; $F_{pk} = 2F_k$, $k = 1,255$; $F_{pk} = 0$, $k = 256,511$.

If we require only the amplitude modulation $a(t)$, we can simply take the inverse FFT of the coefficients F_{pk}, $k = 0,511$, to obtain samples of the analytic signal $f_p(t)$ over the base interval, from which samples of $a(t) = |f_p(t)|$ follow at once. The result is shown in Fig. 3.9b, after interchanging f_{pi}, $i = 0,255$ with f_{pi}, $i = 256,511$. It is indistinguishable from the original function Eq. (3.7).

However, for illustration we will describe the procedure for obtaining the phase modulation as well. With a time span $T = 512/f_s$, the frequency spacing of the coefficients F_{pk} is $\delta f = f_s/512$. To shift left by 10 Hz

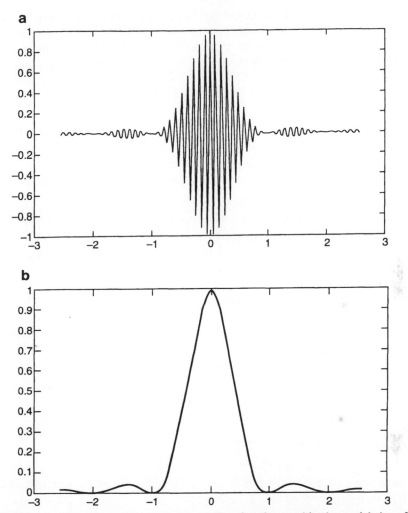

FIGURE 3.9 Waveforms of Example 3.7. (a) Time function resulting by modulation of a carrier with a bandlimited signal. (b) Magnitude of complex envelope recovered by subsampling.

amounts to reindexing k by $(512)(10)/f_s = 51.2$. Since this must be an integer, we will use 51. Thus we take the complex envelope coefficients to be $\tilde{F}_k = \{\text{left circular shift by 51}\}F_{pk}$, which is to say

$$\tilde{F}_k = F_{p,k+51}, \qquad k = 0,460,$$
$$\tilde{F}_k = F_{p,k+51-512} = F_{p,k-461}, \qquad k = 461,511.$$

The inverse FFT of $\tilde{F}_k, k = 0,511$ again yields a complex envelope with amplitude $a(t) = |\tilde{f}(t)|$ shown in Fig. 3.9b. However, the phase

$\phi(t)$ computed from this $\tilde{f}(t)$ will not be zero, whereas the phase of Eq. (3.7) is identically zero. This is because of the slight offset of our down-shift by 51 samples from the desired 51.2 samples. This means our calculated envelope $\tilde{f}(t)$ is riding on a carrier of frequency $0.2\delta f = 0.2/T = (0.2)f_s/512 = 20/512$ Hz. The calculated phase $\phi'(t)$ must then be corrected by subtracting an angle $(40\pi/512)t$ radians:

$$\phi_i = \phi_i' - (40\pi/512)(i/f_s), \qquad i = -256,255.$$

Except for end effects and near $\tilde{f}(t) = 0$, this is found to be zero. In an actual application, we might choose the time span of the analysis to be such that the carrier is an integral multiple of frequency intervals: $f_c = n\delta f = n/T$.

EXAMPLE 3.8 Consider again Example 3.7. Let us now apply the method of Eq. (3.98) to find the real envelope of the modulated wave in Fig. 3.9a. With a carrier f_c of 10 Hz and a half-band $B/2$ of 1 Hz, from Eq. (3.103) we must have $k < 9/4$. Thus we can insert only one replication of the base band $|f| < 1$ between the lower edge of the carrier band and the upper edge of the base band. Then from Eq. (3.102) we choose $f_s = 40/9$ Hz. For convenience we will use 32 samples, spanning $-3.6 \leq t \leq 3.375$, which is far enough out on the waveform Fig. 3.9a that the amplitude is nearly zero.

From the 32 samples of the modulated signal $f_i = f(9i/40)$, $i = -16,15$, as in Eq. (3.98) we form the 32-point complex sequence

$$\tilde{f}_m = (-1)^m [f_{2m} - if_{2m+1}], \qquad m = -8,7,$$

from which we calculate

$$a_m = |\tilde{f}_m|, \qquad m = -8,7.$$

These are exactly the samples of the original envelope function Eq. (3.7).

3.7 FOURIER SERIES REPRESENTATION

Very often in signal processing, it is convenient and natural to proceed in the frequency domain. This is the case because many of the systems to be analyzed are at least approximately stationary, and many random processes are wide-sense stationary over reasonable extents of time. Accordingly, we often have to deal with the Fourier coefficients calculated from sample functions of a random process. In this section we present some results having to do with the properties of those Fourier coefficients, considered as random variables.

We will mostly consider bandlimited functions $f(t)$, with $|F(\omega)| = 0$, $|f| > B/2$. These will be assumed available over some time span T and sampled at the Nyquist rate, $f_s = B$. Thereby we have available $N = BT$ samples f_i, $i = 0, N - 1$. We will consider the FFT coefficients:

$$F_k = \sum_{i=0}^{N-1} f_i \exp\left(\frac{-j2\pi ki}{N}\right), \qquad k = 0, N - 1, \qquad (3.104)$$

$$f_i = \frac{1}{N} \sum_{k=0}^{N-1} F_k \exp\left(\frac{j2\pi ki}{N}\right), \qquad i = 0, N - 1, \qquad (3.105)$$

where we show also the inverse FFT. For such a bandlimited signal, the discrete coefficients Eq. (3.104) relate to the transform $F(\omega)$ of the continuous time signal $f(t)$ as

$$F_k = f_s F(2\pi k/T), \qquad (3.106)$$

over the band $|f| < f_s/2$ where the right side of Eq. (3.106) is nonzero.

We will mention in passing the treatment of a narrowband modulated signal such as in Eq. (3.10). One might sample such a signal at a high rate corresponding to about double the carrier frequency. (Precisely, $f_s > 2f_c + B$, where B is the two-sided bandwidth of the modulating signal.) This would lead to a large number of points for an envelope time span of interest, and a corresponding large number of Fourier coefficients to cover the band $0 \le f < f_s$. Most of these would be found to be zero, specifically those outside the band of the modulated signal. Accordingly, one might as well calculate just those coefficients from the complex envelope of the modulated signal and shift them to the appropriate regions. That is, for the modulated wave the nonzero coefficients are

$$F_k = \tilde{F}_{k-m} = \sum_{i=0}^{N-1} \tilde{f}_i \exp\left[\frac{-j2\pi(k - m)i}{N}\right],$$

where m corresponds to the carrier frequency, and similarly for the replication at frequency $f_s - f_c$. Usually, however, we are presented with the modulated waveform and seek its envelope, so this procedure is not much needed.

Let us consider now the FFT coefficients Eq. (3.104) of the samples $f(i/f_s)$ of a random process $f(t)$ whose sample functions are bandlimited to $|f| < B/2$. The sampling rate is $f_s = B$. The mean of the coefficient sequence is just

$$\mathcal{E}(F_k) = \text{FFT}\{\mathcal{E}(f_i)\},$$

using the indicated notation for the FFT operation. That is, the mean of the FFT sequence is the FFT of the mean of the process.

From Eq. (3.104), the mean square of any coefficient (the variance, if the random process is zero mean) is

$$\mathscr{E}\,|F_k|^2 = \sum_{m=0}^{N-1}\sum_{n=0}^{N-1} \mathscr{E}(f_m f_n^*)\,\exp\left[\frac{-j2\pi(m-n)k}{N}\right]$$

$$= \sum_{m=0}^{N-1}\sum_{i=m-(N-1)}^{m} R_i\,\exp\left(\frac{-j2\pi ik}{N}\right),$$

where $R_i = R_f(i/f_s)$ are the samples of the autocorrelation function $R_f(\tau)$, and we make the change of variable in the second sum that $i = m - n$. Figure 3.10 indicates how to rearrange the double sum to obtain

$$\mathscr{E}\,|F_k|^2 = \left\{\sum_{i=-(N-1)}^{0}\sum_{m=0}^{i+(N-1)} + \sum_{i=1}^{N-1}\sum_{m=i}^{N-1}\right\} R_i\,\exp\left(\frac{-j2\pi ik}{N}\right)$$

$$= \sum_{i=-(N-1)}^{0} (i+N)R_i\,\exp\left(\frac{-j2\pi ik}{N}\right) + \sum_{i=1}^{N-1} (N-i)R_i\,\exp\left(\frac{-j2\pi ik}{N}\right)$$

$$= \sum_{i=-(N-1)}^{N-1} N\left(1-\frac{|i|}{N}\right) R_i\,\exp\left(\frac{-j2\pi ik}{N}\right). \tag{3.107}$$

Then

$$\left(\frac{1}{Nf_s}\right)\mathscr{E}\,|F_k|^2 = \left(\frac{1}{f_s}\right)\sum_{i=-(N-1)}^{N-1}\left(1-\frac{|i|}{N}\right) R_i\,\exp\left(\frac{-j2\pi ik}{N}\right). \tag{3.108}$$

Using Eq. (3.106), this results in

$$\lim_{T\Rightarrow\infty} (1/Nf_s)\mathscr{E}\,|F_k|^2 = (1/f_s)\text{FFT}\{R_f(\tau)\} = S_f(2\pi k\,\delta f). \tag{3.109}$$

Here we indicate that $N \Rightarrow \infty$ must be the case for $T \Rightarrow \infty$, since we assume a fixed sampling rate, and we note $S_f(\omega) = \mathscr{F}\{R_f(\tau)\}$ is the power spectral

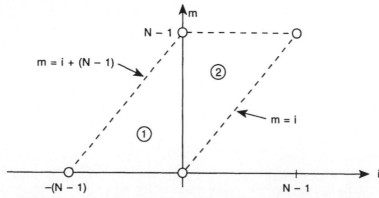

FIGURE 3.10   Relative to change of summation variable in Eq. (3.107).

density of the random process $f(t)$. Equation (3.109) is a more precise statement of Eq. (2.48).

Equation (3.109) suggests using the approximation

$$S_f(\omega_k, \zeta) \cong (1/Nf_s) |F_k(\zeta)|^2 \tag{3.110}$$

for whatever span T of data $f(t)$ is available, where $F_k(\zeta)$ are the FFT coefficients of the samples of $f(t)$. The expression Eq. (3.110) is a random variable, since the particular function $f(t)$ used in its construction is a random process $f(t, \zeta)$. That is why we write $S_f(\omega_k, \zeta)$ in Eq. (3.110), where ζ is the elementary outcome of the abstract experiment which produced the particular sample function $f(t)$ used. Equation (3.109) states that the random variable in Eq. (3.110) has a mean which approaches the quantity we want, if the measurement interval becomes large enough:

$$\lim_{T \Rightarrow \infty} (1/Nf_s)\mathscr{E} |F_k|^2 = S_f(\omega_k). \tag{3.111}$$

However, in addition to Eq. (3.111) we need to be assured that the random fluctuations around that mean, from one sample function to another, will be small—ideally, vanishing as $T \Rightarrow \infty$. That is, we would like to have

$$\lim_{T \Rightarrow \infty} (1/Nf_s) \text{std}\{S_f(\omega_k, \zeta)\} = 0, \tag{3.112}$$

where we indicate the standard deviation of the random variable. A general analysis of the question is difficult. However, for a Gaussian random process it is a fact that, for any T, the standard deviation

$$\text{std}(|F_k|^2/Nf_s) \geq \mathscr{E}(|F_k|^2/Nf_s),$$

so that

$$\lim_{T \Rightarrow \infty} \text{std}(|F_k|^2/Nf_s) \geq S_f(\omega_k). \tag{3.113}$$

The way to fix the problem indicated by Eq. (3.113) is to separate the available data into multiple subintervals and to compute a better approximation to the expected value Eq. (3.109) by using more than a one-term average in Eq. (3.110). There is a trade-off between stability, as evidenced by low variance in the resulting estimator, and the ability to resolve details in the spectrum $S_f(\omega)$ by using a smaller δf, requiring a larger T and thereby leading to fewer data spans over which to average, for a fixed amount of data. We will not pursue the matter further.

In addition to the mean square Eq. (3.107) of the sequence of FFT coefficients, the first joint moment (the covariance, if the process is zero mean) is of interest. Using the same manipulations that led to Eq. (3.107), this is

$$\mathscr{E}(F_k F_l^*) = \exp[-j\pi(1 - 1/N)(k - 1)]$$

$$\times \sum_{i=-(N-1)}^{N-1} R_i \exp\left[\frac{-j\pi i(k + l)}{N}\right]$$

$$\times \sin[\pi(1 - |i|/N)(k - l)]/\sin[\pi(k - l)/N].$$

For $k = l$, this expression recovers Eq. (3.107). For $k \neq l$ and finite N, this correlation is not zero.

Let us examine a fixed frequency separation

$$\Delta f = (k - l)\,\delta f = (k - l)/(N\,\delta t),$$

where δt is the time sampling interval, which we assume to be fixed at some value. Then for the quantity analogous to Eq. (3.111) we have

$$\mathscr{E}(F_k F_l^*)/Nf_s = (1/Nf_s)\exp[-j\pi(N - 1)\,\delta t\,\Delta f]$$

$$\times \sum_{i=-(N-1)}^{N-1} R_i \exp\left[\frac{-j\pi i(k + l)}{N}\right]\frac{\sin[\pi(N - |i|)\,\delta t\,\Delta f]}{\sin(\pi\,\delta t\,\Delta f)}.$$

The sum is finite for all N, so that we have

$$\lim_{T \to \infty} \mathscr{E}(F_k F_l^*)/Nf_s = 0, \qquad k \neq l, \text{fixed } \Delta f. \tag{3.114}$$

That is, for a fixed frequency separation, the corresponding Fourier coefficients are uncorrelated in the limit of infinite observation time. In a later chapter, we will develop the Karhunen–Loève expansion. That is an alternative to the Fourier expansion of the random process, in which the representation basis is such that the coefficients are uncorrelated. The penalty to be paid, however, is that the basis changes for each random process ensemble.

It is sometimes convenient to consider the real and imaginary parts of the complex Fourier coefficients F_k separately. That is, let

$$F_k = F_{ck} + jF_{sk},$$

where

$$F_{ck} = \sum_{i=0}^{N-1} f_i \cos(2\pi ki/N),$$

$$F_{sk} = -\sum_{i=0}^{N-1} f_i \sin(2\pi ki/N).$$

For convenience of notation, let us consider the matrix

$$M = \mathscr{E} \begin{bmatrix} F_{ck} \\ F_{sk} \end{bmatrix} [F_{ck} \quad F_{sk}]$$

$$= \sum_{m=0}^{N-1} \sum_{n=0}^{N-1} \mathscr{E}(f_m f_n) \begin{bmatrix} \cos(2\pi km/N) \\ -\sin(2\pi km/N) \end{bmatrix} [\cos(2\pi kn/N) \quad -\sin(2\pi kn/N)].$$

Inserting $\mathscr{E}(f_m f_n) = R_f(m - n)$ and using the change of variable $i = m - n$ in the second sum leads to

$$M = \frac{1}{2} \sum_{m=0}^{N-1} \sum_{i=m-(N-1)}^{m} R_f(i) \begin{bmatrix} \cos(2\pi ki/N) & \sin(2\pi ki/N) \\ -\sin(2\pi ki/N) & \cos(2\pi ki/N) \end{bmatrix}$$

$$+ \frac{1}{2} \sum_{m=0}^{N-1} \sum_{i=m-(N-1)}^{m} R_f(i) \begin{bmatrix} \cos[2\pi k(2m - i)/N] & -\sin[2\pi k(2m - i)/N] \\ -\sin[2\pi k(2m - i)/N] & -\cos[2\pi k(2m - i)/N] \end{bmatrix}.$$

Using the rearrangement of the sums as indicated in Fig. 3.10, the second term of this last is found to vanish identically, while the first term yields

$$M = \frac{N}{2} \sum_{i=-(N-1)}^{N-1} \left(1 - \frac{|i|}{N}\right) R_f(i) \begin{bmatrix} \cos(2\pi ki/N) & \sin(2\pi ki/N) \\ -\sin(2\pi ki/N) & \cos(2\pi ki/N) \end{bmatrix}.$$

Let us assume that $f(t)$ is real, so that $R_f(\tau)$ is even. Then the off-diagonal terms in this last vanish. That is, the real and imaginary parts of the spectral coefficient are uncorrelated:

$$\mathscr{E} F_{ck} F_{sk} = 0.$$

Comparing the diagonal terms with Eq. (3.108), we have

$$\lim_{T \to \infty} (1/Nf_s)\mathscr{E} |F_{ck}|^2 = \lim_{T \to \infty} (1/Nf_s)\mathscr{E} |F_{sk}|^2 = \tfrac{1}{2} S_f(2\pi k \, \delta f). \quad (3.115)$$

That is, the average power carried by the real and imaginary parts of the spectral coefficients is equal and in turn equals half the total power in the component in question.

In just the same way, we can consider the correlations among the real and imaginary parts of the spectral coefficients of two random processes $f_1(t), f_2(t)$: $F_{1,2}(k) = F_{c1,2}(k) + jF_{s1,2}(k)$. The result is

$$\lim_{T \to \infty} \left(\frac{2}{Nf_s}\right) \mathscr{E} \begin{bmatrix} F_{c1} \\ F_{s1} \\ F_{c2} \\ F_{s2} \end{bmatrix} [F_{c1} \quad F_{s1} \quad F_{c2} \quad F_{s2}]$$

$$= \begin{bmatrix} S_1(k) & 0 & \text{Re } S_{12}(k) & -\text{Im } S_{12}(k) \\ 0 & S_1(k) & \text{Im } S_{12}(k) & \text{Re } S_{12}(k) \\ \hline \text{Re } S_{12}(k) & \text{Im } S_{12}(k) & S_2(k) & 0 \\ -\text{Im } S_{12}(k) & \text{Re } S_{12}(k) & 0 & S_2(k) \end{bmatrix}, \quad (3.116)$$

where $S_{1,2}(k)$ are the power spectral densities of $f_{1,2}(t)$ at frequency $f_k = k\,\delta f$ and $S_{12}(k)$ is the cross-power spectral density between $f_1(t)$ and $f_2(t)$. The partitioning of the matrix will be of interest in Chapter 11 when we discuss a compact notation for complex random variables.

Exercises

3.1 Let $a(t)$, $-\infty < t < \infty$, be a known function with Fourier transform $A(\omega)$. Assume $A(\omega) = 0$ for $\omega > \omega_B$ where $\omega_B \ll \omega_c$.
 (a)   Determine and relate the Fourier transforms of $a(t) \cos \omega_c t$ and $\tfrac{1}{2}a(t)e^{j\omega_c t}$.
 (b)   Determine and relate the Fourier transforms of $a(t) \sin \omega_c t$ and $-j\tfrac{1}{2}a(t)e^{j\omega_c t}$.
 (c)   Relate the Fourier transforms of $a(t) \cos \omega_c t$ and $a(t) \sin \omega_c t$.

3.2 The analytic function or preenvelope of $x(t)$ is

$$z(t) = x(t) + j\hat{x}(t)$$

Show that

$$E\{z(t)z^*(t - \tau)\} = 2[R_x(\tau) + j\hat{R}_x(\tau)]$$

and

$$E\{z(t)z(t - \tau)\} = 0$$

Determine the power spectral density of $z(t)$.

3.3 If a signal $x(t)$ is bandlimited to Ω, show that the magnitude squared of its signal preenvelope is bandlimited to 2Ω.

3.4 For the narrowband stationary process

$$n(t) = x(t) \cos \omega_c t - y(t) \sin \omega_c t$$

derive the relation shown in Eq. (3.75). That is,

$$R_y(\tau) = R_n(\tau) \cos \omega_c \tau + \hat{R}_n(\tau) \sin \omega_c \tau$$

3.5 For the narrowband stationary process

$$n(t) = x(t) \cos \omega_c t - y(t) \sin \omega_c t$$

 (a)   Derive Eq. (3.80):

$$R_n(\tau) = R_x(\tau) \cos \omega_c \tau + R_{xy}(\tau) \sin \omega_c \tau$$

 (b)   If the process is such that Eqs. (3.85), (3.86) are satisfied, what is the envelope of the autocorrelation function of $n(t)$? Show that the power spectral density of $x(t)$ may be expressed

$$S_x(\omega) = S_n(\omega - \omega_c) + S_n(\omega + \omega_c), \qquad -\omega_c < \omega < \omega_c$$

3.6 Consider the *RLC* filter shown in Fig. 3.11. Define $\omega_0^2 = 1/LC$, $\omega_1 = 1/RC$, $\omega_1 \ll \omega_0$. Show that the complex impulse response is given approximately by

$$\bar{h}(t) = \frac{\omega_1}{2} \exp\left(-\frac{\omega_1}{2}t\right), \qquad t \geq 0$$

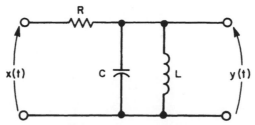

FIGURE 3.11 An *RLC* filter.

3.7 For the *RLC* circuit in Fig. 3.12, assume the input is a voltage, and the output is the current in the circuit. Define $\omega_0^2 = 1/LC$, $\alpha = R/2L$, $\alpha \ll \omega_0$. Show that the complex impulse response is given approximately by $e^{-\alpha t}/2L$, $t \geq 0$.

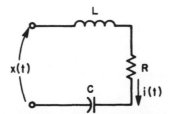

FIGURE 3.12 An *RLC* filter.

4 Gaussian Derived Processes

In the engineering applications of probability theory, of which detection theory is a part, there are three stages. The first is the formulation of a mathematical description of the problem. This must be based on our best understanding of the physical situation and what aspects of that situation are important to the result we seek. The second step, and the only one with which we deal in this book, is the manipulation of that description to obtain a solution to the mathematical problem at hand. Finally, the solution is implemented and the results are observed. If the solution is not satisfactory, more aspects of the problem must be included in the mathematical description and the process iterated.

In the manipulations of the mathematical description of some problem involving randomness, one probability density stands out. For the Gaussian density, to be discussed in this chapter, relatively complete solutions are available for the problems likely to be of interest. As a result, one often seeks to model a problem as Gaussian and then to "tune" the resulting solution to better fit the physical situation. Such a procedure is often successful, and there are three facts about the Gaussian density which contribute to its success.

First, many physical processes are Gaussian, primarily those having to do with the basic noise of thermal origin in systems. Second, the central limit thereom, to be discussed below, states essentially that any process which is the result of a combination of elementary random processes acting in concert will tend to be Gaussian. Finally, linear systems preserve Gaussianity, in that the output of a linear stationary system, excited by a stationary Gaussian process, is itself Gaussian. More pragmatically, a Gaussian density is completely specified by the first two moments of its variables. Since that is often all that we can hope to estimate reliably from a physical situation, the Gaussian is perforce often assumed.

This is not to say that processes of interest are always Gaussian or even approximately so. For example, some noise processes have a burst-like quality, such as "static" noise in radio systems due to relatively nearby lightning storms. Another example is the "clutter" in radar and sonar systems, which is noise in that it results from targets other than those of interest but which is not thermal in origin. Such noise is not as amenable to mathematical treatment as is Gaussian noise. However, relatively complete results are available for the case of Gaussian noise, and here we will discuss the Gaussian process at length.

4.1 THE GAUSSIAN PROBABILITY DENSITY

A scalar random variable z is said to have the (normalized *Gaussian density* if

$$p_z(z) = (2\pi)^{-1/2} \exp(-z^2/2). \qquad (4.1)$$

This is the familiar bell-shaped curve shown in Fig. 4.1. The change of variable $x = \sigma z + m$ ($\sigma > 0$) results in the general one-dimensional Gaussian density:

$$p_x(x) = p_z[z(x)]/(dx/dz)$$
$$= [1/(2\pi)^{1/2}\sigma] \exp[-(x - m)^2/2\sigma^2]. \qquad *(4.2)$$

The parameters m, σ are respectively the mean and standard deviation of x. The density Eq. (4.2) is also called the *normal density* and is often denoted as $N(m, \sigma^2)$.

The distribution function of the density Eq. (4.2) is

$$P_x(X) = \int_{-\infty}^{X} p_x(x)\, dx = \int_{-\infty}^{(X-m)/\sigma} p_z(z)\, dz. \qquad (4.3)$$

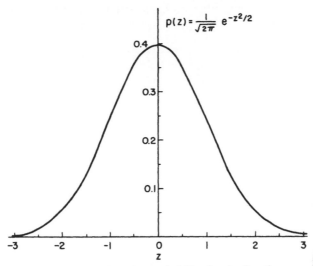

FIGURE 4.1   Gaussian probability density function.

This integral is not expressible in closed form but is tabulated and also provided as a built-in function on many calculators and software systems. Figure 4.2 shows the distribution Eq. (4.3) plotted in terms of $a = (X - m)/\sigma$.

In generalizing the Gaussian density Eq. (4.2), we can consider a finite vector of random variables: $\mathbf{x} = \mathrm{col}(x_i, i = 1, n)$. The multivariate Gaussian density is then defined as

$$p_x(\mathbf{x}) = [(2\pi)^n \, |\mathbf{C}| \,]^{-1/2} \exp[-(\mathbf{x} - \mathbf{m})^{\mathrm{T}}\mathbf{C}^{-1}(\mathbf{x} - \mathbf{m})/2], \qquad *(4.4)$$

where $|\mathbf{C}|$ is the determinant of the $n \times n$ matrix $\mathbf{C}$ and T indicates transpose. By direct integration it can be verified that the parameters $\mathbf{m}$, $\mathbf{C}$ in Eq. (4.4) are the mean of the vector $\mathbf{x}$ and its covariance: $\mathbf{C} = \mathscr{E}(\mathbf{x} - \mathbf{m})(\mathbf{x} - \mathbf{m})^{\mathrm{T}}$. The density is denoted as $N(\mathbf{m}, \mathbf{C})$. It is assumed that the matrix $\mathbf{C}$ is nonsingular, so that $|\mathbf{C}| \neq 0$.

Equation (4.4) is the generalization of the bivariate normal density. For $n = 2$ we have

$$\mathbf{C} = \mathscr{E}\begin{bmatrix} (x_1 - m_1)^2 & (x_1 - m_1)(x_2 - m_2) \\ (x_1 - m_1)(x_2 - m_2) & (x_2 - m_2)^2 \end{bmatrix}$$

$$= \begin{bmatrix} \sigma_1^2 & \rho\sigma_1\sigma_2 \\ \rho\sigma_1\sigma_2 & \sigma_2^2 \end{bmatrix},$$

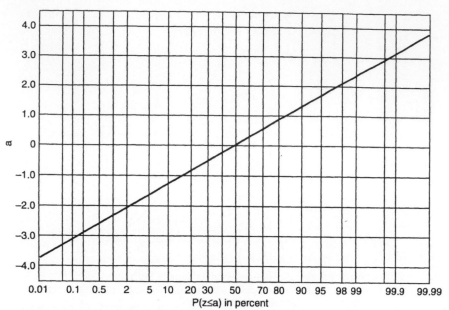

FIGURE 4.2 Gaussian probability distribution function plotted on probability paper.

where ρ is the correlation coefficient Eq. (1.62) of the two random variables. Using this in Eq. (4.4) yields in detail

$$p(x_1, x_2) = 1/[2\pi\sigma_1\sigma_2(1 - \rho^2)^{1/2}]$$
$$\times \exp[-(\tilde{x}_1^2/\sigma_1^2 - 2\rho\tilde{x}_1\tilde{x}_2/\sigma_1\sigma_2 + \tilde{x}_2^2/\sigma_2^2)/2(1 - \rho^2)],$$

where $\tilde{x}_{1,2} = x_{1,2} - m_{1,2}$.

If in Eq. (4.4) the variables x_i are uncorrelated, i.e., if $\mathcal{E}(x_i - m_i)(x_j - m_j) = 0$, $i \neq j$, then we have $\mathbf{C}^{-1} = \text{diag}(1/\sigma_i^2)$ and

$$p_x(\mathbf{x}) = \prod_{i=1}^{n} [(2\pi)^{1/2}\sigma_i]^{-1} \exp\left[\frac{-(x_i - m_i)^2}{2\sigma_i^2}\right], \tag{4.5}$$

so that the variables x_i are independent.

The characteristic function of the density Eq. (4.4) is by definition

$$\Phi_x(\omega) = \int_{-\infty}^{\infty} p_x(\mathbf{x}) \exp(j\omega^T\mathbf{x}) \, d\mathbf{x}.$$

That is, the characteristic function is the conjugate of the Fourier transform of the density. The integral can be evaluated by completing the square in the exponent:

$$\Phi_x(\omega) = (2\pi)^{-n/2} |\mathbf{C}|^{-1/2} \exp(j\mathbf{m}^T\omega - \omega^T\mathbf{C}\omega/2)$$

$$\times \int_{-\infty}^{\infty} \exp(-\mathbf{u}^T\mathbf{C}^{-1}\mathbf{u}/2) \, d\mathbf{u},$$

where $\mathbf{u} = \mathbf{x} - \mathbf{m} - j\mathbf{C}\boldsymbol{\omega}$. Making the change of variable $\mathbf{v} = \mathbf{C}^{-1/2}\mathbf{u}$ in the integral and using the standard result that

$$\int_0^\infty \exp\left(\frac{-v^2}{2}\right) dv = \left(\frac{\pi}{2}\right)^{1/2}$$

yields finally

$$\Phi_x(\boldsymbol{\omega}) = \exp(j\mathbf{m}^T\boldsymbol{\omega} - \boldsymbol{\omega}^T\mathbf{C}\boldsymbol{\omega}/2). \tag{4.6}$$

Linear Transformations

We are often interested in linear systems and their response to a random input function. As we have mentioned above, the output of a linear stationary system excited with a stationary Gaussian input process is also Gaussian. In this section we will give a proof. Suppose that an output m-vector $\mathbf{y}$ is a linear function of an input n-vector $\mathbf{x}$:

$$\mathbf{y} = \mathbf{L}\mathbf{x},$$

where $\mathbf{L}$ is an $m \times n$ matrix ($m \le n$) and is of full rank m. That is, none of the variables y_i is to be deterministically related to the other components of $\mathbf{y}$. Assuming $\mathbf{x}$ is Gaussian, as in Eq. (4.4), we seek the density of $\mathbf{y}$.

Suppose first that $m = n$, so that $\mathbf{L}$ is nonsingular. (If originally $m < n$, we augment $\mathbf{L}$ by defining $y_i = x_i$, $i > m$.) Then from Eq. (1.33) we have at once

$$p_y(\mathbf{y}) = p_x(\mathbf{L}^{-1}\mathbf{y})/\text{abs}(\,|\mathbf{L}|\,).$$

Using Eq. (4.4) this becomes

$$p_y(\mathbf{y}) = (2\pi)^{-n/2}\,|\mathbf{L}^T\mathbf{C}\mathbf{L}|^{-1/2}$$
$$\times \exp[-(\mathbf{y} - \mathbf{L}\mathbf{m})^T(\mathbf{L}\mathbf{C}\mathbf{L}^T)^{-1}(\mathbf{y} - \mathbf{L}\mathbf{m})/2]. \tag{4.7}$$

In this we have used the facts that $|\mathbf{L}^T\mathbf{C}\mathbf{L}| = |\mathbf{L}^T||\mathbf{C}||\mathbf{L}|$, $|\mathbf{L}^T| = |\mathbf{L}|$. Comparison of Eqs. (4.4) and (4.7) identifies $\mathbf{y}$ as a Gaussian random vector with mean and covariance

$$\mathscr{E}\mathbf{y} = \mathbf{L}\mathbf{m}, \qquad \mathbf{Y} = \mathbf{L}\mathbf{C}\mathbf{L}^T. \tag{4.8}$$

In the more usual case that $m < n$, the $n - m$ last of the variables y_i are just the $n - m$ last of the x_i in Eq. (4.7). In that case the excess variables can be integrated out, leaving the marginal density

$$p_y(y_1, \ldots, y_m) = \int_{-\infty}^\infty p_y(\mathbf{y})\, dy_{m+1} \cdots dy_n.$$

We first want to show that this is Gaussian. Let $\mathbf{y}_1$ be the vector of the variables y_i, $i = 1, m$, to be retained, and let $\mathbf{y}_2$ be the vector of the remaining variables y_i, $i = m + 1, n$. Let $\Phi_{y1}(\boldsymbol{\omega}_1)$ be the characteristic function of $\mathbf{y}_1$ and let $\Phi_{y_1}(\boldsymbol{\omega})$ be the characteristic function of the full set of variables $\mathbf{y}$. In $\Phi_y(\boldsymbol{\omega})$, let $\boldsymbol{\omega}_2$ be the variables corresponding to $\mathbf{y}_2$. Then we can write

$$\Phi_{y_1}(\boldsymbol{\omega}_1) = \int_{-\infty}^{\infty} p_{y_1}(\mathbf{y}_1) \, \exp(j\boldsymbol{\omega}_1^T\mathbf{y}_1) \, d\mathbf{y}_1$$

$$= \int_{-\infty}^{\infty} \left[\int_{-\infty}^{\infty} p_y(\mathbf{y}) \, d\mathbf{y}_2\right] \exp(j\boldsymbol{\omega}^T\mathbf{y}) \, d\mathbf{y}_1 \Big|_{\omega_2=0}$$

$$= \int_{-\infty}^{\infty} p_y(\mathbf{y}) \, \exp(j\boldsymbol{\omega}^T\mathbf{y}) \, d\mathbf{y} \Big|_{\omega_2=0} = \Phi_y(\boldsymbol{\omega})\big|_{\omega_2=0}. \qquad (4.9)$$

In order to show from this that $\mathbf{y}_1$ is Gaussian, let us partition the covariance matrix of $\mathbf{y}$ as

$$\mathbf{C}_y = \begin{bmatrix} \mathbf{C}_{11} & \mathbf{C}_{12} \\ \mathbf{C}_{12}^T & \mathbf{C}_{22} \end{bmatrix}.$$

Using this with Eqs. (4.6), (4.9), we have

$$\Phi_{y_1}(\boldsymbol{\omega}_1) = \exp(j\mathbf{m}_1^T\boldsymbol{\omega}_1 - \boldsymbol{\omega}_1^T\mathbf{C}_{11}\boldsymbol{\omega}/2).$$

This is the characteristic function of a Gaussian with mean $\mathbf{m}_1$ and covariance matrix $\mathbf{C}_{11}$. Therefore the marginal density of a Gaussian is itself Gaussian.

In the case of interest, the general $n \times n$ transformation of Eq. (4.7) has the special form

$$\mathbf{L} = \begin{bmatrix} \mathbf{L}_1 \\ \hline \mathbf{0} \mid \mathbf{I} \end{bmatrix},$$

where $\mathbf{L}_1$ is the original $m \times n$ transformation, $\mathbf{I}$ is the $(n - m) \times (n - m)$ unit matrix, and $\mathbf{0}$ is an $(n - m) \times m$ matrix of zeros. We now need only note that the mean and covariance of the variables $\mathbf{y}_1$ are

$$\mathbf{m}_1 = \mathbf{L}_1\mathcal{E}(\mathbf{x}) = \mathbf{L}_1\mathbf{m},$$
$$\mathbf{C}_{11} = \mathcal{E}[\mathbf{L}_1(\mathbf{x} - \mathbf{m})][\mathbf{L}_1(\mathbf{x} - \mathbf{m})]^T = \mathbf{L}_1\mathbf{C}\mathbf{L}_1^T. \qquad (4.10)$$

These lead to a Gaussian of just the form Eq. (4.7) with $\mathbf{L}$ replaced by $\mathbf{L}_1$, so that Eq. (4.7) is a general result for any linear transformation $\mathbf{L}$ of full rank.

Linear Transformation to Decorrelate Random Variables

Let us consider again a vector random variable $\mathbf{x}$, with mean $\mathbf{m}$ and covariance $\mathbf{C} = \mathcal{E}(\mathbf{x} - \mathbf{m})(\mathbf{x} - \mathbf{m})'$. In this section we allow the components of the vector **x** to be complex random variables for generality, in which case the prime symbol that we use indicates the conjugate of the transpose.

In developments concerning multidimensional random variables, i.e., random vectors, it is often convenient to assume that the vector components x_i are uncorrelated, so that the covariance matrix is the unit matrix: **C** = **I**. In the case of Gaussian random variables, the x_i are then independent, because the multivariate density Eq. (4.4) factors as in Eq. (4.5). In this section we describe a procedure for finding a linear transformation of the random vector **x** into a random vector **y** with uncorrelated components y_i.

The problem to be solved is determination of a transformation matrix **T** such that the random variable

$$y = T'x \qquad (4.11)$$

has a covariance matrix

$$Y = \mathscr{E}[T'(x - m)][T'(x - m)]' = T'CT = \Lambda, \qquad (4.12)$$

with Λ a diagonal matrix, with the variances of the new variables y_i on the diagonal. One procedure is to determine the transformation **T** in terms of the normalized eigenvectors of the matrix **C**.

A covariance matrix **C** has two important special properties. First, it is *Hermitian;* that is, it is equal to the conjugate transpose of itself. This is because

$$C' = \mathscr{E}[(x - m)(x - m)']' = \mathscr{E}(x - m)(x - m)' = C.$$

Second, it is *positive semidefinite,* in that, for any vector **v**,

$$r = v'Cv = \mathscr{E}[v'(x - m)][v'(x - m)]'$$
$$= \mathscr{E}|v'(x - m)|^2 \geq 0.$$

If always $r > 0$, except for **v** = **0**, then **C** is called *positive definite.* If a vector **v** ≠ **0** exists with $r = 0$, then the matrix **C** is singular, and at least one of the components x_m of **x** can be expressed deterministically in terms of the others. A degree of freedom then needs to be removed from the problem, and we assume that has been done if necessary.

Suppose then that **C** is any Hermitian positive definite matrix. Consider the eigenproblem for **C**, that is, determination of a constant λ and a vector **t** such that

$$Ct = \lambda t. \qquad (4.13)$$

That is, we seek solutions of the homogeneous set of linear equations

$$(C - \lambda I)t = 0.$$

Certainly one solution is **t** = **0**, which we are not interested in. If the matrix **C** − λ**I** has an inverse, then **0** is the only solution: **t** =

$(\mathbf{C} - \lambda\mathbf{I})^{-1}\mathbf{0} = \mathbf{0}$. Thus, in order to make progress, we must select the number λ to be such that $\mathbf{C} - \lambda\mathbf{I}$ does not have an inverse. Accordingly, we want λ to be such that

$$\det(\mathbf{C} - \lambda\mathbf{I}) = 0. \qquad (4.14)$$

Equation (4.14) is the characteristic equation associated with the matrix $\mathbf{C}$. If $\mathbf{C}$ is an $n \times n$ matrix, and if the determinant in Eq. (4.14) is considered to be expanded by rows, it will be seen that the equation is equivalent to a polynomial equation of degree n, with certain coefficients which depend on the elements of $\mathbf{C}$:

$$\sum_{i=1}^{n} a_i \lambda^i = 0.$$

Such an equation, the *characteristic equation* of the matrix $\mathbf{C}$, has exactly n roots λ_m, $m = 1, n$, called the *eigenvalues* of the matrix $\mathbf{C}$. (In general, there may be repeated roots; that is, some of the λ_m may be equal to one another.) In general, the eigenvalues are complex, although if $\mathbf{C}$ is real, so that the a_i are real numbers, they occur in conjugate pairs. However, if $\mathbf{C}$ is Hermitian, then the eigenvalues are all real numbers. If $\mathbf{C}$ is also positive definite (our case), the roots λ_i of Eq. (4.14) are all positive real numbers.

We therefore have at least one distinct value λ and perhaps as many as n distinct values λ_m from which to choose in considering Eq. (4.13). Now we need to consider whether the resulting equation

$$\mathbf{C}\mathbf{t}_m = \lambda_m \mathbf{t}_m \qquad (4.15)$$

does or does not have solutions. Because the matrix $\mathbf{C}$ of interest to us has the property of being Hermitian positive definite, the answer is unambiguous. If the number λ_m is a simple root of Eq. (4.14), then there is exactly one solution $\mathbf{t}_m$ of Eq. (4.15), called the *eigenvector* belonging to the eigenvalue λ_m. If λ_m is a repeated root of Eq. (4.14) of order k, then there are exactly k corresponding solutions (eigenvectors) of Eq. (4.15), and, furthermore, they are linearly independent. That is, no one of them can be written as a linear combination of the others.

It is also a fact that eigenvectors $\mathbf{t}_{l,m}$ belonging to numerically different eigenvalues are orthogonal, i.e., $\mathbf{t}_l'\mathbf{t}_m = 0$, so that they are (*a fortiori*) linearly independent. Therefore, we can collect together exactly n linearly independent eigenvectors $\mathbf{t}_m$ of the matrix $\mathbf{C}$. Let us arrange them as columns of an nxn matrix $\mathbf{T}$:

$$\mathbf{T} = [\mathbf{t}_1, \ldots, \mathbf{t}_n]. \qquad (4.16)$$

The matrix $\mathbf{T}$, by its construction, has the property that

$$\mathbf{CT} = \mathbf{C}[\mathbf{t}_1, \ldots, \mathbf{t}_n] = [\mathbf{Ct}_1, \ldots, \mathbf{Ct}_n]$$
$$= [\lambda_1 \mathbf{t}_1, \ldots, \lambda_n \mathbf{t}_n] = [\mathbf{t}_1, \ldots, \mathbf{t}_n]\Lambda = \mathbf{T}\Lambda, \tag{4.17}$$

where Λ is the diagonal matrix with the numbers λ_m on the diagonal.

Since the columns of the matrix $\mathbf{T}$ are linearly independent, no column can be written as a linear combination of the others, hence $\det(\mathbf{T}) \neq 0$, and hence $\mathbf{T}$ has an inverse. From Eq. (4.17) we then have

$$\mathbf{T}^{-1}\mathbf{CT} = \Lambda. \tag{4.18}$$

With one additional adjustment, this can be brought to the form of Eq. (4.12). Recall that the columns of $\mathbf{T}$ are orthogonal if the corresponding eigenvectors arose from numerically different eigenvalues. In the case of multiple eigenvectors arising from a repeated root of Eq. (4.14), the eigenvectors are linearly independent. We can always find linear combinations of them that are orthogonal and are still eigenvectors of $\mathbf{C}$. As a result, we can arrange $\mathbf{T}$ to have orthogonal columns. Further dividing through each column of $\mathbf{T}$ by its length $\|\mathbf{t}_m\| = (\mathbf{t}_m'\mathbf{t}_m)^{1/2}$, we obtain a matrix whose columns make up an *orthonormal* set of vectors: $\mathbf{t}_l'\mathbf{t}_m = \delta_{lm}$, the Kronecker delta. That is, $\mathbf{T}$ is a *unitary* matrix (or an *orthogonal* matrix, if $\mathbf{T}$ is real). That is,

$$\mathbf{T}'\mathbf{T} = \begin{bmatrix} \mathbf{t}_1' \\ \vdots \\ \mathbf{t}_n' \end{bmatrix} [\mathbf{t}_1 \ldots \mathbf{t}_n] = \{\mathbf{t}_l'\mathbf{t}_m\}$$

$$= \{\delta_{lm}\} = \mathbf{I}.$$

Hence $\mathbf{T}' = \mathbf{T}^{-1}$, so that Eq. (4.18) becomes just Eq. (4.12).

EXAMPLE 4.1 Consider the symmetric matrix

$$\mathbf{C} = \frac{1}{33} \begin{bmatrix} 82 & 7 & -28 \\ 7 & 34 & -4 \\ -28 & -4 & 49 \end{bmatrix}.$$

It is positive definite and is therefore the covariance matrix of some set of random variables x_i. The characteristic equation Eq. (4.14) is

$$\left(\frac{1}{33}\right)^3 \begin{vmatrix} 33\lambda - 82 & -7 & 28 \\ -7 & 33\lambda - 34 & 4 \\ 28 & 4 & 33\lambda - 49 \end{vmatrix} = \lambda^3 - 5\lambda^2 + 7\lambda - 3 = 0.$$

The roots are $\lambda = 1, 1, 3$. Using the first ($\lambda = 1$) in Eq. (4.15), we seek solutions of

$$49x_1 + 7x_2 - 28x_3 = 0,$$
$$7x_1 + x_2 - 4x_3 = 0,$$
$$-28x_1 - 4x_2 + 16x_3 = 0.$$

The matrix of these equations has rank 1 (the determinant and the determinant of all the second-order minors vanish), so that there are two linearly independent solutions. Since the rank of the system is 1, we can choose any two of the x_i at will. We can then find the two solutions by, say, letting $x_1 = 1$, $x_2 = 0$, so that $x_3 = 49/28$. Then letting $x_1 = 0$, $x_2 = 1$ to obtain a linearly independent solution, we find $x_3 = 1/4$. Although these are independent, they are not orthogonal. Hence we seek, say, a linear combination of the two which is orthogonal to the second:

$$\begin{bmatrix} 0 & 1 & 1/4 \end{bmatrix} \begin{bmatrix} 1 + a(0) \\ 0 + a(1) \\ 49/28 + a/4 \end{bmatrix} = 0,$$

leading to $a = -7/17$. Substituting the value $\lambda = 3$ into Eq. (4.15) and proceeding similarly, now choosing any one of the x_i arbitrarily (but not zero), say $x_1 = 1$, yields the single solution $(1, 1/7, -4/7)$. Finally, normalizing the three vectors to unit length yields the transformation matrix

$$\mathbf{T} = \frac{1}{\sqrt{1122}} \begin{bmatrix} 17 & 0 & 7\sqrt{17} \\ -7 & 4\sqrt{66} & \sqrt{17} \\ 28 & \sqrt{66} & -4\sqrt{17} \end{bmatrix}.$$

Mixed Central Moments

We will now derive a useful expression for the central moments of a Gaussian density function. These are the same as the moments of a zero mean Gaussian density. The generalization of Eq. (1.63) to the multidimensional case yields

$$\mathcal{E}(x_1^{b_1} \cdots x_n^{b_n}) = (-j)^B [(\partial^B / \partial\omega_1^{b_1} \cdots \partial\omega_n^{b_n}) \Phi_x(\omega)]_{\omega=0}, \qquad (4.19)$$

where $B = \Sigma \, b_i$ and the b_i are any nonnegative integers. It is at once clear that all moments for odd B must vanish, because then, the Gaussian density being an even function, the integrand in the defining formula for the moment Eq. (4.19) will be an odd function.

In the case of even B, we can proceed as follows. In the characteristic function Eq. (4.6) we take the mean $\mathbf{m} = \mathbf{0}$ and recognize that the quadratic form $\boldsymbol{\omega}^T \mathbf{C} \boldsymbol{\omega}$ is a scalar. The exponential can then be expanded in a Taylor series around $\boldsymbol{\omega} = \mathbf{0}$ to yield

$$\Phi_x(\boldsymbol{\omega}) = \sum_{p=0}^{\infty} \frac{(-1)^p}{2^p p!} \left[\sum_{k,l=1}^{n} C_{kl} \omega_k \omega_l \right]^p. \tag{4.20}$$

The terms of this sum involve various cross-products of the pairwise products $\omega_k \omega_l$ of the components ω_i of the vector $\boldsymbol{\omega}$. If in some term a particular variable ω_i appears to a power less than b_i, then the operation of taking the first b_i partial derivatives in Eq. (4.19) will yield zero for that term of the sum. On the other hand, if ω_i appears to a power greater than b_i in some term, then after taking b_i partial derivatives ω_i will survive to at least the first power. Setting ω_i to zero in the result then causes that term of the sum to vanish in Eq. (4.19). Hence the only terms of $\Phi_x(\boldsymbol{\omega})$ which can contribute to Eq. (4.19) are those in which ω_i appears to the same power b_i as the order of its partial derivative in Eq. (4.19).

The multinomial expansion now applies. This is the generalization of the binomial expansion:

$$(x_1 + \cdots + x_m)^p = \sum_{k_1,\ldots,k_m=0}^{p} \binom{p}{k_1, \ldots, k_m} x_1^{k_1} \cdots x_m^{k_m}. \tag{4.21}$$

Here the indicated multinomial coefficient is the number

$$\binom{p}{k_1, \ldots, k_m} = \frac{p!}{k_1! \cdots k_m!}$$

and is taken for $k_1 + \cdots + k_m = p$. Consider now that Eq. (4.20) contains terms involving $\omega_r \omega_s$ for $r, s = 1, n$. These are the x_i in Eq. (4.21). In Eq. (4.21) the terms $\omega_r \omega_s$ are taken to certain powers $k_i = k_{rs}$. A particular ω_q appears in terms with $r = q$ but $s \neq q$, with $r \neq q$ but $s = q$, and with r, $s = q$. The resulting power of ω_q in the product Eq. (4.21) is

$$2k_{qq} + \sum_{\substack{s=1 \\ s \neq q}}^{n} k_{qs} + \sum_{\substack{r=1 \\ r \neq q}}^{n} k_{rq} = \sum_{s=1}^{n} k_{qs} + \sum_{r=1}^{n} k_{rq} = b_q,$$

where the restriction to b_q is necessary in order that the term contribute to the moment Eq. (4.19). Since we must also have

$$\sum_{r,s=1}^{n} k_{rs} = p,$$

we have

$$B = \sum_{q=1}^{n} b_q = \sum_{q,s=1}^{n} k_{qs} + \sum_{q,r=1}^{n} k_{rq} = 2p.$$

Thus only a term for which $p = B/2$ can contribute in Eq. (4.19). Remembering that B is even, using only the potentially contributing term $p = B/2$ in Eq. (4.20) and substituting into Eq. (4.19) yields our sought result, in which there may also be terms which do not contribute:

$$\mathscr{E}(x_1^{b_1} \cdots x_n^{b_n}) = [2^{B/2}(B/2)!]^{-1}$$

$$\times (\partial^B / \partial \omega_1^{b_1} \cdots \partial \omega_n^{b_n}) \left[\sum_{k,l=1}^{n} C_{kl} \omega_k \omega_l \right]_{\omega=0}^{B/2}. \qquad *(4.22)$$

As a specific application of Eq. (4.22), for the four-variable *zero-mean* joint Gaussian density we have that

$$\mathscr{E}(x_1 x_2 x_3 x_4) = (1/8)(\partial^4 / \partial \omega_1 \cdots \partial \omega_4) \sum_{ijkl=1}^{n} R_{ij} R_{kl} \omega_i \omega_j \omega_k \omega_l \bigg|_{\omega=0}.$$

In this only those terms contribute in which each ω_r occurs exactly once. There are 24 of them and the derivative of each is the constant $R_{ij} R_{kl}$. Taking account that **R** is a symmetric matrix, we obtain

$$\mathscr{E}(x_1 x_2 x_3 x_4) = R_{12} R_{34} + R_{13} R_{24} + R_{14} R_{23}. \qquad *(4.23)$$

The formula will be considered for complex random variables in Chapter 11.

EXAMPLE 4.2   Suppose $y(t) = x^2(t)$ and that $x(t)$ is a zero-mean stationary Gaussian process with correlation function $R_x(\tau)$. Then

$$R_y(\tau) = \mathscr{E}[x^2(t) x^2(t - \tau)]$$
$$= [\mathscr{E}x^2(t)][\mathscr{E}x^2(t - \tau)] + 2[\mathscr{E}x(t)x(t - \tau)]^2$$
$$= R_x^2(0) + 2R_x^2(\tau).$$

The Gaussian Random Process

A random process $x(t)$ defines random variables $x(t_i)$ for any set of instants t_i, $i = 1, n$. A random process is called Gaussian if the random variables $x(t_i)$ have an n-dimensional Gaussian density Eq. (4.4) for every choice of n and instants t_i. The covariance matrix of each such density has elements

$$C_{ij} = \mathscr{E}[x(t_i) - m(t_i)][x(t_j) - m(t_j)]$$
$$= R_x(t_i, t_j) - m(t_i)m(t_j),$$

where $R_x(t, t')$ is the correlation function of the process and $m(t)$ is the mean. In the common case of a discrete process sampled at uniformly spaced intervals δt, the resulting covariance matrix is a Toeplitz matrix: $C_{ij} = C_x[(i - j)\,\delta t]$.

4.2 THE CENTRAL LIMIT THEOREM

Some support for the common use of the Gaussian probability density as a model of random processes is provided by the central limit theorem. This states that, under modest restrictions, the sum of a number n of independent random variables has a distribution that tends to Gaussian as the number of variables in the sum becomes large. Although various versions of the theorem can be proved under different conditions, one form of the theorem is the following [Grimmett and Stirzaker, 1992, Ch. 5].

Let X_k be a set of independent random variables each having the same distribution, with finite mean m and finite variance σ^2. For convenience we temporarily take $m = 0$. Consider the normalized variables $Y_k = X_k/\sigma\sqrt{n}$, for $n > 0$. We want to show that the sum

$$S = \sum_{k=1}^{n} Y_k = \frac{1}{\sqrt{n}} \sum_{k=1}^{n} \frac{X_k}{\sigma} \tag{4.24}$$

has a density which tends to the normalized Gaussian for large n:

$$\lim_{n \to \infty} p_S(s) = (2\pi)^{-1/2} \exp(-s^2/2). \tag{4.25}$$

As shown in Sect. 1.6, the characteristic function of a random variable always exists. Thus $\Phi_X(\omega)$ and $\Phi_Y(\omega)$ both exist. The densities are related by

$$p_y(y) = (\sigma\sqrt{n})p_x(\sigma y\sqrt{n}).$$

Hence, where $\mathcal{F}$ is the Fourier transform operation, the characteristic functions are related by

$$\Phi_y(\omega) = [\mathcal{F}\{p_y(y)\}]^* = [\Phi_x(-\omega/\sigma\sqrt{n})]^*$$
$$= \Phi_x(\omega/\sigma\sqrt{n}).$$

Then the characteristic function of the sum Eq. (4.24) is

$$\Phi_S(\omega) = [\Phi_y(\omega)]^n = [\Phi_x(\omega/\sigma\sqrt{n})]^n. \tag{4.26}$$

Since we have assumed that X has a finite variance σ^2, the characteristic function $\Phi_x(\omega)$ of X has at least a second derivative, and we can write it as a Taylor series expansion around $\omega = 0$:

$$\Phi_X(\omega) = 1 - \omega^2\sigma^2/2 + o(\omega^2),$$

where we have used Eq. (1.63) to write the first two derivatives in terms of the moments of X. The remainder $o(\omega^2)$ after two terms is a quantity that decreases faster than ω^2 as $\omega \Rightarrow 0$. Using this in Eq. (4.26) then yields

$$\Phi_S(\omega) = [1 - \omega^2/2n + o(1/n)]^n. \tag{4.27}$$

Taking the logarithm of this last yields

$$\ln \Phi_S(\omega) = n \ln[1 - \omega^2/2n + o(1/n)],$$

so that, for large enough n,

$$\Phi_S(\omega) \cong \exp(-\omega^2/2).$$

That is,

$$\lim_{n\Rightarrow\infty} \Phi_S(\omega) = \lim_{n\Rightarrow\infty} \int_{-\infty}^{\infty} p_S(s) \exp(j\omega s)\, ds = \exp\left(\frac{-\omega^2}{2}\right). \tag{4.28}$$

It is a fact that, if a sequence of characteristic functions [such as the sequence Eq. (4.27) for increasing n] converges to a limit which is continuous [as in Eq. (4.28)], then the limit is the characteristic function of the limit of the sequence of corresponding distributions. We therefore conclude from Eq. (4.28) that Eq. (4.25) holds true.

Considering variables X_k with mean m different from zero, the corresponding result is that

$$\lim_{n\Rightarrow\infty} p_S(s) = (2\pi)^{-1/2} \exp(-s^2/2), \tag{4.29}$$

where S is the sum

$$S = \frac{1}{\sqrt{n}} \sum_{k=1}^{n} \frac{X_k - m}{\sigma}. \tag{4.30}$$

Equation (4.29) with S of this latter form is the central limit theorem for independent variables with common distribution.

The same result holds true in the multidimensional case. Specifically, if $\mathbf{X}_k$ is a sequence of m-dimensional column vectors of random variables, independent over k, and if

$$\mathscr{E}\mathbf{X}_k = \mathbf{m}, \qquad \mathscr{E}(\mathbf{X}_k - \mathbf{m})(\mathbf{X}_k - \mathbf{m})^{\mathrm{T}} = \mathbf{C}_x$$

are the mean and covariance of the $\mathbf{X}_k$, then for the sum

$$\mathbf{S} = \frac{1}{\sqrt{n}} \sum_{k=1}^{n} (\mathbf{X}_k - \mathbf{m})$$

we have

$$\lim_{n \Rightarrow \infty} p_S(\mathbf{s}) = (2\pi)^{-m/2} |\det(\mathbf{C}_x)|^{-1/2} \exp[-(\mathbf{s}^T \mathbf{C}_x^{-1} \mathbf{s})/2]. \qquad (4.31)$$

If we introduce new variables $\mathbf{Y}_k = \mathbf{C}_x^{-1/2}(\mathbf{X}_k - \mathbf{m})$, where $\mathbf{C}_x^{-1/2}$ is any symmetric solution of $A\mathbf{C}_x A = I$ (we will see later that such a solution exists), Eq. (4.31) becomes

$$\lim_{n \Rightarrow \infty} p_S(\mathbf{s}) = (2\pi)^{-m/2} \exp(-\mathbf{s}^T \mathbf{s}/2),$$

where

$$\mathbf{S} = \frac{1}{\sqrt{n}} \sum_{k=1}^{n} \mathbf{Y}_k = \frac{1}{\sqrt{n}} \sum_{k=1}^{n} \mathbf{C}_x^{-1/2}(\mathbf{X}_k - \mathbf{m}). \qquad (4.32)$$

This last is the direct analog of Eq. (4.30).

The central limit theorem holds true under more general conditions than assumed above. For example, suppose that the X_k are independent (not necessarily having the same distributions) with finite means m_k and finite variances σ_k^2, and that (the Lindeberg condition), for every $t > 0$,

$$\lim_{n \Rightarrow \infty} s_n^{-2} \sum_{k=1}^{n} \int_{|x| \geq ts_n} x^2 p_k(x)\, dx = 0. \qquad (4.33)$$

In Eq. (4.33),

$$s_n^2 = \sum_{k=1}^{n} \sigma_k^2. \qquad (4.34)$$

Provided the condition Eq. (4.33) holds, the central limit theorem is true for the X_k [Stuart and Ord, 1987, Ch. 7]:

$$\lim_{n \Rightarrow \infty} p_S(s) = (2\pi)^{-1/2} \exp(-s^2/2), \qquad (4.35)$$

where

$$S = \frac{1}{s_n} \sum_{k=1}^{n} (X_k - m_k). \qquad (4.36)$$

Equation (4.36) generalizes Eq. (4.30). Correspondingly, for identically distributed variables, (4.33) holds provided only that $\mathbf{C}_x$ exists.

The condition Eq. (4.33) can be shown to imply that

$$\sigma_k^2 / s_n^2 < \varepsilon,$$

for sufficiently large n, where $\varepsilon > 0$ is any small number. That is, the contribution of any particular random variable X_k to the variance of the sum should be vanishingly small, provided the sum S contains an adequately large number of terms. Roughly speaking, the sum S is the result of a very large number of terms, each of which makes a vanishingly small contribution.

As noted at the beginning of this section, if the X_k have identical distributions, it is necessary only that they have finite mean and variance in order that the central limit theorem hold true. For this special case the Lindeberg condition is satisfied. Substituting $s_n^2 = n\sigma^2$ into Eq. (4.33) in this case yields

$$\lim_{n \Rightarrow \infty} \left(\frac{1}{n\sigma^2}\right) n \int_{|x| \geq t\sigma\sqrt{n}} x^2 p_X(x)\, dx = 0. \qquad (4.37)$$

Although the preceding limit results are important as indicators of the wide applicability of the Gaussian density, for computational work it is useful to know in addition how rapidly the limit density is approached as n increases. Consider again the case that the mutually independent random variables X_k being summed have the same density. Let their mean and variance be finite, and suppose that the third moment exists: $m_3 = \mathscr{E}|X - m|^3 < \infty$. Also assume the characteristic function $\Phi_x(\omega)$ of the X_k is absolutely integrable. Then it can be shown that [Feller, 1966, Ch. 16]

$$p_S(s) = \varphi(s)[1 + (m_3/6\sigma^3\sqrt{n})s(s^2 - 3)] + o(1/\sqrt{n}), \qquad (4.38)$$

where $\varphi(s)$ is the normalized Gaussian density. This is the first of a sequence of results (the Edgeworth series) which we will discuss at more length in Chapter 8.

The remarkable *Berry–Esséen theorem* [Feller, 1966; Papoulis, 1991] also allows an approximation by the normal distribution. Consider again independent random variables X_k with means m_k and variances σ_k^2. Suppose that, for some constant M and for all k,

$$\mathscr{E}(|X_k - m_k|^3) < M\sigma_k^2. \qquad (4.39)$$

Consider the sum Eq. (4.36) with Eq. (4.34). Then, where $\mathscr{G}(s)$ is the normalized Gaussian distribution function,

$$|P_S(s) - \mathscr{G}(s)| < (33/4)M/s_n. \qquad (4.40)$$

Often we are interested in the "tail" of a distribution, particularly in determining the false alarm rate of a detection algorithm. In such a case we are usually interested in the proportional error in approximating some density by the Gaussian, rather than in the error itself. One result is the following [Feller, 1966]. Let X_k be a set of mutually independent random variables with means m_k and variances σ_k^2. Suppose that there exists a finite interval $-a \leq \omega \leq a$ over which the characteristic functions $\Phi_{Xk}(\omega)$ are all analytic. Suppose further that there exists a constant M such that, for all k,

$$\mathscr{E}(|X_k|^3) \leq M\sigma_k^2.$$

Suppose that the variances σ_k^2 are such that $s_n \Rightarrow \infty$ as $n \Rightarrow \infty$, where s_n is as in Eq. (4.34). Then, for fixed s which may $\Rightarrow \infty$ as $n \Rightarrow \infty$, but more slowly than $\sqrt{s_n}$,

$$\ln\{[1 - P_S(s)]/[1 - \mathscr{G}(s)]\} \sim \mu_3^{(n)} s^3/6s_n^3,$$

$$\mu_3^{(n)} = \sum_{k=1}^n \mathscr{E}(X_k - m_k)^3. \tag{4.41}$$

If the X_k have identical distributions, the right side of Eq. (4.41) is $\mu_3 s^3/6(\sigma^3)\sqrt{n}$, where σ^2, μ_3 are the variance and third central moment of the X_k.

EXAMPLE 4.3 Suppose the random variables X_k are uniformly distributed over $(-1/2, 1/2)$. Then $\mathscr{E}(X_k) = 0$ and $\sigma_k^2 = 1/12$. As indicated in Eq. (4.37), since the X_k have a common distribution with finite mean and variance, the central limit theorem holds true. We expect that, as $n \Rightarrow \infty$, the distribution of $(1/\sigma\sqrt{n}) \sum X_k$ tends to the normalized Gaussian. Figure 4.3 compares the sum for $n = 4$ to the Gaussian with mean zero and variance $1/3$. The approximation is quite good.

Example 4.3 makes clear that the central limit theorem is an analytical result having to do with repeated convolution of certain functions. For identically distributed variables X_k, the theorem states that the repeated convolution $p_X[(x - m)/\sigma](*n)p_X[(x - m)/\sigma]$ tends to the normalized Gaussian curve. The result itself makes no reliance on probabilistic notions.

The central limit theorem also applies for the case of discrete random variables and specifically for *lattice variables*, which take on only the

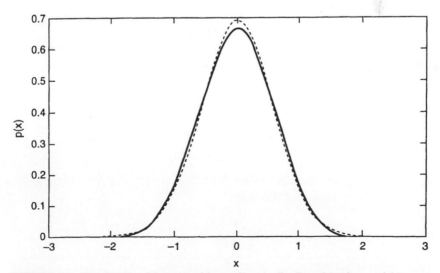

FIGURE 4.3 The density of the sum of four uniformly distributed random variables compared with the Gaussian density of the same mean and variance.

specific values $X_k = a + i_k h$, where i_k is an integer and h is the lattice spacing. Let the X_k have common means m and variances σ^2 and let

$$S = \sum_{k=1}^{n} \frac{X_k - m}{\sigma}. \tag{4.42}$$

Then [Grimmett and Stirzaker, 1992, Ch. 5]

$$\Pr(S = s) \sim (h/\sigma\sqrt{n})_{\mathcal{G}}(s/\sqrt{n}), \tag{4.43}$$

where $_{\mathcal{G}}(x)$ is the normalized Gaussian density.

EXAMPLE 4.4   (DeMoivre–Laplace Theorem) A Bernoulli random variable is a discrete random variable taking only two values, with probabilities say p and $q = 1 - p$. If we label the values as (respectively) 1 and 0, then out of n trials the number of trials resulting in the event with probability p ("success") is

$$K = \sum_{k=1}^{n} X_k = \sigma S + nm,$$

where S is as in Eq. (4.42). The parameters in this are

$$m = \mathcal{E}X_k = (1)p + (0)q = p,$$
$$\sigma^2 = (1 - p)^2 p + (0 - p)^2 q = pq(p + q) = pq.$$

Noting that the lattice constant is $h = 1$, Eq. (4.43) indicates that

$$\Pr(K = k) \sim (2\pi n\sigma^2)^{-1/2} \exp[-(k - nm)^2/2n\sigma^2]$$
$$= (2\pi npq)^{-1/2} \exp[-(k - np)/2npq]. \qquad *(4.44)$$

This result, indicating the asymptotic probability of k successes in n Bernoulli trials, is the *DeMoivre–Laplace theorem*.

EXAMPLE 4.5   Suppose that the independent random variables X_k have the normalized *Cauchy density*, that is,

$$p_{X_k}(x) = (1/\pi)/(1 + x^2).$$

The corresponding characteristic function is

$$\Phi_{X_k}(\omega) = \exp(-|\omega|),$$

which is easily verified by taking the Fourier transform of the right-hand side. Now consider the sum

$$S = \frac{1}{n} \sum_{k=1}^{n} X_k.$$

Since the characteristic function of X_k/n is $\Phi_{X_k}(\omega/n)$, we have

$$\Phi_S(\omega) = \exp(-n|\omega/n|) = \exp(-|\omega|),$$

so that, even for $n \Rightarrow \infty$, S also has the normalized Cauchy density, not the Gaussian density. That the central limit theorem does not apply to this case is a consequence of the fact that a random variable with the Cauchy density has no expected value:

$$\mathscr{E}(X) = \frac{1}{\pi} \lim_{A,B \to \infty} \int_{-A}^{B} \left[\frac{x}{(1 + x^2)} \right] dx$$

$$= (1/2\pi) \lim_{A,B \to \infty} [\ln(1 + B^2) - \ln(1 + A^2)] = \infty.$$

(That the principal value of the improper integral exists is not enough.)

4.3 SUM OF A SINE WAVE AND A GAUSSIAN PROCESS

We will often be interested in the following chapters in deciding the presence or absence of a signal in additive noise. To that end, we will want to calculate the densities of the received data with and without the signal. The simplest such case is that of a constant-amplitude sinusoid of random phase added to zero-mean Gaussian noise. Specifically, we will find the first-order probability density of the received signal

$$r(t) = A \cos(\omega_c t + \theta) + n(t), \tag{4.45}$$

where A and ω_c are specified constants, θ is a uniform random variable on $0 \le \theta < 2\pi$, and $n(t)$ is a zero-mean stationary Gaussian process with variance σ^2. The phase and the noise are assumed independent.

From Eq. (4.45), at a particular value of t the signal $r(t)$ is the sum of two independent random variables. Therefore from Eq. (1.46) we have the characteristic function of the random variable $r(t)$ as

$$\Phi_r(\omega) = \Phi_s(\omega)\Phi_n(\omega),$$

where $s(t) = A \cos(\omega_c t + \theta)$ is the signal of interest. From Eq. (4.6) we have

$$\Phi_n(\omega) = \exp(-\sigma^2 \omega^2/2).$$

By the definition Eq. (1.37), we also have

$$\Phi_s(\omega) = \frac{1}{2\pi} \int_0^{2\pi} \exp[j\omega A \cos(\omega_c t + \theta)] \, d\theta$$

$$= J_0(A\omega), \tag{4.46}$$

where $J_0(u)$ is the Bessel function of first kind and order zero. Then

$$\Phi_r(\omega) = J_0(A\omega) \exp(-\sigma^2 \omega^2/2). \tag{4.47}$$

The (real) density is given by the inverse Fourier transform of this (real) characteristic function. If we replace $J_0(A\omega)$ in Eq. (4.47) with its defining integral:

$$J_0(A\omega) = \frac{1}{\pi} \int_0^\pi \exp[jA\omega \cos(\theta)] \, d\theta,$$

it is easy to complete the square in the exponent of the Fourier transform and make a change of variable to obtain

$$p_r(r) = \frac{1}{2\pi^2} \int_0^\pi \exp\left[\frac{-(A\cos\theta - r)^2}{2\sigma^2}\right]$$

$$\times \int_{-\infty}^\infty \exp\left(\frac{-\sigma^2 u^2}{2}\right) du \, d\theta.$$

Let us change to the normalized density by defining $v = r/\sigma$. Then $p_v(v) = \sigma p_r(\sigma v)$. Evaluating the Gaussian integral in the preceding equation then yields

$$p_v(v) = \frac{1}{\pi} \int_0^\pi \varphi[v - \sqrt{2}\,\alpha \cos(\theta)] \, d\theta. \tag{4.48}$$

In this, $\varphi(x)$ is the normalized Gaussian density $N(0, 1)$, and $\alpha^2 = A^2/2\sigma^2$ is the signal-to-noise ratio, that is, the ratio of the root-mean-square power of the sinusoid to the expected power $\mathscr{E}n^2 = \sigma^2$ of the noise.

The form Eq. (4.48) is convenient for computation, because the Gaussian function is so readily available. However, for use we are more likely to be interested in the variable $z = r/A$. This is because we usually have the case that the signal-to-noise ratio grows by the noise becoming zero (in which case $v = \infty$) rather than by the signal becoming infinite. In Fig. 4.4 we show the density $p_z(z) = (A/\sigma)p_v(Az/\sigma)$ parametrized by the signal-to-noise ratio $\alpha = A/\sigma\sqrt{2}$.

A power series can be obtained for the density Eq. (4.48) as follows. The Gaussian function in the integrand can be expanded in a Taylor series in v, in which the coefficients involve the derivatives $\varphi^{(k)}(-\sqrt{2}\alpha \cos \theta)$ of the Gaussian. Using [Abramowitz and Stegun, 1965, #26.2.38], these can be written in terms of the confluent hypergeometric function $_1F_1(a; b; x)$ (also known as Kummer's function), defined as

$$_1F_1(a; b; x) = 1 + \frac{a}{b}\frac{x}{1!} + \frac{a(a+1)}{b(b+1)}\frac{x^2}{2!}$$

$$+ \frac{a(a+1)(a+2)}{b(b+1)(b+2)}\frac{x^3}{3!} + \cdots . \tag{4.49}$$

The result is

$$p_v(v) = (2/\pi)(2\pi)^{-1/2}$$

$$\times \sum_{k=0}^\infty \frac{(-v^2)^k}{2^k k!} \int_0^{\pi/2} {}_1F_1(k + \tfrac{1}{2}; \tfrac{1}{2}; -\alpha^2 \sin^2 \phi) \, d\phi.$$

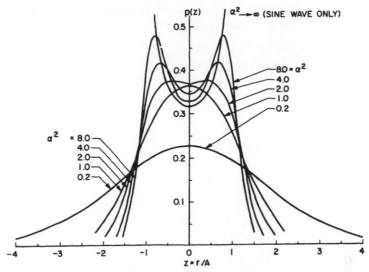

FIGURE 4.4 Amplitude distribution of a sine wave of amplitude A in additive Gaussian noise parametrized by the signal-to-noise ratio $\alpha^2 = A^2/2\sigma^2$.

With some manipulation, and using [Gradshteyn and Ryzhik, 1965], the integral can be evaluated with the result

$$p_v(v) = (2\pi)^{-1/2} \sum_{k=0}^{\infty} \frac{(-v^2)^k}{2^k k!} {}_1F_1(k + \tfrac{1}{2}; 1; -\alpha^2). \qquad (4.50)$$

4.4 DISTRIBUTION OF THE ENVELOPE OF A NARROWBAND PROCESS

Many systems of interest in radar, sonar, and communications involve modulation of the amplitude or phase angle of a sinusoidal carrier by an information waveform. Upon reception, we want to process the information-bearing attributes of the signal. It is usually convenient to suppress the irrelevant carrier waveform in computations. In this section we determine the probability density of the envelope of a narrowband noise signal.

Let us then consider a zero-mean narrowband wide-sense stationary Gaussian random process $n(t)$, with variance σ_n^2. By narrowband, we mean that the power spectrum of $n(t)$ has the form of that in Fig. 3.1, with a bandwidth B such that $B/2 < f_c$. Often we will have $B/2 \ll f_c$, although that is not necessary for the developments either here or in Chapter 3.

As in Eq. (3.11), any sample function of the random process $n(t)$ can be represented in terms of its complex envelope $\tilde{n}(t)$ as

$$n(t) = \text{Re}\{\tilde{n}(t) \exp(j\omega_c t)\}$$
$$= x(t) \cos(\omega_c t) - y(t) \sin(\omega_c t), \qquad (4.51)$$

where

$$\tilde{n}(t) = a(t) \exp[j\phi(t)] = x(t) + jy(t).$$

With this, the signal itself is

$$n(t) = \text{Re}\{a(t) \exp[j(\omega_c t + \phi(t))]\}$$
$$= a(t) \cos[\omega_c t + \phi(t)]. \qquad (4.52)$$

We seek the probability density of the noise envelope $a(t)$.

As indicated in Eq. (3.72), the complex envelope can be represented in terms of the signal $n(t)$ and its Hilbert transform $\hat{n}(t)$ as

$$\tilde{n}(t) = [n(t) + j\hat{n}(t)] \exp(-j\omega_c t). \qquad (4.53)$$

Since the Hilbert transformation is a linear operation on the signal $n(t)$, which is Gaussian, $\hat{n}(t)$ is also a (zero-mean stationary) Gaussian random process. Therefore the complex envelope $\tilde{n}(t)$ is a Gaussian process, and so also are $x(t)$ and $y(t)$. From Eq. (3.82), the variances of $x(t)$ and $y(t)$ are equal to the variance of the original signal $n(t)$: $\sigma_x^2 = \sigma_y^2 = \sigma_n^2$. From Eq. (3.84), the random variables $x = x(t)$, $y = y(t)$ are uncorrelated. Being Gaussian, they are therefore also independent. Their joint density is then

$$p_{xy}(x, y) = p_x(x)p_y(y)$$
$$= (2\pi\sigma_n^2)^{-1} \exp[-(x^2 + y^2)/2\sigma_n^2].$$

Now consider the transformation of variables

$$z(t) = a(t) = [x^2(t) + y^2(t)]^{1/2},$$
$$\phi(t) = \tan^{-1}[y(t)/x(t)], \qquad 0 \le \phi < 2\pi.$$

(We write z for a to accord with the notation for a later more general case.) For convenience using the inverse transformation $x = a \cos(\phi)$, $y = a \sin(\phi)$, and calculating as in Eq. (1.32), results in

$$p_{z,\phi}(z, \phi) = (z/2\pi\sigma_n^2) \exp(-z^2/2\sigma_n^2). \qquad (4.54)$$

Since this is independent of ϕ, we conclude that ϕ has the uniform density over $[0, 2\pi)$. (The density p_ϕ would result by integrating $p_{z,\phi}$ over z; the result would be a constant.) From Eq. (4.54) we then obtain:

$$p_z(z) = \int_0^{2\pi} p_{z,\phi}(z, \phi) \, d\phi$$
$$= (z/\sigma_n^2) \exp(-z^2/2\sigma_n^2). \qquad *(4.55)$$

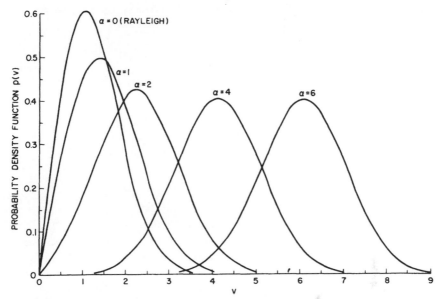

FIGURE 4.5 The Rician density function Eq. (4.66) parametrized by the signal-to-noise ratio α^2. The noise-only case ($\alpha = 0$) is the Rayleigh density Eq. (4.56).

Equation (4.55) is the *Rayleigh density function*. It is plotted in Fig. 4.5 in terms of the normalized variable $v = z/\sigma_n$, having density

$$p_v(v) = \sigma_n p_z(\sigma_n v) = v \exp(-v^2/2). \qquad (4.56)$$

(The particular curve showing Eq. (4.56) is that labeled with the parameter $\alpha = 0$.) The density Eq. (4.55) has mean, second moment, and variance

$$\mathscr{E}(z) = (\pi/2)^{1/2}\sigma_n,$$
$$\mathscr{E}(z^2) = 2\sigma_n^2,$$
$$\sigma_z^2 = (2 - \pi/2)\sigma_n^2. \qquad (4.57)$$

From Eq. (4.54) it is evident that $p_{z,\phi}(z, \phi) = p_z(z)p_\phi(\phi)$, so that $z(t)$ and $\phi(t)$ are independent random variables if taken at the same instant of time. For $z(t)$ and $\phi(t)$ to be independent random processes, however, it is necessary that the random variables $z(t)$, $\phi(t')$ be independent for arbitrary instants t, t'. This is not the case.

The investigation of the joint properties of the envelope and phase processes proceeds as follows [Davenport and Root, 1958]. The in-phase and quadrature components $x(t)$, $x(t')$, $y(t)$, $y(t')$ at arbitrary instants t, t' are jointly Gaussian. Their joint density is expressible in terms of the correlation functions $R_x(\tau)$, $R_y(\tau)$, $R_{xy}(\tau)$, which are given in Eqs. (3.74), (3.75), (3.76). The result is the four-dimensional Gaussian density with correlation matrix ($t - t' = \tau$):

$$\mathbf{R} = \mathscr{E} \begin{bmatrix} x \\ y \\ x' \\ y' \end{bmatrix} [x \quad y \quad x' \quad y']$$

$$= \begin{bmatrix} \sigma_n^2 & 0 & R_x(\tau) & R_{xy}(\tau) \\ 0 & \sigma_n^2 & -R_{xy}(\tau) & R_x(\tau) \\ R_x(\tau) & -R_{xy}(\tau) & \sigma_n^2 & 0 \\ R_{xy}(\tau) & R_x(\tau) & 0 & \sigma_n^2 \end{bmatrix}, \tag{4.58}$$

where we write $x = x(t)$, $x' = x(t')$, etc., and also use Eq. (3.77). Using the transformation of variables from x, y, x', y' to the envelope and phase samples z, ϕ, z', ϕ', the density of the envelope and phase follows easily. Because of the special form of the matrix Eq. (4.58), the computations are eased by use of the formulas for the determinant and inverse of a partitioned matrix (Chapter 11). The result is

$$p(z, z', \phi, \phi') = (zz'/4\pi^2 |\mathbf{R}|^{1/2}) \exp(-u/2|\mathbf{R}|^{1/2}), \tag{4.59}$$

where the determinant of the matrix Eq. (4.58) is such that

$$|\mathbf{R}|^{1/2} = \sigma_n^4 - [R_x^2(\tau) + R_{xy}^2(\tau)]$$

and the quantity in the exponent is

$$\begin{aligned} u = &\ \sigma_n^2(z^2 + z'^2) \\ &- 2zz'[R_x(\tau)\cos(\phi' - \phi) + R_{xy}(\tau)\sin(\phi' - \phi)]. \end{aligned}$$

Integrating the density Eq. (4.59) first with respect to the phase variables and then with respect to the amplitude variables results in the densities

$$\begin{aligned} p_{zz'}(z, z') = &\ (zz'/|\mathbf{R}|^{1/2}) \\ &\times I_0\{zz'[R_x^2(\tau) + R_{xy}^2(\tau)]^{1/2}/|\mathbf{R}|^{1/2}\} \\ &\times \exp[-\sigma_n^2(z^2 + z'^2)/2|\mathbf{R}|^{1/2}], \end{aligned} \tag{4.60}$$

where $I_0(x)$ is the modified Bessel function of the first kind and order zero, and

$$\begin{aligned} p_{\phi\phi'}(\phi, \phi') = &\ (|\mathbf{R}|^{1/2}/4\pi^2\sigma_n^4)[(1 - \beta^2)^{1/2} \\ &+ \beta(\pi - \cos^{-1}(\alpha)]/(1 - \beta^2)^{3/2} \end{aligned}$$

where the inverse cosine takes values in $[0, \pi)$ and β is defined as

$$\beta = [R_x(\tau)\cos(\phi' - \phi) + R_{xy}(\tau)\sin(\phi' - \phi)]/\sigma_n^2.$$

From these it is clear that

$$p_{zz'\phi\phi'}(z, z', \phi, \phi') \neq p_{zz'}(z, z')p_{\phi\phi'}(\phi, \phi'),$$

so that the envelope and phase processes are not independent.

4.5 ENVELOPE OF A NARROWBAND SIGNAL PLUS NARROWBAND NOISE

We now want to determine the probability density for the envelope of a narrowband signal in additive narrowband Gaussian noise. Therefore we consider

$$r(t) = a(t)\cos(\omega_c t + \theta) + n(t), \tag{4.61}$$

where $a(t)$ is a known deterministic narrowband envelope function, the phase angle θ is uniformly distributed over $[0, 2\pi)$, and the noise $n(t)$ is as considered in Sect. 4.4.

Using the representation Eq. (4.51) of $n(t)$ in terms of its in-phase and quadrature components $x(t)$, $y(t)$, the signal Eq. (4.61) becomes

$$\begin{aligned}
r(t) &= [a(t)\cos(\theta) + x(t)]\cos(\omega_c t) \\
&\quad - [a(t)\sin(\theta) + y(t)]\sin(\omega_c t) \\
&= r_c\cos(\omega_c t) - r_s\sin(\omega_c t), \tag{4.62}
\end{aligned}$$

where $r_c(t)$, $r_s(t)$ are so defined and are the in-phase and quadrature components of $r(t)$. Following Sect. 4.4, define new variables

$$\begin{aligned}
z &= [r_c^2 + r_s^2]^{1/2}, \tag{4.63}\\
\phi &= \tan^{-1}(r_s/r_c), \qquad 0 \le \phi < 2\pi.
\end{aligned}$$

These are the amplitude and phase of the envelope of $r(t)$. We seek their densities.

First consider a particular fixed value of the random phase θ. From their definitions Eq. (4.62), $r_c(t)$ and $r_s(t)$ are Gaussian variables for specified θ and with zero covariance [since $x(t)$, $y(t)$ are uncorrelated at a common instant]. Their means are $a(t)\cos(\theta)$, $a(t)\sin(\theta)$ and they have variances σ_n^2. Hence they are independent and have the density

$$\begin{aligned}
p_{r_c r_s}(r_c, r_s \mid \theta) &= (2\pi\sigma_n^2)^{-1} \\
&\quad \times \exp\{-[(r_c - a\cos\theta)^2 + (r_s - a\sin\theta)^2]/2\sigma_n^2\}.
\end{aligned}$$

Transforming to the new variables z, ϕ yields

$$p_{z\phi}(z, \phi \mid \theta) = (z/2\pi\sigma_n^2)\exp\{-[z^2 + a^2 - 2za\cos(\phi - \theta)]/2\sigma_n^2\}. \tag{4.64}$$

Integrating ϕ out of this density yields

$$p_z(z \mid \theta) = (z/2\pi\sigma_n^2) \exp[-(z^2 + a^2)/2\sigma_n^2]$$

$$\times \int_0^{2\pi} \exp\left[\left(\frac{az}{\sigma_n^2}\right)\cos(\phi - \theta)\right] d\phi$$

$$= (z/\sigma_n^2) \exp[-(z^2 + a^2)/2\sigma_n^2]I_0(az/\sigma_n^2) = p_z(z), \qquad *(4.65)$$

where $I_0(x)$ is the modified Bessel function of the first kind and order zero. The last equality in Eq. (4.65) results from the observation that the density conditioned on θ is in fact independent of θ and is therefore the unconditioned density as well.

The density Eq. (4.65) is called the *Rician density function* (after S. O. Rice). It is conveniently parametrized by the ratio of the instantaneous power $a^2(t)$ of the deterministic amplitude of the signal envelope in Eq. (4.61) to the stochastic average of the noise power: $\mathcal{E}[n^2(t)] = \sigma_n^2$. That is, the signal-to-noise ratio (SNR) $\alpha = a/\sigma_n$. The normalized Rician density results by transforming to the variable $v = z/\sigma_n$. The resulting density

$$p_v(v) = \sigma_n p_z(\sigma_n v) = v \exp[-(v^2 + \alpha^2)/2]I_0(\alpha v) \qquad (4.66)$$

is shown in Fig. 4.5 for various values of the SNR α.

The moments of the density Eq. (4.65) are

$$\mathcal{E}(z^n) = (2\sigma_n^2)^{n/2}\,\Gamma(n/2 + 1)\,{}_1F_1(-n/2; 1; -\alpha^2/2), \qquad (4.67)$$

where Γ is the gamma function, ${}_1F_1$ is the confluent hypergeometric function, and $\alpha = a/\sigma_n$ is the SNR. The mean and standard deviation are shown in Fig. 4.6.

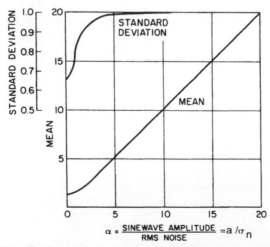

FIGURE 4.6 Normalized mean $\mathcal{E}z/\sigma_n$ and standard deviation σ_z/σ_n for Rician distribution.

In detection problems, it is the probability that the envelope z exceeds some threshold β which is usually of interest, rather than the density $p_z(z)$ itself. Accordingly, we introduce the *Marcum Q function*, which is one minus the distribution function of a standardized Rician variable. That is, from Eq. (4.66),

$$Q(\alpha, \beta) = 1 - \int_0^\beta v \exp\left[\frac{-(v^2 + \alpha^2)}{2}\right] I_0(\alpha v)\, dv$$

$$= \int_\beta^\infty v \exp\left[\frac{-(v^2 + \alpha^2)}{2}\right] I_0(\alpha v)\, dv. \tag{4.68}$$

This is shown in Figs. 4.7 and 4.8. The diagrams are on so-called normal probability paper, on which the distribution of a Gaussian appears as a straight line. The form of the plots in Fig. 4.8 indicates that the Rician density is well approximated by a Gaussian for moderately large signal-to-noise ratios. Helstrom (1968) gives the useful expansions

$$Q(\alpha, \beta) = \exp\left[-\frac{1}{2}(\alpha^2 + \beta^2)\right] \sum_{n=0}^\infty \left(\frac{\alpha}{\beta}\right)^n I_n(\alpha \beta)$$

$$= 1 - \exp\left[-\frac{1}{2}(\alpha^2 + \beta^2)\right] \sum_{n=1}^\infty \left(\frac{\beta}{\alpha}\right)^n I_n(\alpha \beta),$$

where $I_n(x)$ is the modified Bessel function of first kind and order n.

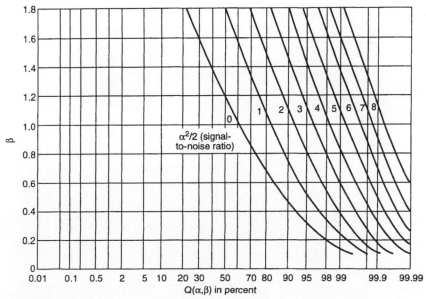

FIGURE 4.7 The Marcum Q-function parametrized by the signal-to-noise ratio $\alpha^2/2$.

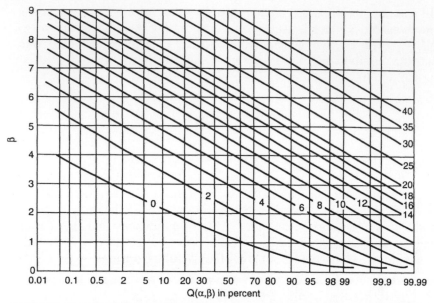

FIGURE 4.8   The Marcum Q-function parametrized by the signal-to-noise ratio $\alpha^2/2$.

From the joint conditional density Eq.(4.64) it is simple to obtain the marginal conditional density of the phase $\phi(t)$ of the received signal $r(t)$ of Eq. (4.61). Completing the square in the exponent of Eq. (4.64) and making a change of variable, we have

$$p_\phi(\phi \mid \theta) = \int_0^\infty p_{z\phi}(z, \phi \mid \theta) \, dz$$

$$= (2\pi\sigma_n^2)^{-1} \exp[-a^2 \sin^2(\phi - \theta)/2\sigma_n^2]$$

$$\times \int_0^\infty [u + \left(\frac{a}{\sigma_n}\right) \cos(\phi - \theta)] \exp\left(\frac{-u^2}{2}\right) du$$

$$= (1/2\pi) \exp(-a^2/2\sigma_n^2)$$

$$\times \{1 + (\pi/2)^{1/2}(a/\sigma_n) \cos(\phi - \theta) \exp[a^2 \cos^2(\phi - \theta)/2\sigma_n^2]$$

$$\times [1 + 2 \operatorname{erf}[a \cos(\phi - \theta)/\sigma_n]\}. \tag{4.69}$$

In this the *error function* is defined as

$$\operatorname{erf}(u) = (2\pi)^{-1/2} \int_0^u \exp\left(\frac{-x^2}{2}\right) dx.$$

The density Eq. (4.69) is plotted in Fig. 4.9 with the instantaneous signal-to-noise ratio $\alpha = a/\sigma_n$ as a parameter.

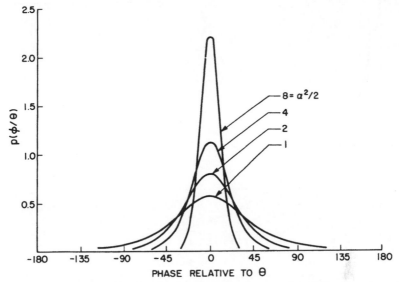

FIGURE 4.9 The density function for the phase of a sine wave in additive Gaussian noise, parametrized by the signal-to-noise ratio $\alpha^2 = a^2/\sigma_n^2$.

4.6 SQUARED ENVELOPE OF A NARROWBAND NOISE PROCESS

In Sect. 4.4 we considered the envelope $a(t)$ of a narrowband noise process Eq. (4.52). The result was the Rayleigh density Eq. (4.55). The density of the square of the envelope is also important. Using the one-to-one (for $z \geq 0$) transformation $q = z^2$, the density of the squared envelope

$$q = x^2(t) + y^2(t) \qquad (4.70)$$

follows at once from Eq. (4.55) as

$$p_q(q) = p_z(z)/(dq/dz) = p_z(\sqrt{q})/(2\sqrt{q})$$
$$= (2\sigma_n^2)^{-1} \exp(-q/2\sigma_n^2). \qquad *(4.71)$$

This is the (unnormalized) *chi-squared density* with two degrees of freedom. The mean and variance are

$$\mathscr{E}(q) = 2\sigma_q^2,$$
$$\sigma_q^2 = 4\sigma_n^4. \qquad (4.72)$$

The density Eq. (4.71) is plotted in Fig. 4.10 (2 degrees of freedom curve) for the normalized variable $v = q/\sigma_n^2$, so that

$$p_v(v) = \tfrac{1}{2} \exp(-v/2).$$

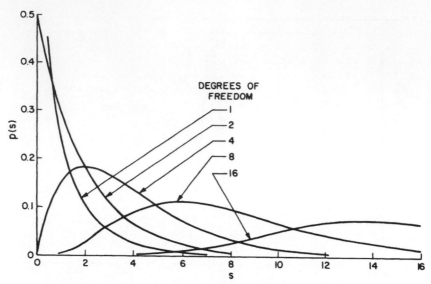

FIGURE 4.10   The chi-squared probability density function Eq. (4.76).

This last density is the normalized chi-squared density with two degrees of freedom. Figure 4.11 shows the corresponding distribution.

4.7 THE CHI-SQUARED DENSITY

The squared envelope of the narrowband signal Eq. (4.52) is the sum of two zero-mean independent Gaussian variables having the same variance:

$$q = z^2 = x^2 + y^2,$$

where x and y have variance σ_n^2. In Sect. 4.4 the variables x and y arose as the in-phase and quadrature components of a narrowband waveform. In this section we will generalize the quantity q and consider the sum of the squares of any number of zero-mean independent Gaussian variables having a common variance.

Accordingly, let x_i, $i = 1, \ldots, n$, be independent Gaussian random variables with zero mean and variance σ_x^2. Consider the sum

$$s = \sum_{i=1}^{n} x_i^2. \tag{4.73}$$

As indicated in Eq. (1.46), the characteristic function of a sum of independent random variables is the product of the characteristic functions of the summands.

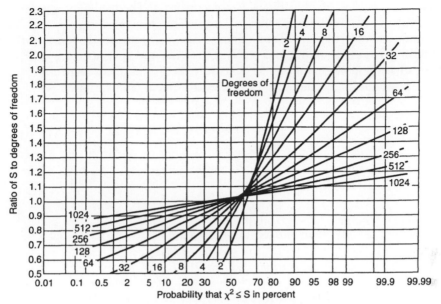

FIGURE 4.11 The chi-squared distribution function.

From Eq. (1.28), the density of $y = x^2$ for Gaussian x is

$$p_y(y) = p_x(\sqrt{y})/\sqrt{y} = (2\pi\sigma_x^2 y)^{-1/2}\exp(-y/2\sigma_x^2), \qquad y \geq 0.$$

The corresponding characteristic function, the conjugate of the Fourier transform, is

$$\Phi_y(\omega) = (1 - 2j\omega\sigma_x^2)^{-1/2}. \tag{4.74}$$

Hence the characteristic function of s is

$$\Phi_s(\omega) = (1 - 2j\omega\sigma_x^2)^{-n/2}.$$

The (real) density of the sum s is the inverse Fourier transform of the conjugate of this. The result is

$$p_s(s) = [(2\sigma_x^2)^{n/2}\Gamma(n/2)]^{-1}s^{n/2-1}\exp(-s/2\sigma_x^2). \qquad \text{*(4.75)}$$

Equation (4.75) is the *gamma density*, and the expression is valid for any n, integer or not. The density is more commonly called the *chi-squared density with n degrees of freedom*. In many applications, the density arises from a problem in which n is naturally an integer, such as the case Eq. (4.73). The normalized form of the density results from the change of variable $v = s/\sigma_x^2$. The result is

$$\begin{aligned} p_v(v) &= \sigma_x^2 p_S(v\sigma_x^2) \\ &= [2^{n/2}\Gamma(n/2)]^{-1}v^{n/2-1}\exp(-v/2). \end{aligned} \tag{4.76}$$

Equation (4.76) is plotted in Fig. 4.10 for various n. The tendency toward a Gaussian in accord with the central limit theorem is evident.

The mean and variance of the density Eq. (4.75) are easily found for integer n. From Eqs. (1.63) and (4.74), the mean and variance of $y = x^2$, where x has the unnormalized Gaussian density $N(0, \sigma_x)$, are respectively σ_x^2, $2\sigma_x^4$. Accordingly, the sum Eq. (4.73) has mean and variance

$$\mathscr{E}(s) = n\sigma_x^2,$$
$$\mathrm{Var}(s) = 2n\sigma_x^4. \tag{4.77}$$

From these, the number of degrees of freedom to use in fitting a chi-squared density to a random variable can be judged from the ratio of the measured mean and variance:

$$n = 2[\mathscr{E}(s)]^2/\mathrm{Var}(s).$$

This will not usually be an integer, but that is not necessary for the general chi-squared density, that is, for the gamma density.

The distribution of the normalized density Eq. (4.76) is

$$P_v(v) = \int_0^v p_v(u)\, du$$

$$= \frac{1}{\Gamma(n/2)} \int_0^{v/2} w^{n/2-1} \exp(-w)\, dw$$

$$= I[v(2n)^{-1/2}, n/2-1]. \tag{*4.78}$$

In this last, we have introduced Pearson's form of the *incomplete gamma function*:

$$I(u, p) = \frac{1}{\Gamma(p+1)} \int_0^{u(p+1)^{1/2}} w^p \exp(-w)\, dw. \tag{4.79}$$

The normalized distribution Eq. (4.78) is plotted in Fig. 4.11. The distribution of the unnormalized variable Eq. (4.73) is $P_s(s) = P_v(s/\sigma_x^2)$.

It is worth noting that the sum of a variable with chi-squared density with M degrees of freedom and one with N degrees of freedom is a variable with chi-squared density having $M + N$ degrees of freedom, provided the underlying variances σ_x^2 of the two variables of the form Eq. (4.73) are the same. Otherwise, the density of the sum is not chi-squared.

4.8 SQUARED ENVELOPE OF A SINE WAVE PLUS A NARROWBAND PROCESS

In Sect. 4.4 we considered the envelope of a narrowband noise process. In Sect. 4.5 signal was added. Section 4.6 considered the squared envelope of the noise alone. In this section, we consider the squared envelope of signal plus noise. From Eq. (4.63), we therefore consider the variable

$$q = z^2(t) = [a(t)\cos(\theta) + x(t)]^2 + [a(t)\sin(\theta) + y(t)]^2, \quad (4.80)$$

where $a(t)$ is deterministic, θ is uniform over $[0, 2\pi)$, and $x(t)$, $y(t)$ are independent, zero-mean Gaussian variables with common variance σ_n^2. Applying the transformation $q = z^2$ to the density Eq. (4.65) yields

$$
\begin{aligned}
p_q(q \mid \theta) &= p_z(q^{1/2} \mid \theta)/(2q^{1/2}) \\
&= (2\sigma_n^2)^{-1} \exp[-(q + a^2)/2\sigma_n^2]\, I_0(aq^{1/2}/\sigma_n^2) = p_q(q). \quad *(4.81)
\end{aligned}
$$

The last equality follows by observing that the conditional density is independent of the random variable θ.

The density Eq. (4.81) is the (unnormalized) *noncentral chi-squared density* with two degrees of freedom. It is the generalization of the chi-squared density Eq. (4.71) to the case that a signal of deterministic amplitude $a(t)$ is present. The characteristic function of the density Eq. (4.81) is

$$
\begin{aligned}
\Phi_q(\omega) &= \int_0^\infty p_q(q) \exp(j\omega q)\, dq \\
&= \exp[j\omega a^2/(1 - j2\omega\sigma^2)](1 - j2\omega\sigma^2)^{-1}. \quad (4.82)
\end{aligned}
$$

This is generalized in the next section. The calculation leading to the expression Eq. (4.82) follows from a special case of [Gradshteyn and Ryzhik, 1965, #6.643.2]:

$$\int_0^\infty \exp(-\alpha q) I_0(2\beta q^{1/2})\, dq = \frac{1}{\alpha} \exp\left(\frac{\beta^2}{\alpha}\right). \quad (4.83)$$

4.9 NONCENTRAL CHI-SQUARED DENSITY

Finally, we determine the density of the quantity which generalizes Eq. (4.75) in the way that the quantity Eq. (4.81) generalizes Eq. (4.71). That is, we now consider

$$s = \sum_{i=1}^{n} (A_i + x_i)^2, \quad (4.84)$$

where A_i are constants and the x_i are independent zero-mean Gaussian variables with common variance σ^2. The variables $y_i = A_i + x_i$ are independent Gaussians with means A_i and variances σ^2: $p_{yi}(y_i) = N(A_i, \sigma^2)$. The density of one term $s_i = (A_i + x_i)^2$ of the sum Eq. (4.84) is therefore, from Eq. (1.28),

$$p_{si}(s_i) = (2\pi\sigma^2 s_i)^{-1/2} \exp[-(s_i + A_i^2)/2\sigma^2] \cosh(A_i s_i^{1/2}/\sigma^2).$$

Using [Gradshteyn and Ryzhik, 1965, #3.546.2], the characteristic function follows as

$$\Phi_{si}(\omega) = \exp[j\omega A_i^2/(1 - j2\omega\sigma^2)]\,(1 - j2\omega\sigma^2)^{-1/2}.$$

Then

$$\Phi_s(\omega) = \exp[j\omega S/(1 - j2\omega\sigma^2)]\,(1 - j2\omega\sigma^2)^{-n/2} \qquad (4.85)$$

where we define

$$S = \sum_{i=1}^{n} A_i^2. \qquad (4.86)$$

Equation (4.82) is the special case of this with $n = 2$, $A_1 = a\cos(\theta)$, $A_2 = a\sin(\theta)$, so that $S = a^2$.

Using [Gradshteyn and Ryzhik, 1965, #6.643.2], it can be verified that the following density has the characteristic function Eq. (4.85):

$$p_s(s) = (2\sigma^2)^{-1}(s/S)^{(n-2)/4}$$
$$\times \exp[-(s + S)/2\sigma^2]\,I_{n/2-1}[(Ss)^{1/2}/\sigma^2]. \qquad *(4.87)$$

This is the (unnormalized) *noncentral chi-squared density* with n degrees of freedom, and it is the generalization of Eq. (4.81). The density Eq. (4.87)

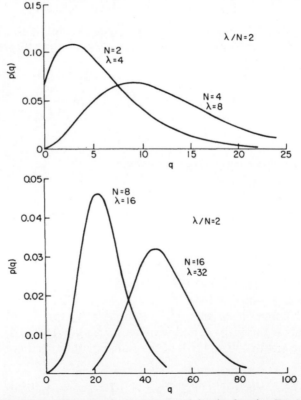

FIGURE 4.12   The noncentral chi-squared density function Eq. (4.89).

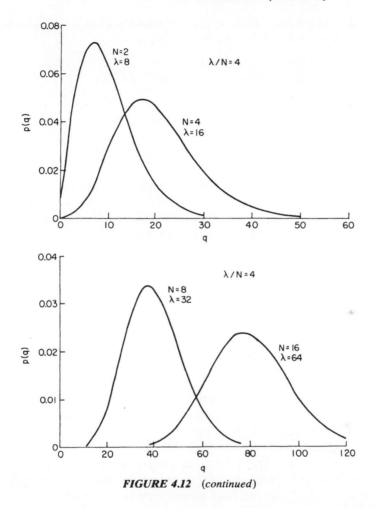

FIGURE 4.12   (*continued*)

stands on its own for noninteger n and need not arise from consideration of the sum Eq. (4.84). As before, $I_m(x)$ is the modified Bessel function of first kind and order m. The mean and variance of the density Eq. (4.87) are

$$\mathscr{E}(s) = S + n\sigma^2,$$
$$\text{Var}(s) = 4\sigma^2 S + 2n\sigma^4. \tag{4.88}$$

Let us define the normalized variable $v = s/\sigma^2$ and the parameter $\lambda = S/\sigma^2$. The variable v has the density

$$p_v(v) = \sigma^2 p_s(\sigma^2 v)$$
$$= \tfrac{1}{2}(v/\lambda)^{(n-2)/4} \exp[-(v + \lambda)/2]I_{n/2-1}[(\lambda v)^{1/2}]. \tag{4.89}$$

This is the standard form of the noncentral chi-squared density. The quantity $\lambda \geq 0$ is called the *noncentrality parameter*. Figure 4.12 shows the density Eq. (4.89) for various n and λ. From Eq. (4.88), the mean and variance of the normalized density Eq. (4.89) are

$$\mathcal{E}(v) = \mathcal{E}(s)/\sigma^2 = S/\sigma^2 + n = \lambda + n,$$
$$\text{Var}(v) = \mathcal{E}(s^2)/\sigma^4 - (\lambda + n)^2 = 4\lambda + 2n. \qquad (4.90)$$

From Eq. (4.84), it is clear that, if s_1 and s_2 are normalized noncentral chi-squared variables with n_1 and n_2 degrees of freedom and noncentrality parameters λ_1, λ_2, then their sum is normalized noncentral chi-squared with degrees of freedom $n_1 + n_2$ and noncentrality parameter $\lambda_1 + \lambda_2$.

A common application of the noncentral chi-squared density involves the system sketched in Fig. 4.13. There a signal is received in additive noise and passed through a narrowband filter to reject noise outside the signal band. The filter output is a narrowband signal $a(t) \cos [\omega_c t + \theta(t)]$ in additive narrowband noise $n(t)$ with zero mean and constant variance σ_n^2. As in Eq. (4.51), the noise can be represented in terms of zero-mean low-pass signals $x(t)$, $y(t)$, having common variance σ_n^2, as

$$n(t) = x(t) \cos (\omega_c t) - y(t) \sin (\omega_c t).$$

Therefore the filter output can be written

$$r(t) = \{a(t) \cos[\theta(t)] + x(t)\} \cos(\omega_c t)$$
$$- \{a(t) \sin[\theta(t)] + y(t)\} \sin(\omega_c t).$$
$$= \text{Re}\{[(a \cos \theta + x) + j(a \sin \theta + y)] \exp(j\omega_c t)\}.$$

In this last, the slowly varying complex envelope $\tilde{r}(t)$ is evident. That quantity is recovered by the second stage of the system of Fig. 4.13, using some method such as one of those indicated in Chapter 3.

In many applications to radar and sonar, the phase θ of the envelope is not reliably measured. This is because a slight change in the time t (the two-way travel time from transmitter to target) causes a large change in the phase $\omega_c t + \theta$ of the received signal, due to the $\omega_c t$ term and the relatively large value of ω_c. Consequently, it is usually the amplitude of the envelope, or its square, which is calculated. The complex envelope can be sampled at an interval $\delta t = 1/B$ appropriate to its two-sided bandwidth B to produce time samples $\tilde{r}_i$ which are uncorrelated from one sample to another. This is because the power spectrum of $\tilde{r}(t)$ considered as a random process is bandlimited to B, so that the autocorrelation function falls essentially to zero after a time $1/B = \delta t$. If the noise is assumed Gaussian, then the samples $\tilde{r}_i$ are independent.

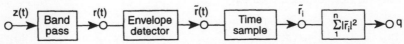

FIGURE 4.13 Formation of the test statistic for detection of a narrowband signal in narrowband Gaussian noise.

The squaring device in Fig. 4.13 then computes at its output the numbers

$$|\tilde{r}_i|^2 = (a_i \cos \theta_i + x_i)^2 + (a_i \sin \theta_i + y_i)^2.$$

The summing device finally produces an output

$$q = \sum_{i=1}^{n} |\tilde{r}_i|^2.$$

Comparing these last with Eq. (4.84), it is clear that the output q of the system of Fig. 4.13 has the noncentral chi-squared density with $2n$ degrees of freedom and with the quantity S of Eq. 4.86 such that

$$S = \sum_{i=1}^{n} [(a_i \cos \theta_i)^2 + (a_i \sin \theta_i)^2] = \sum_{i=1}^{n} a_i^2.$$

From Eq. (4.89), the density of the normalized quantity $v = q/\sigma_n^2$ is then

$$p_v(v) = \tfrac{1}{2}(v/\lambda)^{(n-1)/2} \exp[-(v + \lambda)/2]I_{n-1}[(\lambda v)^{1/2}], \qquad (4.91)$$

where the noncentrality parameter is $\lambda = S/\sigma_n^2$.

The energy in the signal over the observation interval $n \, \delta t$ is

$$E_s = \sum_{i=1}^{n} [a_i \cos(\omega_c t_i + \theta_i)]^2 \, \delta t \cong S \, \delta t/2,$$

where the sum of the double frequency terms is taken as approximately zero. The average power of the signal is therefore $S \, \delta t/(2n \, \delta t) = S/2n$. The average power of the noise is σ_n^2. We can then define the output signal-to-noise (power) ratio of the system of Fig. 4.13 as the quantity

$$\alpha^2 = S/2n\sigma_n^2 = \lambda/2n.$$

This identifies the noncentrality parameter λ of the density governing the output of the system of Fig. 4.13 as the product of the input power signal-to-noise ratio α^2 and the number $2n$ of degrees of freedom.

4.10 STUDENT'S t-DENSITY

In Chapter 5 we will consider ways of estimating whether or not some signal of interest is present in a received waveform. A simple case is that in which a specified constant A has or has not been added to zero-mean Gaussian noise $n(t)$ before we receive the signal. That is, we seek to decide between the alternatives that $r(t) = n(t)$ and $r(t) = A + n(t)$. A simple strategy, using a single sample $r(t_i)$, is to determine whether or not the value of r_i is consistent with a sample of a random variable with mean zero and variance σ_n^2. If it is, then we might conclude that the signal A is not present in the sample r_i. The problem is normalized by dividing by σ_n, so that we examine the quantity $v = r_i/\sigma_n$, where σ_n is a measure of the noise amplitude. The quantity v has unit variance under both hypotheses, and we seek to assess whether its mean is zero.

In analyzing the performance of such a strategy, we need to know the density of the random variable v. If σ_n is known, then v is a Gaussian with mean either 0 or A/σ_n and with unit variance. However, in many cases σ_n is not known, perhaps because it changes with time. Then we must make an estimate, based on samples of $r(t)$ over times when we know signal is not present. We might then consider the random variable

$$v = \frac{r_i}{\left[(1/n) \sum_{k=1}^{n} n_k^2 \right]^{1/2}}.$$

In the case of noise only, the numerator of v is Gaussian with mean zero and unknown variance σ_n^2, while the denominator involves a variable with the chi-squared density with n degrees of freedom.

In general, then, we are interested in the variable

$$t = x/(y/n)^{1/2}, \tag{4.92}$$

where x is zero-mean Gaussian and y is chi-squared with n degrees of freedom. (We assume the samples n_k entering into y are taken far enough apart that they are uncorrelated random variables. That will be the case if the sampling interval $\delta t > 1/B$, where B is the two-sided bandwidth of the noise spectrum.) From the Gaussian density and the density Eq. (4.75) of y, we have

$$p_{xy}(x, y) = [(2\pi \sigma_n^2)^{1/2} (2\sigma_n^2)^{n/2} \Gamma(n/2)]^{-1}$$
$$\times y^{n/2-1} \exp[-(x^2 + y)/2\sigma_n^2].$$

Introduce the new variables $t = x/(y/n)^{1/2}$, $w = y$. The Jacobian of the transformation is $|det[\partial(t, w)/\partial(x, y)]| = (n/y)^{1/2}$. Using Eq. (1.33), we then have

$$p_{tw}(t, w) = [(2\pi n)^{1/2} 2^{n/2} \sigma_n^{n+1} \Gamma(n/2)]^{-1} w^{(n-1)/2}$$
$$\times \exp[-(1 + t^2/n)w/2\sigma_n^2].$$

Integrating this over $0 \le w \le \infty$ yields

$$p_t(t) = (n\pi)^{-1/2} \Gamma[(n + 1)/2]/\Gamma(n/2)$$
$$\times (1 + t^2/n)^{-(n+1)/2}. \tag{*(4.93)}$$

This is the Student t density, named for W. S. Gosset, who wrote under the pseudonym "Student." We shall have occasion in a later chapter to make use of it.

The distribution of the density Eq. (4.93) is

$$P_t(t) = \int_{-\infty}^{t} p_t(v)\, dv = 1/2 + \int_{0}^{t} p_t(v)\, dv.$$

Inserting Eq. (4.93) into this and making the change of variable $u = (v^2/n)/(1 + v^2/n)$ leads to

$$P_t(t) = \frac{1}{2} + \frac{\Gamma[(n+1)/2]}{2\pi^{1/2}\Gamma(n/2)}$$

$$\times \int_{0}^{U} u^{-1/2}(1-u)^{\frac{n}{2}-1}\, du, \qquad (4.94)$$

where $U = (t^2/n)/(1 + t^2/n)$. The integral is expressible in terms of the incomplete beta function, defined as

$$I_U(a, b) = \frac{1}{B(a,b)} \int_{0}^{U} u^{a-1}(1-u)^{b-1}\, du, \qquad 0 \le U \le 1. \quad (4.95)$$

In this, $B(a, b)$ is the (complete) beta function:

$$B(a, b) = \Gamma(a)\Gamma(b)/\Gamma(a+b).$$

Finally, using the relation

$$I_U(a, b) = 1 - I_{1-U}(b, a),$$

the distribution Eq. (4.94) of a normalized Student t variable becomes

$$P_t(t) = 1 - \tfrac{1}{2} I_u(n/2, 1/2), \qquad \qquad *(4.96)$$

where $u = (1 + t^2/n)^{-1}$. The distribution Eq. (4.96) is tabulated in [3].

For $n = 1$, Eq. (4.93) is the Cauchy density. The distribution is

$$P_t(t) = \tfrac{1}{2} + (1/\pi)\tan^{-1}(t).$$

4.11 SNEDECOR'S F-DENSITY

In the previous section we considered briefly the question of detecting the presence of a known constant signal A in additive zero-mean Gaussian noise with unknown variance σ_n^2. We were led to examine the quantity r_i/σ_n where r_i are the received signal samples. This has the same variance but different means under the two hypotheses. Using an estimate for the unknown noise variance led to the quantity Eq. (4.92), and its density Eq. (4.93). However, in other cases of interest, the signal may itself be a random process, having different statistical properties from the noise process. Accordingly, we are led to examine the quantity $v = \sigma_r^2/\sigma_n^2$, where the numerator is calculated from the region being interrogated for signal and the denominator is taken in regions known to be free of signal.

If v is near unity, we conclude that signal is not present. If v is significantly greater than unity, we are led to conclude that there is more variance in the numerator than attributable to noise alone, and signal is present.

Again we must use estimates of the variances σ_r^2, σ_n^2. If we know that the processes in question are zero mean, we can reasonably examine the quantity

$$F = \frac{(1/m)\sum_{i=1}^{m} r_i^2}{(1/n)\sum_{i=1}^{n} n_i^2}.$$

The sum in the numerator of this has the chi-squared density with m degrees of freedom, based on the (unknown) variance σ_r^2, and the denominator sum is chi-squared with n degrees of freedom and σ_n^2, also unknown. If signal is not present, then $\sigma_r^2 = \sigma_n^2$. We will seek the properties of the ratio F in the case of signal not present in order to use them to assess some measured value of F.

Accordingly, we write in general

$$F = (x/m)/(y/n), \tag{4.97}$$

where x and y are independent chi-squared, and seek the density of F. To F we adjoin the variable $G = y$. The Jacobian of the transformation is $|det[\partial(F, G)/\partial(x, y)]| = n/(my)$. Using this with Eq. (4.75), in the case that the two chi-squared quantities are based on the same variance σ_n^2, we have

$$p_{FG}(F, G) = (m/n)\,[(2\sigma_n^2)^{(m+n)/2}\,\Gamma(m/2)\Gamma(n/2)]^{-1}$$
$$\times (mFG/n)^{m/2-1}\,G^{n/2}\exp[-(mF/n + 1)G/2\sigma_n^2].$$

Integrating this over $0 \le G \le \infty$ leads to

$$p_F(F) = m^{m/2}\,n^{n/2}\,\frac{\Gamma[(m + n)/2]}{\Gamma(m/2)\Gamma(n/2)}$$
$$\times F^{m/2-1}(mF + n)^{-(m+n)/2}, \qquad F \ge 0. \qquad *(4.98)$$

The distribution of F is

$$P_F(F) = \int_0^F p_F(f)\,df = 1 - I_U\left(\frac{n}{2}, \frac{m}{2}\right), \qquad *(4.99)$$

where $U = (1 + mF/n)^{-1}$. The result Eq. (4.99) follows by making the change of variable $u = (1 + n/mf)^{-1}$ in the integral and introducing the incomplete beta function Eq. (4.95). The distribution Eq. (4.99) is tabulated by Bendat and Piersol (1986).

For completeness, we will mention the noncentral F-distribution. Suppose that

$$r_i = A_i + x_i,$$

where the A_i are known and the x_i are zero-mean independent Gaussian variables, with common variance σ_r^2. Then, as in Eq. (4.84), the quantity

$$s = \sum_{i=1}^{m} r_i^2$$

has the noncentral chi-squared density Eq. (4.87), with m degrees of freedom. As before, let n_i be zero-mean independent Gaussian variables with common variance σ_n^2. Then

$$t = \sum_{i=1}^{n} n_i^2$$

has the (central) chi-squared density Eq. (4.75) with n degrees of freedom.
Now define

$$F = (s/m\sigma_r^2)/(t/n\sigma_n^2).$$

Then [Hodges, 1955] F has the noncentral F-distribution.

The noncentral F-distribution involves two indexes a, b and the noncentrality parameter λ. These are given by

$$a = m/2,$$

$$b = n/2,$$

$$\lambda = (1/\sigma_r^2) \sum_{i=1}^{m} A_i^2.$$

With these, the distribution function, written for $x = 1/(1 + F)$, is

$$P_x(x) = \exp(-\lambda x)\{I_{1-x}(a, b) + (1 - x)^a$$

$$\times \sum_{j=1}^{b-1} [x(1 - x)\lambda]^j P_j/j!\}, \qquad (4.100)$$

where

$$P_j = \sum_{k=0}^{b-j-1} (-1)^k \binom{b - j - 1}{k}$$

$$\times \frac{(a + b - 1)(a + b - 2) \cdots (a + j)}{(b - j - 1)!(a + j + k)}(1 - x)^k. \qquad (4.101)$$

Here $I_U(a, b)$ is the incomplete beta function. Beyer (1968) gives plots of Eq. (4.100) and Robertson (1976) has given a table and a calculation method. In Chapter 11 we will make some use of the rather formidable expression Eq. (4.101) and give a way to calculate it conveniently.

Exercises

4.1 Assume the Gaussian vector $\mathbf{x}$ is zero mean and has a covariance matrix $\mathbf{R}_x$. Show that the characteristic function of the random variable $q = \mathbf{x}'\mathbf{Q}\mathbf{x}$ is equal to

$$\frac{1}{|\mathbf{I} - j2\omega\,\mathbf{R}_x\,\mathbf{Q}|^{1/2}}$$

Hint: It is obvious if $\mathbf{x}$ is n-dimensional that

$$\int_{\text{all }\mathbf{x}} \frac{1}{(2\pi)^{n/2}|\mathbf{R}_x|^{1/2}} \exp(-\tfrac{1}{2}\mathbf{x}'\mathbf{R}_x^{-1}\mathbf{x})\, d\mathbf{x} = 1$$

4.2 Assume white Gaussian noise of power spectral density $N_0/2$ is inserted into a filter with transfer function

$$H(j\omega) = \frac{1}{1 + j\omega/\omega_1}$$

Find the probability density functions of the output and the envelope of the output.

4.3 What is the probability density function of $z(t) = A \cos \omega_c t + n(t)$? Assume A and ω_c constant, and $n(t)$ zero-mean Gaussian with power σ^2.

4.4 Consider the following orthogonal transformation of the correlated zero-mean variables y_1 and y_2

$$\begin{bmatrix} \cos\theta & -\sin\theta \\ \sin\theta & \cos\theta \end{bmatrix} \begin{bmatrix} y_1 \\ y_2 \end{bmatrix}$$

Define $E\{y_1^2\} = \sigma_{y_1}^2$, $E\{y_2^2\} = \sigma_{y_2}^2$, and $E\{y_1 y_2\} = \rho\sigma_{y_1}\sigma_{y_2}$ with $\sigma_{y_1} = \sigma_{y_2}$. Show that to make the transformed variables uncorrelated θ must be chosen such that

$$\tan 2\theta = \frac{2\rho\sigma_{y_1}\sigma_{y_2}}{\sigma_{y_2}^2 - \sigma_{y_1}^2}$$

4.5 Consider the narrowband Gaussian process

$$n(t) = x(t) \cos \omega_c t - y(t) \sin \omega_c t$$

and assume that the spectral density is symmetric about the carrier frequency ω_c. Determine the multivariate probability density function $p(x_t, x_{t-\tau}, y_t, y_{t-\tau})$.

4.6 Consider the narrowband signal

$$z(t) = A \cos(\omega_c t + \theta) + n(t)$$

with $n(t)$ Gaussian and

$$n(t) = x(t) \cos \omega_c t - y(t) \sin \omega_c t$$

(a) Show that the correlation function of the envelope squared of $z(t)$ is

$$R(\tau) = A^4 + 4A^2\sigma^2 + 4\sigma^4 + 4(A^2 R_x(\tau) + R_x^2(\tau) + R_{xy}^2(\tau))$$

(b) What is the power spectral density for the noise only case if the noise spectral density is symmetric about ω_c for $\omega > 0$?

4.7 Assume that the variables x and y are jointly Gaussian with zero mean. Denote $E\{x^2\} = \sigma_x^2, E\{y^2\} = \sigma_y^2, E\{xy\} = \rho\sigma_x\sigma_y$. Show that the conditional density function of y given x is

$$p(y \mid x) = \frac{1}{(2\pi)^{1/2}\sigma_y(1 - \rho^2)^{1/2}} \exp\left[\frac{(y - \rho(\sigma_y/\sigma_x)x)^2}{-\sigma_y^2(1 - \rho^2)}\right]$$

4.8 This exercise demonstrates that RC filtered white Gaussian noise (see Example 2.8) is a Markov process (Exercise 2.4). Such a process, denote it $y(t)$, will have an exponential correlation function such as $e^{-\omega_1|\tau|}$. Define $t_3 - t_2 = t_2 - t_1 = \Delta$, $e^{-\omega_1|\Delta|} \triangleq \rho$, $y(t_3) = y_3$, $y(t_2) = y_2$, and $y(t_1) = y_1$ where $t_3 > t_2 > t_1$. Show that

$$p(y_1 \mid y_2, y_3) = \frac{1}{(2\pi)^{1/2}(1 - \rho^2)^{1/2}} \exp\left[\frac{(y_1 - \rho y_2)^2}{-2(1 - \rho^2)}\right]$$

and therefore that

$$p(y_1 \mid y_2, y_3) = p(y_1 \mid y_2)$$

4.9 Determine the eigenvectors and eigenvalues of $\mathbf{N} = \sigma_w^2\mathbf{I} + \mathbf{R}_n$ in terms of the eigenvectors and eigenvalues of $\mathbf{R}_n$.

4.10 Show that
(a) The eigenvalues of the matrix

$$\mathbf{R} = \begin{bmatrix} 2 & 3^{1/2}/2 \\ 3^{1/2}/2 & 3 \end{bmatrix}$$

are 1.5 and 3.5, and that the matrix of eigenvectors is

$$\mathbf{A} = \begin{bmatrix} 3^{1/2}/2 & 1/2 \\ -1/2 & 3^{1/2}/2 \end{bmatrix}$$

(b) Verify that the matrix $\mathbf{A}$ is orthogonal.
(c) Verify by direct calculation that $\mathbf{RA} = \mathbf{A\Lambda}$.

(d)   If samples x_1 and x_2 have the covariance matrix $\mathbf{R}$, find a transformation on x_1 and x_2 such that the transformed variables have the unit covariance matrix $\mathbf{I}$. Relate the answer to both $\mathbf{A}$ and Λ.

4.11   What are the eigenvalues of the Markov covariance matrix

$$\begin{bmatrix} 1 & \rho & \rho^2 \\ \rho & 1 & \rho \\ \rho^2 & \rho & 1 \end{bmatrix}$$

4.12   From Fig. 4.8 it is clear that the Rician density function is approximately normal over a reasonable range of β for large values of signal-to-noise. Show that this is to be expected using the expression for the density function.

4.13   The chi variable is the square root of the chi-squared variable.

(a)   Show that the probability density function for the chi variable with n degrees of freedom is

$$p(\chi) = \frac{\chi^{n-1}e^{-\chi^2/2}}{2^{(n-2)/2}\Gamma(n/2)}$$

(b)   Do the same for the noncentral chi variable and show that

$$p(\chi) = \chi \left(\frac{\chi^2}{\lambda} \right)^{(n-2)/4} \exp\left(-\frac{\lambda}{2} - \frac{\chi^2}{2} \right) I_{n/2-1}((\chi^2\lambda)^{1/2})$$

4.14   Show that the mth-order central moment of the central chi-squared variable with n degrees of freedom is

$$2^m \left(\frac{n}{2} \right) \left(\frac{n}{2} + 1 \right) \cdots \left(\frac{n}{2} + m - 1 \right)$$

4.15   Suppose x is central chi-square with N degrees of freedom, and y is noncentral with M degrees of freedom and noncentral parameter λ. Assume x and y statistically independent. What is the probability distribution of $z = x + y$? (We assume that x and y were generated by Gaussian variables of unit variance.)

4.16   The average power, defined as $y = (1/T) \int_0^T x^2(t)\, dt$, of a zero-mean Gaussian process is often approximated by a central chi-square probability distribution.

(a)   Using the mean and variance of y, show that the approximate number of degrees of freedom of y is

$$\nu \approx R_x^2(0) \left[\frac{1}{T} \int_{-T}^{T} \left(1 - \frac{|v|}{T} \right) R_x^2(v)\, dv \right]^{-1}$$

and that for large T

$$\nu \approx R_x^2(0) \left[\frac{1}{T} \int_{-\infty}^{\infty} R_x^2(v) \, dv \right]^{-1}$$

where $R_x(\tau)$ is the autocorrelation function of $x(t)$.

(b)   Suppose the power spectral density of $x(t)$ is bandlimited in frequency to $|\omega| \leq 2\pi B$. Within this range assume $S_x(\omega)$ is constant, that is, bandlimited white noise. For large T, show that $\nu \approx 2BT$.

4.17   Suppose we were to approximate, by matching the means and variances, a probability distribution of mean μ and variance σ^2 to a noncentral chi-square distribution with ν degrees of freedom and a noncentral parameter λ. Show that we would choose

$$\lambda = \frac{\sigma^2}{2} - \mu, \qquad \nu = 2\mu - \frac{\sigma^2}{2}$$

4.18   When chi-squared variables are added the result is chi-squared because of the normalization. This exercise shows that the same result does not necessarily hold for the gamma distribution. Assume that

$$z_1 = x_1^2 + x_2^2, \qquad z_2 = y_1^2 + y_2^2, \qquad \text{and} \qquad q = z_1 + z_2$$

where x_1, x_2 are statistically independent zero-mean Gaussian with variance σ_1^2 and y_1, y_2 are similar except with variance σ_2^2, and $\sigma_1 \neq \sigma_2$.

(a)   Show that the distribution of q is

$$p(q) = \frac{1}{2(\sigma_1^2 - \sigma_2^2)} \left[\exp\left(-\frac{q}{2\sigma_1^2} \right) - \exp\left(-\frac{q}{2\sigma_2^2} \right) \right]$$

Is this the gamma distribution?

(b)   What is the distribution if $\sigma_1 = \sigma_2$?

5 Hypothesis Testing

In this chapter, we introduce the general problem with which we will be concerned in the remainder of the book. Given some data and some number of probability distributions from which the data might have been sampled, we want to determine, as best we can, which distribution was in effect at the time the data were taken. This is the problem of statistical hypothesis testing, and its solution rests on the branch of mathematics called decision theory. In engineering, this is often called detection theory, because the prototypical problem was that of detecting the presence of a target in some region of a radar surveillance area.

Problems in detection theory also arise in the context of digital communication systems. In that case, some source of message information produces a bit 0 or 1, which is to be transmitted to us. An encoder generates a corresponding signal $s_0(t)$ or $s_1(t)$, which is distorted in transmission and corrupted by noise in the channel and the receiver. The task is to design a receiver (a signal processor) which will determine, as best possible, whether a 0 or a 1 was the intended message. The design rests on a model of the system which specifies, in more or less completeness, the probability distributions of the received signal in the two message cases. The task of the receiver is to select between these two distributions, given the received data. There are three related problems to be solved. First, specify the model of the system. Second, design the receiver, under some mathematization of the desire for the "best" receiver. Finally, evaluate its performance "on the average."

In this chapter, we consider the second and third of these three tasks, that is, determination of some general algorithms for the receiver, given the distributions of the data under the various hypotheses which might be operating, and assessment of their performance. In subsequent chapters we will apply those results to specific cases of data models which are both tractable and useful. We will also in Chapter 10 give consideration to the closely related question of estimation. That is, rather than deciding among some distributions based on data, we may want to estimate as well as possible the values of some parameters of a particular distribution. We will also consider to a lesser extent the combined problem of detection and estimation, in which we first want to determine which of several distributions was in effect at the time some data were generated and then estimate the values of some parameters of that distribution.

As an introduction, we will set the general framework of problems in decision theory. We will then consider a simple specific detection problem which has many of the elements we will develop in more detail in this and the following chapters. In the remainder of this chapter we will discuss and illustrate the most common procedures available for solving various of the general problems in decision theory.

5.1 INTRODUCTION

Decision theory is the branch of probability theory dealing with the following situation. We have available some data which we model as a random process because some elements in the source of the data are not describable with certainty. For example, we might be viewing the output voltage waveform of a radar receiver. At some particular range, a target might or might not be present, and the receiver voltage at the corresponding time will depend on which is the case. Also, any actual data are corrupted with noise of one kind or another, which is not deterministically predictable.

Suppose that in such a case we are interested in determining which of a number of situations gave rise to the data at hand. We will specify some number of hypotheses H_i, $i = 0, m - 1$, among which we believe there is one which describes the state of affairs at the time the data were produced. Let these hypotheses specifically refer to m probabilistic models. By processing the data set y at hand, we want to determine, as best we can, which of the models H_i was in effect to produce the data in question. The result of such processing will be a decision D_j that the data should be associated with hypothesis H_j. Given the hypotheses H_i, $i = 0, m - 1$, we want to determine how to arrive at the decision D_j which best fits the case and evaluate how well that strategy performs on average.

Let us introduce some terminology and indicate broadly some of the variants of the problem. First, if only two hypotheses are involved, usually numbered H_0, H_1, we have a *binary* hypothesis-testing problem. Traditionally, the hypothesis numbered H_0 is called the *null hypothesis*, and H_1 is the *alternative hypothesis*. The data y might be scalar or multidimensional. In any case, we want to divide the data space into two regions R_0, R_1, called the *acceptance region* (we accept H_0) and the *critical region,* or the *rejection region*, in which we reject H_0 and instead accept H_1 as true. That is, we make decision D_0, that hypothesis H_0 was in effect, if the data point y lies in region R_0, and similarly for decision D_1. These two regions must include every point of the y-space, since in any event we require that a decision be made. Also, they must have no point of the y-space in common, else there is ambiguity about what decision is made for such a point. That is, the regions R_0, R_1 must dichotomize the data space.

We will deal only with *parametric* decision theory, in which we assume that the probability distributions corresponding to the hypotheses H_i are certain functions of known forms, possibly with parameters having unknown values. For any particular hypothesis H_i two cases are possible. The hypothesis might have no parameters with unknown values, in which case it is called *simple,* or it might have one or more parameters whose values are unspecified, in which case it is called *composite*. For example, in the next section we treat the case of two hypotheses, which are that the mean of a scalar Gaussian random variable with specified variance has one of two specified values m_0, m_1. Since the distribution corresponding to each hypothesis is of known form (a Gaussian density), we have a parametric problem. It is binary, since we admit only two possibilities. Both hypotheses are simple, because we associate a single specified value of the mean with each hypothesis. If the variance were unknown, each hypothesis would be composite, because each would involve a parameter (the variance) with unknown value.

In case we are interested in sorting among more than two hypotheses, the problem becomes one of *multiple* hypothesis testing. In particular, if we introduce hypotheses H_i, $i = 0, m - 1$ we have the m-ary hypothesis-testing problem.

Our first aim in solving any of these problems is to describe how best to process the data to arrive at a decision. We must describe mathematically what is meant by "best"in order to proceed. This is done in terms of the errors which might be made in any particular procedure. In the binary problem, for example, if hypothesis H_0 was in effect to generate the data, and we make decision D_1 that H_1 was in effect, we have made a so-called *Type I error*, or *error of the first kind*. Depending on how we arrive at the decision, we will make that error with some probability $P_f = P(D_1 \mid H_0)$, called the *size* of the test in statistical work. (We use the notation P_f because, in radar, traditionally the null hypothesis H_0 is taken as "no target present," in which case the size of the test is reasonably called the probability of *false alarm*.) On the other hand, if H_1 was in effect to generate the data, and we decide D_0, then we have committed a *Type II error*, or *error of the second kind*. (In radar a Type II error is called a *missed detection*.) We do that with some probability $P(D_0 \mid H_1)$. Since in this binary problem we must decide either D_0 or D_1 in any event, we have $P_d = P(D_1 \mid H_1) = 1 - P(D_0 \mid H_1)$. This is called the *power* of the test in statistical work. The notation P_d recalls that this is the *probability of detection* in radar.

Continuing with the binary problem, there are a number of ways to describe the "best" decision strategy in terms of the probabilities of detection and false alarm. One of the most useful is the *Neyman–Pearson strategy*, which applies to the case that both hypotheses are simple. In this, we choose the decision rule such that the false alarm probability P_f is no larger than some specified upper bound and, within that constraint, maximize the probability of detection. This is perhaps the most important criterion in radar detection problems. Although the detection of targets is ultimately the "name of the game," an uncontrolled rate of false alarms can lead to a situation in which the overall system is overwhelmed and ceases to function effectively.

Perhaps the most general mathematization of the problem of building a best detection strategy is based on Bayes' rule. In this case, costs (positive or negative) are assigned to the various responses a receiver could indicate, such as correctly detecting a target or indicating a false alarm. The total cost is then averaged over the underlying probabilities of the hypotheses, and the detection strategy is chosen which minimizes the average cost. This is a useful strategy in the cases of both simple and composite hypotheses. In particular, in the case of a composite hypothesis, some probability density can be assumed for the unknown parameters of the underlying density and those parameters averaged away in the course of computing the average cost of a strategy.

As we will see, most schemes for building a receiver require the designer to state the underlying probability densities. A central problem in implementing such strategies is that often some needed density is not known completely. We will also discuss some ways of dealing with that problem. For example, the underlying unknown density can be chosen to be that which has the most deleterious effect on performance. Thereby, we design against the worst that nature can do and at least limit our losses. As such a strategy can lead to performance which is very much worse than might be the case with a less pessimistic view of the world, it is perhaps more common to parametrize the unknown densities and estimate the needed parameter values by processing some data. In a later chapter on estimation (Chapter 10) we will describe ways of doing that. Finally, since the state of the world may change as time goes on, it is usually necessary to update such estimates from time to time. In Chapter 11 we will discuss some ways of doing that.

In the rest of this chapter, we discuss the simpler receiver strategies available.

5.2 A SIMPLE DETECTION PROBLEM

Suppose that a message m takes the value m_0 (e.g., 0) or the value m_1 (e.g., 1). Since we presumably don't know in advance which message is to be sent, we model this as a random variable m taking values m_0, m_1 with respective probabilities P_0, P_1. The transmission channel adds a zero-mean Gaussian random noise variable n to the message. This results in a value of the random variable $y = m + n$ being presented as data. By processing a particular value y, we want to decide whether the message was m_0 or m_1. We also want to evaluate how well the decision strategy works over the ensemble of messages and channel noise.

Suppose the noise variance is σ_n^2. Then the two cases of interest, m_0 and m_1, correspond to two different densities for the received data y. In the two cases, y is Gaussian with variance σ_n^2 and with mean m respectively m_0 or m_1. That is,

$$p_y(y \mid m) = (2\pi\sigma_n^2)^{-1/2} \exp[-(y - m)^2/2\sigma_n^2]. \tag{5.1}$$

We need to state what we mean by making the best choice between the two possible message values. One strategy is to choose the value of m which has the higher probability of having been in effect, given the particular value of y which was received. That is, we choose m_1 in the case that

$$p_m(m_1 \mid y) > p_m(m_0 \mid y). \tag{5.2}$$

This is called the *maximum a posteriori probability* (MAP) criterion. Here we have arbitrarily awarded a tie to $m = m_0$.

Using the prior probabilities P_0, P_1 of the two messages, Bayes' rule yields

$$p_m(m_1 \mid y) = p_y(y \mid m_1)P_1/p_y(y),$$

and similarly for $p_m(m_0 \mid y)$. Then the decision rule Eq. (5.2) becomes: Choose $m = m_1$ provided

$$\lambda(y) = p_m(m_1 \mid y)/p_m(m_0 \mid y) = p_y(y \mid m_1)P_1/p_y(y \mid m_0)P_0 > 1. \quad (5.3)$$

The quantity $\lambda(y)$ is called the *likelihood ratio*, and the rule Eq. (5.3) which has resulted from our criterion of goodness is a *likelihood ratio test*. Since the logarithm is a monotonically increasing function, this test can also be written in the form involving the log likelihood ratio:

$$\ln(\lambda) > 0. \quad (5.4)$$

Substituting the two assumed Gaussian densities Eq. (5.1) into this last, and assuming for definiteness $m_1 > m_0$, leads to the rule to choose m_1 provided

$$y > y_t = [\sigma_n^2/(m_1 - m_0)]\ln(P_0/P_1) + (m_1 + m_0)/2. \quad (5.5)$$

(For $m_1 < m_0$, the sense of the inequality reverses.) That is, we compare the received data value y to a threshold. For this particular problem, the data y itself appears as a *test statistic,* i.e., the quantity to be computed by the receiver. Figure 5.1 shows the two conditional densities in question and the decision rule, for the special case $m_0 = 0$, $m_1 = 1$, $\sigma_n = 1$. The region R_1 of the test statistic y is that in which Eq. (5.5) is satisfied and for which we choose $m = 1$ as the receiver output.

Equation (5.5) specifies the form of the maximum *a posteriori* probability receiver for this communication system. To assess its performance, we might compute the probability of error. There are two ways to make an error. Either the message is m_0 and we declare it to be m_1, or the message is m_1 and we declare it to be m_0. Relative to the hypothesis that the message is m_0, the first of these is an error of type I, or a *false alarm*, and the second is an error of type II, or a *missed detection*. The terminology reverses if we choose to regard the hypothesis as sending message m_1.

In our binary communication system, the probabilities of the two types of error are $P(D_1 \mid m_0)$, $P(D_0 \mid m_1)$. Using Eq. (5.5), and still assuming $m_1 > m_0$, the two errors have probabilities.

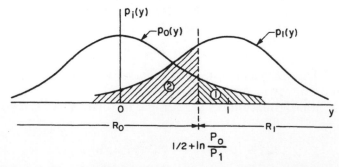

FIGURE 5.1 Conditional probability density functions and dichotomy of sample space for binary hypothesis test. Regions 1 and 2 show the error probabilities $P(D_1 \mid H_0)$ and $P(D_0 \mid H_1)$.

$$P(D_1 \mid H_0) = P(y > y_t \mid H_0)$$

$$= \int_{y_t}^{\infty} (2\pi \sigma_n^2)^{-1/2} \exp\left[\frac{-(y - m_0)^2}{2\sigma_n^2}\right] dy$$

$$= \int_{(1/\alpha) \ln(P_0/P_1) + \alpha/2}^{\infty} (2\pi)^{-1/2} \exp\left(\frac{-u^2}{2}\right) du,$$

$$P(D_0 \mid H_1) = P(y \le y_t \mid H_1)$$

$$= \int_{-\infty}^{y_t} (2\pi \sigma_n^2)^{-1/2} \exp\left[\frac{-(y - m_1)^2}{\sigma_n^2}\right] dy$$

$$= \int_{-\infty}^{(1/\alpha) \ln(P_0/P_1) - \alpha/2} (2\pi)^{-1/2} \exp\left(\frac{-u^2}{2}\right) du, \qquad (5.6)$$

where

$$\alpha = (m_1 - m_0)/\sigma_n.$$

If $m_1 < m_0$, Eqs. (5.6) hold as written, but with $\alpha = (m_0 - m_1)/\sigma_n$. Both cases are included by taking Eqs. (5.6) with the definition

$$\alpha = |m_1 - m_0|/\sigma_n. \qquad (5.7)$$

It is noteworthy that, for given prior probabilities P_0, P_1, the error probabilities Eq. (5.6) depend on the single parameter Eq. (5.7). The parameter α of Eq. (5.7) can be taken as the *signal-to-noise ratio* (SNR) for this problem. Since one could equally well define the SNR as any function of α, it is always important to link an expression such as Eq. (5.7) to the expression in which it is used, such as Eqs. (5.6).

Figure 5.1 shows the two error probabilities Eqs. (5.6). Figure 5.2 shows the quantities Eq. (5.6) plotted with the SNR Eq. (5.7) as parameter. This figure is one form of *receiver operating characteristic* (ROC) curve for the test Eq. (5.5).

In a communication system, we are not usually interested in the separate probabilities of errors of the two types, Eq. (5.6). Rather, we are interested in the total probability of error. Since the events $m = m_0$ and $m = m_1$ are mutually exclusive, this can be written

$$P_e = P(D_1 \mid m_0)P_0 + P(D_0 \mid m_1)P_1. \qquad (5.8)$$

Figure 5.3 shows this quantity plotted as a function of the SNR Eq. (5.7) with P_0 as parameter. The function Eq. (5.8) is symmetric in P_0, P_1, so that the parameter in Fig. 5.3 can equally well be read as P_1. Equivalently, the curve for some value $P_0 = p$ is also the curve for $P_0 = 1 - p$. The curves accord with intuition. In particular, with $\alpha = 0$ the noise overwhelms the signal and the receiver chooses according to the prior probabilities P_0,

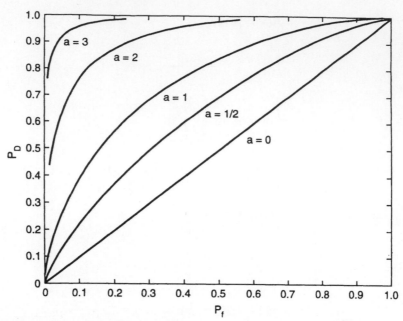

FIGURE 5.2   Receiver operating characteristic curve for binary detection as in Eqs. (5.6), (5.7), with $P_0 = P_1$. The SNR is a.

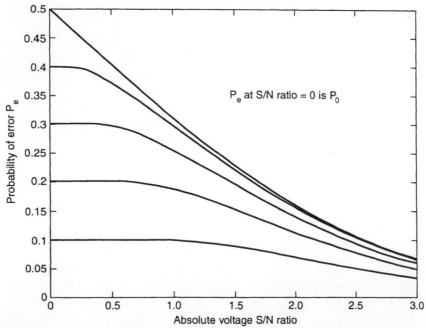

FIGURE 5.3   Error probability Eq. (5.8) for binary Gaussian channel parametrized by signal-to-noise ratio Eq. (5.7).

P_1. Then if $P_0 < P_1$ we always choose H_1, and the probability of error is P_0. On the other hand, if $P_1 \leq P_0$, we always choose H_0 and the probability of being in error is P_1.

In the particular case that $P_0 = P_1 = 1/2$, we have from Eqs. (5.6), (5.8) that

$$P_e = P(D_1 \mid m_0) = P(D_0 \mid m_1) = \int_{\alpha/2}^{\infty} (2\pi)^{-1/2} \exp(-u^2/2) \, du. \quad (5.9)$$

This is the uppermost curve in Fig. 5.3.

In the remainder of this chapter, we discuss hypothesis testing in more general situations than indicated above. In later chapters we will apply the general formalism to specific problem models which have proved to be useful in applications.

5.3 THE NEYMAN–PEARSON CRITERION

In Sect. 5.2, we took as criterion of goodness for the receiver that the signal selected should be that with maximum probability of having been sent, given the particular data that were actually received. That is, the criterion Eq. (5.2) was the basis for the design. In radar and sonar problems, for example, another criterion is more appropriate, the Neyman–Pearson criterion. With this approach, we choose a probability of type I error (false alarm) as large as we are willing to tolerate and seek to minimize the probability of type II error (missed detection) under that limitation. That is, for some given probability P_f we work the problem:

$$\min_{D(y)} P(D_0 \mid H_1), \qquad P(D_1 \mid H_0) \leq P_f \quad (5.10)$$

where $D(y)$ symbolizes the decision rule used to process the data y. We have also generalized the notation to any hypotheses H_0, H_1. In Sect. 5.2, these were taken as, e.g., $H_0 = $ "the message sent was m_0."

In radar and sonar P_f is the false alarm probability, if the hypothesis is "no target," and $1 - P(D_0 \mid H_1) = P(D_1 \mid H_1) = P_d$ is the probability of detection. The Neyman–Pearson criterion is therefore also phrased as: "Maximize P_d while maintaining the false alarm probability at most at the specified level P_f." In this form, we have to solve the problem

$$\max P(D_1 \mid H_1) = \max \int_{R_1} p_1(y) \, dy,$$

$$P(D_1 \mid H_0) = \int_{R_1} p_0(y) \, dy \leq P_f, \quad (5.11)$$

by choosing the region R_1 of y-space. Here y symbolizes whatever (perhaps multidimensional) data are available, R_1 is the region of the y-space in

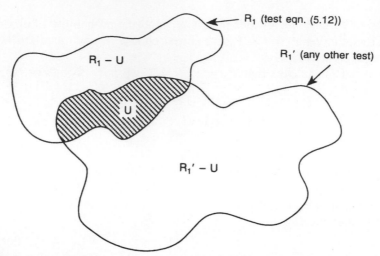

FIGURE 5.4 Two regions of data space, each of which results in false alarm probability $P(D_1 \mid H_0) \leq P_f$. The intersection U might be empty.

which we will decide D_1, and $p_0(y)$, $p_1(y)$ are the densities of y conditioned on H_0, H_1, respectively.

Since the entire y-space is to be separated into the region R_1 (the critical region, in which we reject H_0) and the remainder (the acceptance region), the question is, what points of y should be included in R_1? Clearly, the answer is: those points y for which $p_1(y)$ contributes proportionately more to increasing the first integral of Eq. (5.11) than does $p_0(y)$ to increasing the second integral. That suggests including in R_1 those points y for which $p_1(y) > \lambda_0 p_0(y)$, where λ_0 is some constant to be found.

The precise statement of this strategy for finding R_1 is the *Neyman–Pearson lemma*. Assume there to be only two hypotheses H_0, H_1; that is, we are concerned with a binary hypothesis-testing problem. Both hypotheses are to be simple. [The densities $p_0(y)$, $p_1(y)$ are to have no parameters whose values are unknown.] The data y may be multidimensional. Then, for a false alarm probability $P(D_1 \mid H_0) \leq P_f$, with $P_f > 0$, the test with greatest detection probability P_d is one for which the critical region R_1 is found from a likelihood ratio test, which we phrase as follows, to allow for the case that $p_0(y) = 0$:

If $p_1(y) < \lambda_0 p_0(y)$,   choose H_0
If $p_1(y) > \lambda_0 p_0(y)$,   choose H_1
If $p_1(y) = \lambda_0 p_0(y)$,   choose H_1 in accord with a Bernoulli trial
  with probability of success β.   *(5.12)

In these, $\lambda_0 \geq 0$ and $\beta > 0$ are constants to be determined. In the unrealistic case that the desired $P_f = 0$, then choose H_0 unless $p_0(y) = 0$, in which case choose H_1.

To see that Eq. (5.12) solves the problem Eq. (5.11), consider as follows. First, note that any test satisfying Eq. (5.11) must be such that $P(D_1 \mid H_0) = P_f$. This is because any region R_1 for which $P(D_1 \mid H_0) < P_f$ can be improved by including more points of y in R_1, since then both $P(D_1 \mid H_1)$ and $P(D_1 \mid H_0)$ increase. This process continues until $P(D_1 \mid H_0)$ is at its limit. The question is, exactly which points of y are most effective in increasing $P(D_1 \mid H_1)$ relative to increase in $P(D_0 \mid H_1)$?

To find the answer to this question, let R_1 be a region in which H_1 is chosen in accord with Eq. (5.12). Let R_1' be any other region with false alarm probability $P_f' = P'(D_1 \mid H_0) \le P(D_1 \mid H_0) = P_f$. Let these be as sketched in Fig. 5.4, possibly having an intersection U. Consider the integral

$$I = \left(\int_{R_1 - U} - \int_{R_1' - U} \right) [p_1(y) - \lambda_0 p_0(y)] \, dy.$$

By assumption, the integrand in this last is nonnegative over the region $R_1 - U$, since Eq. (5.12) is to hold, and on R_1 we choose H_1. On the other hand, on $R_1' - U$ the integrand is nonpositive, since U is the region of commonality with R_1 and hence contains all points of R_1' with nonnegative integrand. Hence $I > 0$. Adding and subtracting the integrals over the common region U results in

$$\left(\int_{R_1} - \int_{R_1'} \right) p_1(y) \, dy \ge \lambda_0 \left(\int_{R_1} - \int_{R_1'} \right) p_0(y) \, dy.$$

That is,

$$P_d - P_d' \ge \lambda_0 (P_f - P_f') \ge 0.$$

That is, the test defined by Eq. (5.12) has at least as large a detection probability as any other test meeting the requirement on false alarm probability. It is therefore the optimum test.

It remains to determine the values of λ_0 and β. We require that

$$P(D_1 \mid H_0) = P[p_1(y) > \lambda_0 p_0(y) \mid H_0] \\ + \beta P[p_1(y) = \lambda_0 p_0(y) \mid H_0] = P_f, \qquad (5.13)$$

for the given value P_f. If y is a continuous random variable, as implied by the integrals above, it suffices to take $\beta = 0$ and solve Eq. (5.13) for λ_0. On the other hand, if y is a discrete random variable, there may not be a region R_1 which matches P_f exactly. In that case, the largest available region is used, and β is selected to make up the deficit in P_f.

A number of points should be noted here. First, for the Neyman–Pearson criterion to be solved by the strategy Eq. (5.12), which we now write somewhat loosely as the likelihood ratio test

$$\lambda(y) = p_1(y)/p_0(y) > \lambda_0, \qquad \qquad *(5.14)$$

we must have a problem with only two alternatives to be chosen between, since we have divided the data space into only two decision regions, R_0 and R_1. Second, the hypotheses H_0, H_1 must be simple; that is, there must be no unknown parameters in the conditional densities $p_0(y)$, $p_1(y)$. Otherwise, we will not be able to compute the value of $\lambda(y)$ to compare to the threshold λ_0. Importantly, we have not made any assumptions about the forms of the densities $p_0(y)$, $p_1(y)$. The Neyman–Pearson procedure is completely general in that respect. Finally, there is no need to specify the *a priori* probabilities of the two hypotheses. On the other hand, if those probabilities are known, then the Neyman–Pearson test makes no use of that information.

EXAMPLE 5.1   Let us consider again the simple problem of Sect. 5.2. that is, we want to decide between two Gaussian densities with means m_0, m_1 and common variance σ_n^2. The likelihood ratio test Eq. (5.14) is

$$\begin{aligned} \lambda(y) &= p_1(y)/p_0(y) \\ &= \exp\{-[(y - m_1)^2 - (y - m_0)^2]/2\sigma_n^2\} > \lambda_0. \end{aligned}$$

In the case $m_1 > m_0$, taking the logarithm and rearranging yields Eq. (5.5),

$$y > y_t,$$

except that now the threshold is some value to be found. (If $m_1 < m_0$, the sense of the inequality reverses.) The requirement on the threshold is

$$P(D_1 \mid H_0) = \int_{y_t}^{\infty} p_0(y)\, dy$$

$$= \int_{(y_t - m_0)/\sigma_n}^{\infty} (2\pi)^{-1/2} \exp\left(-\frac{u^2}{2}\right) du = P_f. \qquad (5.15)$$

From this equation, the threshold y_t can be determined for any specified size P_f, numerically or from a table of the Gaussian error function. With that threshold in hand, the likelihood ratio test attains a probability of detection (power)

$$P_d = P(D_1 \mid H_1) = \int_{y_t}^{\infty} p_1(y)\, dy$$

$$= \int_{(y_t - m_1)/\sigma_n}^{\infty} (2\pi)^{-1/2} \exp\left(\frac{-u^2}{2}\right) du$$

$$= \int_{(y_t - m_0)/\sigma_n - (m_1 - m_0)/\sigma_n}^{\infty} (2\pi)^{-1/2} \exp\left(\frac{-u^2}{2}\right) du. \qquad (5.16)$$

If $m_1 < m_0$, Eqs. (5.15), (5.16) hold with the signs of the lower limits reversed.

In either case ($m_1 < m_0$ or $m_1 > m_0$), with a given P_f the value of P_d is again determined by the value of the single parameter Eq. (5.7), i.e.,

$$\alpha = |m_1 - m_0|/\sigma_n, \qquad (5.17)$$

which appears as an appropriate signal-to-noise ratio for the problem. Figure 5.2 shows the probability of detection P_d as a function of P_f, parametrized by the required signal-to-noise ratio α of Eq. (5.17). This is the ROC curve for this problem and is a universal curve applicable to many situations. That is, the curve Fig. 5.2 simply portrays the result of solution, for P_d given α and P_f, of the two equations

$$\int_U^\infty (2\pi)^{-1/2} \exp\left(\frac{-u^2}{2}\right) du = P_f,$$

$$\int_{U-\alpha}^\infty (2\pi)^{-1/2} \exp\left(\frac{-u^2}{2}\right) du = P_d. \qquad *(5.18)$$

The form of the SNR α in terms of the parameters of the problem may of course be different for different problems and must always be stated in references to the universal curves Fig. 5.2.

Figure 5.5 is a replotting of Fig. 5.2 with P_f as parameter. Figure 5.6 replots Fig. 5.5, with P_d shown using the *normal probability scale.*

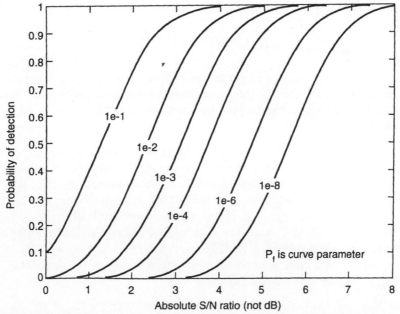

FIGURE 5.5 Universal receiver operating characteristic curve for binary Gaussian problem, Eq. (5.18).

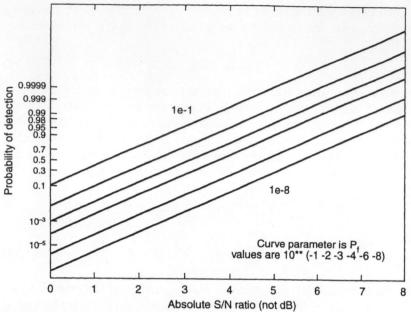

FIGURE 5.6   Receiver operating characteristic curve plotted on probability paper in terms of absolute signal to noise ratio.

This displays the ordinate distance y labeled with the numbers

$$P_d = \int_{-\infty}^{y} (2\pi)^{-1/2} \exp\left(\frac{-u^2}{2}\right) du.$$

With this last relation, y is linear with SNR: $y = \alpha - U$, where U depends on P_f. Figure 5.7 shows another presentation, with the SNR α in dB (i.e., $20 \log_{10} \alpha$) and P_d on the normal probability scale, parametrized by P_f.

EXAMPLE 5.2   Suppose that we require a false alarm probability no larger than $P_f = 10^{-6}$. Suppose also that the conditions of the problem, whatever it might be, present us with a signal-to-noise ratio of $\alpha = 5$, in absolute units. Then Fig. 5.5 shows that we can at best obtain a detection probability $P_d = 0.6$. If the specific problem involves deciding between two Gaussian densities with means m_0 and m_1 and common variance σ_n, then from Eq. (5.15) we obtain $(y_t - m_0)/\sigma_n = 4.75$. If for example, we have a problem with $m_0 = 1$, $\sigma_n = 1/4$, we obta⁻ finally $y_t = 2.19$ and a detector $y > 2.19$.

EXAMPLE 5.3   Because of the prevalence in engineering of te⁻ simple hypotheses in a Gaussian framework, it is perhaps easy

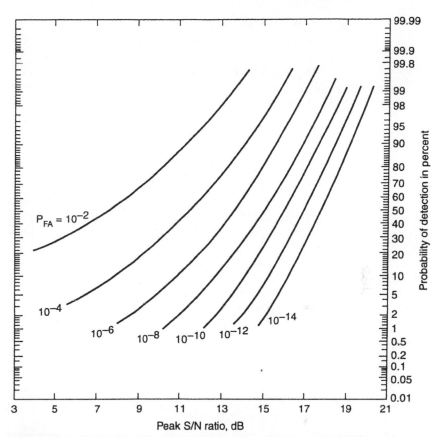

FIGURE 5.7   Receiver operating characteristic curve plotted on probability paper in terms of signal-to-noise ratio in dB. (From *Radar Detection*, by J. V. DiFranco and W. L. Rubin © 1980, Artech House, Norwood, MA.)

one hand to overlook the generality of the Neyman–Pearson detector and on the other to forget that curves such as Figs. 5.5–5.7 apply only to the particular case of Gaussian hypotheses. To emphasize the generality of the Neyman–Pearson procedure, suppose that we want to test the hypothesis H_0 that a single observation y is a sample of a random variable with the normalized Gaussian density

$$p_0(y) = (2\pi)^{-1/2} \exp(-y^2/2),$$

against the hypothesis H_1 that the governing density is exponential with parameter a, where a is some specified value:

$$p_1(y) = a \exp(-y/a), \qquad y \geq 0.$$

The Neyman–Pearson test Eq. (5.14) is to choose H_1 whenever (for $y \geq 0$)

$$\lambda(y) = a(2\pi)^{1/2}\exp(y^2/2 - y/a) > \lambda_t,$$
$$f(y) = y^2 - 2y/a - 2\ln[\lambda_t/a(2\pi)^{1/2}] > 0. \qquad (5.19)$$

Whenever $y < 0$ we choose H_0, because negative values of y have zero probability of occurring under H_1. For $y \geq 0$, the detector Eq. (5.19) prescribes the critical region of y to be that shown in Fig. 5.8, where $\gamma^2 = 1 + 2a^2\ln(\lambda_t/a\sqrt{2\pi})$. The false alarm and missed detection probabilities are

$$P_f = P(D_1 \mid H_0) = \left[\int_0^{(1-\gamma)/a} + \int_{(1+\gamma)/a}^{\infty}\right](2\pi)^{-1/2}\exp\left(\frac{-y^2}{2}\right)dy,$$

$$1 - P_d = P(D_0 \mid H_1) = \int_{(1-\gamma)/a}^{(1+\gamma)/a} a\exp\left(\frac{-y}{a}\right)dy.$$

For given values of P_f and a, the first of these could be solved for γ and hence λ_t. Using that in the second yields P_d. From those, the ROC curve could be constructed to show corresponding values of P_f and P_d with a as parameter, which then would appear as a reasonable definition of signal-to-noise ratio for the problem. Such a ROC curve would not be found in standard texts, however, because of the limited interest of the problem.

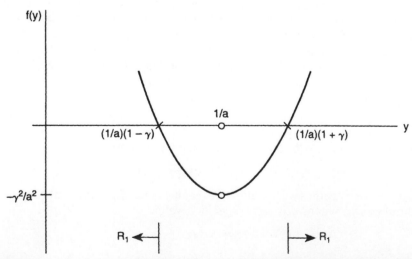

FIGURE 5.8   Critical region R_1 in Example 5.3 for deciding in favor of y having exponential density vs. y having Gaussian density. The parameter γ is determined by P_f.

5.4 BAYES' CRITERION

A useful and general criterion for goodness of a detection strategy can be set up in terms of costs of the various outcomes of the strategy. Such costs must be assigned by the user of the decision process. Once assigned, an average cost of using a particular decision procedure can be calculated and that procedure chosen which minimizes the average cost.

Let the decision process make decision D_i, that hypothesis H_i was in effect when the data were generated, when in fact hypothesis H_j is the correct decision. If $i = j$, we have made a correct decision, whereas if $i \neq j$ we have made an error. In general, let C_{ij} be the *cost* of choosing H_i when H_j is true. The cost might be positive or negative; a negative (or zero) cost would reasonably accompany a correct decision. Suppose that this cost assignment is such that an incorrect decision has a cost which is algebraically greater than the cost of a correct decision. That is, we assume

$$C_{i \neq j, j} - C_{jj} > 0, \quad \text{all } i, j. \tag{5.20}$$

In a later section we will consider the case of m-ary hypothesis testing. For clarity, in this section we will suppose that the problem is binary, that is, that there are only two hypotheses from which to select. As in the previous section, we will seek a region R_1 of the data space, the critical region, in which we decide that H_1 is true and reject the hypothesis that H_0 is true. The average cost, or the *risk*, of each decision is then

$$
\begin{aligned}
C = &\ P_0[C_{00}P(D_0 \mid H_0) + C_{10}P(D_1 \mid H_0)] \\
&+ P_1[C_{01}P(D_0 \mid H_1) + C_{11}P(D_1 \mid H_1),
\end{aligned}
\tag{5.21}
$$

where P_0, P_1 are the *a priori* probabilities of the two hypotheses, which are assumed to be specified. Since we decide D_1 whenever y is in the sought region R_1, we have

$$P(D_1 \mid H_0) = \int_{R_1} p_0(y)\, dy,$$

$$P(D_1 \mid H_1) = \int_{R_1} p_1(y)\, dy. \tag{5.22}$$

Using Eq. (5.22) in Eq. (5.21) and recalling that $P(D_0 \mid H_0) = 1 - P(D_1 \mid H_0)$, $P(D_0 \mid H_1) = 1 - P(D_1 \mid H_1)$, we obtain the average cost in the form

$$
C = P_0 C_{00} + P_1 C_{01} \\
+ \int_{R_1} [P_0(C_{10} - C_{00})p_0(y) - P_1(C_{01} - C_{11})p_1(y)]\, dy. \tag{5.23}
$$

We want to choose the region R_1 in such a way as to minimize the average cost Eq. (5.23). That is, we want to decide which points y of the data space should be included in the region of integration R_1. Note that the coefficients of $p_0(y)$ and $p_1(y)$ in the integrand are positive, by the assumption Eq. (5.20). Therefore, to minimize the value of the integral, we include in R_1 all points y for which the second term in the integrand is larger than the first, because that will make the value of C smaller than would be the case if we did not include the point. That is, we put into R_1 all points y for which:

$$\lambda(y) = p_1(y)/p_0(y) > P_0(C_{10} - C_{00})/P_1(C_{01} - C_{11}) = \lambda_t. \quad *(5.24)$$

That is, the decision rule is to include in R_1, and hence include among those data points for which we decide that H_1 is true, all those points y for which Eq. (5.24) is true.

The Bayes criterion of minimizing the average risk has thus led us again to a likelihood ratio test. The test Eq. (5.24) is the same as the earlier tests Eq. (5.3) using maximum *a posteriori* probability and Eq. (5.14) using the Neyman–Pearson criterion, except that the threshold is different in each case. In fact, the maximum *a posteriori* probability test Eq. (5.3) is obtained as a special case of the Bayes test Eq. (5.24) if the costs are chosen such that $C_{10} - C_{00} = C_{01} - C_{11}$.

5.5 MINIMUM ERROR PROBABILITY CRITERION

In a binary communication system, we usually have no particular reason to penalize an incorrect choice of H_0 differently from an incorrect choice of H_1. Therefore, a reasonable criterion is to minimize the cost of errors Eq. (5.21) using the particular choice $C_{00} = C_{11} = 0$, $C_{01} = C_{10} = c$. That is, we consider Eq. (5.21) in the form

$$C = c[P_0 P(D_1 \mid H_0) + P_1 P(D_0 \mid H_1)] = cP_e, \quad (5.25)$$

where we indicate the probability of error P_e. Since c is just a constant, the strategy for minimizing cP_e is the same as the strategy for minimizing the probability of error P_e itself. From Eq. (5.24), the answer is to choose message H_1 for received data y such that

$$\lambda(y) = p_1(y)/p_0(y) > P_0/P_1. \quad *(5.26)$$

This is just the test Eq. (5.3) which maximizes the *a posteriori* probability of the two hypotheses.

EXAMPLE 5.4   Usually in a communication system we have no reason to select different *a priori* probabilities of the various symbols. Therefore, suppose in a binary communication system we choose $P_0 = P_1 = 1/2$. We model the system as selection of a symbol value $m = 0$ or $m = 1$, which is corrupted during transmission by zero-mean additive Gaussian noise with variance $\sigma_n^2 = 1$. The received datum $y = m + n$ is therefore a Gaussian random variable with mean 0 or 1 under the two hypotheses and with unit variance. The criterion of minimum probability of error prescribes the receiver Eq. (5.26). Upon substituting for $p_0(y)$ and $p_1(y)$ the corresponding Gaussian densities, the receiver becomes the reasonable rule:

$$\text{Choose} \quad m = 1 \quad \text{if } y > y_t = \tfrac{1}{2}. \tag{5.27}$$

For this receiver the false alarm and detection probabilities are

$$P_f = \int_{1/2}^{\infty} (2\pi)^{-1/2} \exp\left(\frac{-u^2}{2}\right) du = 0.309,$$

$$P_d = \int_{-1/2}^{\infty} (2\pi)^{-1/2} \exp\left(\frac{-u^2}{2}\right) du = 1 - 0.309 = 0.691.$$

The probabilities of errors of the two kinds are equal:

$$P(D_1 \mid H_0) = P_f = 0.309,$$
$$P(D_0 \mid H_1) = 1 - P_d = 0.309,$$

and they are each equal to the probability of error:

$$P_e = \tfrac{1}{2} P(D_1 \mid H_0) + \tfrac{1}{2} P(D_0 \mid H_1) = 0.309. \tag{5.28}$$

For this particular example, the appropriate signal-to-noise ratio is that of Eq. (5.17) and has value $|\Delta m|/\sigma_n = 1$. The value $P_e = 0.31$ is found on the chart in Fig. 5.3 using SNR $\alpha = 1$.

5.6 MINIMAX CRITERION

So far, we have discussed three criteria by which the "best" detector can be selected. Using the Bayes criterion, we needed both the costs of decisions and the *a priori* probabilities of the hypotheses. With the criterion of maximum *a posteriori* probability, only the *a priori* probabilities were required. Finally, using the Neyman–Pearson strategy, neither costs nor prior probabilities were needed. The fourth combination, to be investigated in this section, is that in which we wish to assign costs to the decisions but do not have enough information to assign *a priori* probabilities to the hypotheses. The strategy will be to use the Bayes criterion nonetheless and assign prior probabilities, but to do it in such a way that the deleterious effects of a wrong assignment are minimized. That is, we admit our ignorance but act in such a way as to minimize its potential consequences.

We will consider here the case of two simple hypotheses. The Bayes strategy for this case led to the receiver Eq. (5.24) and attained the minimum average cost Eq. (5.21) in which the probabilities Eq. (5.22) have been substituted, with R_1 the region of y-space described by Eq. (5.24). Let us now consider how the attained minimum cost varies as a function of the *a priori* probability P_0 that H_0 is true. For illustration, consider again the problem of Sect. 5.2, that of detecting the mean of a unit variance Gaussian, but with a general prior probability P_0 of hypothesis H_0 that $m = 0$ and its companion probability $P_1 = 1 - P_0$ that $m = 1$. Also, to avoid a nongeneral symmetry in the problem, let us now choose costs $C_{00} = C_{11} = 0$ for correct decisions and $C_{01} = c$, $C_{10} = 1$ for incorrect decisions. The detector Eq. (5.24) for this example of selecting between the two Gaussians then becomes

$$y > y_t = \tfrac{1}{2} + \ln(P_0/cP_1). \tag{5.29}$$

This results in the attained minimum average cost

$$C_{\min} = P_0 P(D_1 \mid H_0) + cP_1 P(D_0 \mid H_1)$$

with

$$P(D_1 \mid H_0) = \int_{y_t}^{\infty} (2\pi)^{-1/2} \exp\left(\frac{-y^2}{2}\right) dy,$$

$$P(D_0 \mid H_1) = \int_{-\infty}^{y_t} (2\pi)^{-1/2} \exp\left[\frac{-(y-1)^2}{2}\right] dy. \tag{5.30}$$

Since $P_1 = 1 - P_0$ in this binary problem, the attained minimum cost Eq. (5.30) involves only the parameters P_0 and c. Figure 5.9a shows this cost $C_{\min}$ as a function of P_0, for the particular case that $c = 2$.

Figure 5.9a illustrates the general fact that the minimum of the cost Eq. (5.21) gets less the closer the problem gets to one of the deterministic situations $P_0 = 0$ (always choose H_1 and we are right with probability one) or $P_0 = 1$ (always choose H_0). The curve has a single relative maximum at a value of P_0 which can be found by setting the derivative of the attained minimum of the cost Eq. (5.21) equal to zero. For the example Eq. (5.30) the result, found as we will describe shortly, is $P_0 = 0.60$, for which the attained cost is $C_{\min} = 0.42$. This is the greatest minimum cost which can be associated with this problem, and it occurs only if nature presents us with the least fortunate situation, namely $P_0 = 0.60$. For any other state of nature, the incurred cost of using the optimum receiver will be less, as shown in Fig. 5.9a.

However, the situation changes if we have not designed the receiver using in Eq. (5.29) the value of P_0 which nature has in effect. Suppose, for example, that we assume nature is using $P_0 = 0.3$ in this example and design the receiver accordingly. That is, in Eqs. (5.29), (5.30) we use $P_0 = 0.3$, $P_1 = 0.7$. Suppose, however, that we are wrong and that nature is using some other value P_0'. Then the attained cost is

$$C' = P'_0 P(D_1 \mid H_0, P_0) + c(1 - P'_0) P(D_0 \mid H_1, P_0), \qquad (5.31)$$

where we indicate that the error probabilities depend only on the prese-
lected P_0 and not on the actual P'_0. Equation (5.31) is linear in P'_0 and is
plotted in Fig. 5.9b for the case of $c = 2$ and $P_0 = 0.3$. Since this (nonoptimal) cost function must have a value larger than the optimal Bayes cost
everywhere except at the point $P'_0 = P_0$, the two curves are tangent at
$P_0 = 0.3$ as shown.

As seen in Fig. 5.9b, the penalty for guessing wrong about nature can
be small, as for $P'_0 < 0.5$ in this example. It can also be large, as for
$P'_0 > 0.6$. We want to avoid the possibly large penalty which might result
from our ignorance about nature. The *minimax strategy* is that which
minimizes the maximum penalty we might pay for our mistake in using
the wrong value of P_0. It corresponds to arranging matters so that the
nonoptimal loss curve in Fig. 5.9b is horizontal. That is, we make the
average cost a constant, regardless of what value nature might choose for
P'_0. To do that, we place the point of tangency of the two curves at the
maximum of the optimal Bayes risk curve Fig. 5.9a. That is, we choose
for the design value of P_0 that value which causes the optimal Bayes risk
to be a maximum.

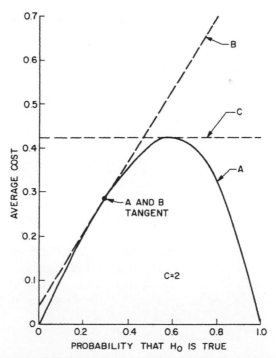

FIGURE 5.9 Example of minimax solution. (a) Bayes minimum cost; (b) cost for particular
decision threshold; (c) minimax cost.

Let us now consider the minimax strategy for the general binary problem with simple hypotheses. The solution to the problem has interesting features which would become apparent during a deeper discussion than we will give. We will only point them out as we proceed with a simple discussion. Consider the cost in the form Eq. (5.23). It is minimized by the choice of critical region R_1 as indicated by Eq. (5.24). We want to find the maximum of the minimum Bayes risk Eq. (5.23) as P_0 varies. Recalling that $P_1 = 1 - P_0$, the differential of the cost Eq. (5.23) is

$$
dC_{min} = \left\{ C_{00} - C_{01} + \int_{R_1} [(C_{10} - C_{00})p_0(y) \right.
$$

$$
\left. + (C_{01} - C_{11})p_1(y)] \, dy \right\} dP_0
$$

$$
+ \int_{\delta R_1} [P_0(C_{10} - C_{00})p_0(y) - P_1(C_{01} - C_{11})p_1(y)] \, dy. \qquad (5.32)
$$

The second integral in Eq. (5.32) is over a differential region of y-space around the optimal region R_1 generated by the variation dP_0. The value of that integral, to first order, is the integrand evaluated on the boundary of R_1, times the signed volume of the region δR_1. Now from Eq. (5.24), everywhere on the boundary of R_1 we have the integrand as

$$
P_0(C_{10} - C_{00})p_0(y) - P_1(C_{01} - C_{11})p_1(y) = 0.
$$

Therefore, the integral over δR_1 in Eq. (5.32) vanishes, and we have

$$
dC_{min}/dP_0 = C_{00} - C_{01} + (C_{10} - C_{00})P_{10} + (C_{01} - C_{11})P_{11},
$$

where $P_{ij} = P(D_i \mid H_j)$ Setting this to zero, the value of P_0 which maximizes the minimum Bayes risk Eq. (5.21) is found to be just that value for which

$$
C_{00}P_{00} + C_{10}P_{10} = C_{01}P_{01} + C_{11}P_{11}. \qquad (5.33)
$$

According to Eq. (5.33), the optimal Bayes risk is maximized by choosing the prior probability P_0 such that the average cost of a decision, if H_0 is true, is equal the average cost if H_1 is true. This is the condition for choice of P_0 under the minimax criterion.

It is now easy to find the average cost attained by the minimax receiver. Substituting Eq. (5.33) into the cost Eq. (5.21) yields

$$
\begin{aligned}
C_{mm} &= P_0(C_{00}P_{00} + C_{10}P_{10}) + P_1(C_{01}P_{01} + C_{11}P_{11}) \\
&= (P_0 + P_1)(C_{00}P_{00} + C_{10}P_{10}) \\
&= C_{00}P_{00} + C_{10}P_{10} = C_{01}P_{01} + C_{11}P_{11}.
\end{aligned}
$$

That is, the attained average cost is equal the average cost if hypothesis H_0 is true, which is in turn the average cost if hypothesis H_1 is true.

EXAMPLE 5.5 In a binary communication system we choose costs as $C_{00} = C_{11} = 0$, $C_{01} = C_{10} = 1$. The prior probabilities P_0, P_1 are unknown. From Eq. (5.33), the minimax philosophy prescribes building the receiver assuming a value of P_0 such that

$$P(D_1 \mid H_0) = P(D_0 \mid H_1). \tag{5.34}$$

The average probability of error will then be

$$P_e = P(D_1 \mid H_0) = P(D_0 \mid H_1).$$

In the particular case of the system of Example 5.4, in which the two symbols correspond to a Gaussian data number y with mean either 0 or 1, and with unit variance, according to Eq. (5.34) the value of P_0 is taken to be that for which

$$\int_{y_t}^{\infty} (2\pi)^{-1/2} \exp\left(\frac{-u^2}{2}\right) du = \int_{-\infty}^{y_t-1} (2\pi)^{-1/2} \exp\left(\frac{-u^2}{2}\right) du,$$

which yields $y_t = \frac{1}{2}$ as the minimax receiver. Since from Eq. (5.29) we have $y_t = \frac{1}{2} + \ln(P_0/P_1)$, this corresponds to the choice $P_0 = \frac{1}{2}$.

5.7 MULTIPLE MEASUREMENTS

Nothing in the foregoing discussions is restricted to a one-dimensional data element y. If y is multidimensional, the various appearances of y above are simply replaced by a data vector $\mathbf{y}$. Since the likelihood ratio involves the densities $p_0(\mathbf{y})$ and $p_1(\mathbf{y})$, it is necessary to know the joint probabilities among all the elements of $\mathbf{y}$. For example, if the elements of $\mathbf{y}$ arise from multiple performances of the same experiment, it may be reasonable to assume the components y_i of $\mathbf{y}$ are independent. On the other hand, if the elements y_i arise as sequential samples $y(t_i)$ of some random process, we will need to make some assumptions about the statistics of the process $y(t)$. An example will suffice to indicate the procedure.

EXAMPLE 5.6 A constant signal, either 0 or A, is added to a zero-mean stationary Gaussian noise process to form a data waveform $y(t)$. We want to decide which signal is present. The data waveform is sampled over some interval to produce n numbers $y(t_i)$. The power spectrum of the noise is that of low-pass bandlimited white noise, say $S_n(f) = N_0/2$, $|f| < B$, and $S_n(f) = 0$, $|f| > B$. The sampling interval is $\delta t = 1/2B$. In this case, the noise process has correlation function

$$R_n(\tau) = \mathcal{F}^{-1}[S_n(f)] = (N_0 B)\,[\sin(2\pi B\tau)]/(2\pi B\tau).$$

The n noise samples $n_i = n(t_i)$ are therefore zero-mean Gaussian random variables with correlation

$$\mathscr{E} n_i n_j = R_n(\,|i - j|\, \delta t) = (N_0 B)\, \delta_{ij}.$$

That is, they are uncorrelated and with constant variance $\sigma_{n_i}^2 = N_0 B$.

The data samples y_i are therefore uncorrelated random variables with mean either 0 (hypothesis H_0) or A (hypothesis H_1) and with variance $\sigma_n^2 = N_0 B$ under either hypothesis. Since the y_i are Gaussian, they are independent. Then we have

$$p_1(\mathbf{y}) = (2\pi)^{-n/2}\, \sigma_n^{-n} \exp\left[-\sum_{i=1}^{n} \frac{(y_i - A)^2}{2\sigma_n^2} \right],$$

and similarly for $p_0(\mathbf{y})$. The likelihood ratio test is

$$\lambda(\mathbf{y}) = p_1(\mathbf{y})/p_0(\mathbf{y})$$

$$= \exp\left[-\sum_{i=1}^{n} \frac{(y_i - A)^2}{2\sigma_n^2} + \sum_{i=1}^{n} \frac{y_i^2}{2\sigma_n^2} \right] > \lambda_t,$$

where the threshold λ_t will depend on the detector criterion we wish to use. Rearranging this, the test becomes

$$s = \frac{1}{n}\sum_{i=1}^{n} y_i > \frac{\sigma_n^2 [\ln(\lambda_t)]}{An} + \frac{A}{2} = s_t. \tag{5.35}$$

That is, the sample mean of the data is to be thresholded.

The performance of the detector Eq. (5.35) can be presented as a ROC curve as follows. Since the *detection statistic s* is a sum of Gaussian variables (the y_i), it is itself Gaussian. To describe it completely we therefore need only its mean and variance. The mean is

$$\mathscr{E}(s) = \frac{1}{n}\sum_{i=1}^{n} \mathscr{E}(y_i) = 0 \text{ or } A,$$

that is, $\mathscr{E}(s \mid H_0) = 0$, $\mathscr{E}(s \mid H_1) = A$. The variance under the two hypotheses is

$$\sigma_s^2 = \mathscr{E}[(s - \mathscr{E}s)^2 \mid H_i] = \sigma_n^2/n.$$

The false alarm and detection probabilities are

$$P_f = \int_{s_t/\sigma_s}^{\infty} (2\pi)^{-1/2} \exp\left(\frac{-u^2}{2} \right) du,$$

$$P_d = \int_{(s_t - A)/\sigma_s}^{\infty} (2\pi)^{-1/2} \exp\left(\frac{-u^2}{2} \right) du.$$

That is, the performance can be described by the standard ROC curves Figs. 5.5–5.7, with the SNR taken as $\alpha = A/\sigma_s = nA/\sigma_n$. There is a *coherent integration gain* of n, i.e., $20 \log_{10} n$ dB, due to processing of n time samples rather than one.

5.8 MULTIPLE ALTERNATIVE HYPOTHESIS TESTING

There are situations in which more than two alternative hypotheses are of interest concerning the generator of some observed set of data. For example, a communication system may use an alphabet of more than two characters, so that during each character interval we must determine as best we can which of more than two characters constituted the message. In other cases, we may be interested in the value of some parameter which was in effect during generation of the data. That value would normally be quantized into more than two levels, with each level corresponding to a hypothesis about the data source. With this last point of view, problems in hypothesis testing become problems in parameter estimation, a subject we will deal with in some detail in Chapter 10.

In this section, we pursue the multiple (m-ary) hypothesis-testing problem by extending the Bayes strategy developed in Sect. 5.4. That is, we consider m hypotheses, from among which we are to choose one, based on some data set y, which might be multidimensional. Each decision D_i (that hypothesis H_i was in effect) is assigned a cost C_{ij}, which depends on which hypothesis H_j was in fact in effect. Depending on the decision rule, we will make the decisions D_i with probabilities $P(D_i \mid H_j)$ which depend on the hypothesis H_j which was in effect to produce the data. The hypotheses are assumed to have specified *a priori* probabilities P_j of being in effect. Then we seek the decision rule which minimizes the average cost of making decisions. That cost is

$$C = \sum_{i=1}^{m} \sum_{j=1}^{m} C_{ij} P(D_i \mid H_j) P_j. \tag{5.36}$$

Let R_i be the region of the space of the data y in which we will choose decision D_i, that hypothesis H_i is true. Since these decision regions must include the entire data space R (that is, we demand that some decision be made) and they are disjoint regions (each data point y may correspond to only one decision), we can write

$$R = \sum_{i=1}^{m} R_i,$$

$$R_j = R - \sum_{\substack{i=1 \\ i \neq j}}^{m} R_i.$$

Using this, we have

$$P(D_j \mid H_j) = \int_{R_j} p_j(y)dy = \int_R p_j(y)\,dy - \sum_{\substack{i=1 \\ i\neq j}}^{m} \int_{R_i} p_j(y)\,dy.$$

Since the first term of this is unity, the cost Eq. (5.36) can be written

$$C = \sum_{j=1}^{m} \left\{ C_{jj}P_j + \sum_{\substack{i=1 \\ i\neq j}}^{m} (C_{ij} - C_{jj})P_j \int_{R_i} p_j(y)\,dy \right\}.$$

In the sum over i, the term with $i = j$ can be formally included, since its value is zero. Then we can rearrange this as

$$C = \sum_{j=1}^{m} C_{jj}P_j + \sum_{i=1}^{m} \int_{R_i} \left\{ \sum_{j=1}^{m} P_j(C_{ij} - C_{jj})p_j(y)\,dy \right\}. \tag{5.37}$$

Let us assume that the cost of a wrong decision is always more than the cost of a right decision, that is, $C_{ij} > C_{jj}$, $i \neq j$. The first sum in Eq. (5.37) is fixed. We minimize the second sum by allocating each data point y to the region R_i to which it contributes least. That is, we assign any particular y to that region R_i for which the following quantity is a minimum:

$$\lambda_i = \sum_{j=1}^{m} P_j(C_{ij} - C_{jj})p_j(y). \tag{5.38}$$

A special case of Eq. (5.36) is often of interest, that in which correct decisions have zero cost and all incorrect decisions have equal cost, which can be taken to be unity. Then from Eq. (5.38) we choose that hypothesis for which

$$\lambda_i = \sum_{\substack{j=1 \\ j\neq i}}^{m} p_j(y)P_j = \sum_{\substack{j=1 \\ j\neq i}}^{m} p(y, H_j) = p(y) \sum_{\substack{j=1 \\ j\neq i}}^{m} P(H_j \mid y)$$

is minimum. But since $p(y)$ is independent of i, this corresponds to choosing that H_i for which

$$\sum_{\substack{j=1 \\ j\neq i}}^{m} P(H_j \mid y) = 1 - P(H_i \mid y)$$

is minimum, which is to say we choose that H_i for which the *a posteriori* probability $P(H_i \mid y)$ is maximum. That is, the Bayes criterion leads to the same receiver as that which realizes the maximum *a posteriori* probability criterion in this case. In this case the cost Eq. (5.36) is just

$$C = \sum_{j=1}^{m} P_j \sum_{i\neq j}^{m} P(D_i \mid H_j) = \sum_{j=1}^{m} P_j P_e(H_j) = P_e,$$

where $P_e(H_j)$ is the probability of error if H_j is true, and P_e is the total probability of error in the process. That is, the maximum *a posteriori* receiver minimizes the probability of error in this *m*-ary system.

EXAMPLE 5.7   Consider again Example 5.6, but suppose that there are four hypotheses, namely that the mean of the process being observed is 1, 2, 3, or 4, with equal *a priori* probabilities $P_j = \frac{1}{4}$. Then the *a posteriori* probabilities are

$$P(m_j \mid \mathbf{y}) = P_j p_j(\mathbf{y})/p(\mathbf{y}) = \left[\frac{P_j}{p(\mathbf{y})}\right] (2\pi)^{-n/2} \sigma_n^{-n} \exp\left[-\sum_{i=1}^{n} \frac{(y_i - m_j)^2}{2\sigma_n^2}\right].$$

To maximize this we choose that hypothesis m_j for which $p_j(\mathbf{y})$ is maximum, which corresponds to choosing m_j such that the quantity

$$S = \frac{2m_j}{n} \sum_{i=1}^{n} y_i - m_j^2$$

is an algebraic maximum. In our case of $m_j = 1, 2, 3, 4$, we compute the sample mean

$$\bar{y} = \frac{1}{n} \sum_{i=1}^{n} y_i$$

and choose the maximum from among the quantities

$$g(\bar{y}) = 2\bar{y} - 1, 4\bar{y} - 4, 6\bar{y} - 9, 8\bar{y} - 16, \qquad (5.39)$$

corresponding respectively to the decisions that $m = 1, 2, 3, 4$.

Each of the four quantities Eq. (5.39) is a maximum over a particular span of the sample mean $\bar{y}$. Those regions are shown in Fig. 5.10 together with the probability density functions $p_j(\bar{y})$.

5.9 COMPOSITE HYPOTHESIS TESTING WITH MINIMUM COST

In the discussions so far, we have considered only cases in which the various hypotheses of interest were simple. That is, the probability density

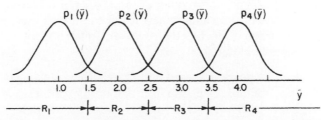

FIGURE 5.10 Conditional density functions and acceptance regions for Example 5.7.

of the data y under every hypothesis H_i was completely specified. However, it is common that this is not the situation. For example, we may be interested in testing H_0, that $p(y)$ is Gaussian with mean 0 and some variance σ_y^2, against the hypothesis H_1 that $p(y)$ is Gaussian with mean 1 and the same variance σ_y^2, but it might be the case that we don't know an appropriate value of σ_y^2 to use. If we want to use a likelihood ratio test, for example, we have to compute $\lambda(y) = p_1(y)/p_0(y)$ for the data y we want to test. This ratio generally involves the unknown σ_y^2, and hence we can't compute its value. Hypotheses such as this H_0, H_1, depending on unknown parameters, are called *composite hypotheses*.

In Chapter 11 we will consider detection strategies which amount to using some of the available data to estimate the unknown parameters in composite hypotheses. In this section and in Sect. 5.11 we will summarize some of the classical procedures used in statistics. The subject of testing general composite hypotheses is not easy. In this book we will treat only some special cases which are particularly applicable to problems in engineering. We introduce the subject using an example.

EXAMPLE 5.8   At a certain instant t_i, a signal $\sin(\omega_c t + \theta)$ is (H_1) or is not (H_0) present in additive zero-mean unit variance Gaussian noise. We want to decide the situation based on a single sample $y(t_i) = y_i$. The detector which minimizes the probability of error is the maximum *a posteriori* detector Eq. (5.26). For equal *a priori* probabilities, the corresponding log likelihood detector is

$$y_i \sin(\omega_c t_i + \theta) > (\tfrac{1}{2}) \sin^2(\omega_c t_i + \theta).$$

If the value of θ is known, this is a problem of the type considered earlier (simple hypotheses). However, if θ is unknown, there is difficulty both in the threshold and with the sign of the coefficient of y_i on the left, since they both depend on θ. The hypothesis H_1 is composite, since the mean of the underlying density is $\sin(\omega_c t + \theta)$, which depends on the unknown parameter θ. This is the common situation in radar and communications, as the carrier frequency ω_c is so high that slight perturbations in the sampling time t_i will result in large changes in $\omega_c t_i$. For example, a radar operating at X-band may have $\omega_c = 2\pi \times 10^9$. A range deviation of $\delta R = 0.1$ m amounts to a two-way timing difference of $\delta t = 2\delta R/c = 0.67$ ns, for a value $\omega_c\, \delta t = 4.2$ rad. Clearly, even if we knew the phase of the signal when it left the transmitter, we may not know it well when it reaches the receiver. In later chapters we will pursue this example in detail.

For definiteness, suppose we have a binary hypothesis-testing problem, with H_0, H_1 the possibilities. Suppose first that both hypotheses are composite; that is, suppose the densities underlying the data under the two hypotheses contain unknown parameters: $H_0 = H_0(\phi)$, $H_1 = H_1(\theta)$. Either or both of the parameter sets ϕ, θ might be multidimensional, and in general there might be some parameters which belong to both sets. In this section we will assume that they do not and that ϕ, θ are sets of distinct variables. We will also assume that we are regarding these as random variables and that we are willing to specify probability densities for them. In the preceding example, in which we are essentially in complete ignorance of the signal phase, we might assume that θ was a random variable with uniform density: $p_1(\theta) = (2\pi)^{-1}$, $0 \le \theta < 2\pi$.

With these assumptions, we can deal with the problem of unknown parameters by using the Bayes criterion of Sect. 5.4. Now the average cost Eq. (5.21), or the risk, of each decision also requires averaging over the parameters ϕ, θ. For generality for the moment we can assume also that the costs C_{ij} depend on the unknown parameters in that C_{00}, C_{10} depend on ϕ and C_{01}, C_{11} depend on θ. Then we have the average cost of decision making as

$$
\begin{aligned}
C &= \mathscr{E}_\phi\{P_0[C_{00}(\phi)P(D_0 \mid H_0(\phi)) + C_{10}(\phi)P(D_1 \mid H_0(\phi))]\} \\
&\quad + \mathscr{E}_\theta\{P_1[C_{01}(\theta)P(D_0 \mid H_1(\theta)) + C_{11}(\theta)P(D_1 \mid H_1(\theta))]\} \\
&= P_0\mathscr{E}_\phi\left\{C_{00}(\phi) + [C_{10}(\phi) - C_{00}(\phi)]\int_{R_1} p_0(y \mid \phi)\, dy]\right\} \\
&\quad + P_1\mathscr{E}_\theta\left\{C_{01}(\theta) + [C_{11}(\theta) - C_{01}(\theta)]\int_{R_1} p_1(y \mid \theta)\, dy]\right\} \\
&= P_0\mathscr{E}_\phi\{C_{00}(\phi)\} + P_1\mathscr{E}_\theta\{C_{01}(\theta)\} \\
&\quad + \int_{R_1} \{P_0\mathscr{E}[(C_{10}(\phi) - C_{00}(\phi))p_0(y \mid \phi)] \\
&\quad\quad + P_1\mathscr{E}[(C_{11}(\theta) - C_{01}(\theta))p_1(y \mid \theta)]\}\, dy.
\end{aligned}
\tag{5.40}
$$

Here we regard R_1 as a nonrandom region of y-space. We want to allocate the points y to R_1 in such a way as to minimize the average cost Eq. (5.40). We again make the reasonable assumption that errors cost more than correct decisions:

$$
C_{10}(\phi) > C_{00}(\phi), \qquad C_{01}(\theta) > C_{11}(\theta).
$$

Then the strategy is to include in R_1 any point for which the integrand is negative, which leads to the criterion of choosing H_1 provided

$$
g(y) = \frac{\mathscr{E}\{[C_{01}(\theta) - C_{11}(\theta)]p_1(y \mid \theta)\}}{\mathscr{E}\{[C_{10}(\phi) - C_{00}(\phi)]p_0(y \mid \phi)\}} > \frac{P_0}{P_1}.
\tag{5.41}
$$

In the common case that the costs are independent of the parameter values ϕ, θ, the Bayes detector Eq. (5.41) is just

$$g(y) = \mathscr{E}p_1(y \mid \theta)/\mathscr{E}p_0(y \mid \phi) > P_0(C_{10} - C_{00})/P_1(C_{01} - C_{11}). \qquad (5.42)$$

Since, for example,

$$\mathscr{E}p_1(y \mid \theta) = \int p_1(y \mid \theta)p_\theta(\theta)\, d\theta = \int p_1(y, \theta)\, d\theta = p_1(y), \qquad (5.43)$$

Eq. (5.42) is just

$$\lambda(y) = p_1(y)/p_0(y) > P_0(C_{10}' - C_{00})/P_1(C_{01} - C_{11}). \qquad *(5.44)$$

This is just the Bayes test Eq. (5.24). That is, in the case of composite hypotheses with parameter-independent costs and with known *a priori* density functions for the parameters in the hypothesis densities, the Bayes test for simple hypotheses is recovered.

In the case of composite hypotheses, the attained probabilities of errors of any test are

$$P_f = P(D_1 \mid H_0) = \int_{R_1} p_0(y \mid \phi)\, dy = P_f(\phi),$$

$$P_d = P(D_1 \mid H_1) = \int_{R_1} p_1(y \mid \theta)\, dy = P_d(\theta). \qquad *(5.45)$$

In the common case that hypothesis H_0 is simple, P_f is a constant, while in general the probability of detection P_d depends on the parameters of the composite hypothesis. The function $P_d(\theta)$ is called the *power function* of the test.

EXAMPLE 5.9   Let us consider testing hypothesis H_0 that a scalar observation y is from a zero-mean Gaussian density with unit variance, against hypothesis H_1, that the mean is some value $m \neq 0$. For specified m, the Bayes detector is that of Eq. (5.24):

$$p_1(y \mid m)/p_0(y) = \exp[-(y - m)^2/2 + y^2/2] > 1, \qquad (5.46)$$

where to be specific we chose costs corresponding to the criterion of minimum probability of error and equal prior probabilities. That is, the test is

$$my > m^2/2. \qquad (5.47)$$

If $m > 0$, this becomes

$$y > m/2,$$

whereas for $m < 0$ the test is

$$y < m/2.$$

It is easy to see that the probability of error of the first kind is the same in both cases:

$$P_f = \int_{|m|/2}^{\infty} (2\pi)^{-1/2} \exp(-y^2/2) \, dy. \tag{5.48}$$

Similarly, in both cases the probability of detection is

$$P_d = \int_{-|m|/2}^{\infty} (2\pi)^{-1/2} \exp(-y^2/2) \, dy. \tag{5.49}$$

Therefore there is no difficulty in constructing a ROC curve, with $|m|$ as the SNR. However, unless we know the sign of m, we don't know which test to apply.

If we have some reason to believe that, for example, the unknown mean is the value of a Gaussian random variable with zero mean and unit variance, then we can apply the Bayes test discussed above. The test itself is that in Eq. (5.44). However, we need to calculate from Eq. (5.43) that

$$p_1(y) = \mathscr{E}_m p_1(y \mid m) = \int_{-\infty}^{\infty} (2\pi)^{-1} \exp\left[\frac{-(y-m)^2}{2} - \frac{m^2}{2}\right] dm$$

$$= (4\pi)^{-1/2} \exp(-y^2/4).$$

The test Eq. (5.44) is then to choose H_1 if

$$\lambda(y) = p_1(y)/p_0(y) = 2^{-1/2} \exp(y^2/4) > 1,$$
$$y^2 > 2 \ln(2) = 1.386 = (1.177)^2, \quad |y| > 1.177. \tag{5.50}$$

This test no longer involves the unknown value of m. Since the test is now different from Eq. (5.46), the error probabilities are no longer given by Eqs. (5.48), (5.49), but rather

$$P_f = 2 \int_{1.177}^{\infty} (2\pi)^{-1/2} \exp\left(\frac{-y^2}{2}\right) dy = 0.239,$$

$$P_d = \left(\int_{1.177-m}^{\infty} + \int_{1.177+m}^{\infty}\right) (2\pi)^{-1/2} \exp\left(\frac{-u^2}{2}\right) du. \tag{5.51}$$

The power P_d now depends on the unknown parameter m. The function $P_d(m)$ is the power function of the test and ranges from $P_d = P_f = 0.239$ at $m = 0$ to $P_d = 1$ at $m = \pm\infty$. The important fact here is that we can actually apply the test Eq. (5.50) without knowing the value of m.

Although the attained P_d, Eq. (5.51), depends on the unknown m, with this Bayes strategy we are guaranteed that on the average, over m, the total error probability is minimum. Since $P_e = P_0 P_f + P_1(1 - P_d)$, in this example we have

$$\text{aver}(P_e) = \frac{P_f}{2} + \frac{1}{2} \int_{-\infty}^{\infty} (2\pi)^{-1/2} [1 - P_d(m)] \exp\left(\frac{-m^2}{2}\right) dm$$

$$= \frac{P_f}{2} + \frac{1}{4\pi} \int_{-\infty}^{\infty} \int_{m-1.177}^{m+1.1177} \exp\left[\frac{-(u^2 + m^2)}{2}\right] dm \, du = 0.417.$$

(The integral is evaluated using a 45° coordinate rotation in the m, u plane.) This is only a little better than guessing in accord with the prior probabilities, using a coin toss.

5.10 SUFFICIENT STATISTICS

In Example 5.6, a multidimensional measurement set y_i, upon which we wanted to base some decision, was replaced in the test procedure by its sample mean. The dimension of the likelihood calculation, and the labor in calculating the test performance, were thereby much reduced. In this section we discuss in what generality such a fortunate circumstance comes about. The discussion involves the idea of a sufficient statistic.

Suppose that we are interested in a random variable y, whose density $p(y \mid \theta)$ depends on a parameter θ. (Either y or θ or both could be multidimensional.) Suppose that we find a function $s(y)$, defining a random variable s, which is such that

$$p_y(y \mid \theta, s) = p_y(y \mid s). \tag{5.52}$$

Then $s(y)$ is called a *sufficient statistic* for θ [Stuart and Ord, 1991, Ch. 17]. Using Eq. (5.52) we have equivalently

$$p_\theta(\theta \mid y, s) = p_y(y \mid \theta, s)p_\theta(\theta \mid s)/p_y(y \mid s) = p_\theta(\theta \mid s). \tag{5.53}$$

This indicates that the information about the parameter θ of interest is carried entirely by the random variable $s(y)$. In processing the data to infer θ, we can just as well process values $s(y)$ as values y. The usefulness of introducing s is that, in many problems, such a function $s(y)$ exists with the dimension of s smaller than that of y.

Using Eq. (5.52), we have

$$p_y(y, s \mid \theta) = p_y(y \mid \theta, s)p_s(s \mid \theta) = p_y(y \mid s)p_s(s \mid \theta).$$

Then, formally using both y and $s(y)$ as data, the likelihood ratio to discriminate between two hypotheses H_0, H_1, that $\theta = \theta_0$ or $\theta = \theta_1$, becomes

$$\lambda(y, s) = p(y, s \mid \theta_1)/p(y, s \mid \theta_0) = p_s(s \mid \theta_1)/p_s(s \mid \theta_0) = \lambda(s).$$

That is, the derived data $s(y)$ can substitute for the measured data y in the hypothesis-testing procedure using a likelihood ratio.

Sometimes a sufficient statistic arises naturally in a problem. In Example 5.6 we sought to decide whether the mean m of a unit variance white Gaussian sequence was 0 or A. The test Eq. (5.35) involved only the sample mean $s = \bar{y}$ of the data y, rather than the individual samples y_i themselves. This indicates that the sample mean is a sufficient statistic s for the sequence means $\theta = 0$ or A.

There is a result which helps in recognizing when a sufficient statistic exists without having to carry through the calculation of the receiver as in Example 5.6. In order that $s(y)$ be a sufficient statistic for the parameter θ of a density $p_y(y \mid \theta)$, it is necessary that the density factor in the form

$$p_y(y \mid \theta) = g[s(y), \theta]h(y). \tag{5.54}$$

If the range of y is independent of θ, condition Eq. (5.54) is also sufficient. In the case of a discrete random variable, the probability of the random variable should factor as in Eq. (5.54) [Stuart and Ord, 1991].

EXAMPLE 5.10 Suppose we make n independent determinations y_i of a discrete random variable governed by the Poisson distribution. We want to make inferences about the parameter μ of the distribution. We have

$$P(y \mid \mu) = \prod_{i=1}^{n} \left[\exp(-\mu)\frac{\mu^{y_i}}{y_i!} \right] = \left[\exp(-n\mu)\, \mu^{\Sigma\, y_i} \right] \left[\frac{1}{\prod_{i=1}^{n} y_i!} \right], \tag{5.55}$$

where the sum in the exponent is also over $1 \le i \le n$. Equation (5.55) indicates a factorization of the form Eq. (5.54), provided we choose

$$s = \sum_{i=1}^{n} y_i.$$

We can implement any likelihood ratio test using $P(s \mid \mu)$ rather than $P(y \mid \mu)$.

The criterion Eq. (5.54) has a useful formulation in terms of densities of a particular form. Specifically, in order that Eq. (5.54) hold, it is necessary that the density $p_y(y \mid \theta)$ be from the *exponential family* of densities. That is, in order that a sufficient statistic $s(y)$ exist for a parameter vector θ of the same dimension it is necessary that [Stuart and Ord, 1991]

$$p_y(y \mid \theta) = C(\theta)h(y) \exp\left[\sum_{i=1}^{n} a_i(\theta)g_i(y) \right]. \qquad *(5.56)$$

Comparing this with Eq. (5.54) it is seen that the functions $g_i(y)$ can be taken as the sufficient statistics $s_i(y)$. Provided that the ranges of the variables y_i do not depend on the θ_j, condition Eq. (5.56) is also sufficient for Eq. (5.52) to hold. Furthermore, the set s_i is the smallest dimension set which is sufficient for the set θ_j. (It is a minimal set of sufficient statistics for θ.)

EXAMPLE 5.11 Suppose that we want to make inferences about the mean m and standard deviation σ of a Gaussian density. The data are n independent samples y_i. We have

$$p(\mathbf{y}) = (2\pi\sigma^2)^{-n/2} \exp\left[-\sum_i (y_i - m)^2/2\sigma^2\right]$$

$$= (2\pi\sigma^2)^{-n/2} \exp\left\{\left[-\sum_i y_i^2 + 2m \sum_i y_i - nm^2\right]/2\sigma^2\right\}. \quad (5.57)$$

This is of the form Eq. (5.56), where we can take specifically

$$C(m, \sigma) = (2\pi\sigma^2)^{-n/2} \exp(-nm^2/2\sigma^2),$$
$$h(\mathbf{y}) = 1,$$
$$a_1(m, \sigma) = -1/2\sigma^2,$$
$$a_2(m, \sigma) = m/\sigma^2,$$
$$g_1(\mathbf{y}) = \sum_i y_i^2,$$
$$g_2(\mathbf{y}) = \sum_i y_i.$$

We can take g_1, g_2 as sufficient statistics, or equivalently the sample mean and variance:

$$s_1 = \bar{y} = g_2/n,$$

$$s_2 = \overline{\sigma^2} = \frac{1}{n} \sum_i (y_i - \bar{y})^2 = \frac{g_1}{n} - \left(\frac{g_2}{n}\right)^2.$$

EXAMPLE 5.12 We wish to discriminate the two situations that data samples y_i are drawn from an uncorrelated Gaussian sequence with mean 1 and variance 1 (H_0) from the case of mean 1 and variance 2 (H_1). From the form of Eq. (5.57), we can conclude that, since the mean $m = 1$ is known, a sufficient statistic for σ^2 is $s = \sum y_i^2 - 2m \sum y_i$ with $a_1(\sigma) = -1/2\sigma^2$, $g_1(y) = s$. Equivalently, a sufficient statistic is

$$s = \frac{1}{n} \sum_{i=1}^n (y_i - m)^2. \quad (5.58)$$

We can see this directly by using the densities:

$$p_0(\mathbf{y}) = (2\pi\sigma_0^2)^{-n/2} \exp\left[-\sum_i (y_i - m)^2/2\sigma_0^2\right],$$

$$p_1(\mathbf{y}) = (2\pi\sigma_1^2)^{-n/2} \exp\left[-\sum_i (y_i - m)^2/2\sigma_1^2\right].$$

With these, the log likelihood ratio test is

$$\ln[p_1(\mathbf{y})/p_0(\mathbf{y})] = -n \ln(\sigma_1/\sigma_0)$$

$$+ \frac{1}{2}\left(\frac{1}{\sigma_0^2} - \frac{1}{\sigma_1^2}\right)\sum_{i=1}^{n}(y_i - m)^2 > t,$$

which rearranges as

$$s = \frac{1}{n}\sum_{i=1}^{n}(y_i - m)^2$$

$$> 2(1/\sigma_0^2 - 1/\sigma_1^2)^{-1}[t/n + \ln(\sigma_1/\sigma_0)] = s_t.$$

Note that s is not the sample variance as usually defined, since we have used the actual known mean of the process rather than the sample mean.

The performance of this test is described as always by

$$P_f = \int_{s_t}^{\infty} p_0(s)\, ds,$$

$$P_d = \int_{s_t}^{\infty} p_1(s)\, ds.$$

In these, $p(s)$ is not Gaussian. Rather, from Eq. (4.73), the quantity $u = ns$ has the chi-squared density with n degrees of freedom, so that

$$p_s(s) = np_u(ns) = n[(2\sigma_y^2)^{n/2}\Gamma(n/2)]^{-1}(ns)^{n/2-1}\exp(-ns/2\sigma_y^2).$$

Then

$$P_f = \int_{s_t}^{\infty} n\left[(2\sigma_0^2)^{n/2}\Gamma\left(\frac{n}{2}\right)\right]^{-1}(ns)^{n/2-1}\exp\left(\frac{-ns}{2\sigma_0^2}\right)ds,$$

$$P_d = \int_{s_t}^{\infty}\left(\frac{\sigma_0}{\sigma_1}\right)^n n\left[(2\sigma_0^2)^{n/2}\Gamma\left(\frac{n}{2}\right)\right]^{-1}$$

$$\times (ns)^{n/2-1}\exp[-(\sigma_0/\sigma_1)^2\, ns/2\sigma_0^2]. \tag{5.59}$$

From this it is clear that an appropriate definition of signal-to-noise ratio, for specified n, is the quantity $\alpha = \sigma_1/\sigma_0$. A ROC curve in terms of α and n could be computed for this problem from Eq. (5.59). However, it is often assumed that n is large enough that the central limit theorem has taken over, in which case the test statistic s itself is assumed to be Gaussian. From Eq. (4.77),

$$\mathscr{E}(s) = \mathscr{E}(u)/n = \sigma_y^2,$$
$$\mathrm{Var}(s) = \mathrm{Var}(u)/n^2 = (2/n)\sigma_y^4.$$

Then approximately

$$P_f = \int_{s_t}^{\infty} \left(\frac{4\pi\sigma_0^4}{n}\right)^{-1/2} \exp\left[\frac{-(s - \sigma_0^2)^2 n}{4\sigma_0^4}\right] ds$$

$$= \int_{u_0}^{\infty} (2\pi)^{-1/2} \exp\left(\frac{-u^2}{2}\right) du,$$

$$P_d = \int_{u_1}^{\infty} (2\pi)^{-1/2} \exp\left(\frac{-u^2}{2}\right) du,$$

where the lower limits are

$$u_{0,1} = (n/2)^{1/2} (s_t/\sigma_{0,1}^2 - 1).$$

From this,

$$u_1 = (\sigma_0/\sigma_1)^2 u_0 - (n/2)^{1/2}(1 - \sigma_0^2/\sigma_1^2).$$

In case H_0 corresponds to noise only, and H_1 corresponds to signal plus independent noise, we have $\sigma_0^2 = \sigma_n^2$, $\sigma_1^2 = \sigma_n^2 + \sigma_s^2$. Then, for $\sigma_s \ll \sigma_n$ (the small-signal assumption), we have

$$u_1 \cong u_0 - (n/2)^{1/2}(\sigma_s/\sigma_n)^2. \tag{5.60}$$

Then the ROC curves in Figs. 5.5–5.7 apply, where, from Eqs. (5.18), (5.60), the SNR is to be taken as

$$\alpha = (n/2)^{1/2}(\sigma_s/\sigma_n)^2. \tag{5.61}$$

5.11 UNIFORMLY MOST POWERFUL TESTS

In Sect. 5.9, we assumed that the parameters in any composite hypothesis were regarded as random variables and that density functions were available for them. In a Bayes philosophy, it was then possible to average them away and obtain computable tests. Sometimes, however, there is no evident *a priori* density available for one or more of the hypotheses. In this section we discuss cases in which we can proceed nonetheless.

Let us first consider more generally the case of Example 5.9. That is, suppose the hypothesis H_0 is simple but that the alternative $H_1(\theta)$ is composite and depends on some parameters θ whose values are not known. To be specific, suppose we want to proceed under the Neyman–Pearson criterion. For any specific value of θ, we could in principle determine a test

$$\lambda(y \mid \theta) = p_1(y \mid \theta)/p_0(y) > \lambda_t. \tag{5.62}$$

The threshold does not depend on θ, because whatever the value of θ might be, the threshold will be set to yield the desired false alarm probability P_f. Since the false alarm probability involves only $p_0(y)$, which does not depend on θ, neither the threshold nor the attained P_f of the test depends on θ.

However, for the test $\lambda(y \mid \theta)$, the detection probability (power) in general will depend on the parameter θ, because $p_1(y \mid \theta)$ is involved: $P_d = P_d(\theta)$. Any test (optimal or not) attains a certain power P_d. Relative to the Neyman–Pearson criterion, the test Eq. (5.62) has the largest power (for the given P_f) and is called the *most powerful* test for the particular values of parameters θ involved. It may happen that, in fact, the test Eq. (5.62) does not depend on the unknown parameters θ. In that case, it is called a *uniformly most powerful* (UMP) test. In that highly desirable situation, we can construct and carry out the test Eq. (5.62) without needing to know the values of the unknown parameters. Fortunately, for many problems of interest with one composite hypothesis, that situation does occur. In this section we give some examples and some general criteria. Usually, the power $P_d(\theta)$ of even a uniformly most powerful test does depend on θ, so we don't necessarily know how well the test is performing. But at least we know it is doing as well as could be done, even if we did have information about θ.

EXAMPLE 5.13   Consider again Example 5.9. Hypothesis H_0 is simple (y is zero mean unit variance Gaussian), but H_1 is composite (y is unit variance Gaussian with unknown mean m). Suppose now that we have at least the information that m is positive. Then the test Eq. (5.47) is the obvious one:

$$y > y_t, \tag{5.63}$$

where now we will use the Neyman–Pearson criterion and set y_t to attain the desired P_f:

$$P_f = \int_{y_t}^{\infty} (2\pi)^{-1/2} \exp\left(\frac{-y^2}{2}\right) dy. \tag{5.64}$$

The test Eq. (5.63) does not involve the unknown m and is therefore uniformly most powerful. The attained power is

$$P_d(m) = \int_{y_t - m}^{\infty} (2\pi)^{-1/2} \exp\left(\frac{-u^2}{2}\right) du. \tag{5.65}$$

In at least one important case, a uniformly most powerful test always exists [Lehmann, 1991]. Suppose that in the case of interest the densities of the data samples depend on a single unknown scalar parameter: $p_y(y) = p_y(y \mid \theta)$. Suppose that the hypotheses are $H_0: \theta \le \gamma$, $H_1: \theta > \gamma$, for some specified γ. Suppose that the density $p_y(y \mid \theta)$ is such that it has a *monotone likelihood ratio*. This last means that there is to exist a scalar function $s(y)$ of the data which is independent of θ and which is such that, for every pair of values θ_0, $\theta_1 > \theta_0$, the likelihood ratio $\lambda(y) = p(y \mid \theta_1)/p(y \mid \theta_0)$ is a nondecreasing function of s. Then thresholding s is equivalent to thresholding λ. Since s does not depend on the unknown parameter θ, the test is uniformly most powerful. That is, if $\lambda(y)$ is monotone in the problem of discriminating H_0 from H_1, the threshold test $s > s_t$ is uniformly most powerful. The function $s(y)$ is a minimal sufficient statistic, and the threshold is determined from $P_f = P_0(s > s_t)$. In the case of a discrete random variable, the result holds true with the densities replaced by probabilities.

EXAMPLE 5.14   [Lehmann, 1991]. A lot of N items contains a number D of defective items, which we want to estimate by sampling the lot. We take $n \le N$ samples and note that $d \le D$ items of the n are defective. It seems reasonable to use d as data to decide whether $D > D_0$, where D_0 is some tolerance level. We hope that a test exists which we can apply without having to know the value of D for the particular lot of

N items. Of the D defective items, there are $\binom{D}{d}$ ways in which we

could have gotten the d defective ones we drew. Since we drew exactly d defectives, we must draw exactly $n - d$ good items, and there are

$\binom{N - D}{n - d}$ ways to do that. We could have drawn our total n items in

$\binom{N}{n}$ ways. Thus, the probability of drawing d defects in a sample of

n is

$$P_d(d \mid D) = \frac{\binom{D}{d}\binom{N - D}{n - d}}{\binom{N}{n}}. \tag{5.66}$$

This is the *hypergeometric distribution,* so called because the probabilities $P_d(d)$ appear as coefficients in the Taylor series of a certain case of the hypergeometric function.

Now let $D = \theta_0$ be any value of the unknown parameter D, and let $D + a = \theta_1$ be any larger value ($a > 0$). We want to show that the resulting likelihood ratio is a monotonically increasing function of d. Hence we consider

$$\lambda(d) = \frac{P_d(d \mid D + a)}{P_d(d \mid D)} = \frac{\binom{D + a}{d}\binom{N - D - a}{n - d}}{\binom{D}{d}\binom{N - D}{n - d}}.$$

Thus we have

$$\lambda(d + 1)/\lambda(d) = [1 + a/(D - d)]/[1 - a/(N - D - n + d + 1)] > 1.$$

Hence $\lambda(d)$ is monotonically increasing, and d itself is an appropriate function $s(d)$. Therefore the test $d > d_t$ is uniformly most powerful for testing for $D > D_0$. The threshold would need to be set from a specified probability P_f of improperly rejecting a lot and from summation of the density Eq. (5.66) using $D = D_0$.

In addition to possessing sufficient statistics, the exponential density Eq. (5.56) has the following desirable property [Lehmann, 1991]. If the density Eq. (5.56) has a single parameter, that is, if the density is of the form

$$p(\mathbf{y} \mid \boldsymbol{\theta}) = C(\theta)h(\mathbf{y}) \exp[a(\theta)s(\mathbf{y})], \tag{5.67}$$

and if $a(\theta)$ is monotone increasing, then the test $s > s_t$ is uniformly most powerful for testing $\theta > \theta_0$. Correspondingly, if $a(\theta)$ is monotone decreasing, then $s < s_t$ is UMP for testing $\theta > \theta_0$.

EXAMPLE 5.15. Consider again Example 5.10, concerning inference about the parameter μ in a Poisson distribution, given n samples y_i. We have Eq. (5.55) as

$$p(\mathbf{y} \mid \mu) = \exp(-n\mu) \left(\prod_i y_i! \right)^{-1} \exp[\ln(\mu) \sum_i y_i].$$

This is of the form Eq. (5.67), where $a(\theta) = \ln(\mu)$ is monotonically increasing with μ. Accordingly, the test

$$s = \sum_{i=1}^{n} y_i > s_t$$

is uniformly most powerful for testing $\mu > \mu_0$. The threshold s_t is set from the distribution of s.

Since the y_i are Poisson variables, the characteristic function of their distribution is

$$\Phi_Y(\omega) = \mathcal{E}[\exp(j\omega k)] = \sum_{k=0}^{\infty} \exp(-\mu) \left(\frac{\mu^k}{k!} \right) \exp(j\omega k)$$

$$= \exp(-\mu) \sum_{k=0}^{\infty} \frac{[\mu \exp(j\omega)]^k}{k!} = \exp\{\mu[\exp(j\omega) - 1]\}. \tag{5.68}$$

Accordingly, the sum s of n independent Poisson variables has characteristic function

$$\Phi_s(\omega) = \exp\{n\mu[\exp(j\omega) - 1]\}, \tag{5.69}$$

so that the test statistic s is Poisson with parameter $n\mu$. Then we set the threshold from

$$P_{\mathrm{f}} = P(s > s_{\mathrm{t}} \mid \mu = \mu_0) = \sum_{s=s_{\mathrm{t}}+1}^{\infty} \exp(-n\mu_0)\frac{(n\mu_0)^s}{s!}.$$

It may be that there is no integer solution of this equation. Then one can simply accept that P_{f} which results from the integer s_{t} coming closest to a solution. More precisely, a randomized test can be used:

If $\quad s < s_{\mathrm{t}}, \quad$ decide H_0;
If $\quad s > s_{\mathrm{t}}, \quad$ decide H_1;
If $\quad s = s_{\mathrm{t}}, \quad$ decide H_1 with probability
β and H_0 with probability $1 - \beta$.

Then the threshold is set by

$$P_{\mathrm{f}} = \sum_{s > s_{\mathrm{t}}+1} \exp(-n\mu_0)\frac{(n\mu_0)^s}{s!} + \beta \exp(-n\mu_0)\frac{(n\mu_0)^{s_{\mathrm{t}}+1}}{(s_{\mathrm{t}} + 1)!}.$$

This is solved by using the value s_{t} closest to the solution without the β term and then choosing β to supply the needed small increment in P_{f}.

5.12 UNKNOWN A PRIORI INFORMATION AND NONOPTIMAL TESTS

We have just seen that, for a wide class of densities [the one-parameter exponential family Eq. (5.67)] and for a one-sided test of the parameter $(\theta > \theta_0)$, a uniformly most powerful test always exists. On the other hand, for a two-sided test, even if the density is single parameter, a UMP test does not usually exist. Specifically, suppose we want to test a single-parameter density for the alternatives H_0: $\theta = \theta_0$ versus the two-sided alternative H_1: $\theta = \theta_1 \neq \theta_0$. No uniformly most powerful test exists in this case [Stuart and Ord, 1991, Ch. 21].

To see that this is true, expand the densities in the likelihood ratio as

$$p(y \mid \theta) = p(y \mid \theta_0) + p'(y \mid \theta_*)(\theta - \theta_0),$$

where θ_* is some unknown value between θ and θ_0. Then the likelihood ratio test becomes

$$\lambda(y) = p(y \mid \theta_1)/p(y \mid \theta_0) = 1 + p'(y \mid \theta_*)(\theta_1 - \theta_0)/p(y \mid \theta_0) > \lambda_{\mathrm{t}}.$$

In this we would like to divide out the term $\theta_1 - \theta_0$ to obtain a test not involving the unknown value θ_1. However, depending on whether $\theta_1 > \theta_0$ or $\theta_1 < \theta_0$, the sense of the inequality reverses, and we don't know what test to apply. That is, there is no UMP test.

EXAMPLE 5.16 Consider again Example 5.9, in which we sought to test the mean m of a Gaussian density with known variance. The alternatives are $m = 0$ and $m \neq 0$. The log likelihood ratio test is

$$\ln[\lambda(y)] = (2my - m^2)/2\sigma^2 > t.$$

If we assume (without information that it is true) that $m > 0$, then we have the test

$$y > (2\sigma^2 t + m^2)/2m = y_t,$$

whereas if we assume $m < 0$, the test is

$$y < y_t.$$

The corresponding detection probabilities (powers) $P(y > y_t \mid m)$, $P(y < y_t \mid m)$ are shown by the solid lines in Fig. 5.11, for the case $P_f = 0.1$. If we guess right about the sign of m we do adequately well, but if we guess wrong we do very badly.

The nonoptimal procedure involving *confidence intervals* to calculate a test is designed to cope with situations such as that of the last example. In that example, it produces a test which is not optimal in either case of the sign of m but which is also not bad whichever case may be true. The test applies to cases of a one-parameter density with H_0: $\theta = \theta_0$, H_1: $\theta \neq \theta_0$. We can phrase this loosely as testing whether the given data y might reasonably have occurred for a value θ "not too far away from" the value θ_0 of interest. If the answer is no, then we choose H_1. To do this, we use the test with acceptance region

$$a_1 \leq s(y) \leq a_2, \tag{5.70}$$

where $s(y)$ is a sufficient statistic for the parameter θ of interest. The performance of the test is described by

$$P_f = \int_{-\infty}^{a_1} p(s \mid \theta_0)\, ds + \int_{a_2}^{\infty} p(s \mid \theta_0)\, ds,$$

with $P_d(\theta)$ being the same expression with θ_0 replaced by the unknown true θ. Although there is no necessity to do so, it is conventional to select

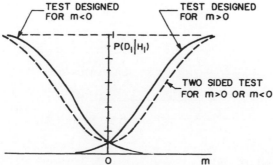

FIGURE 5.11   Performance for tests of the mean with $P(D_1 \mid H_0) = 0.1$ fixed.

a_1, a_2 to be such that the two integrals have equal contribution to P_f. (As we will see below, in fact there are cases in which a better choice can be made.) Figure 5.12 illustrates the difference between a one-sided Neyman–Pearson test and the corresponding two-sided confidence interval test.

EXAMPLE 5.17   Continuing Example 5.9 (testing $m \neq 0$), we consider

$$p(y \mid m = 0) = (2\pi\sigma^2)^{-1/2} \exp(-y^2/2\sigma^2).$$

In this case the data value y is itself a sufficient statistic for the mean. With $a_1 = a_2$, the test Eq. (5.70) is $|y| > y_t$. The threshold for the test is set by

$$P_f = 2 \int_{y_t/\sigma}^{\infty} (2\pi)^{-1/2} \exp\left(\frac{-u^2}{2}\right) du.$$

The power is then computed as

$$P_d(m) = \left(\int_{-\infty}^{(-y_t-m)/\sigma} + \int_{(y_t-m)/\sigma}^{\infty} \right) (2\pi)^{-1/2} \exp\left(\frac{-u^2}{2}\right) du.$$

This latter is plotted as the dashed line in Fig. 5.11, for the same value $P_f = 0.1$ as used for the Neyman–Pearson tests shown by the solid lines. The test Eq. (5.70) always has less power than does the Neyman–Pearson test if we were to guess right about the sign of m. However, it always has greater power than the optimal test if the optimal test was designed using an incorrect assumption about the sign of m.

In standard terminology, the test Eq. (5.70) establishes a *confidence interval* for the parameter θ. With some confidence, using the test we can

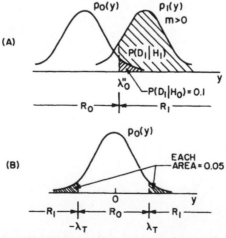

FIGURE 5.12   Likelihood functions. (a) One-sided test for $m > 0$. (b) Two-sided test for $m > 0$ or $m < 0$, with $P(D_1 \mid H_0) = 0.1$.

establish that the unknown value θ is near θ_0 if Eq. (5.70) holds. We are right with probability

$$P(a_1 \leq s \leq a_2 \mid \theta = \theta_0) = P_d = 1 - P_f.$$

The quantity P_f is called the *level of significance* of the test, and $1 - P_f$ is called the *confidence coefficient*.

The test Eq. (5.70) is an example of a class of procedures which are useful in cases in which a uniformly most powerful test does not exist, the *uniformly most powerful unbiased* (UMPU) tests [Lehmann, 1991]. In this context, the term "unbiased" means that, for any value θ of the unknown parameter, the probability of false alarm is less than the probability of detection. Example 5.16 and Fig. 5.11 show that this obviously desirable situation is not necessarily true for a test if there is a composite hypothesis involved. In Fig. 5.11, if the true value of m is of sign opposite to what was assumed in building the test, the probability of detection $P_d = P(D_1 \mid H_1)$ is always less than the false alarm probability $P_f = 0.1$. On the other hand, the confidence interval test in Example 5.17, from which the dashed curve in Fig. 5.11 resulted, is such that P_d is always greater than P_f (or equal, on the boundary). An UMPU test is that one from among the class of tests which are unbiased, in the sense just described, which has the largest power (detection probability) among all such tests with the specified P_f. A general discussion of such tests is lengthy, and we will give only some examples here.

For the exponential family of densities, Eq. (5.56), such UMPU tests often exist if a scalar parameter is of interest. That is, suppose we are interested in the scalar parameter θ and the data $\mathbf{y}$ are described by

$$p(\mathbf{y} \mid \theta, \varphi) = C(\theta, \varphi) \, h(\mathbf{y}) \, \exp[a(\theta)s(\mathbf{y}) + \sum_i t_i(\mathbf{y})b_i(\phi)]. \qquad (5.71)$$

Here φ is a vector of density parameters of no interest to us (*nuisance parameters*), and s and the t_i are sufficient statistics for θ and ϕ. Then a UMPU test exists for a wide variety of hypotheses about the parameter θ.

EXAMPLE 5.18 Consider again Example 5.11, and suppose we are interested in the variance σ^2 of the Gaussian density with unknown (and uninteresting) mean m. We have as before

$$p(\mathbf{y} \mid m, \sigma^2) = (2\pi\sigma^2)^{-n/2} \exp\left[-\sum_i \frac{(y_i - m)^2}{2\sigma^2} \right]$$

$$= (2\pi\sigma^2)^{-n/2} \exp\left(\frac{-nm^2}{2\sigma^2} \right)$$

$$\times \exp\left[-\left(\frac{1}{\sigma^2} \right) \sum_i \frac{y_i^2}{2} + \left(\frac{m}{\sigma^2} \right) \sum_i y_i \right].$$

This is of the form Eq. (5.71), with the identification of $\Sigma\, y_i^2$ as statistic sufficient for $\theta = 1/\sigma^2$ and with m/σ^2 as nuisance parameter. Suppose we want to test $H_0: \sigma^2 = \sigma_0^2$ versus $H_1: \sigma^2 \neq \sigma_0^2$. We are assured there exists a UMPU test, that is, an unbiased test for which $P_d \geq P_f$, regardless of m and such that it has the highest P_d of any such test. However, we need to find what it is.

It happens that the appropriate test is the obvious one: choose H_0 whenever

$$a_1 \leq \sum_{i=1}^{n} y_i^2 \leq a_2. \tag{5.72}$$

The limits a_1, a_2 of the acceptance region Eq. (5.72) can be expected to depend on both σ_0 and on the sample mean $\bar{y}$, but they will not depend on the unknown actual mean m. Figure 5.11 gives a clue to use in searching for these limits. If the test is unbiased, then $P_d(\theta)$ dare not go below P_f for any value of the unknown parameter. But for every value of θ, P_f is constant and fixed at the required level. Furthermore, at the value $\theta = \theta_0$, P_d and P_f are equal. The conclusion is that, as indicated in Fig. 5.11, $P_d(\theta)$ must be a minimum at $\theta = \theta_0$.

In the current example, in order to determine the boundary points a_1, a_2 in Eq. (5.72), we therefore need to investigate the density of the test statistic. If as above we base the test on $\Sigma\, y_i^2$, then the noncentral chi-squared density Eq. (4.87) applies. It is equivalent, and neater, to use instead the test

$$a_1 \leq s = \frac{1}{\sigma_0^2} \sum_{i=1}^{n} (y_i - \bar{y})^2 \leq a_2.$$

Now s has the chi-squared density because $\mathscr{E}(y_i - \bar{y}) = 0$. However, the number of degrees of freedom is $n - 1$, rather than n, because there exists one deterministic relation among the variables, namely $\Sigma(y_i - \bar{y}) = 0$. Therefore the n variables indicated are not independent, and the sum could be expressed in terms of $n - 1$ independent zero-mean variables. Then the requirement

$$P_f = \int_0^{a_1} p(s \mid \sigma_0)\, ds + \int_{a_2}^{\infty} p(s \mid \sigma_0)\, ds \tag{5.73}$$

in principle determines a_2 as a function of a_1, with the known value σ_0 as parameter. Then

$$P_d(\sigma) = \int_0^{a_1} p(s \mid \sigma)\, ds + \int_{a_2(a_1)}^{\infty} p(s \mid \sigma)\, ds.$$

Setting the derivative of this last with respect to σ equal to zero at the point $\sigma = \sigma_0$ in principle finally determines a_1 and hence a_2. Carrying all that through would yield the condition

$$(a_1/a_2)^{(n-1)/2} = \exp[(a_1 - a_2)/2].$$

Solving this together with Eq. (5.73) yields finally a_1 and a_2. For example, with $n = 11$ samples and $P_f = 0.05$ there results $a_1 = 3.52, a_2 = 21.73$.

Another approach to the problem of designing a test when some parameters of the hypotheses are unknown is to assume a distribution for them, as done in Sect. 5.9, but to guard against an unexpectedly large Bayes risk by choosing for them the least favorable distribution. That is, we assume nature will act in such a way that we incur the largest possible average cost of making decisions. We then use that least favorable distribution in the design of the Bayes test. The problem to be solved is determination of that least favorable distribution in any specific problem.

Suppose, for example, that H_0 is a simple hypothesis but that $H_1(\boldsymbol{\theta})$ is composite, depending on a vector of unknown parameters. The Bayes test is that of Eq. (5.44):

$$\lambda(\mathbf{y}) = \mathscr{E}_\theta p_1(\mathbf{y} \mid \boldsymbol{\theta})/p_0(\mathbf{y}) > \lambda_t.$$

If the cost C is the expected error probability, for example, then

$$C = \mathscr{E}_\theta[P(D_1 \mid H_0) + 1 - P(D_1 \mid H_1)].$$

Since H_0 does not depend on $\boldsymbol{\theta}$, the cost C is maximized by choosing $p(\boldsymbol{\theta})$ to minimize the expected power $P(D_1 \mid H_1)$ of the test:

$$\mathscr{E}_\theta P(D_1 \mid H_1) = \int_\theta \left\{ \int_{R_1} p_1(\mathbf{y} \mid \boldsymbol{\theta}) \, d\mathbf{y} \right\} p(\boldsymbol{\theta}) \, d\boldsymbol{\theta},$$

where R_1 is the critical region of the data space, in which we reject H_0. Such a least favorable distribution can be found for many specific problems. It is often clear from context what one should choose. For example, we will often encounter the case in which the phase θ of a received signal is an unknown parameter, in which case selection of the uniform density $p(\theta) = 1/2\pi$ expresses our lack of any information and is clearly the least favorable choice.

We will finally mention the useful alternative of using the data $\mathbf{y}$ to estimate the values of the unknown parameters and then using those estimates in the likelihood ratio of the test of interest. Suppose, for example, both hypotheses in a binary problem are composite: $H_0(\phi), H_1(\theta)$. One class of good estimators of the unknown parameters ϕ, θ is that of the maximum likelihood estimates. Suppose that certain data y are governed by a density $p(y \mid \theta)$. Then the maximum likelihood estimate $\hat{\theta}(y)$ of θ is that value which maximizes the likelihood function $p(y \mid \theta)$. That is, $\hat{\theta}$ is that value of θ for which the data in hand had the highest probability of having occurred:

$$\hat{\theta}(y) \Leftarrow \max_{\theta} p(y \mid \theta).$$ *(5.74)

Either θ or y or both might be multidimensional. Using the maximum likelihood estimate Eq. (5.74) for both hypotheses, the likelihood ratio test becomes

$$\lambda(y) = p_1(y \mid \hat{\theta})/p_0(y \mid \hat{\phi}) > \lambda_t.$$ *(5.75)

The quantity $\lambda(y)$ in Eq. (5.75) is called the *generalized likelihood ratio* for the problem in question. The generalized likelihood ratio test Eq. (5.75) is not necessarily optimum under any criterion but is often not a bad test.

EXAMPLE 5.19   Consider again Example 5.4, but suppose now that n samples are taken of a Gaussian process with mean m and known variance σ^2. We wish to discriminate H_0: $m = 0$ from H_1: $m \neq 0$. The likelihood function for the composite hypothesis H_1 is

$$p(y \mid m) = (2\pi\sigma^2)^{-n/2} \exp[-\textstyle\sum(y_i - m)^2/2\sigma^2].$$

The *likelihood equation* for determining $\hat{m}$ is

$$dp(y \mid m)/dm = \sigma^{-2}(2\pi\sigma^2)^{-n/2} \exp[-\textstyle\sum(y_i - m)^2/2\sigma^2]\textstyle\sum(y_i - m) = 0,$$

leading to

$$\hat{m} = \bar{y} = \frac{1}{n}\sum_{i=1}^{n} y_i,$$ (5.76)

the sample mean. The generalized likelihood ratio Eq. (5.75) is then:

$$\lambda(y) = \exp[-\textstyle\sum(y_i - \bar{y})^2/2\sigma^2 + \textstyle\sum y_i^2/2\sigma^2].$$

This leads to the log likelihood ratio test

$$(2\sigma^2/n) \ln(\lambda) = \bar{y}^2 > t.$$

This is just the two-sided test which would be arrived at by the procedure of Example 5.18.

Exercises

5.1   Using just a single observation, what is the likelihood ratio receiver to choose between the hypotheses that for H_0 the sample is zero-mean Gaussian with variance 1, and for H_1 the sample is zero-mean Gaussian with variance 2.

   (a)   In terms of the observation, what are the decision regions R_0 and R_1?
   (b)   What is the probability of choosing H_1 when H_0 is true?

5.2 Based on N independent samples design a likelihood ratio test to choose between

H_0: $r(t) = n(t)$
H_1: $r(t) = 1 + n(t)$
H_2: $r(t) = -1 + n(t)$

where $n(t)$ is zero-mean Gaussian with variance σ^2. Assume equal *a priori* probabilities for each hypothesis, no cost for correct decisions, and equal costs for any error. Show that the test statistic may be chosen to be $\bar{r} = (1/N) \sum_{i=1}^{N} r_i$. Find the decision regions for $\bar{r}$.

5.3 Design a likelihood ratio test to choose between

$$H_1: \quad p_1(y) = \frac{1}{(2\pi)^{1/2}} \exp\left(-\frac{y^2}{2\sigma^2}\right), \quad -\infty < y < \infty$$

$$H_0: \quad p_0(y) = \begin{cases} \frac{1}{2}, & -1 \leq y \leq 1 \\ 0, & \text{otherwise} \end{cases}$$

(a) Assume that the threshold $\lambda_0 = 1$. In terms of y, and as a function of σ^2, what are the decision regions?

(b) Use a Neyman–Pearson test with $P(D_1 \mid H_0) = \alpha$. What are the decision regions?

5.4 (a) Design a likelihood ratio test to choose between the hypotheses

H_1: The sample x is chi-squared distributed with n degrees of freedom.
H_0: The sample x is chi-squared distributed with two degrees of freedom.

Assume equal *a priori* probabilities and costs of error.

(b) Suppose the number of degrees of freedom in H_1 is a discrete random variable such that

$$P(n = N) = \tfrac{1}{2}, \qquad P(n = M) = \tfrac{1}{2}$$

What is the likelihood ratio test for this case?

5.5 Based on a single observation, use a minimax test to decide between

H_1: $r(t) = n(t)$
H_0: $r(t) = 1 + n(t)$

Assume $n(t)$ is Gaussian with zero mean and average power σ^2. Assume $C_{00} = C_{11} = 0$, $C_{10} = C_{01} = 1$. What is the threshold in terms of the observation? What *a priori* probability for each hypothesis is implied by the solution?

5.6   For the preceding problem, assume $C_{10} = 3$ and $C_{01} = 6$.

   (a)   What *a priori* probability of each hypothesis would limit the maximum possible cost?

   (b)   What is the decision region in terms of the single observation?

5.7   Consider the receiver shown in Fig. 5.13. The hypotheses are

H_1: $s(t) = A \cos(\omega_c t + \theta)$

H_0: $s(t) = 0$

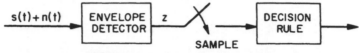

FIGURE 5.13   Single sample receiver.

where A and ω_c are constants. The narrowband Gaussian noise has zero mean and variance σ_n^2.

   (a)   Can the decision rule be implemented by comparing a single sample, z, of the envelope to a threshold? Why?

   (b)   In closed form, what is the probability of choosing H_1 when H_0 is true? Using a Neyman–Pearson test with a fixed value of $P(D_1 \mid H_0)$, show that the threshold value against which z is compared is

$$[-2\sigma_n^2 \ln P(D_1 \mid H_0)]^{1/2}$$

   (c)   Determine the decision rule if M independent samples of z are used.

5.8   Consider the receiver shown in Fig. 5.14. The hypotheses are

H_1: $s(t) = A \cos(\omega_c t + \theta)$

H_0: $s(t) = 0$

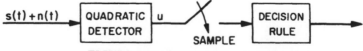

FIGURE 5.14   Single sample receiver.

where A and ω_c are constants. The narrowband Gaussian noise has zero mean and variance σ_n^2.

   (a)   Can the decision rule be implemented by comparing a single sample of u to a threshold? Why? How would the performance of this system compare with that of part (a) of the preceding problem?

(b) Determine the decision rule if M independent samples of u are used.

5.9 For the receiver in Fig. 5.14, assume that the noise is zero-mean Gaussian with variance σ_n^2 and, in addition, the signal is also zero-mean Gaussian with variance σ_s^2. The null hypothesis is noise only.

(a) For a single sample, show that the decision rule can be implemented by comparing z to a threshold.

(b) Show that

$$P(D_1 \mid H_1) = [P(D_1 \mid H_0)]^{\sigma_n^2/(\sigma_n^2 + \sigma_s^2)}$$

(c) Assume that M statistically independent samples of z are used. Show that the likelihood ratio receiver can be implemented by comparing $\lambda_T = \sum_{i=1}^{M} z_i^2$ to a threshold.

(d) Denote this threshold as V_T. Show that

$$P(D_1 \mid H_0) = \int_{V_T/\sigma_n^2}^{\infty} \frac{y^{M-1} e^{-y/2}}{2^M \gamma(M)} \, dy$$

and

$$P(D_1 \mid H_1) = \int_{V_T/(\sigma_n^2 + \sigma_s^2)}^{\infty} \frac{y^{M-1} e^{-y/2}}{2^M \gamma(M)} \, dy$$

5.10 Assume that the four hypotheses to be tested are

H_0: chi-squared distribution with two degrees of freedom
H_1: chi-squared distribution with four degrees of freedom
H_2: chi-squared distribution with six degrees of freedom
H_3: chi-squared distribution with eight degrees of freedom.

Assume equal *a priori* probabilities, no cost for correct decisions, and equal cost for any error.

(a) Based on a single sample x show that the result of a likelihood ratio test is the following: choose

H_0 if $0 \le x < 2$
H_1 if $2 < x < 4$
H_2 if $4 < x < 6$
H_3 if $6 < x$

(Neglect the problem of what to do if x is equal to 2, 4, or 6. It is not important in this example.)

(b) Assume M statistically independent samples x_i, $i = 1, \ldots, M$ are used. Show that the optimum test is the same as in part (a) except that x is replaced by $(\prod_{i=1}^{M} x_i)^{1/M}$.

5.11   Design a likelihood ratio test to choose between the hypotheses

H_0:   The samples are chi-squared distributed with two degrees of freedom

H_1:   The samples are chi-squared distributed with either four, six, or eight degrees of freedom.

Assume M statistically independent samples are used. Assume $P(H_0) = \frac{1}{4}$ and $P(H_1) = \frac{3}{4}$. Furthermore, the *a priori* probability of the degrees of freedom ν is $P(\nu = 4) = P(\nu = 6) = P(\nu = 8) = \frac{1}{3}$.

6 Detection of Known Signals

*I*n the preceding chapter, we surveyed many of the general types of problems with which hypothesis testing, or detection theory, deals. In this chapter we return to the simplest of these, the problem of choosing among simple hypotheses. Our specific aim will be to specify the design of receivers which discriminate noise-corrupted signals from noise only or which distinguish among different known signals in the presence of noise. The signals and the noise will be assumed to have known means and variances. The signal variance may be zero, in which case the signal is a completely specified waveform.

Our primary intent will be to specify the design of optimum receivers. By an optimum receiver, we mean one whose performance, measured in some way, satisfies some mathematically specified criterion of goodness, under some particular set of assumptions about the signals and noise which will be processed. If either the criterion of goodness or the assumptions about the signals and noise change, in general the corresponding receiver will change. If a receiver is designed under one optimality criterion and set of assumptions, it may perform poorly with respect to a different criterion or in the presence of signals or noise which is not in accord with the assumptions made during the receiver design. It is therefore important in any practical design to investigate the robustness of a receiver's performance to deviations from the assumptions used in its design. In any event, the calculated performance of an optimum receiver, under some criterion, provides an indication of the performance which might be expected of a practical receiver.

Although a number of criteria of optimality could be used, we will consider mainly the criterion of minimum probability of error, the most reasonable for a communication system, and the Neyman–Pearson criterion for radar and sonar systems. This latter allows us to specify a maximum probability of false alarm for the system, which can be translated into a rate of false alarms, and to maximize the probability of detection under that restriction. Throughout the chapter, we will be applying the results of Chapter 5, but in the context of specific problems of engineering interest.

6.1 TWO COMPLETELY KNOWN SIGNALS IN ADDITIVE GAUSSIAN NOISE

Suppose that we are to process a received data signal $r(t)$ which consists of one or the other of two completely specified signals $s_0(t)$, $s_1(t)$ in additive zero-mean Gaussian noise. Both of the signals and the noise are bandlimited to $|f| \leq B$. The received data are sampled at the Nyquist rate $f_s = 2B = 1/\delta t$ over some time interval $0 \leq t < T$ to produce $m = T/\delta t = 2BT$ samples $r(t_k) = r_k$, $0 \leq k \leq m - 1$. We will consider these to be arranged as an m-dimensional column vector $\mathbf{r}$.

We now consider the following two hypotheses:

$$H_0: \quad r(t) = s_0(t) + n(t),$$
$$H_1: \quad r(t) = s_1(t) + n(t).$$

Equivalently, in terms of the vector of time samples to be processed, we have

$$H_0: \quad \mathbf{r} = \mathbf{s}_0 + \mathbf{n},$$
$$H_1: \quad \mathbf{r} = \mathbf{s}_1 + \mathbf{n}. \tag{6.1}$$

Since both signals are given time functions, the vectors $\mathbf{s}_0$, $\mathbf{s}_1$ are given numerical vectors. Since $n(t)$ is a zero-mean Gaussian random process, the vector $\mathbf{n}$ is an m-dimensional zero-mean Gaussian random vector. From Eq. (6.1) we have $\mathscr{E}(\mathbf{r} \mid H_i) = \mathbf{s}_i$.

With these assumptions taken into account, the probability density of the data vector $\mathbf{r}$ is completely known for each of the hypotheses H_0, H_1.

We are therefore dealing with the problem of deciding between two simple hypotheses. From Chapter 5, we know that the optimum receiver, under any of the definitions of optimality considered in Chapter 5, should calculate the likelihood ratio and compare it to a threshold. Depending on whether we want to use the maximum *a posteriori* (MAP) criterion of optimality, a Bayes cost criterion, or the Neyman–Pearson criterion, the threshold will change, but the receiver will otherwise be the same. That is, we have the design:

$$\text{Choose } H_1 \text{ if:} \quad \lambda(\mathbf{r}) = p_1(\mathbf{r})/p_0(\mathbf{r}) > \lambda_t, \tag{6.2}$$

where λ_t is some constant determined by whichever criterion we choose to use.

The likelihood ratio follows at once from Eq. (4.4), which specifies the density of the m-dimensional data vector $\mathbf{r}$:

$$p_i(\mathbf{r}) = (2\pi)^{-m/2} [\det(\mathbf{C})]^{-1/2}$$
$$\times \exp[-\tfrac{1}{2}(\mathbf{r} - \mathbf{s}_i)^{\mathrm{T}}\mathbf{C}^{-1}(\mathbf{r} - \mathbf{s}_i)]. \tag{6.3}$$

Then

$$\lambda(\mathbf{r}) = \det(\mathbf{C}_0\mathbf{C}_1^{-1}) \exp[-\tfrac{1}{2}(\mathbf{r} - \mathbf{s}_1)^{\mathrm{T}}\mathbf{C}_1^{-1}(\mathbf{r} - \mathbf{s}_1)$$
$$+ \tfrac{1}{2}(\mathbf{r} - \mathbf{s}_0)^{\mathrm{T}}\mathbf{C}_0^{-1}(\mathbf{r} - \mathbf{s}_0)], \tag{6.4}$$

where the superscript T means transpose and $\mathbf{C}_0$, $\mathbf{C}_1$ are the covariance matrices of $\mathbf{r}$ under the two hypotheses. For these latter we have

$$\mathbf{C}_i = \mathscr{E}(\mathbf{r}\mathbf{r}^{\mathrm{T}} \mid H_i) - [\mathscr{E}(\mathbf{r} \mid H_i)][\mathscr{E}(\mathbf{r} \mid H_i)]^{\mathrm{T}}$$
$$= (\mathbf{N} + \mathbf{s}_i\mathbf{s}_i^{\mathrm{T}}) - \mathbf{s}_i\mathbf{s}_i^{\mathrm{T}} = \mathbf{N},$$

where $\mathbf{N}$ is the covariance of the noise $\mathbf{n}$.

Taking logarithms in Eq. (6.4) and rearranging, the receiver Eq. (6.2) becomes

$$S(\mathbf{r}) = (\mathbf{s}_1 - \mathbf{s}_0)^{\mathrm{T}}\mathbf{N}^{-1}\mathbf{r}$$
$$> \ln(\lambda_t) + \tfrac{1}{2}(\mathbf{s}_1^{\mathrm{T}}\mathbf{N}^{-1}\mathbf{s}_1 - \mathbf{s}_0^{\mathrm{T}}\mathbf{N}^{-1}\mathbf{s}_0) = S_t. \tag{*(6.5)}$$

The scalar quantity $S(\mathbf{r})$ appears as a sufficient statistic for the problem at hand. The receiver computes S from the data $\mathbf{r}$ and compares the result to a threshold, as shown in Fig. 6.1.

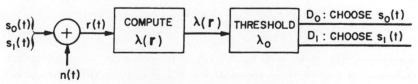

FIGURE 6.1 Optimum receiver.

The performance of the receiver Eq. (6.5) can be expressed in terms of the probability density of the sufficient statistic $S(\mathbf{r})$ under each of the two hypotheses H_0, H_1. Since by Eq. (6.5) S is a linear combination of the Gaussian random variables r_k, $0 \le k < m - 1$, it is itself Gaussian, and we need only its mean and variance to specify its density. From Eq. (6.5) we have

$$\mathscr{E}(S \mid H_i) = (\mathbf{s}_1 - \mathbf{s}_0)^T \mathbf{N}^{-1} \mathscr{E}(\mathbf{r} \mid H_i) = (\mathbf{s}_1 - \mathbf{s}_0)^T \mathbf{N}^{-1} \mathbf{s}_i, \tag{6.6}$$

$$\begin{aligned} \operatorname{Var}(S \mid H_i) = \sigma_S^2 &= \mathscr{E}\{[S - \mathscr{E}(S \mid H_i)]^2 \mid H_i\} \\ &= \mathscr{E}_i[(\mathbf{s}_1 - \mathbf{s}_0)^T \mathbf{N}^{-1}(\mathbf{r} - \mathbf{s}_i)][(\mathbf{s}_1 - \mathbf{s}_0)^T \mathbf{N}^{-1}(\mathbf{r} - \mathbf{s}_i)]^T \\ &= (\mathbf{s}_1 - \mathbf{s}_0)^T \mathbf{N}^{-1}[\mathscr{E}_i(\mathbf{r} - \mathbf{s}_i)(\mathbf{r} - \mathbf{s}_i)^T]\mathbf{N}^{-1}(\mathbf{s}_1 - \mathbf{s}_0) \\ &= (\mathbf{s}_1 - \mathbf{s}_0)^T \mathbf{N}^{-1} \mathbf{N} \mathbf{N}^{-1}(\mathbf{s}_1 - \mathbf{s}_0) = (\Delta\mathbf{s})^T \mathbf{N}^{-1}(\Delta\mathbf{s}). \end{aligned} \tag{6.7}$$

In this last, we write $\Delta\mathbf{s} = \mathbf{s}_1 - \mathbf{s}_0$. With Eqs. (6.6), (6.7), we have

$$p(S \mid H_i) = (2\pi\sigma_S^2)^{-1/2} \exp[-(S - \Delta\mathbf{s}^T \mathbf{N}^{-1}\mathbf{s}_i)^2/2\sigma_S^2].$$

The probabilities of false alarm and of detection of the receiver Eq. (6.5) are

$$P_f = P(S > S_t \mid H_0) = \int_{S_t}^{\infty} p(S \mid H_0)\, dS$$

$$= \int_{(S_t - \Delta\mathbf{s}^T \mathbf{N}^{-1}\mathbf{s}_0)/\sigma_S}^{\infty} (2\pi)^{-1/2} \exp\left(\frac{-u^2}{2}\right) du,$$

$$P_d = P(S > S_t \mid H_1) = \int_{S_t}^{\infty} p(S \mid H_1)\, dS$$

$$= \int_{(S_t - \Delta\mathbf{s}^T \mathbf{N}^{-1}\mathbf{s}_1)/\sigma_S}^{\infty} (2\pi)^{-1/2} \exp\left(\frac{-u^2}{2}\right) du.$$

Letting $U = (S_t - \Delta\mathbf{s}^T \mathbf{N}^{-1}\mathbf{s}_0)/\sigma_S$, these can be written

$$P_f = \int_U^{\infty} (2\pi)^{-1/2} \exp\left(\frac{-u^2}{2}\right) du,$$

$$P_d = \int_{U-\alpha}^{\infty} (2\pi)^{-1/2} \exp\left(\frac{-u^2}{2}\right) du, \tag{6.8}$$

where

$$\begin{aligned} \alpha &= (\Delta\mathbf{s})^T \mathbf{N}^{-1}(\Delta\mathbf{s})/\sigma_S = (\Delta\mathbf{s})^T \mathbf{N}^{-1}(\Delta\mathbf{s})/[(\Delta\mathbf{s})^T \mathbf{N}^{-1}(\Delta\mathbf{s})]^{1/2} \\ &= [(\Delta\mathbf{s})^T \mathbf{N}^{-1}(\Delta\mathbf{s})]^{1/2}. \end{aligned} \tag{*6.9}$$

By comparing Eq. (6.8) with Eq. (5.18), it is seen that the performance of the receiver Eq. (6.5), in terms of P_f, P_d, is described by the curves of Fig. 5.2, with the quantity α of Eq. (6.9) as signal-to-noise ratio.

Equations (6.5), (6.9) are completely general for the case of known signals in additive zero-mean Gaussian noise. Let us now further assume that the noise is white, with constant power spectral density over the band:

$$S_n(f) = N_0/2, \qquad |f| \le B.$$

Then from Eq. (2.45) the correlation function of the noise is

$$R_n(\tau) = \mathcal{F}^{-1}\{S_n(f)\} = \int_{-B}^{B} (N_0/2) \exp(j2\pi f\tau) \, df$$

$$= N_0 B[\sin(2\pi B\tau)]/(2\pi B\tau).$$

With sampling at the Nyquist rate, so that $\delta t = 1/2B$, the correlation coefficient of the noise samples $n(t_k) = n_k$ is

$$\mathcal{E} n_j n_k = R_n(|j - k|\delta t) = N_0 B \delta_{jk},$$

where δ_{jk} is the Kronecker delta. That is, the vector $\mathbf{n}$ in Eq. (6.1) is a zero-mean random vector with correlation matrix

$$\mathbf{N} = (N_0 B)\mathbf{I},$$

where $\mathbf{I}$ is the $m \times m$ unit matrix. The noise samples n_k therefore have variance $\sigma_n^2 = N_0 B$.

In this special case of uncorrelated data samples, the receiver Eq. (6.5) becomes

$$\begin{aligned} S(\mathbf{r}) &= (\mathbf{s}_1 - \mathbf{s}_0)^{\mathrm{T}}\mathbf{r}/\sigma_n^2 = (\Delta\mathbf{s})^{\mathrm{T}}\mathbf{r}/\sigma_n^2 \\ &> \ln(\lambda_t) + (\mathbf{s}_1^{\mathrm{T}}\mathbf{s}_1 - \mathbf{s}_0^{\mathrm{T}}\mathbf{s}_0)/2\sigma_n^2 \\ &= \ln(\lambda_t) + (\|\mathbf{s}_1\|^2 - \|\mathbf{s}_0\|^2)/2\sigma_n^2. \end{aligned} \qquad (6.10)$$

In this last we indicate the length of a vector as $\| \cdots \|$. The signal-to-noise ratio Eq. (6.9) becomes

$$\alpha = [(\Delta\mathbf{s})^{\mathrm{T}}(\Delta\mathbf{s})/\sigma_n^2]^{1/2} = \|\mathbf{s}_1 - \mathbf{s}_0\|/\sigma_n. \qquad (6.11)$$

It is sometimes convenient to consider these developments in the case that the sampling interval $\delta t \Rightarrow 0$. That is, we consider the case of infinite bandwidth. Then the foregoing formulas in terms of sums go over to formulas involving integrals of functions in continuous time. We emphasize that the formulas Eqs. (6.5), (6.9) are general in the case of a Gaussian noise, white or not. However, in the case of uncorrelated noise samples we obtain Eqs. (6.10), (6.11), which involve simple sums that pass over to integrals in the limit of infinite bandwidth.

Specifically, in Eqs. (6.10), (6.11), as a prelude to taking the limit as $\delta t \Rightarrow 0$, let us replace σ_n^2 by $\sigma_n^2 = N_0 B = N_0/2\delta t$. Then we have

$$S(\mathbf{r}) = \frac{2}{N_0} \sum_{k=0}^{m-1} (s_{1k} - s_{0k}) r_k \, \delta t$$

$$> \ln(\lambda_t) + \sum_{k=0}^{m-1} \frac{(s_{1k}^2 - s_{0k}^2) \delta t}{N_0} = S_t, \tag{6.12}$$

$$\alpha^2 = \frac{2}{N_0} \sum_{k=0}^{m-1} (s_{1k} - s_{0k})^2 \, \delta t. \tag{6.13}$$

Now let $\delta t \Rightarrow 0$ while maintaining the observation interval fixed at $0 \le t < T$. Consequently $m = T/\delta t \Rightarrow \infty$. In the limit, from Eq. (6.12) we obtain

$$S = \frac{2}{N_0} \int_0^T [s_1(t) - s_0(t)] r(t) \, dt$$

$$> \ln(\lambda_t) + \frac{1}{N_0} \int_0^T [s_1^2(t) - s_0^2(t)] \, dt = S_t. \tag{*6.14}$$

For neatness in later formulas, we will write this as

$$S' = (N_0/2)S - (1/2) \int_0^T [s_1^2(t) - s_0^2(t)] \, dt$$

$$> (N_0/2) \ln(\lambda_t) = S_t'. \tag{6.15}$$

This is called a *correlation receiver*, because the time cross-correlation Eq. (2.28) of the data waveform $r(t)$ with a receiver template is computed and compared with a threshold. The performance of the receiver Eq. (6.15) is parametrized by the limit of the signal-to-noise ratio (SNR) Eq. (6.13):

$$\alpha^2 = \frac{2}{N_0} \int_0^T [s_1(t) - s_0(t)]^2 \, dt. \tag{6.16}$$

It is revealing to rewrite the SNR Eq. (6.16) as follows:

$$\alpha^2 = \frac{2}{N_0} \int_0^T [s_1^2(t) + s_0^2(t)] \, dt - \frac{4}{N_0} \int_0^T s_0(t) s_1(t) \, dt$$

$$= (4E_{av}/N_0)(1 - \rho).$$

In this last we define

$$E_{av} = \frac{1}{2} \int_0^T [s_1^2(t) + s_0^2(t)] \, dt, \tag{6.17}$$

which is the average energy in the two signals $s_1(t)$, $s_0(t)$, and

$$\rho = \frac{1}{E_{av}} \int_0^T s_0(t) s_1(t) \, dt, \tag{6.18}$$

which is the normalized time cross-correlation between the two signals.

Then the performance of the receiver Eq. (6.15) is described by the curves of Fig. 5.2 parametrized by the SNR

$$\alpha = [4E_{av}(1 - \rho)/N_0]^{1/2}. \tag{6.19}$$

Similarly, the probabilities of false alarm and detection of the receiver Eq. (6.15) are given by Eq. (6.8), where

$$U = (S_t' - \mathscr{E}_0 S')/\sigma_{S'}. \tag{6.20}$$

From Eqs. (6.14), (6.15), for the white noise case we have

$$\mathscr{E}_0(S') = \left(\frac{N_0}{2}\right)\mathscr{E}_0(S) - \frac{1}{2}\int_0^T [s_1^2(t) - s_0^2(t)]\, dt$$

$$= -\frac{1}{2}\int_0^T [s_1(t) - s_0(t)]^2\, dt = -E_{av}(1 - \rho),$$

and similarly

$$\mathscr{E}_1(S') = E_{av}(1 - \rho).$$

Also,

$$\sigma_{S'}^2 = (N_0/2)^2\sigma_S^2 = (N_0/2\sigma_n)^2\, (\Delta s)^T(\Delta s)$$

$$\Rightarrow \frac{N_0}{2}\int_0^T [s_1(t) - s_0(t)]^2\, dt.$$

$$= N_0 E_{av}(1 - \rho). \tag{6.21}$$

With these, the performance of the receiver is described by

$$P_f = \int_U^\infty (2\pi)^{-1/2} \exp\left(\frac{-u^2}{2}\right) du,$$

$$P_d = \int_{U-\alpha}^\infty (2\pi)^{-1/2} \exp\left(\frac{-u^2}{2}\right) du, \tag{6.22}$$

where

$$U = S_t'[N_0 E_{av}(1 - \rho)]^{-1/2} + [E_{av}(1 - \rho)/N_0]^{1/2},$$
$$\alpha = 2[E_{av}(1 - \rho)/N_0]^{1/2}. \tag{*6.23}$$

This last recovers the SNR of Eq. (6.19).

It is important to note that the performance of the correlation receiver Eq. (6.15) depends only on the average energy of the waveforms, their time correlation coefficient, and the noise variance. It is independent of the detailed waveform structure of the signals.

It is easy to see that the time correlation coefficient ρ is bounded as $|\rho| \leq 1$. We have

$$0 \le \int_0^T [s_0(t) \pm s_1(t)]^2 \, dt$$

$$= \int_0^T [s_0^2(t) + s_1^2(t)] \, dt \pm 2 \int_0^T s_1(t) s_0(t) \, dt$$

$$= 2E_{av} \pm 2E_{av}\rho = 2E_{av}(1 \pm \rho).$$

Since $E_{av} > 0$, this is to say that

$$1 \pm \rho \ge 0,$$
$$\pm \rho \ge -1,$$
$$-1 \le \rho \le 1, \qquad |\rho| \le 1.$$

6.2 APPLICATION TO RADAR

In the case of radar (or active sonar), the hypotheses are H_0: no echo of the transmitted waveform is present in a specific time interval of the received signal (a "range bin"), and H_1: there is an echo of the transmitted signal in the specific interval. Then we appropriately take $s_0(t) = 0$ in the developments of the previous section. The signal $s_1(t)$ should be whatever form the transmitted signal takes after passage through the medium to a potential target, reflection, and passage back to the receiver. In the case of radar, it is usual to assume that $s_1(t)$ is just the transmitted signal delayed in time, whereas in sonar it is in general a difficult problem to compute what the received signal due to a target would look like. Fortunately, as noted at the end of the previous section, for performance prediction we are concerned only with the energy in the received signal, the computation of which is much easier.

Accordingly, suppose that, over a segment $0 \le t \le T$ of the received signal $r(t)$ (beginning at the range time in question), we investigate the hypotheses:

$$H_0: r(t) = n(t),$$
$$H_1: r(t) = s(t) + n(t).$$

From Eq. (6.14), the receiver can be taken as

$$S'' = \int_0^T s(t) r(t) \, dt > S_t'', \qquad \qquad *(6.24)$$

where S_t'' is a threshold set in accordance with the optimality criterion of interest. In radar and sonar, that is usually the Neyman–Pearson criterion, so that S_t'' is set to yield whatever false alarm probability is requested.

The performance of the receiver Eq. (6.24) is described by the receiver operating characteristic (ROC) curves of Fig. 5.2. From Eq. (6.17), we have

$$E_{av} = \frac{1}{2} \int_0^T s^2(t) \, dt = \frac{E}{2},$$

where E is the energy of the signal received from a target. Also, from Eq. (6.18) we have $\rho = 0$. The result from Eq. (6.19) is a signal-to-noise ratio

$$\alpha = (2E/N_0)^{1/2}. \qquad\qquad *(6.25)$$

The threshold S''_t in Eq. (6.24) is determined from the probability density of S''. From Eq. (6.24) we have

$$\mathscr{E}(S'' \mid H_0) = 0,$$

$$\mathscr{E}(S'' \mid H_1) = \int_0^T s^2(t) \, dt = E,$$

while from Eq. (6.21)

$$\sigma_{S''}^2 = N_0 E/2.$$

Hence with the Neyman–Pearson criterion the threshold S''_t would be set from

$$P_f = \int_{S''_t(2/N_0 E)^{1/2}}^{\infty} (2\pi)^{-1/2} \exp\left(\frac{-u^2}{2}\right) du. \qquad (6.26)$$

6.3 APPLICATION TO BINARY COMMUNICATIONS

In the case of a binary communication system, we are interested in discriminating between message bits 0 and 1, encoded as the known waveforms $s_0(t)$, $s_1(t)$. It is not necessary, or even usual, that $s_0(t) = 0$, as is the case in radar. Also, the Neyman–Pearson criterion is not appropriate. Rather, a Bayesian criterion or MAP criterion will be used. Specifically, here we will assume that we are interested in minimum probability of error. In that case, from Eq. (5.26), we use the MAP criterion, resulting in the likelihood ratio test Eq. (6.2) as

$$\lambda(\mathbf{r}) > \lambda_t = P_0/P_1, \qquad (6.27)$$

where P_0, P_1 are the prior probabilities of the message bits 0, 1. Usually these are taken equal; however, there is no reason not to carry the more general case forward, and we will do that here.

With the threshold of the likelihood ratio test as in Eq. (6.27), the corresponding continuous time receiver is that of Eq. (6.14):

$$R = \int_0^T [s_1(t) - s_0(t)]r(t)\, dt > \left(\frac{N_0}{2}\right) \ln\left(\frac{P_0}{P_1}\right)$$

$$+ \frac{1}{2}\int_0^T [s_1^2(t) - s_0^2(t)]\, dt = R_t. \tag{6.28}$$

The probabilities of the errors are

$$P(D_1 \mid H_0) = P_f = \int_{\beta+\alpha/2}^{\infty} (2\pi)^{-1/2} \exp\left(\frac{u^2}{2}\right) du,$$

$$P(D_0 \mid H_1) = 1 - P_d = \int_{-\beta+\alpha/2}^{\infty} (2\pi)^{-1/2} \exp\left(\frac{u^2}{2}\right) du, \tag{6.29}$$

where

$$\beta = (1/\alpha) \ln(P_0/P_1) \tag{6.30}$$

and α is the SNR:

$$\alpha = 2[E_{av}(1 - \rho)/N_0]^{1/2} \tag{6.31}$$

With these the total probability of error is

$$P_e = P_0 P(D_1 \mid H_0) + P_1 P(D_0 \mid H_1)$$

$$= \int_{\alpha/2}^{\infty} (2\pi)^{-1/2} \exp\left(\frac{-u^2}{2}\right) du$$

$$+ \left(P_1 \int_{\alpha/2-\beta}^{\alpha/2} - P_0 \int_{\alpha/2}^{\alpha/2+\beta}\right) (2\pi)^{-1/2} \exp\left(\frac{-u^2}{2}\right) du. \tag{6.32}$$

In the usual case that $P_1 = P_0$, the parameter β of Eq. (6.30) vanishes, and the error probability Eq. (6.32) is

$$P_e = \int_{[E_{av}(1-\rho)/N_0]^{1/2}}^{\infty} (2\pi)^{-1/2} \exp\left(\frac{-u^2}{2}\right) du. \qquad *(6.33)$$

Figure 6.2 shows the probability of error Eq. (6.33) as a function of the quantity E_{av}/N_0. This is related to the signal-to-noise ratio Eq. (6.31) through the parameter ρ. The curves of Fig. 6.2 are parametrized by the correlation ρ, and two values are shown: $\rho = 0, -1$. Recall that here E_{av} is the average energy of the two signals $s_0(t)$, $s_1(t)$ and N_0 is the one-sided power spectral density of the white noise, i.e., $S_n(f) = N_0/2$, $|f| < B$. In the particular case of Fig. 6.2, with $P_0 = P_1 = \frac{1}{2}$, from Eq. (6.17) the quantity E_{av} is the stochastic average energy of the received waveform, so that the average power in the received waveform is E_{av}/T. For that reason E_{av}/N_0 is a more useful parameter in this application than is the signal-to-noise ratio α.

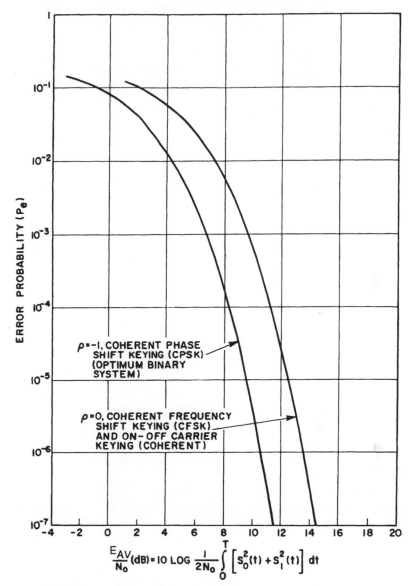

FIGURE 6.2 Error performance for binary communication systems.

On–Off Carrier Keyed System

Suppose that the communication system simply turns the carrier on for a duration T if the message bit is 1 and leaves it off if the message is 0. Then we have

$$s_0(t) = 0,$$
$$s_1(t) = A \cos(\omega_c t + \theta), \qquad 0 \le t \le T.$$

We assume that the amplitude and phase of $s_1(t)$ are known. Because of the known phase angle, the system is said to be *coherent*. In this case, from Eqs. (6.17), (6.18) we have

$$\rho = 0,$$

$$E_{av} = \frac{1}{2} \int_0^T s_1^2(t) \, dt = E/2,$$

where E is the energy of $s_1(t)$. Then from Eq. (6.31) the SNR is

$$\alpha = (4E_{av}/N_0)^{1/2} = (2E/N_0)^{1/2}, \tag{6.34}$$

or

$$\alpha_{dB} = 10 \log(2E/N_0).$$

This is just the SNR of Fig. 5.7, and Fig. 6.2 for $\rho = 0$ is a replotting of the information of Fig. 5.7.

In this particular system, assuming equal prior probabilities of the bits 0, 1, the threshold Eq. (6.28) becomes

$$R_t = E/2,$$

while from Eq. (6.33)

$$P_e = \int_{(E_{av}/N_0)^{1/2}}^{\infty} (2\pi)^{-1/2} \exp\left(\frac{-u^2}{2}\right) du. \tag{6.35}$$

Antipodal Binary System

In this case, we choose signals such that the SNR Eq. (6.31) is maximized for a given average energy. That is, we want to arrange that $\rho = -1$. (Recall that we must have $|\rho| \le 1$.) From Eqs. (6.17), (6.18), this results by the choice $s_1(t) = -s_0(t)$, hence the terminology *antipodal*. This situation is commonly realized in the coherent phase shift keying system by the choice

$$s_0(t) = A \cos(\omega_c t + \theta),$$
$$s_1(t) = -A \cos(\omega_c t + \theta) = A \cos(\omega_c t + \theta + \pi).$$

Again this is a coherent system because we assume the phase angle θ is known, and the phase difference π is attained.

For this system, from Eqs. (6.28), (6.17), (6.31), and (6.33) we have respectively

$$R_t = 0,$$
$$E_{av} \cong A^2 T/2,$$
$$\alpha = (8E_{av}/N_0)^{1/2},$$

$$P_e = \int_{(2E_{av}/N_0)^{1/2}}^{\infty} (2\pi)^{-1/2} \exp\left(\frac{-u^2}{2}\right) du.$$

Figure 6.2 for $\rho = -1$ shows P_e as a function of E_{av}/N_0 for this case. The performance is 3 dB better than that of the on–off system of the previous section.

Coherent Frequency Shift Keying

In this system, the choice is

$$s_0(t) = A \cos(\omega_0 t),$$
$$s_1(t) = A \cos(\omega_1 t), \qquad 0 \le t \le T.$$

In the particular case that $|\omega_1 - \omega_0|$ is an integer multiple of π/T, it is easily seen that $\rho \approx 0$. This system therefore is governed by Fig. 6.2 with $\rho = 0$. For other choices of frequencies it is possible to attain values $\rho < 0$ with performance intermediate to those depicted in the two cases of Fig. 6.2.

6.4 THE LIKELIHOOD FUNCTIONS

The multidimensional Gaussian density Eq. (6.3) provides the starting point for discussion of many detection problems. In the case of a bandlimited white process $r(t)$, with power spectral density $S_r(f) = N_0/2$, $|f| \le B$, sampling at the Nyquist rate $1/\delta t = 2B$ yields independent samples, as indicated in Sec. 6.1. With a fixed time interval of length T and $m = T/\delta t$ samples, the density Eq. (6.3) is then

$$p(\mathbf{r}) = (\delta t/\pi N_0)^{m/2} \exp[-\Sigma(r_i - \bar{r}_i)^2 \, \delta t/N_0],$$

where the sum is over the m samples and $\bar{r}(t) = \mathscr{E}[r(t)]$. Letting $B \Rightarrow \infty$ so that $\delta t \Rightarrow 0$ results in the density

$$p(\mathbf{r}) = A \exp\left\{-(1/N_0) \int_0^T [r(t) - \bar{r}(t)]^2 \, dt\right\}. \qquad *(6.36)$$

In the limit, the constant $A \Rightarrow 0$, which is a consequence of the passage to white noise. Since such a process has correlation function

$$R_n(\tau) = (N_0/2) \, \delta(\tau),$$

each of its samples has infinite variance, and we might expect zero probability of being in any particular small cell of the space of the variables.

Since it is usually ratios of densities such as Eq. (6.36) that are of interest, i.e., likelihood ratios, the constant A cancels. For example, suppose that we are interested in the likelihood ratio for continuous sampling in the problem of discriminating zero-mean white Gaussian noise $n(t)$ from known signal plus noise: $s(t) + n(t)$. From Eq. (6.36) the log likelihood ratio is

$$\ln[\lambda(\mathbf{r})] = \frac{1}{N_0}\left\{2\int_0^T s(t)r(t)\,dt - \int_0^T s^2(t)\,dt\right\}. \qquad (6.37)$$

In another case, we might want to decide which of n known signals $s_i(t)$ was present in additive noise. Using the MAP criterion, we would choose that signal which maximized the probability Eq. (6.36). That is, we would choose that signal which had minimum value of the integrated squared difference

$$\Delta = \int_0^T [r(t) - s_i(t)]^2\,dt.$$

In subsequent calculations, we will use various such ratios of probabilities Eq. (6.36) which individually vanish in the limit. The ratios, however, all have well-defined values. The use of such continuous time sampling is conventional and often convenient in theoretical developments. However, practical calculations will always be carried out in the discrete domain, for which such vanishing probability densities do not arise.

6.5 MATCHED FILTERS

In this section we will reinterpret the correlation receiver of (for example) Eq. (6.14) in terms of the so-called matched filter. We will consider both discrete data and continuous data and both white noise and colored noise. We will also derive the matched filter from a criterion for quality of detection different from those we have considered to this point. We begin with the simplest situation.

Known Signals in White Noise

Suppose we are interested in the radar problem of Sect. 6.2. We seek to determine whether or not a completely specified signal $s(t)$ is present in zero-mean white additive Gaussian noise over a specified interval $0 \leq t \leq T$. From Eq. (6.24) the receiver is to correlate the data $r(t)$ with the known $s(t)$ and threshold the result:

$$S = \int_0^T s(t)r(t)\,dt > S_t. \qquad (6.38)$$

Let us now define a filter by its impulse response $h(t)$ and take

$$h(t) = s(-t), \qquad -T \le t \le 0, \qquad \qquad *(6.39)$$

with $h(t)$ zero elsewhere. This is illustrated in Fig. 6.3. The filter Eq. (6.39) is called the *matched filter* corresponding to the signal $s(t)$. Since the impulse response $h(t)$ is nonzero for negative time, the filter is not physically realizable. (It responds before the impulse happens). This is not of any difficulty, unless true real time operation is required, an unusual situation.

The response of the matched filter Eq. (6.39) to the signal $r(t)$ to be processed over $0 \le t \le T$ is (Fig. 6.3)

$$g(t) = \int_{-\infty}^{\infty} h(t-u)r(u)\,du = \int_{-\infty}^{\infty} s(u-t)r(u)\,du$$

$$= \int_t^{t+T} s(u-t)r(u)\,du.$$

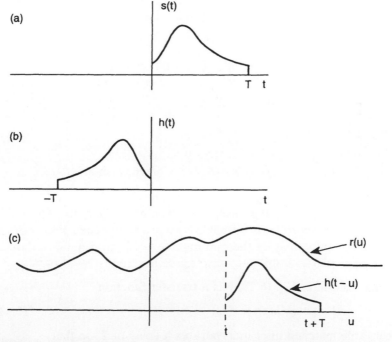

FIGURE 6.3 Illustrating a signal and its matched filter. (a) The signal; (b) the corresponding matched filter impulse response; (c) convolution of the filter response with the signal.

This is such that

$$g(0) = \int_0^T s(u)r(u)\, du = S,$$

where S is the test statistic in Eq. (6.38). That is, the correlation receiver output S results by sampling the matched filter output at the time $t = 0$ when the observation interval begins.

EXAMPLE 6.1   A signal $s(t) = \sin(8\pi t/T)$, $0 \le t \le T$, is to be detected in additive white zero-mean Gaussian noise. We define a running time correlator output as

$$S(t) = \int_{t-T}^t s(u)r(u)\, du.$$

The matched filter has impulse response $h(t) = -\sin(8\pi t/T)$, $-T \le t \le 0$. The matched filter output is

$$g(t) = \int_t^{t+T} \sin\left[\frac{8\pi(u-t)}{T}\right] r(u)\, du.$$

Thereby the output of the matched filter at $t = 0$ equals the output of the running correlator at $t = T$. If the nonrealizability of the matched filter in this case is to be avoided, we can introduce a delay T into the system and define a realizable matched filter as having impulse response

$$h_r(t) = h(t - T) = s(T - t) \qquad 0 \le t \le T. \tag{6.40}$$

With this, the outputs of the filter $h_r(t)$ and the running correlator $S(t)$ are equal at the time $t = T$ at which the observation interval ends. Let the input be the noiseless signal: $r(t) = \sin(8\pi t/T)$, $0 \le t \le T$. The corresponding correlator and realizable matched filter outputs are shown in Fig. 6.4:

$$S(t) = (T - |t'|)/2 + (T/36\pi) \sin(16\pi |t'|/T),$$
$$g_r(t) = \tfrac{1}{2}(T - |t'|) \cos(8\pi |t'|/T) + (T/16\pi) \sin(8\pi |t'|/T),$$

where $t' = t - T$. If the signal were buried in additive zero-mean noise, Fig. 6.4 would portray the ensemble mean of the outputs. The point here is that the correlator output $S(t)$ and the matched filter output $g_r(t)$ are the same only at the one instant $t = T$, in general, although in this particular example they also agree at other isolated points.

The matched filter Eq. (6.39) has a transfer function

$$H(f) = \mathscr{F}\{h(t)\} = \mathscr{F}\{s(-t)\} = S(-f) = S^*(f). \qquad *(6.41)$$

The realizable matched filter incorporates a delay of T, so that

$$H_r(f) = \exp(-j\omega T)H(f) = \exp(-j\omega T)S^*(f).$$

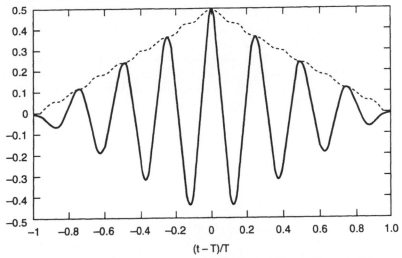

(t – T)/T

FIGURE 6.4 Outputs of correlator and matched filter in Example 6.1.

Maximizing Signal-to-Noise Ratio—White Noise

In various discussions up to here, relative to receiver operating characteristics, we have discussed the signal-to-noise ratio for specific problems. So far, that has been taken as whatever quantity α appears in the expressions, such as Eq. (5.18), which describe the performance of one detection scheme or another. In this section we take another view. Consider a scalar random variable y under two hypotheses, H_0 and H_1. Relative to the two hypotheses, the signal-to-noise ratio of the random variable y is defined as a quantity α, with:

$$\alpha^2 = [\mathcal{E}(y \mid H_1) - \mathcal{E}(y \mid H_0)]^2/\text{Var}(y \mid H_0). \qquad (6.42)$$

The prototypical case is that in which H_0 is zero-mean noise only and H_1 is constant signal A added to the noise. Then

$$\alpha^2 = A^2/\sigma_n^2.$$

This is the ratio of the signal power to the ensemble average noise power, if we consider the variable y to represent a voltage.

Now suppose that we want to work a detection problem, for example, that of constant signal $s(t)$ in additive zero-mean white noise, but in the case that we do not know the probability density of the noise. As we discussed in Chapter 5, we could proceed by seeking a least favorable density for the noise, for example, and using the Neyman–Pearson criterion. It is much more usual, however, to forsake precisely specified optimality criteria and simply assume that, if we can process the data $r(t)$ in such a way as to maximize the signal-to-noise ratio Eq. (6.42), we will make the signal "stand out" well in the noise background, if it is present. This is usually not a bad strategy.

Suppose then that we want to discriminate a known signal $s(t)$ in zero-mean additive white noise $n(t)$ from the case of noise only. For simplicity and general predictability of the process, we will use a linear filter to process the data $r(t)$. Let the impulse response of the filter be $h(t)$. We will allow the filter to be noncausal in general and observe its output at some instant T of interest. That is, our processor forms a filter output

$$g(t) = \int_{-\infty}^{\infty} h(t - u)r(u)\, du, \tag{6.43}$$

and makes a threshold comparison:

$$g(T) > g_t,$$

where g_t is a threshold to be determined. We want to find $h(t)$ such that the random variable $g(T)$ has greatest signal-to-noise ratio.

The signal-to-noise ratio to be maximized is

$$\alpha^2 = \{\mathscr{E}[g(T) \mid H_1] - \mathscr{E}[g(T) \mid H_0]\}^2/\mathrm{Var}[g(T) \mid H_0]. \tag{6.44}$$

Using Eq. (6.43) in this last, with $H_0 : r = n$ and $H_1 : r = s + n$, yields

$$\alpha^2 = \frac{\left[\int_{-\infty}^{\infty} h(T - u)s(u)\, du\right]^2}{\int_{-\infty}^{\infty}\int_{-\infty}^{\infty} h(T - u)h(T - v)\mathscr{E}[n(u)n(v)]\, dv\, du}$$

$$= \frac{(2/N_0)\left[\int_{-\infty}^{\infty} h(T - u)s(u)\, du\right]^2}{\int_{-\infty}^{\infty} h^2(T - u)\, du}. \tag{6.45}$$

In this last, we have used the fact that, for white noise with two-sided power spectral density $N_0/2$, we have

$$\mathscr{E}[n(u)n(v)] = R_n(u - v) = (N_0/2)\, \delta(u - v),$$

where $\delta(u)$ is the Dirac delta function (impulse).

We now need an important result, the *Schwartz inequality*. In its general form, it applies to any two abstract vectors x, y in a linear space with an inner product $\langle x, y \rangle$ defined. The result is that

$$|\langle x, y \rangle|^2 \le \|x\|^2 \|y\|^2, \tag{*6.46}$$

with equality holding if and only if $x = \beta y$, where β is an arbitrary constant. (The squared length $\|x\|^2$ of a vector x is defined as $\langle x, x \rangle$.) Without going too far afield from our current discussion, we will only mention three specific versions of the general relation Eq. (6.46). The vectors x, y might be scalar random variables, in which case the appropriate inner product is

$$\langle x, y \rangle = \mathscr{E}(xy),$$

the (unnormalized) correlation. Again, x, y might be the usual column vectors of real numbers $\mathbf{x}$, $\mathbf{y}$, in which case the inner product is the dot product of analytic geometry: $\langle x, y \rangle = \mathbf{x}^T\mathbf{y}$. Finally, the case at hand, x, y might be real or complex scalar time functions, in which case

$$\langle x, y \rangle = \int_a^b x^*(t) y(t) \, dt,$$

where $[a, b]$ is any convenient time interval. We will prove the relation Eq. (6.46) for this last case, although, with some standard postulates about the properties of abstract vectors and inner products, the proof of the general relation is no harder.

Consider then the relation, valid for arbitrary possibly complex functions $x(t)$, $y(t)$:

$$0 \le \int_a^b |x(t) + \gamma y(t)|^2 \, dt$$

$$= \int_a^b |x(t)|^2 \, dt + 2 \operatorname{Re}\left[\gamma \int_a^b x^*(t) y(t) \, dt\right] + |\gamma|^2 \int_a^b |y(t)|^2 \, dt. \qquad (6.47)$$

This is necessarily true for all complex γ. In particular, for any given $x(t)$, $y(t)$, let us choose $\gamma = A \exp(-j\phi)$, where ϕ is the angle of the complex number $\int x^*y \, dt$. Then we have

$$0 \le \int_a^b |x(t)|^2 \, dt + 2A \left|\int_a^b x^*(t) y(t) \, dt\right| + A^2 \int_a^b |y(t)|^2 \, dt. \qquad (6.48)$$

The right side of this is a quadratic equation in A. If the relation is to be true for all A, then the quadratic must have no real roots, other than possibly a single real root of multiplicity two. That is, the discriminant of the quadratic must be zero or negative. That observation yields just Eq. (6.46) in the form

$$\left|\int_a^b x^*(t) y(t) \, dt\right|^2 \le \int_a^b |x(t)|^2 \, dt \int_a^b |y(t)|^2 \, dt. \qquad *(6.49)$$

This relation obviously holds with equality if $x(t) = \beta y(t)$, for any constant β. On the other hand, if the expression holds with equality then there is a root A_0 of Eq. (6.48) for which $\gamma_0 = A_0 \exp(-j\phi)$ satisfies Eq. (6.47) with equality. Then $x(t) = -\gamma_0 y(t)$, except possibly at discrete points.

We now apply the Schwartz inequality in the form Eq. (6.49) to the SNR Eq. (6.45). We obtain

$$\alpha^2 \le \frac{2}{N_0} \int_{-\infty}^{\infty} s^2(u)\, du = 2E/N_0.$$

Equality holds if $h(T - u) = \gamma s(u)$. Since Eq. (6.45) is homogeneous in γ, we can take $\gamma = 1$ and obtain

$$h(t) = s(T - t). \tag{6.50}$$

In the case that the signal $s(t)$ extends over $0 \le t \le T$, this is just the impulse response of the realizable form of the matched filter, Eq. (6.40). That is, we have determined that the realizable matched filter maximizes the output signal-to-noise ratio Eq. (6.44) in the case of a known signal in additive zero-mean white noise, if we threshold the filter output at the time T at which the signal ends.

Using the matched filter Eq. (6.50), the attained maximum of the SNR is

$$\alpha_{\max}^2 = \frac{2}{N_0} \int_{-\infty}^{\infty} s^2(u)\, du = \frac{2E}{N_0}, \tag{6.51}$$

where E is the energy of the signal $s(t)$. This is the same as the SNR Eq. (6.25) which resulted from a likelihood ratio test in the same problem, with underlying Gaussian noise.

Maximizing Signal-to-Noise Ratio—Colored Noise

We now want to consider the more general case of the problem considered in the preceding section, that in which the noise is not white. That is, we now assume zero mean noise with a general covariance function $R_n(\tau)$ which is not necessarily of the form $(N_0/2)\, \delta(\tau)$. We will proceed by converting the problem to that considered in the preceding section. That is, we whiten the noise and then apply the result Eq. (6.50). The result will be the system of Fig. 6.5.

Let us then seek a filter (perhaps nonrealizable) with impulse response $w(t)$ such that, if its input is noise $n(t)$ with covariance $R_n(\tau)$, then its output is white noise with covariance $A\, \delta(\tau)$. Using Eq. (2.65), we convert this into a problem involving the power spectral densities of the noise.

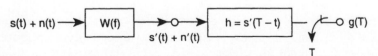

FIGURE 6.5 Noise whitener and matched filter.

That is, we seek a transfer function $W(j\omega)$ such that

$$|W(j\omega)|^2 S_n(f) = A,$$

where $S_n(f) = \mathscr{F}\{R_n(\tau)\}$ is the power spectrum of the input noise. The answer is clearly to choose $W(j\omega)$ such that

$$|W(j\omega)|^2 = A/S_n(f).$$

In this we might as well take $A = 1$, since scale factor is irrelevant to the ratio of signal power to noise power at the filter output. Note that the phase spectrum of the filter $W(j\omega)$ is arbitrary.

With one assumption, the solution of the problem of finding $W(j\omega)$ is immediate. We assume that the noise power spectrum $S_n(f)$ is a rational function, that is, a ratio of finite polynomials in the variable f. This is no real restriction in practice, because we can approximate any given spectrum by a rational spectrum with as small error as we want. However, the order of the rational spectrum may grow large. Furthermore, for analytical calculations it may not be convenient to make such an approximation. In a later section we will consider a procedure for working this problem without the assumption of a rational spectrum.

In any event, since the covariance function $R_n(\tau)$ is necessarily an even function of τ, we must have $S_n(f) = S_n(-f)$. Furthermore, since $R_n(\tau)$ is real, $S_n(s)$ is necessarily a real function, so the roots of its numerator and denominator polynomials, i.e., its poles and zeros, must occur in conjugate pairs in the complex plane $s = \sigma + j\omega$. Taking these properties together, we conclude that the poles and zeros of $S_n(s)$ must have the quadrantal symmetry illustrated in Fig. 6.6.

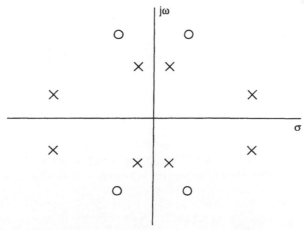

FIGURE 6.6 Locations of poles and zeros of power spectral density, showing quadrantal symmetry.

It is now clear how to "pull apart" $1/S_n(f) = |W(j\omega)|^2$ to find $W(j\omega)$. Since the impulse response $w(t)$ should be real, we want to have $W(-j\omega) = W^*(j\omega)$. Then we want $1/S_n(s) = W(-s)W(s)$. We can therefore take any pole s_{pi} in Fig. 6.6 and associate it with $W(s)$ and assign its mirror image in the origin, i.e., $-s_{pi}$, to $W(-s)$. The same procedure applies to the zeros. If we want $W(s)$ to be physically realizable, then we assign the left plane zeros of $S_n(s)$ to $W(s)$. If we further want $W(s)$ to be minimum phase (that is, at any frequency its phase shift is to be the least among all filters with the same amplitude characteristic), then we assign the left plane poles to $W(s)$. The noise-whitening filter is then obtained.

At the output of the noise-whitening filter in Fig. 6.5, the noise $n'(t)$ is white, with spectrum $S_{n'}(f) = 1$. Therefore, we apply the results of the preceding section with $N_0 = 2$. The filter $h(t)$ in Fig. 6.5 is now matched to the signal $s(t)$ as distorted by the whitening filter, i.e., to $s'(t)$. The result [using the realizable filter Eq. (6.40)] is

$$H(j\omega) = W(-j\omega)S(-j\omega)\exp(-j\omega T), \qquad (6.52)$$

since in the time domain we reverse the signal $w(t)*s(t)$ and delay the result by T. The cascade of the noise whitener with the filter Eq. (6.52) results in an overall system function for nonwhite noise

$$H_{nw}(j\omega) = W(j\omega)H(j\omega) = W(j\omega)W(-j\omega)S(-j\omega)\exp(-j\omega T)$$
$$= |W(j\omega)|^2 S(-j\omega)\exp(-j\omega T) = S(-j\omega)\exp(-j\omega T)/S_n(f).$$
$$(6.53)$$

EXAMPLE 6.2   A signal $s(t) = \exp(-2t)$, $0 \le t \le 4$, is buried in zero mean noise with covariance function $R_n(\tau) = 2\exp(-|\tau|)\cos(2\tau) + \exp(-3|\tau|)$. (Note that it is not easy to tell in advance that this is a legitimate covariance function. It is, because its spectrum happens to be positive for every frequency f.) We seek the filter Eq. (6.53). The noise spectrum is

$$S_n(s) = \frac{-4s^2 + 20}{(s^2 + 2s + 5)(s^2 - 2s + 5)} - \frac{6}{(s-3)(s+3)}$$
$$= \frac{10s^4 - 20s^2 + 330}{(s^4 + 6s^2 + 25)(-s^2 + 9)}.$$

This has poles and zeros with quadrantal symmetry: $s_p = \pm(-1 \pm j2)$, ± 3; $s_z = \pm(1.836 \pm j1.540)$. Should we want it for some additional purpose, the minimum phase causal whitening filter transfer function is ($s = j\omega$)

$$W(j\omega) = (10)^{-1/2}\frac{(s^2 + 2s + 5)(s + 3)}{s^2 + 3.672s + 5.742}.$$

As might be expected, this involves differentiation to decorrelate the noise. With $S(-s) = [-1/(s-2)]\{1 - \exp[4(s-2)]\}$, the system function $H_{nw}(j\omega)$ follows at once. In Fig. 6.7 we sketch the various time functions involved in this example. It is worth noting that, even though the signal is of limited time extent and we delay the matched filter by the appropriate amount, the filter $h_{nw}(t)$ is still noncausal, i.e., $h_{nw}(t)$ does not vanish for $t < 0$, because $w(t)$ does not vanish for $t > T$.

Maximizing SNR—Colored Noise and Finite Data Span

In the preceding section, we indicated how to find a filter which maximizes signal-to-noise ratio for the problem of a known signal in colored noise. The result was the filter $H_{nw}(j\omega)$ of Eq. (6.53). That filter, however, has a corresponding impulse response $h_{nw}(t)$ which is nonzero over $-\infty \leq t < 0$, even though the whitening filter $w(t)$ itself is causal by construction. Therefore, we need to process an infinite span of data $r(t)$, even though the sought signal $s(t)$ might be nonzero only over some finite range $a \leq t \leq b$. This is because the noise outside that interval tells us something

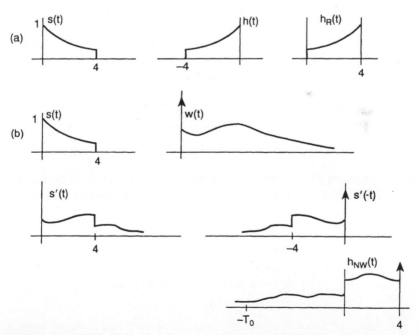

FIGURE 6.7 Matched filter impulse responses: (a) Realizable filter response $h_R(t)$ for finite-length signal in white noise; (b) nonrealizable filter $h_{nw}(t)$ with use of noise-whitening filter $w(t)$ of Fig. 6.5.

about the noise inside the interval, since the noise is not white. We must process noise either into the infinite future (causal whitening filter, Fig. 6.7b) or into the infinite past [were we to choose the noncausal whitening filter $w(-t)$]. This is because the noise spectrum $S_n(f)$ is assumed rational, so that the corresponding correlation function $R_n(\tau)$ is nonzero for all τ.

In the (usual) case of non-real-time processing, we have access to all the data, both future and past. Then the question of causality (physical realizability) disappears, and we have just the problem that we can process only a finite amount of data. In practice, the signal $s(t)$ of Fig. 6.7b always has some finite end point T ($T = 4$ in the figure). Then the signal $s'(t)$ at the output of the whitening filter falls off to zero as $t \Rightarrow \infty$, and we simply truncate it at some appropriately large time. Then the system response $h_{\text{nw}}(t)$ is at least of finite span.

With real-time processing, however, we can process only past and present data. Then it is essential that $h_{\text{nw}}(t)$ vanish for $t < 0$. With the choice of whitening filter we made in the preceding section, as in Fig. 6.7b the impulse response $h_{\text{nw}}(t)$ decays to zero as $t \Rightarrow -\infty$. We can therefore truncate $h_{\text{nw}}(t)$ at some appropriate time $-T_0$ and introduce additional delay T_0. The truncated and delayed response $h_{\text{nw}}(t - T_0)$ then vanishes for $t < 0$. The penalty is that we now attain maximum signal-to-noise ratio at a time T_0 later than the time T when the signal turns off, but at least we have an acceptable approximation to the solution.

With another approach, it is sometimes possible to calculate an exact solution to the problem of real-time processing. We build into the optimization problem the constraint that we may process data only over some span $a \leq t \leq b$ including the signal $s(t)$. In this approach, we return to the starting point, Eq. (6.43), and consider only filters such that

$$g(T) = \int_a^b h(T - u)r(u)\, du,$$

where T is some time at which we want the SNR to be maximum. The quantity to be maximized, in the general case of nonwhite noise, rather than Eq. (6.45) is now

$$\alpha^2 = \frac{\left[\int_a^b h(T - u)s(u)\, du\right]^2}{\int_a^b \int_a^b h(T - u)h(T - v)R_n(u - v)\, du\, dv}. \tag{6.54}$$

Here we have assumed that the noise is stationary and that we seek a stationary filter.

Whereas Eq. (6.45) involved only inner products, allowing use of the Schwartz inequality, the denominator of Eq. (6.54) involves a *quadratic form*. In general, if x, y are abstract vectors in a linear space and if $A(\cdot)$

is an operator, converting a vector into a vector, then the inner product $\langle x, A(y) \rangle$ is the quadratic form based on the operator A. In terms of vectors which are columns of n real numbers, the manifestation of this is just $\mathbf{x}^T A \mathbf{y}$, where A is a specified $n \times n$ matrix. In terms of time functions, the denominator of Eq. (6.54) is the quadratic form based on $R_n(t, t')$ with the abstract vectors being time functions. Although there is a powerful analogy here, its development and exploitation would take us farther afield than we want to go at the moment. Therefore, we will work the problem Eq. (6.54) by another method, the *calculus of variations*.

It is no more effort to work the general problem of nonstationary noise and a possibly nonstationary filter. Then the quantity Eq. (6.54) becomes

$$\alpha^2 = \frac{\left[\int_a^b h(T, u)s(u) \, du \right]^2}{\int_a^b \int_a^b h(T, u)h(T, v)R_n(u, v) \, du \, dv}. \tag{6.55}$$

We will minimize this by constraining the numerator to be constant while minimizing the denominator. The constraint is adjoined to the quantity to be minimized by a Lagrange multiplier μ, so that we consider the quantity

$$Q = \int_a^b \int_a^b h(T, u)h(T, v)R_n(u, v) \, du \, dv - \mu \int_a^b h(T, u)s(u) \, du. \tag{6.56}$$

Now we hypothesize that a function $h_0(T, t)$ minimizing Q exists, and write the general $h(T, t)$ as

$$h(T, t) = h_0(T, t) + \varepsilon f(T, t), \tag{6.57}$$

where $f(T, t)$ is an arbitrary function.

We now seek a necessary condition that must be satisfied by $h_0(T, t)$. If $h_0(T, t)$ is the function that minimizes Q, then Q, considered as a function of ε, must have a stationary point for $\varepsilon = 0$; i.e., we must have

$$\partial Q(\varepsilon)/\partial \varepsilon|_{\varepsilon=0} = 0.$$

Substituting Eq. (6.57) into Eq. (6.56) and differentiating, this last yields the condition

$$\int_a^b \left[\int_a^b h_0(T, v)R_n(v, u) \, dv - (\mu/2)s(u) \right] f(T, u) \, du = 0.$$

Now the function $f(T, u)$ is entirely arbitrary. In particular, we could take it to be a very narrow pulse, of width δu and height $1/\delta u$, centered at any arbitrary value $u = u_0$, with $a \leq u_0 \leq b$. (That is, we use an impulse at u_0.) Then the last equation yields the necessary condition

$$\int_a^b h_0(T, v)R_n(v, u_0) \, dv = (\mu/2)s(u_0), \qquad a \leq u_0 \leq b.$$

Finally we change the arbitrary u_0 in this last to an arbitrary u, scale the optimal $h_0(T, v)$ by the constant $\mu/2$ (since SNR is independent of a constant multiplier in the filter), and drop the subscript from the optimal function. The result is the necessary condition for an optimal filter, maximizing SNR at time T:

$$\int_a^b h(T, v)R_n(v, u)\, dv = s(u), \qquad a \le u \le b. \qquad *(6.58)$$

This is the *Wiener-Hopf equation*.

The general solution of Eq. (6.58) is not easy, and we will postpone that discussion until Chapter 9. Here we will only note the special case of a stationary noise process, so that $R_n(v, u) = R_n(v - u)$, and a stationary filter, so that $h(T, v) = h(T - v)$. With the changes of variable $v' = T - v$, $u' = T - u$, after erasing the primes Eq. (6.58) becomes

$$\int_{T-b}^{T-a} R_n(u - v)h(v)\, dv = s(T - u), \qquad T - b \le u \le T - a. \qquad (6.59)$$

Now finally letting the observation time become doubly infinite, so that $a \Rightarrow -\infty$, $b \Rightarrow \infty$, Eq. (6.59) becomes

$$\int_{-\infty}^{\infty} R_n(u - v)h(v)\, dv = s(T - u), \qquad -\infty \le u \le \infty.$$

We can take the Fourier transform of both sides of this. Recognizing the convolution on the left, that yields

$$S_n(f)H(j\omega) = S(-j\omega)\exp(-j\omega T),$$

where $S_n(f)$ is the power spectrum of the colored noise and $S(j\omega)$ is the transform of the signal $s(t)$. The optimal filter transfer function is then

$$H(j\omega) = [S(-j\omega)/S_n(f)]\exp(-j\omega T), \qquad (6.60)$$

which agrees with the earlier result Eq. (6.53). In the further case of white noise and $T = 0$, this is

$$H(j\omega) = (2/N_0)S(-j\omega),$$

so that

$$h(t) = (2/N_0)s(-t).$$

This recovers Eq. (6.39), except for the constant $2/N_0$, which we retain here for use in the next calculation.

Finally, it is easy to calculate the (maximum) SNR attained by the filter specified in the condition Eq. (6.58). Using Eq. (6.58) in the denominator of Eq. (6.55), and recalling that $R_n(v, u) = R_n(u, v)$, yields

$$\alpha_{max}^2 = \frac{\left[\int_a^b h(T, u)s(u)\, du \right]^2}{\int_a^b h(T, u)s(u)\, du}$$

$$= \int_a^b h(T, u)s(u)\, du. \tag{6.61}$$

In the case of infinite observation interval and white noise, with a stationary filter and $T = 0$ this becomes

$$\alpha_{max}^2 = \int_{-\infty}^{\infty} h(-u)s(u)\, du = \frac{2}{N_0} \int_{-\infty}^{\infty} s^2(u)\, du = \frac{2E}{N_0}, \tag{6.62}$$

where E is the signal energy. This recovers Eq. (6.51).

The general expression for the attained α_{max}^2 of Eq. (6.61) in terms of $s(t)$ and $R_n(u, v)$ is found in Chapter 9. In the special case of a stationary noise process and doubly infinite observation interval, we can use *Parseval's relation*. For real functions $f_1(t)$, $f_2(t)$ that is

$$\int_{-\infty}^{\infty} f_1(t)f_2(t)\, dt = \int_{-\infty}^{\infty} F_1(j\omega)F_2(-j\omega)\, df.$$

In this, take

$$f_1(t) \Rightarrow h(T - u),$$
$$f_2(t) \Rightarrow s(u),$$

so that, from Eq. (6.60),

$$F_1(f) = H(-j\omega) \exp(-j\omega T) = S(j\omega)/S_n(-f) = S(j\omega)/S_n(f).$$

Then

$$\alpha_{max}^2 = \int_{-\infty}^{\infty} \left\{ \frac{|S(j\omega)|^2}{S_n(f)} \right\} df. \tag{6.63}$$

With $S_n(f) = N_0/2$, this again recovers Eq. (6.62).

6.6 THE GENERAL DISCRETE MATCHED FILTER

In this section we return to the hypotheses

$$H_1: \quad \mathbf{r} = \mathbf{s}_1 + \mathbf{n},$$
$$H_0: \quad \mathbf{r} = \mathbf{s}_0 + \mathbf{n}.$$

Here $\mathbf{s}_1$, $\mathbf{s}_0$ are specified m-dimensional column vectors of real numbers and $\mathbf{n}$ is a zero-mean random vector. Without loss of generality we will take s_0 to be zero and write s for s_1. In the case that the noise is Gaussian, in general not white, Eq. (6.5) shows the filter which realizes the likelihood ratio test between the two hypotheses, and Eq. (6.9) gives the attained signal-to-noise ratio to be used with a general Gaussian receiver operating characteristic such as Fig. 5.7.

In this section we will not assume a density for the noise, but only that its covariance matrix $C_n = \mathscr{E}(nn')$ is given. (For generality in this section, we will use the notation that a prime indicates conjugate transposition.) We will seek a linear operator H which maximizes the signal-to-noise ratio of the filter output. We take H to be an $m \times m$ matrix, and the filter output to be

$$g = Hr. \tag{6.64}$$

This is analogous to the complete time history of the output of a continuous time nonstationary filter. We will specify a particular instant at which the SNR is to be maximum by specifying a vector e_i which has unity in its ith place and zeros elsewhere.

With these arrangements, the squared SNR to be maximized by choice of the matrix H is

$$\alpha^2 = |\mathscr{E}(e_i'Hr \mid H_1) - \mathscr{E}(e_i'Hr \mid H_0)|^2/\mathrm{Var}(e_i'Hr \mid H_0)$$
$$= |h's|^2/h'C_nh. \tag{6.65}$$

In the last step we have written $H'e_i = h$, and we will seek h.

We first assume that the noise sequence is white, with $C_n = \sigma_n^2 I$, where I is the $m \times m$ unit matrix. Then Eq. (6.65) becomes

$$\alpha^2 = |h's|^2/\sigma_n^2 h'h. \tag{6.66}$$

The Schwartz inequality Eq. (6.46) shows that the maximum value of Eq. (6.66) is $\|s\|^2/\sigma_n^2$, which is attained by the choice

$$h = s. \tag{6.67}$$

This is the matched filter in this discrete version of the problem of maximizing SNR in the case of white noise.

Note in Eq. (6.67) that the vector h is not in any sense a time-reversed (reindexed) version of the signal s. This is because the application of the filter to the data is a simple data weighting, carried out by the inner produce $g = h'r$. In earlier sections, in discussing the continuous time formulations, we sought an impulse response $h(t)$. That is applied to the data by a convolution, which involves reversing the impulse response before using it to weight the data samples. That reversal is anticipated by taking the impulse response to be the reversal of the signal, so that the weights applied to the data are the signal samples themselves.

In order to solve the general problem Eq. (6.65), let us suppose that we can find a (square) nonsingular matrix $\mathbf{A}$ such that

$$\mathbf{C}_n = \mathbf{A}\mathbf{A}'. \tag{6.68}$$

Were we to succeed in doing that, then the problem Eq. (6.65) becomes that of maximizing

$$\alpha^2 = |\mathbf{h}'\mathbf{s}|^2/\mathbf{h}'\mathbf{A}\mathbf{A}'\mathbf{h} = |\mathbf{h}'\mathbf{A}\mathbf{A}^{-1}\mathbf{s}|^2/\mathbf{h}'\mathbf{A}\mathbf{A}'\mathbf{h}.$$

The Schwartz inequality Eq. (6.46) shows us that the maximum value of α^2 is attained by

$$\mathbf{A}^{-1}\mathbf{s} = \mathbf{A}'\mathbf{h},$$

and is

$$\alpha^2_{\max} = \|\mathbf{A}^{-1}\mathbf{s}\|^2.$$

The optimal filter is

$$\mathbf{h} = (\mathbf{A}')^{-1}\mathbf{A}^{-1}\mathbf{s} = (\mathbf{A}\mathbf{A}')^{-1}\mathbf{s} = \mathbf{C}_n^{-1}\mathbf{s}. \tag{6.69}$$

This processor results in the output of the optimal filter being

$$g = \mathbf{h}'\mathbf{r} = \mathbf{s}'\mathbf{C}_n^{-1}\mathbf{r}. \tag{6.70}$$

This is the matched filter and noise whitener. The attained maximum SNR is then

$$\alpha^2_{\max} = \mathbf{s}'\mathbf{C}_n^{-1}\mathbf{s}. \tag{6.71}$$

That Eq. (6.70) is a matched filter and whitener follows from the observation that the operator $\mathbf{A}^{-1}$ applied to the noise $\mathbf{n}$ results in an output $\mathbf{f}$ which is white. We have

$$\begin{aligned}
\mathscr{E}(\mathbf{f}\mathbf{f}') &= \mathscr{E}[\mathbf{A}^{-1}\mathbf{n}\mathbf{n}'(\mathbf{A}^{-1})'] = \mathbf{A}^{-1}\mathscr{E}(\mathbf{n}\mathbf{n}')(\mathbf{A}^{-1})' \\
&= \mathbf{A}^{-1}\mathbf{C}_n(\mathbf{A}^{-1})' = \mathbf{A}^{-1}\mathbf{A}\mathbf{A}'(\mathbf{A}')^{-1} = \mathbf{I}.
\end{aligned}$$

Passing the signal $\mathbf{s}$ through this filter $\mathbf{A}^{-1}$ yields a filtered signal $\mathbf{A}^{-1}\mathbf{s}$ in white noise. The corresponding matched filter is $\mathbf{h} = \mathbf{A}^{-1}\mathbf{s}$, so that the cascade of the whitener with the matched filter is an operator with output

$$g = \mathbf{s}'(\mathbf{A}^{-1})'\mathbf{A}^{-1}\mathbf{r} = \mathbf{s}'(\mathbf{A}\mathbf{A}')^{-1}\mathbf{r} = \mathbf{s}'\mathbf{C}_n^{-1}\mathbf{r}.$$

We now want to consider the factorization Eq. (6.68). Such a nonsingular $\mathbf{A}$ always exists, although it is not unique. We will indicate two possibilities, one involving the eigenvalues and eigenvectors of $\mathbf{C}_n$ and the other using essentially the *Gram–Schmidt procedure* for orthogonalizing a set of independent vectors. Recall that the matrix $\mathbf{C}_n$ is nonsingular $[\det(\mathbf{C}_n) \neq 0]$, provided there is no deterministic relation among the components n_i of the noise vector $\mathbf{n}$, which we assume. Because, for any vector $\mathbf{v}$, we have

$$\mathbf{v}'\mathbf{C}_n\mathbf{v} = \mathbf{v}'(\mathscr{E}\mathbf{nn}')\mathbf{v} = \mathscr{E}(|\mathbf{v}'\mathbf{n}|^2) \geq 0,$$

the matrix $\mathbf{C}_n$ is in fact positive definite. Since it is symmetric, it has n orthonormal eigenvectors and all its eigenvalues λ_i are positive numbers (Sect. 4.1). We can write

$$\mathbf{C}_n = \sum_{i=1}^{n} \lambda_i \mathbf{t}_i \mathbf{t}_i',$$

where the $\mathbf{t}_i$ are the (orthonormal, possibly complex) eigenvectors corresponding to the real λ_i. Now take

$$\mathbf{A} = \sum_{i=1}^{n} (\lambda_i)^{1/2} \mathbf{t}_i \mathbf{t}_i'. \tag{6.72}$$

Then $\mathbf{A}$ is also nonsingular. (All its eigenvalues $\lambda_i^{1/2}$ are nonzero and in fact positive.) Also,

$$\mathbf{A}\mathbf{A}' = \sum_{i=1}^{n}\sum_{j=1}^{n} (\lambda_i\lambda_j)^{1/2} \mathbf{t}_i (\mathbf{t}_i'\mathbf{t}_j)\mathbf{t}_j'$$

$$= \sum_{i=1}^{n} \lambda_i \mathbf{t}_i \mathbf{t}_i' = \mathbf{C}_n. \tag{6.73}$$

Here we have used the orthonormality of the eigenvectors: $\mathbf{t}_i'\mathbf{t}_j = \delta_{ij}$. With the construction Eq. (6.72), $\mathbf{A}$ is symmetric. Then we have $\mathbf{C}_n = \mathbf{A}^2$, so that $\mathbf{A}$ is the positive square root of the matrix $\mathbf{C}_n$.

Equation (6.72) can be written

$$\mathbf{A} = \mathbf{T}\boldsymbol{\Lambda}^{1/2}\mathbf{T}', \tag{6.74}$$

where $\mathbf{T}$ is the matrix whose columns are the eigenvectors $\mathbf{t}_i$ of $\mathbf{C}_n$ and $\boldsymbol{\Lambda}$ is the diagonal matrix of the corresponding eigenvalues. In the whitening transformation

$$\mathbf{f} = \mathbf{A}^{-1}\mathbf{n},$$

we may then take

$$\mathbf{A}^{-1} = \mathbf{T}\boldsymbol{\Lambda}^{-1/2}\mathbf{T}', \tag{6.75}$$

using the fact that $\mathbf{T}$ is *unitary* (*orthogonal*, in the case of a real matrix), so that $\mathbf{T}^{-1} = \mathbf{T}'$.

More directly, writing Eq. (6.73) as

$$\mathbf{A}\mathbf{A}' = \mathbf{T}\boldsymbol{\Lambda}\mathbf{T}' = \mathbf{C}_n$$

indicates that we may also take

$$\mathbf{A} = \mathbf{T}\mathbf{\Lambda}^{1/2},$$
$$\mathbf{A}^{-1} = \mathbf{\Lambda}^{-1/2}\mathbf{T}'. \tag{6.76}$$

In this case, we have another (nonsymmetric) solution for the whitening transformation. In Sect. 9.1 we discuss the solution Eq. (6.76) as the *singular value decomposition* of the matrix $\mathbf{C}$.

The matrix Eq. (6.72) corresponds to a noncausal system. Suppose we think of the elements r_i of the data vector $\mathbf{r}$ as ordered time samples. Then any element y_j of the output $\mathbf{A}^{-1}\mathbf{r}$ of the whitening filter depends on all elements of $\mathbf{r}$. That is, the output at time j of the filter depends on inputs for $i > j$. The filter $\mathbf{A}^{-1}$ is noncausal. There is no particular reason to object to this, except in the rare cases that true real-time processing is required. Nonetheless, it is a circumstance that can be avoided by another solution $\mathbf{A}$ of Eq. (6.68). If we require a causal filter, in the sense just described, we must have

$$y_j = \sum_{i=1}^{n} (\mathbf{A}^{-1})_{ji} r_i = \sum_{i=1}^{j} (\mathbf{A}^{-1})_{ji} r_i. \tag{6.77}$$

That is, the matrix $\mathbf{A}^{-1}$ must be lower triangular. Therefore, $\mathbf{A}$ must also be lower triangular.

There is a standard procedure in matrix analysis called *Cholesky decomposition*. It applies to any nonsingular matrix, such as $\mathbf{C}_n$, and produces a lower triangular matrix $\mathbf{A}$ such that Eq. (6.68) holds. That is, the solution of our problem of a causal whitening transformation is provided by a standard algorithm of numerical analysis. It is of independent interest to show how such a decomposition comes about. We will do that in terms of the *Gram–Schmidt orthonormalization procedure* for a set of abstract vectors. In our particular application, the "vectors" will be zero-mean random variables.

In general, suppose we have a set of linearly independent vectors s_i which we want to transform to an orthonormal set t_j. Let the s_i be written as columns of a nonsingular matrix $\mathbf{S}$ as

$$\mathbf{S} = [s_1, \ldots, s_n].$$

We seek new vectors t_j which are orthonormal and related to the s_i by an upper triangular transformation matrix $\mathbf{U}$:

$$\mathbf{T} = [t_1, \ldots, t_n] = \mathbf{S}\mathbf{U},$$

that is,

$$\mathbf{T}' = \begin{bmatrix} t_1' \\ \vdots \\ t_n' \end{bmatrix} = \mathbf{U}'\mathbf{S}' = \mathbf{U}' \begin{bmatrix} s_1' \\ \vdots \\ s_n' \end{bmatrix},$$

or

$$\mathbf{t}_j = \sum_{i=1}^{j} u_{ij}\mathbf{s}_i. \tag{6.78}$$

We proceed as follows. We have

$$\mathbf{t}_1 = u_{11}\mathbf{s}_1.$$

We choose the constant u_{11} such that

$$1 = \mathbf{t}_1'\mathbf{t}_1 = u_{11}^2\mathbf{s}_1'\mathbf{s}_1,$$
$$u_{11} = 1/\|\mathbf{s}_1\|.$$

Now we must have

$$\mathbf{t}_2 = u_{12}\mathbf{s}_1 + u_{22}\mathbf{s}_2.$$

This is to be orthogonal to $\mathbf{t}_1$. From Fig. 6.8, we must have

$$\mathbf{t}_2 = c_2[\mathbf{s}_2 - (\mathbf{t}_1'\mathbf{s}_2)\mathbf{t}_1],$$

where c_2 is whatever constant makes $\mathbf{t}_2$ have unit length, that is,

$$c_2 = 1/\|\mathbf{s}_2 - (\mathbf{t}_1'\mathbf{s}_2)\mathbf{t}_1\|.$$

Since the columns of $\mathbf{S}$ are independent, the denominator of this last cannot vanish.

The process is clear. We construct

$$\mathbf{t}_j = c_j\left[\mathbf{s}_j - \sum_{i=1}^{j-1}(\mathbf{t}_i'\mathbf{s}_j)\mathbf{t}_i\right],$$

where c_j is the inverse of the length of the vector in brackets. Then $\mathbf{t}_j$ is of unit length, and

$$\mathbf{t}_j'\mathbf{t}_{k<j} = c_j\left[\mathbf{s}_j'\mathbf{t}_{k<j} - \sum_{i=1}^{j-1}(\mathbf{s}_j'\mathbf{t}_i)\mathbf{t}_i'\mathbf{t}_{k<j}\right]$$
$$= c_j(\mathbf{s}_j'\mathbf{t}_k - \mathbf{s}_j'\mathbf{t}_k) = 0.$$

The last step follows because, by construction, the $\mathbf{t}_i$ form an orthonormal set for $i < j$. This is the Gram–Schmidt procedure for construction of an orthonormal set of vectors from a linearly independent set.

In the particular application of interest at the moment, we consider the vectors $\mathbf{s}_i$, $\mathbf{t}_j$ above to be zero-mean random variables s_i, t_j. The inner products such as $\mathbf{s}_i'\mathbf{t}_j$ become correlations $\mathscr{E}(s_it_j)$. A squared length is a variance: $\mathbf{s}_i'\mathbf{s}_i = \mathscr{E}(s_i^2) = \sigma_s^2$. Then Eq. (6.78) becomes

$$t_j = \sum_{i=1}^{j} u_{ij}s_i,$$

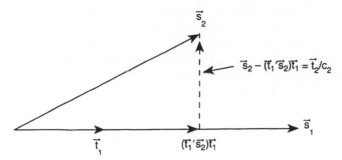

FIGURE 6.8 Construction of a vector t_2 perpendicular to t_1.

or, in terms of the random variables of Eq. (6.77),

$$y_j = \sum_{i=1}^{j} u_{ij} r_i. \tag{6.79}$$

In terms of the column vectors $\mathbf{y}$, $\mathbf{r}$ of the random variables y_j, r_i, Eq. (6.79) becomes

$$\mathbf{y} = \mathbf{U}'\mathbf{r}. \tag{6.80}$$

By construction, $\mathbf{U}$ is upper triangular, so that $\mathbf{U}'$ is lower triangular and the filter is causal. Comparing Eqs. (6.77), (6.80), we then have

$$\mathbf{U}' = \mathbf{A}^{-1} \tag{6.81}$$

as the whitening transformation. In Sect. 9.1 we discuss this solution using the LDU *decomposition* of $\mathbf{C}_n$. The factorization of the noise covariance is

$$\mathbf{C}_n = \mathbf{A}\mathbf{A}' = (\mathbf{U}^{-1})'(\mathbf{U}^{-1}), \tag{6.82}$$

where $\mathbf{A}$ is lower triangular.

6.7 AN m-ARY COMMUNICATION SYSTEM

In Sect. 6.3 we considered the Gaussian binary detection problem in the context of a binary communication system, with bits 0 or 1 encoded as signals $s_0(t)$ or $s_1(t)$. More generally, we can consider an alphabet of m symbols encoded using m completely specified signals $s_i(t)$. Just one symbol is transmitted in each symbol interval and is received as one of the $s_i(t)$ in additive zero-mean Gaussian noise. To be specific, we will assume that the signals at the receiver are orthogonal with equal energies:

$$\rho_{ij} = \int_0^T s_i(t) s_j(t) \, dt = E\delta_{ij}. \tag{6.83}$$

For any linearly independent set of signals, it can always be arranged that Eq. (6.83) is true. Figure 6.9 shows an example of such an orthogonal set.

If we assume the message characters to have equal *a priori* probabilities and if we take as quality criterion minimization of the probability of error, then Sect. 5.8 prescribes that we choose as the message that corresponding to the signal $s_j(t)$ with maximum probability $P[r(t) \mid s_i(t)]$, where $r(t)$ is the received data waveform:

$$s_j(t) \Leftarrow \max_i P[r(t) \mid s_i(t)].$$

Assuming white noise with spectral density $N_0/2$, since

$$r(t) = s_i(t) + n(t),$$

the likelihood function Eq. (6.36) indicates that we should choose that $s_j(t)$ for which

$$j \Leftarrow \min_i \int_0^T [r(t) - s_i(t)]^2 \, dt.$$

That is,

$$j \Leftarrow \min_i \left[\int_0^T r^2(t) \, dt - 2 \int_0^T s_i(t) r(t) \, dt + \int_0^T s_i^2(t) \, dt \right].$$

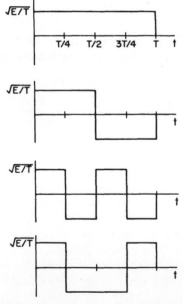

FIGURE 6.9   Example of a set of orthogonal signals.

Since the integral of $s_i^2(t)$ is the same for each $s_i(t)$, we carry out the processing

$$j \Leftarrow \max_i \int_0^T s_i(t)r(t)\,dt. \qquad *(6.84)$$

This last indicates that we should set up a bank of correlation receivers and choose that message corresponding to the receiver with greatest sampled output at the end of the observation interval. Equivalently, using the results of Sect. 6.5, we can use a bank of matched filters, as sketched in Fig. 6.10.

We now want to calculate the probability of error P_e attained by the optimal receiver Eq. (6.84). We need the probabilities $P(D_j \mid H_i)$ of deciding that symbol H_j was sent when in fact H_i was sent. Since $r(t)$ is Gaussian, the quantities being searched over in Eq. (6.84) are Gaussian random variables. Letting them be g_i, we have

$$\mathscr{E}(g_i \mid H_k) = \int_0^T s_i(t)s_k(t)\,dt = E\delta_{ik}, \qquad (6.85)$$

in this special case of orthogonal signals. Furthermore,

$$\mathscr{E}(g_i g_j \mid H_k) = \mathscr{E}\left[\int_0^T s_i(t)[s_k(t) + n(t)]\,dt\right]$$

$$\times \left[\int_0^T s_j(t)[s_k(t) + n(t)]\,dt\right]$$

$$= E^2\delta_{ik}\delta_{jk} + \int_0^T\int_0^T s_i(t)s_j(t')\mathscr{E}[n(t)n(t')]\,dt'\,dt$$

$$= E^2\delta_{ik}\delta_{jk} + (N_0/2)\int_0^T s_i(t)s_j(t)\,dt$$

$$= E^2\delta_{ik}\delta_{jk} + E\delta_{ij}N_0/2. \qquad (6.86)$$

In this last we used the assumption that the noise is white with spectral density $N_0/2$.

Using Eq. (6.85) with Eq. (6.86), we have

$$\mathscr{E}_k\{[g_i - \mathscr{E}_k(g_i)][g_j - \mathscr{E}_k(g_j)]\}$$
$$= E^2\delta_{ik}\delta_{jk} + E\delta_{ij}N_0/2 - E^2\delta_{ik}\delta_{jk} = E\delta_{ij}N_0/2.$$

That is, under every hypothesis H_k, the variables g_i are independent. Also, $\text{Var}(g_i) = EN_0/2$ is independent of hypothesis.

Hence the detector chooses the largest among m independent Gaussian variables with means Eqs. (6.85) and common variance. We make an error

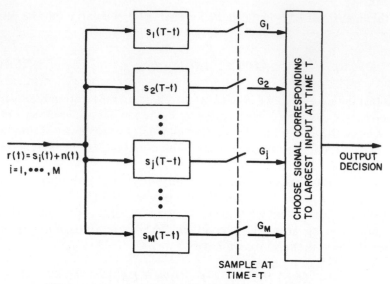

FIGURE 6.10. Matched filter implementation of *m*-ary system.

if the one we choose is not that corresponding to the sent message. If the sent message is H_k, the probability of being correct in our choice is

$$1 - P_e = \int_{-\infty}^{\infty} P\{g_{i \neq k} < g_k \mid H_k\} p(g_k \mid H_k)\, dg_k$$

$$= \int_{-\infty}^{\infty} \left\{ \prod_{i \neq k} P(g_i < g_k \mid H_k) \right\} p(g_k \mid H_k)\, dg_k$$

$$= \int_{-\infty}^{\infty} (2\pi \sigma_g^2)^{-1/2} \exp \left[\frac{-(g_k - E)^2}{2\sigma_g^2} \right]$$

$$\times \left[\int_{-\infty}^{g_k} (2\pi \sigma_g^2)^{-1/2} \exp \left[\frac{-g_i^2}{2\sigma_g^2} \right] dg_i \right]^{m-1} dg_k,$$

where σ_g^2 is the common variance of the decision variables and we take note from Eq. (6.85) that $\mathscr{E}(g_{i \neq k} \mid H_k) = 0$. Then

$$P_e = 1 - \int_{-\infty}^{\infty} (2\pi)^{-1/2} \exp \left(\frac{-u^2}{2} \right)$$

$$\times \left[\int_{-\infty}^{u + (2E/N_0)^{1/2}} (2\pi)^{-1/2} \exp \left(\frac{-v^2}{2} \right) dv \right]^{m-1} du. \quad (6.87)$$

From this last it is clear that $(2E/N_0)^{1/2}$ is an appropriate signal-to-noise ratio for this problem. However, to conform to the usual usage in discussions of communication systems, we will use $(E/N_0)^{1/2}$.

For normalization, it is convenient to express the SNR in terms of the power SNR per bit of information transmitted by the m-ary system in each character interval. With m equally probable signals there are associated $\log_2(m)$ information bits. As discussed in Sect. 7.6, the character error probability P_e can be expressed on a per-bit basis. Figure 6.11 shows that quantity for various m plotted as a function of the per bit SNR, which is

$$\gamma_{b,dB} = 10 \log_{10}[(E/N_0)/\log_2(m)].$$

6.8 THE GENERAL DISCRETE GAUSSIAN PROBLEM

Throughout this chapter to this point we have assumed that the signals among which we wish to discriminate were deterministic and fully specified, but were received in noise. In this section we return to the general binary Gaussian case of deciding between two hypotheses

$$H_1: \quad \mathbf{r} = \mathbf{s} + \mathbf{n},$$
$$H_0: \quad \mathbf{r} = \mathbf{n},$$

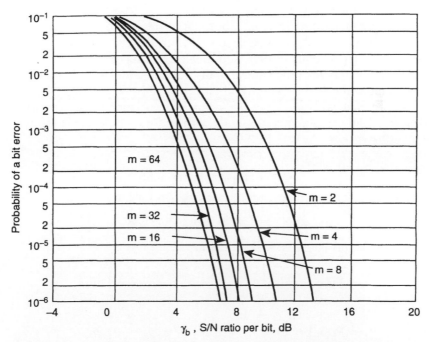

FIGURE 6.11 Probability of bit error for coherent detection of orthogonal signals. (From J. G. Proakis, "Digital Communications", McGraw-Hill, 1963, by permission.)

in the case that both **s** and **n** are Gaussian vectors. We assume zero-mean noise and a signal **s** with mean **m**. The covariance matrices $\mathbf{S} = \mathscr{E}(\mathbf{s} - \mathbf{m})(\mathbf{s} - \mathbf{m})^T$ and $\mathbf{N} = \mathscr{E}\,\mathbf{nn}^T$ are assumed to be known. We take the signal and noise to be independent.

Equation (6.4) is the likelihood ratio for this general binary Gaussian problem. We take $\mathbf{C}_1 = \mathbf{S} + \mathbf{N}$, $\mathbf{C}_0 = \mathbf{N}$, and obtain

$$\ln[\lambda(\mathbf{r})] = (-\tfrac{1}{2})(\mathbf{r} - \mathbf{m})^T(\mathbf{S} + \mathbf{N})^{-1}(\mathbf{r} - \mathbf{m}) + \tfrac{1}{2}\mathbf{r}^T\mathbf{N}^{-1}\mathbf{r}$$
$$= \tfrac{1}{2}\mathbf{r}^T[\mathbf{N}^{-1} - (\mathbf{S} + \mathbf{N})^{-1}]\mathbf{r} + \mathbf{m}^T(\mathbf{S} + \mathbf{N})^{-1}\mathbf{r} + \text{const.}$$

Dropping the constant, which in any event would be incorporated in the threshold, we obtain the receiver

$$S(\mathbf{r}) = \tfrac{1}{2}\mathbf{r}^T[\mathbf{N}^{-1} - (\mathbf{S} + \mathbf{N})^{-1}]\mathbf{r} + \mathbf{m}^T(\mathbf{S} + \mathbf{N})^{-1}\mathbf{r} > S_t, \qquad *(6.88)$$

where S_t is the threshold appropriate to the optimality criterion selected.

Suppose that we know the signal under hypothesis H_1. Then we can model it as a random vector, with mean **m**, but covariance $\mathbf{S} = \mathbf{0}$, the zero matrix. That is, we now have a deterministic situation. Then the receiver Eq. (6.88) recovers the matched filter Eq. (6.5) in the current case of zero-mean additive noise. On the other hand, the general expression Eq. (6.88) allows us to modulate from certain knowledge of the signal ($\mathbf{S} = \mathbf{0}$) to a model in which the signal **s** is a zero-mean random process, just as is the noise, but with different covariance.

The performance of the receiver Eq. (6.88) is described in terms of the probability densities of the sufficient statistic $S(\mathbf{r})$ under the two hypotheses. Because $S(\mathbf{r})$ is a quadratic function of the observables r_i, it is not Gaussian. We will invoke the central limit theorem under the assumption that the dimension of the vector **r** is large. That is, we assume $S(\mathbf{r})$ is Gaussian, so that we need only its mean and variance under the two hypotheses. The calculations are dealt with in Sect. 11.5, and here we only give the results:

$$\mathscr{E}_0(S) = \tfrac{1}{2}\,\text{trace}[\mathbf{S}(\mathbf{S} + \mathbf{N})^{-1}],$$
$$\mathscr{E}_1(S) = \tfrac{1}{2}\,\text{trace}(\mathbf{SN}^{-1}) + \tfrac{1}{2}\mathbf{m}^T[\mathbf{N}^{-1} + (\mathbf{S} + \mathbf{N})^{-1}]\mathbf{m}$$
$$\text{Var}_0(S) = \tfrac{1}{2}\,\text{trace}[\mathbf{S}(\mathbf{S} + \mathbf{N})^{-1}]^2 + \mathbf{m}^T(\mathbf{S} + \mathbf{N})^{-1}\mathbf{N}(\mathbf{S} + \mathbf{N})^{-1}\mathbf{m}$$
$$\text{Var}_1(S) = \tfrac{1}{2}\,\text{trace}(\mathbf{SN}^{-1})^2 + \mathbf{m}^T\mathbf{N}^{-1}(\mathbf{S} + \mathbf{N})\mathbf{N}^{-1}\mathbf{m}$$

As usual, the probabilities of interest are

$$P_f = P(D_1 \mid H_0) = \int_{(S_t - S_0)/\sigma_0}^{\infty} (2\pi)^{-1/2} \exp\left(\frac{-u^2}{2}\right) du,$$

$$P_d = P(D_1 \mid H_1) = \int_{(S_t - S_1)/\sigma_1}^{\infty} (2\pi)^{-1/2} \exp\left(\frac{-u^2}{2}\right) du,$$

where we write $S_{0,1}$ for the mean of the statistic $S(\mathbf{r})$ under hypotheses $H_{0,1}$ and similarly for the standard deviations. However, if the variances under the two hypotheses are not equal, these cannot be related on a simple ROC curve such as Fig. 5.7. Said another way, the SNR will depend on P_f in general. As in Eq. (5.18), let

$$\alpha = (S_t - S_0)/\sigma_0 - (S_t - S_1)/\sigma_1. \tag{6.89}$$

In this we can use

$$(S_t - S_0)/\sigma_0 = \text{erf}^{-1}(\tfrac{1}{2} - P_f),$$

where erf^{-1} is the inverse function to the error function

$$\text{erf}(x) = \int_0^x (2\pi)^{-1/2} \exp\left(\frac{-u^2}{2}\right) du.$$

Then P_d follows.

One special case of the above is that in which $\mathbf{m} = \mathbf{0}$ and the input signal-to-noise ratio, defined as the matrix $\mathbf{SN}^{-1}$, is "small." Here by a small input SNR we mean that the largest eigenvalue of $\mathbf{SN}^{-1}$ is much less than unity. In the more general case that the largest eigenvalue of the input SNR is only less than unity, there is a convergent expansion:

$$(\mathbf{S} + \mathbf{N})^{-1} = \mathbf{N}^{-1}(\mathbf{I} + \mathbf{SN}^{-1})^{-1}$$

$$= \mathbf{N}^{-1} \sum_{i=0}^{\infty} (-\mathbf{SN}^{-1})^i.$$

In the case of small input SNR, this can be truncated to yield

$$(\mathbf{S} + \mathbf{N})^{-1} \cong \mathbf{N}^{-1}(\mathbf{I} - \mathbf{SN}^{-1}).$$

In the case $\mathbf{m} = \mathbf{0}$, the detector Eq. (6.88) then becomes approximately

$$S(\mathbf{r}) = \tfrac{1}{2}\mathbf{r}'\mathbf{N}^{-1}\mathbf{SN}^{-1}\mathbf{r} > S_t \qquad *(6.90)$$

This is called the *threshold detector*, because it is appropriate to signals whose power is much less than the noise power.

Under this zero-mean small-signal assumption, the detector parameters become

$$\mathscr{E}_0(S) = \tfrac{1}{2}\text{trace}[\mathbf{SN}^{-1}(\mathbf{I} - \mathbf{SN}^{-1})],$$

$$\mathscr{E}_1(S) = \tfrac{1}{2}\text{trace}(\mathbf{SN}^{-1}),$$

$$\text{Var}_0(S) = \tfrac{1}{2}\text{trace}[\mathbf{SN}^{-1}(\mathbf{I} - \mathbf{SN}^{-1})]^2 \cong \tfrac{1}{2}\text{trace}(\mathbf{SN}^{-1})^2,$$

$$\text{Var}_1(S) = \tfrac{1}{2}\text{trace}(\mathbf{SN}^{-1})^2.$$

The important thing to note is that the two variances are now approximately equal, so that the threshold terms cancel in Eq. (6.89). The performance of the detector Eq. (6.90) is now decribed by standard ROC curves such as Fig. 5.7 with the SNR Eq. (6.89) being

$$\alpha = (S_1 - S_0)/\sigma_0 = \tfrac{1}{2} \operatorname{trace}(\mathbf{SN}^{-1})^2 / [\tfrac{1}{4} \operatorname{trace}(\mathbf{SN}^{-1})^2]^{1/2}$$
$$= [\tfrac{1}{4} \operatorname{trace} (\mathbf{SN}^{-1})^2]^{1/2}. \qquad *(6.91)$$

That is,

$$\alpha_{\mathrm{dB}} = 20 \log_{10}(\alpha) = 10 \log_{10}[\tfrac{1}{4} \operatorname{trace}(\mathbf{SN}^{-1})^2].$$

From this latter general result for real matrices we can transcribe a result for narrowband Gaussian white noise processes sampled over an interval $0 \le t \le T$ at the Nyquist rate. In that case, with a one-sided bandwidth B and one-sided signal and noise spectral densities S_0, N_0, we have the detector Eq. (6.90) as

$$S(r) = \frac{1}{2} \sum_{i=0}^{m-1} \left(\frac{\sigma_s}{\sigma_n^2}\right)^2 r_i^2 > S_t, \qquad (6.92)$$

where $\sigma_s^2 = S_0 B$, $\sigma_n^2 = N_0 B$. As usual letting $B \Rightarrow \infty$, so that the sampling interval $\delta t = 1/2B \Rightarrow 0$, and noting $\sigma_s^2 = S_0/2\,\delta t$, and similarly for σ_n^2, the detector Eq. (6.92) becomes

$$S(r) = \frac{S_0}{N_0^2} \int_0^T r^2(t)\, dt > S_t,$$

in which the constant scale factor can be dropped. This is a quadratic detector, which forms an estimate of the total energy in the received waveform. If we normalize by the observation interval, we obtain a detector

$$S(r) = \frac{1}{T} \int_0^T r^2(t)\, dt > S_t$$

which estimates the average received power.

In this case, the SNR Eq. (6.91) becomes

$$\alpha^2 = \frac{1}{2} \sum_{i=0}^{m-1} \left(\frac{\sigma_s^2}{\sigma_n^2}\right)^2 = \left(\frac{m}{2}\right)\left(\frac{\sigma_s^2}{\sigma_n^2}\right)^2,$$
$$\alpha = (m/2)^{1/2} (\sigma_s^2/\sigma_n^2). \qquad *(6.93)$$

That is, the SNR at the detector output is the input SNR scaled up by the square root of the number of independent samples used in forming the test statistic.

The detector Eq. (6.92) can be analyzed exactly, because under either hypothesis the r_i are Gaussian random variables, so that $S(r)$ is a chi-squared variable with m degrees of freedom. We will postpone that calculation to Chapter 8, however, where it will arise in a more general context.

One other trancription of the threshold detector Eq. (6.90) can be made. Suppose that the components r_i of the vector of observables $\mathbf{r}$ are complex Fourier coefficients R_i calculated from bandlimited sample functions $r(t)$ sampled in time at the Nyquist rate appropriate to the band. If the data interval is adequately long, then we saw in Eq. (3.114) that the Fourier coefficients R_i are uncorrelated from one another, whether or not they are calculated from white noise time samples. That is, in the frequency domain representation, the covariance matrices $\mathbf{S}$, $\mathbf{N}$ are always approximately diagonal. In the limit of infinite observation time, the result becomes exact. As indicated in Eq. (3.109), the variance of a coefficient R_i calculated for some frequency f_i is proportional to the power spectral density of $r(t)$ evaluated at that frequency.

Now let the signal and noise have general power spectral densities $S(f)$, $N(f)$ over some band B. The threshold detector Eq. (6.90) becomes

$$S = \frac{1}{2mf_s} \sum_{i=0}^{m-1} \left[\frac{S(f_i)}{N^2(f_i)} \right] |R_i|^2. \tag{6.94}$$

In this we have used the fact, which we will also discuss in Chapter 11, that the foregoing expressions involving the transpose hold for complex quantities with the transpose taken as the transpose conjugate operation. We have also used Eq. (3.109) to introduce the scaling between the variances of the Fourier coefficients of the signal and noise and their power spectral densities. The detector Eq. (6.94) incorporates a whitener $1/N(f)$ and a filter $S(f)$ constraining the process to the band of the signal.

The SNR Eq. (6.91) in the case Eq. (6.94) becomes

$$\alpha^2 = \frac{1}{2} \sum_{i=0}^{m-1} \left[\frac{S(f_i)}{N(f_i)} \right]^2. \tag{6.95}$$

In the case of white noise, this recovers Eq. (6.93).

We can pass to the limit of infinite bandwidth in Eqs. (6.94), (6.95). That is, we let the sampling interval $\delta t \Rightarrow 0$ while keeping the observation interval fixed at T. Then the number of samples $m \Rightarrow \infty$. We recall that the Fourier coefficients R_i are

$$R_i = \sum_{k=0}^{m-1} r(t_k) \exp\left(\frac{-j2\pi ik}{m} \right) \cong \frac{R(f_i)}{\delta t},$$

where $R(f)$ is the Fourier transform of $r(t)$. With this, Eq. (6.94) becomes

$$S = \frac{1}{2(m\,\delta t)(f_s\,\delta t)} \sum_{i=0}^{m-1} \left[\frac{S(f_i)}{N^2(f_i)} \right] |R(f_i)|^2$$

$$\Rightarrow \frac{1}{2} \int_{-\infty}^{\infty} \left[\frac{S(f)}{N^2(f)} \right] |R(f)|^2 \, df. \tag{6.96}$$

Here we have used $f_s \, \delta t = 1$ and $m(\delta t) = T = 1/\delta f$. We have also taken account that the zero-frequency term is that with $i = 0$ and changed the sun to run over $-(m/2) + 1 \le i \le m/2$.

Correspondingly, the SNR eq. (6.95) becomes

$$\alpha^2 = \frac{T}{2} \int_{-\infty}^{\infty} \left[\frac{S(f)}{N(f)} \right]^2 df. \qquad \text{*(6.97)}$$

This becomes infinite as the observation interval $T \Rightarrow \infty$, reflecting the fact that we obtain increasingly good estimates of the data average power as the observation interval lengthens. In the limit, there is no question about the average power of the stationary data waveform $r(t)$, after whitening and filtering, and we can decide with arbitrarily small probability of error which of the two variance hypotheses is in effect. In the case of white signal and noise, Eq. (6.97) becomes

$$\alpha^2 = (Tm \, \delta f/2) \, (\sigma_s^2/\sigma_n^2)^2,$$

which recovers Eq. (6.93), since $T \, \delta f = 1$.

Exercises

6.1 Design a receiver to choose between two hypotheses

$$H_1: \quad r(t) = s_1(t) + n(t)$$
$$H_0: \quad r(t) = s_0(t) + n(t)$$

using minimum error probability as the criterion. The signals $s_1(t)$ and $s_0(t)$ are shown in Fig. 6.12. The additive noise is white, Gaussian with power spectral density $N_0/2$. Assume equal *a priori* probabilities. Determine the probability of error for $E/N_0 = 2$.

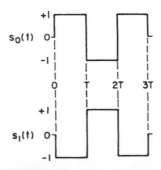

FIGURE 6.12   Three-bit words.

6.2 For the signals discussed in the preceding problem, each signal is a "word," and each word has three "bits." Suppose we detect each bit, one at a time. If at most one of the bits is in error, the word can still be decoded properly.

(a) What is the error probability of each bit?

(b) Given that we can "correct" a single bit error, what is the probability that the word is decoded in error?

(c) Compare the results with the preceding problem.

6.3 Consider a coherent frequency shift keying system with signals

$$s_1(t) = \sin \omega_1 t, \qquad 0 \le t \le T$$
$$s_0(t) = \sin \omega_0 t, \qquad 0 \le t \le T$$

For white Gaussian noise, the optimum receiver was discussed in Sect. 6.3. Denote $\omega_d = \omega_1 - \omega_0$.

(a) Show that the error probability is minimized when the frequency difference is chosen such that

$$\omega_d/2\pi \simeq 0.7/T$$

Assume either $(\omega_1 + \omega_0)T = k\pi$, where k is an integer, or that $\omega_1 + \omega_0 \gg 0$.

(b) In terms of the required signal energy what is the improvement of this system over the system for which $\rho = 0$?

6.4 Consider a four-phase communication system with signals $(0 \le t \le T)$

$$s_0(t) = A \sin \omega_0 t$$
$$s_1(t) = A \sin(\omega_0 t + \pi/2)$$
$$s_2(t) = A \sin (\omega_0 t + \pi)$$
$$s_3(t) = A \sin(\omega_0 t + 3\pi/2)$$

Assume the additive noise is white Gaussian with spectral density $N_0/2$. Assume equal *a priori* probabilities and equal costs for errors.

(a) What is the optimum receiver? How many correlators are required?

(b) Show that the probability of a correct decision is

$$P_c = \left[\int_{-(E/N_0)^{1/2}}^{\infty} \frac{1}{(2\pi)^{1/2}} e^{-z^2/2} \, dz \right]^2$$

(c) How does this compare with the probability of correct decision for the binary coherent case?

6.5 Consider a ternary communication system with equally likely signals $(0 \le t \le T)$

$$H_0: \quad s_0(t) = 0$$
$$H_1: \quad s_1(t) = A \sin \omega_0 t$$
$$H_2: \quad s_2(t) = -A \sin \omega_0 t$$

Assume white Gaussian noise with spectral density $N_0/2$.
- (a) Design a likelihood ratio test to minimize the error probability.
- (b) Show that

$$P(D_0 \mid H_0) = 2 \int_0^{(E/2N_0)^{1/2}} \frac{1}{(2\pi)^{1/2}} e^{-u^2/2} \, du$$

- (c) Show that

$$P(D_1 \mid H_1) = P(D_2 \mid H_2) = \int_{-(E/2N_0)^{1/2}}^{\infty} \frac{1}{(2\pi)^{1/2}} e^{-u^2/2} du$$

and therefore that the probability of a correct decision is

$$P_c = \frac{2}{3}\left[2 \int_0^{(E/2N_0)^{1/2}} \frac{1}{(2\pi)^{1/2}} e^{-u^2/2} \, du + \frac{1}{2} \right]$$

- (d) Compare this ternary system with the binary FSK case, and the four-phase system of the previous exercise.

6.6 Determine the likelihood ratio receiver to choose between the hypotheses

$$H_1: \quad r(t) = A \cos \omega_1 t + B \cos(\omega_2 t + \phi) + n(t)$$
$$H_0: \quad r(t) = B \cos(\omega_2 t + \phi) + n(t)$$

where A, B, ω_1, ω_2, and ϕ are known constants. The noise is white Gaussian, with power spectral density $N_0/2$. How is the performance of the receiver influenced by the signal $B \cos(\omega_2 t + \phi)$?

6.7 Consider the matched filter for a signal

$$s(t) = \begin{cases} A, & 0 \le t \le T \\ 0, & \text{otherwise} \end{cases}$$

in white Gaussian noise.
- (a) What is the peak output signal-to-noise ratio?
- (b) Suppose that instead of the matched filter, a filter

$$h(t) = \begin{cases} e^{-\alpha t}, & 0 \le t \le T \\ 0, & \text{otherwise} \end{cases}$$

is used. What is the peak output signal-to-noise ratio? What would you expect the optimum value of α to be?
- (c) Suppose that a filter

$$h(t) = e^{-\alpha t}, \quad t \ge 0$$

is used. What is the peak output signal-to-noise ratio? Show that the signal-to-noise ratio for this case is always less than or equal to that produced in part (b).

6.8 For the preceding problem, consider a Gaussian filter

$$h(\tau) = (1/\alpha) \exp\left(-\frac{(\tau - \tau_0)^2}{2\alpha^2}\right), \qquad -\infty < \tau < \infty, \qquad \tau_0 > 0$$

(Note if $\tau_0 \gg \alpha$ this can be approximated by a physically realizable filter.)
  (a) At what value of time will the output signal-to-noise ratio be a maximum?
  (b) Derive an expression for the signal-to-noise ratio.

6.9 Consider the signal $s(t) = 1 - \cos \omega_0 t$, $0 \le t \le 2\pi/\omega_0$ and the RC-filtered noise with power spectral density

$$S_n(\omega) = \frac{\omega_1^2}{\omega^2 + \omega_1^2}$$

  (a) Find the generalized matched filter for this example by solving Eq. (6.53) with $T_0 = 2\pi/\omega_0$. The answer is

$$h(t) = 1 - \frac{\omega_0^2 + \omega_1^2}{\omega_1^2} \cos \omega_0 t, \qquad 0 \le t \le 2\pi/\omega_0$$

  (b) What is the resulting maximum signal-to-noise ratio?

6.10 Consider the integral equation Eq. (6.58) for the generalized matched filter for a signal $s(t) = \sin \omega_0 t$, $0 \le t \le 2\pi/\omega_0 = T$, and an autocorrelation function $(\omega_1/2)e^{-\omega_1 |t|}$.
  (a) Verify that the filter

$$h(t) = s(T - t) - \frac{1}{\omega_1^2}\frac{d^2}{dt^2} s(T - t)$$

does not identically solve the integral equation.
  (b) Suppose that the delta functions

$$a\delta(t) + b\delta(t - T)$$

are added to the impulse response $h(t)$. Can the integral equation now be satisfied, and if so for what values of a and b?

6.11 For the preceding exercise,
  (a) Find the maximum signal-to-noise ratio.
  (b) For the same signal and noise conditions as in the previous exercise, assume that the filter

$$z(t) = -\left(\frac{\omega_1^2 + \omega_0^2}{\omega_1^2}\right) \sin \omega_0 t, \qquad 0 \le t \le 2\pi/\omega_0 = T$$

is used instead of the optimum filter. Determine the signal-to-noise ratio for this filter. Compare the optimum and suboptimum results for $\omega_0 = \omega_1$.

6.12 Determine the upper and lower triangular matrices whose product $C'C$ is the Markov covariance matrix

$$\begin{bmatrix} 1 & \rho & \rho^2 \\ \rho & 1 & \rho \\ \rho^2 & \rho & 1 \end{bmatrix}$$

7 Detection of Signals with Random Parameters

In Chapter 5 we discussed detection strategies in two different situations. In the first, we dealt with simple hypotheses, which are such that the probability density $p(y \mid H_i)$ governing the data under each specific hypothesis H_i is entirely known. In Chapter 6 we discussed in detail the simplest such situation, that in which the $p(y \mid H_i)$ were Gaussian with known means and variances. In fact, we dealt mostly with the binary hypothesis-testing problem and even more specifically that in which the densities $p(y \mid H_{0,1})$ had the same variance but different means. The processor which resulted in that case was the matched filter. However, in Chapter 5 we also discussed composite hypotheses. That is, the densities $p(y \mid H_i)$ were allowed to have parameters whose values were not known. The simplest strategy for dealing with such situations was developed in Sect. 5.9. There we considered the unknown parameters in the data densities $p(y \mid H_i)$ to be random variables, for which densities were in turn specified. That allowed us to design receivers which were optimal on average over the ensemble of parameter

247

values. In this chapter, we apply that general idea to specific cases of interest in radar and communications.

Throughout this chapter we will be interested in systems in which a relatively narrowband pulse of energy modulates a sinusoidal carrier. The receiver will process the narrowband data to determine a likelihood ratio. The data will usually be a function $r(t)$ of time, observed over a finite interval $0 \le t \le T$. Often we will assume the procedures of Chapter 3 have been applied as preprocessing to retrieve the complex envelope function $\tilde{r}(t)$ from the raw data $r(t)$. In that case, the receiver processing will be phrased in terms of the complex envelope function.

Four parameters of a received signal will be considered in turn and in various combinations. These are the phase, the amplitude, the time of arrival, and the carrier frequency of the signal. Phase is almost never known in a carrier system, and in this chapter we will assume in every case that it is not. If it is, then the *coherent* processing procedures of Chapter 6 will usually be applicable. Amplitude is often not known, due to ignorance about the precise details of the propagation medium (multipath transmission effects in communications) or about the scattering mechanism of a target in radar and sonar. Frequency is often known, except in the case of unknown Doppler shift from a moving target or from instabilities in the propagation medium. Time of arrival is not known if we don't know the range at which a target might be present. Since we usually lack any other guidance, we will mostly assume these parameters are random variables with the uniform probability density over some range of feasible values. The Bayesian procedures of Sect. 5.9 can then be applied.

Throughout this chapter, we will suppose that the signals of interest, if present, are in additive Gaussian white noise. The more complicated analysis involving colored noise will be pursued in Chapter 9. Also, we will mostly consider the binary hypothesis-testing problem of interest in radar and sonar. We will therefore limit the discussion of communication systems to binary alphabets. That will, however, serve as an introductory discussion of the general procedures.

7.1 PROCESSING NARROWBAND SIGNALS

Let us consider again the simple problem of Sect. 6.1—that of binary hypothesis testing of two known signals in additive zero mean Gaussian white noise. In this chapter, we will further take the signals to be narrowband sinusoids. Then we consider the hypotheses

$$\begin{aligned} H_0\!: \quad r(t) &= s_0(t) + n(t), \\ H_1\!: \quad r(t) &= s_1(t) + n(t), \end{aligned} \tag{7.1}$$

where the signals at the receiver input are

$$s_{0,1}(t) = A_{0,1}(t) \cos[\omega_c t + \phi_{0,1}(t) + \theta_{0,1}]. \tag{7.2}$$

The amplitude and phase modulation functions A_i, ϕ_i and the carrier frequency f_c are initially assumed to be specified. The noise $n(t)$ is zero mean, Gaussian, and white with two-sided power spectral density $N_0/2$. The signals, if present, exist over some interval $0 \le t \le T$, which for the time being we take as known. The signals are to be narrowband in the sense described in Sect. 3.2. That is, the modulated sinusoids $s_0(t)$, $s_1(t)$ in Eq. (7.2) are to be such that their Fourier spectra $S_{0,1}(j\omega) = 0$, $|f \pm f_c| > B/2$, as illustrated in Fig. 3.3. Furthermore, we require $f_c > B/2$, and in fact we will usually have $f_c \gg B/2$.

The applicable densities can be obtained at once from Eq. (6.36). Where α is a constant, we have

$$p_{0,1}(r) = \alpha \exp \left\{ \frac{-1}{N_0} \int_0^T [r(t) - s_{0,1}(t)]^2 \, dt \right\}. \tag{7.3}$$

The appropriate log likelihood ratio follows as in Eq. (6.37). In the radar case that $s_0(t) = 0$ in Eq. (7.1), Eq. (6.37) yields

$$\ln\{\lambda[r(t)]\} = \frac{2}{N_0} \int_0^T A(t) \cos[\omega_c t + \phi(t) + \theta] r(t) \, dt - \frac{E}{N_0} \tag{7.4}$$

where

$$E = \int_0^T A^2(t) \cos^2[\omega_c t + \phi(t) + \theta] \, dt \cong \frac{1}{2} \int_0^T A^2(t) \, dt \tag{7.5}$$

is the energy of the sought signal. In the case $s_0(t) \ne 0$, Eq. (7.3) gives the result generalizing Eq. (7.4):

$$\ln\{\lambda[r(t)]\} = \frac{2}{N_0} \int_0^T [s_1(t) - s_0(t)] r(t) \, dt - \frac{(E_1 - E_0)}{N_0}, \qquad *(7.6)$$

where $E_{0,1}$ are the energies of the signals $s_{0,1}(t)$.

The result Eq. (7.6) is convenient for analog processing. For digital processing, and for generality, it is worth pursuing the development in terms of the modulating functions in Eq. (7.2). That is, we will suppress the irrelevant carrier. As a first step in the development, we will consider the analytic signals and complex envelopes corresponding to the real signals $s_{0,1}(t)$, the noise $n(t)$, and the data $r(t)$, as in Sect. 3.1. That is, from Eq. (3.11) we write the signals in Eq. (7.1) in the form

$$s(t) = \text{Re}\{\tilde{s}(t) \exp(j\omega_c t)\}, \tag{7.7}$$

where

$$\bar{s}(t) = [s(t) + j\hat{s}(t)] \exp(-j\omega_c t) \tag{7.8}$$

is the complex envelope of the real signal $s(t)$, with $\hat{s}(t)$ the Hilbert transform of $s(t)$.

In the case of $s_{0,1}(t)$ as in Eq. (7.2), it is clear from Eq. (7.7) that the complex envelopes are of the form

$$\bar{s}(t) = A(t) \exp[j\phi(t) + \theta]. \tag{7.9}$$

In this the function $A(t)$ is the real envelope of the signal $s(t)$ and $\phi(t)$ is the phase modulation. However, if the carrier frequency is not large with respect to the signal bandwidth B, the envelope $A(t)$ will not be readily apparent by inspection of $s(t)$, even though the mathematical procedure is valid. With a carrier $f_c \gg B$, the real envelope $A(t)$ will look like the envelope of $s(t)$. Figure 3.2 shows an example.

Since a real signal is exactly recoverable from its corresponding complex envelope, as in Eq. (7.7), the transformation between signal and complex envelope is one-to-one. It then makes no difference whether we work the detection problem in terms of the modulated signals such as in Eq. (7.2), or the corresponding complex envelopes Eq. (7.8). If we use the latter, the resulting receiver will need to have an *envelope detector* as preprocessor. [Note the common double terminology. A "detector" is an algorithm for discriminating among hypotheses Eq. (7.1). It is also a circuit or procedure for determining the envelope, often just the real envelope, from a modulated signal.]

Rather than Eq. (7.1), we therefore consider the problem:

$$\begin{aligned} H_0\!: &\quad \bar{r}(t) = A_0(t) \exp\{j[\phi_0(t) + \theta_0]\} + \bar{n}(t), \\ H_1\!: &\quad \bar{r}(t) = A_1(t) \exp\{j[\phi_1(t) + \theta_1]\} + \bar{n}(t), \end{aligned} \tag{7.10}$$

where we deal with complex envelope signals. Since we want to build likelihood ratio tests, we will need the appropriate probability densities for the problem Eq. (7.10). In Eq. (7.10) we need to consider the densities $p(z)$ of complex random processes $z = x + jy$. As we mentioned in Sect. 1.8, this is defined as $p(z) = p(x, y)$. That is, the density of a complex random variable or process is defined as the joint density of its real and imaginary parts. Therefore, in considering the densities relative to Eq. (7.10), we will need the densities of the noise $n(t)$ and of its Hilbert transform $\hat{n}(t)$, as well as the joint density of these two random processes.

Since the noise process $n(t)$ is Gaussian, from Eq. (3.32) so also is its transform $\hat{n}(t)$, because the latter is a linear transformation of the former. Hence so also are the in-phase and quadrature components $n_i(t)$, $n_q(t)$ of the envelope

$$\tilde{n}(t) = n_i(t) + jn_q(t) = [n(t) + j\hat{n}(t)] \exp(-j\omega_c t).$$

We will use a sampling strategy, similar to what was done in Sect. 6.1, in order to write the likelihood ratio functions for the problem Eq. (7.10). Since the signals Eq. (7.2) are assumed to have a bandwidth B such that $B/2 \le f_c$, we can restrict attention to the band $|f| \le 2f_c$ with no loss of detectability. (Affairs outside that band are the same whichever hypothesis is true.) Therefore, we will assume that the processor is preceded by a low-pass filter with passband $|f| \le 2f_c$. [In the usual case of narrowband data $r(t)$, the filtering operation is only conceptual.] The data itself occupies some band $|f \pm f_c| \le B/2$ within the band $|f| \le 2f_c$. The complex envelopes in the problem Eq. (7.10) are then restricted to the band $|f| \le B/2$. We now consider the complex envelope signals to be sampled at the corresponding Nyquist interval $\delta t = 1/B$.

Hence, we replace the problem Eq. (7.10) by the discrete problem with hypotheses of the form

$$H: \quad \bar{\mathbf{r}} = \bar{\mathbf{s}} + \bar{\mathbf{n}} = \mathbf{r}_i + j\mathbf{r}_q, \tag{7.11}$$

involving samples of the complex envelope functions. In order to write down the likelihood ratio for the problem Eq. (7.11), we need the covariance matrix of the vector Eq. (7.11) under both hypotheses. Since we are assuming the signals to be completely known, that amounts to knowing the correlation matrix of the noise envelope in-phase and quadrature samples. That is, we need the matrix elements

$$\mathcal{E} \begin{bmatrix} n_i[(k+l)\,\delta t] \\ n_q[(k+l)\,\delta t] \end{bmatrix} [n_i(k\,\delta t) \quad n_q(k\,\delta t)]$$

$$= \begin{bmatrix} R_{ni}(l\,\delta t) & R_{niq}(l\,\delta t) \\ R_{nqi}(l\,\delta t) & R_{nq}(l\,\delta t) \end{bmatrix}, \tag{7.12}$$

where $\delta t = 1/B$.

In this bandlimited white noise case, the conditions leading to Eq. (3.89) apply, so that

$$R_{niq}(\tau) = 0,$$
$$R_{nqi}(\tau) = R_{niq}(-\tau) = 0. \tag{7.13}$$

Furthermore, we have

$$R_n(\tau) = 2 \int_{f_c - B/2}^{f_c + B/2} \left(\frac{N_0}{2}\right) \cos(\omega\tau)\, df$$

$$= 2N_0 \cos(\omega_c\tau) \int_0^{B/2} \cos(\omega\tau)\, df = \frac{N_0 B \cos(\omega_c\tau)[\sin(\pi B\tau)]}{\pi B\tau}.$$

Comparing with Eq. (3.80) and using Eq. (3.75) then shows

$$R_{ni}(\tau) = R_{nq}(\tau) = \frac{N_0 B [\sin(\pi B \tau)]}{\pi B \tau}. \tag{7.14}$$

With this sampling plan, and $\delta t = 1/B$, we can now write down the likelihood function for the discrete problem Eq. (7.11). That is, for $m = T/\delta t$ samples we seek

$$p(\tilde{\mathbf{r}}) = p(\mathbf{r}_i, \mathbf{r}_q) = p[r_i(1), r_q(1), \ldots, r_i(m), r_q(m)]. \tag{7.15}$$

Using $\delta t = 1/B$, from Eqs. (7.13), (7.14) we have

$$R_{ni}(l \, \delta t) = R_{nq}(l \, \delta t) = N_0 B \delta_{l0},$$
$$R_{niq}(l \, \delta t) = R_{nqi}(l \, \delta t) = 0.$$

Then in Eq. (7.15) we only need to consider the noise sample correlations

$$\mathscr{E} \begin{bmatrix} n_i(l) \\ n_q(l) \end{bmatrix} [n_i(l) \quad n_q(l)] = N_0 B \mathbf{I},$$

where $\mathbf{I}$ is the 2×2 unit matrix. Hence we have

$$p(\tilde{\mathbf{r}}) = \alpha \exp\left\{ -\left(\frac{1}{2N_0 B}\right) \sum_{l=1}^{m} [(r_{il} - s_{il})^2 + (r_{ql} - s_{ql})^2] \right\}$$
$$= \alpha \exp\{(-1/2N_0 B) \|\tilde{\mathbf{r}} - \tilde{\mathbf{s}}\|^2\}, \tag{7.16}$$

where

$$\|\tilde{\mathbf{r}} - \tilde{\mathbf{s}}\|^2 = \sum_{l=1}^{m} |\tilde{r}(l \, \delta t) - \tilde{s}(l \, \delta t)|^2 = (\tilde{\mathbf{r}} - \tilde{\mathbf{s}})'(\tilde{\mathbf{r}} - \tilde{\mathbf{s}}), \tag{7.17}$$

with the prime being the transpose conjugate when complex vectors are concerned.

The log likelihood ratio for the discrete version of the problem Eq. (7.10) is then

$$\ln[\lambda(\tilde{\mathbf{r}})] = \ln[p_1(\tilde{\mathbf{r}})/p_0(\tilde{\mathbf{r}})]$$
$$= (-1/2N_0 B)[(\tilde{\mathbf{r}} - \tilde{\mathbf{s}}_1)'(\tilde{\mathbf{r}} - \tilde{\mathbf{s}}_1) - (\tilde{\mathbf{r}} - \tilde{\mathbf{s}}_0)'(\tilde{\mathbf{r}} - \tilde{\mathbf{s}}_0)]$$
$$= (\delta t/2N_0)\{2 \operatorname{Re}[(\tilde{\mathbf{s}}_1 - \tilde{\mathbf{s}}_0)'\tilde{\mathbf{r}}] - (\|\tilde{\mathbf{s}}_1\|^2 - \|\tilde{\mathbf{s}}_0\|^2)\}. \tag{7.18}$$

The signal-processing part of this is the matched filter corresponding to the samples of the signal envelope difference $\tilde{s}_1(t) - \tilde{s}_0(t)$.

Since all signals involved here are bandlimited and we have sampled at the Nyquist rate, the discrete sums such as Eq. (7.17), scaled by δt, are equal the corresponding integrals. That is, we can write Eqs. (7.16), (7.18) as

$$p_{0,1}[\bar{r}(t)] = \alpha \exp\left[-\int_0^T \frac{|\bar{r}(t) - \bar{s}_{0,1}(t)|^2 \, dt}{2N_0}\right], \qquad (7.19)$$

$$\ln\{\lambda[\bar{r}(t)]\} = \frac{1}{N_0} \operatorname{Re} \int_0^T [\bar{s}_1(t) - \bar{s}_0(t)]^* \, \bar{r}(t) \, dt$$

$$- \left(\frac{1}{2N_0}\right) \left[\int_0^T |\bar{s}_1(t)|^2 \, dt - \int_0^T |\bar{s}_0(t)|^2 \, dt\right]. \qquad *(7.20)$$

(These assume the bandlimited signals approximately vanish outside $0 \le t \le T$.) These last are convenient in theoretical developments, whereas Eq. (7.18) is appropriate for computations. The processor Eq. (7.20) is the correlation receiver Eq. (6.14), written for complex envelopes rather than for real signals.

In the case $f_c = 0$, Eq. (7.20) reduces to Eq. (7.6). This follows from Parseval's relation,

$$\int_{-\infty}^{\infty} f(t)g(t) \, dt = \int_{-j\infty}^{j\infty} F(f) G^*(f) \, df,$$

so that

$$\int_{-\infty}^{\infty} \hat{f}(t)\hat{g}(t) \, dt = \int_{-j\infty}^{j\infty} |H(f)|^2 F(f) G^*(f) \, df$$

$$= \int_{-\infty}^{\infty} f(t)g(t) \, dt,$$

where $H(f)$ is the Hilbert transformer Eq. (3.23). With $f_c = 0$, we are considering the case that the phase modulation $\phi(t)$ has an unknown additive constant θ.

7.2 DETECTION OF SIGNALS WITH UNKNOWN CARRIER PHASE

In Eq. (7.2) we have included constant phase angles $\theta_{0,1}$ in the signals $s_{0,1}$ to be detected. The likelihood ratios Eqs. (7.6), (7.20) assume those angles are known, so that the detection statistic can be computed. However, it is unusual for the phase angles in Eq. (7.2) to be known in a communications problem. Carrier frequencies f_c are usually high enough that slight perturbations in the transmission medium, or the presence of multipath effects, or uncertainties in the scattering properties of a target cause timing errors corresponding to large phase excursions. For example, with a radar signal of 1 GHz (L-band) a range error of 4 cm causes a phase excursion of $\pi/2$ rad. Again, in a sonar system a multipath transmission channel leads to uncertainty in the phase of the signal we wish to detect.

For these reasons, the phase angles $\theta_{0,1}$ in Eq. (7.2) must be assumed to be unknown. We therefore have to deal with the problem Eq. (7.1) as a problem involving composite hypotheses. A realistic model is to assume complete lack of information about the phase angles and to treat them as random variables with the uniform density. Then the procedures of Sect. 5.9 can be applied. The minimum-risk likelihood ratio is given by Eq. (5.42). In the problem Eq. (7.10), which generalizes Eq. (7.1), this becomes

$$\lambda[\bar{\mathbf{r}}(t)] = \mathscr{E}p_1(\bar{r} \mid \theta_1)/\mathscr{E}p_0(\bar{r} \mid \theta_0), \tag{7.21}$$

where the expectations are over the phase angles $\theta_{0,1}$.

Using Eq. (7.19), for the terms in Eq. (7.21) we have

$$\mathscr{E}p[\bar{r}(t) \mid \theta] = \frac{\alpha}{2\pi} \int_{-\pi}^{\pi} \exp\left[-\int_0^T \frac{|\bar{r}(t) - \tilde{s}(t)|^2 \, dt}{2N_0}\right] d\theta. \tag{7.22}$$

For future developments, from here on it will be convenient to take the signals Eq. (7.9) in the form

$$\tilde{s}(t) = Aa(t) \exp\{j[\phi(t) + \theta]\}, \tag{7.23}$$

where the real envelope waveform $a(t)$ is normalized as

$$\frac{1}{T} \int_0^T a^2(t) \, dt = 1. \tag{7.24}$$

The important term in the integral in Eq. (7.22) is then

$$\int_0^T a(t)\bar{r}(t) \exp[-j\phi(t)] \, dt = 2q \exp(j\beta), \tag{7.25}$$

where q and β are so defined. (The factor of 2 is introduced for later convenience.) With this definition, we have

$$\mathscr{E}p[\bar{r}(t) \mid \theta] = \frac{\alpha}{2\pi} \exp\left[-\int_0^T \frac{\{|\bar{r}(t)|^2 + A^2a^2(t)\} \, dt}{2N_0}\right]$$

$$\times \int_{-\pi}^{\pi} \exp\left[\left(\frac{2Aq}{N_0}\right) \cos(\theta - \beta)\right] d\theta. \tag{7.26}$$

We have in general that

$$\int_{-\pi}^{\pi} \exp[\gamma \cos(\theta - \beta)] \, d\theta = \int_{-\pi-\beta}^{\pi-\beta} \exp[\gamma \cos(\theta)] \, d\theta = \int_{-\pi}^{\pi} \exp[\gamma \cos(\theta)] \, d\theta$$

$$= 2 \int_0^{\pi} \exp[\gamma \cos(\theta)] \, d\theta = 2\pi I_0(\gamma), \tag{7.27}$$

where $I_0(\gamma)$ is the modified Bessel function of the first kind and order zero. Using this in Eq. (7.26) yields

$$\mathscr{E}p[\bar{r}(t) \mid \theta] = \alpha I_0(2Aq/N_0) \exp\left[-\left(\frac{1}{2N_0}\right)\int_0^T \{|\bar{r}(t)|^2\right.$$
$$\left. + A^2 a^2(t)\} \, dt\right]. \qquad (7.28)$$

Now we can substitute Eq. (7.28), written for $\bar{s}_0$ and $\bar{s}_1$, into Eq. (7.21) to obtain finally

$$\lambda[\bar{r}(t)] = [I_0(2A_1 q_1/N_0)/I_0(2A_0 q_0/N_0)]$$
$$\times \exp\left[-\left(\frac{1}{2N_0}\right)\int_0^T [A_1^2 a_1^2(t) - A_0^2 a_0^2(t)] \, dt\right]. \qquad (7.29)$$

In the special case that $s_0(t) = 0$, Eq. (7.29) becomes

$$\lambda[\bar{r}(t)] = I_0(2Aq/N_0) \exp(-A^2 T/2N_0), \qquad *(7.30)$$

using Eq. (7.24) and taking account that $I_0(0) = 1$. From Eq. (7.25)

$$q = \frac{1}{2}\left|\int_0^T a(t)\bar{r}(t) \exp[-j\phi(t)] \, dt\right|. \qquad *(7.31)$$

It is worth noting that the signal energy is

$$E = \int_0^T s^2(t) \, dt \cong \frac{A^2}{2}\int_0^T a^2(t) \, dt = A^2 T/2. \qquad (7.32)$$

(The approximation results by dropping the integral of the double frequency term.)

In the case Eq. (7.30), the hypothesis test can be written

$$I_0(2A \, q/N_0) > I_t, \qquad (7.33)$$

where I_t is a threshold appropriate to the error criterion of interest and after absorbing the constant in Eq. (7.30) into the threshold. It happens that the Bessel function $I_0(\gamma)$ is a monotonically increasing function of γ. Hence the test can be carried out in terms of either $I_0(2Aq/N_0)$ or simply q. For convenience, the test is often implemented equivalently in terms of q^2 as

$$q^2 > q_t^2. \qquad *(7.34)$$

This is the so-called quadrature receiver for the problem Eq. (7.10) in the case of testing signal plus noise against noise alone with unknown carrier phase. It is important to note that the test Eq. (7.34) is independent of the signal scale factor A. That is, the test is uniformly most powerful with respect to signal amplitude scale factor.

7.3 THE QUADRATURE RECEIVER AND EQUIVALENT FORMS

In the case of detecting the presence or absence of a signal

$$s(t) = Aa(t)\cos[\omega_c t + \phi(t) + \theta]$$

in zero mean white Gaussian noise, with phase θ a uniform random variable, we have arrived at the *quadrature receiver* Eq. (7.34). From Eq. (7.31), the test statistic is

$$q^2 = \frac{1}{4}\left|\int_0^T a(t)\tilde{r}(t)\exp[-j\phi(t)]\,dt\right|^2. \tag{7.35}$$

Here $\tilde{r}(t)$ is the complex envelope function corresponding to the received signal $r(t)$. That can be determined by any of the methods detailed in Sect. 3.6. In particular, from Eqs. (3.96), (3.97) we have

$$\tilde{r}(t) = 2[r(t)\exp(-j\omega_c t)]_{LP}, \tag{7.36}$$

where the notation LP indicates low-pass filtering of the quantity in brackets. Since the integration in Eq. (7.35) acts as an approximate low-pass filter, the quadrature receiver Eq. (7.34) can be written as

$$q^2 = \left|\int_0^T a(t)r(t)\exp\{-j[\omega_c t + \phi(t)]\}\,dt\right|^2 > q_t^2. \tag{7.37}$$

The receiver Eq. (7.37) is diagrammed in Fig. 7.1.

As discussed in Sect. 6.5.1, the result of the correlation operations in Eq. (7.37) can also be generated by sampling the output of a matched filter. The most useful form of matched filter is that with impulse response

$$\tilde{h}(t)/2 = a(T - t)\exp[-j\phi(T - t)].$$

In terms of this, Eq. (7.35) can be written

$$q = (1/2)|\tilde{y}(T)|,$$
$$\tilde{y}(t) = [\tilde{h}(t)/2]*\tilde{r}(t), \tag{7.38}$$

where we indicate sampling of the filter output $\tilde{y}(t)$ at the time T ending the observation interval. The tilde sign on $\tilde{r}(t)$ indicates that this is the complex envelope corresponding to the available real data signal $r(t)$. The tildes on $\tilde{h}(t)$ and $\tilde{y}(t)$ are to emphasize that these are base band signals. In the form Eq. (7.38), the computations are carried out at base band, as would be appropriate for a digital receiver. However, the same manipulations could be carried out equally well in the carrier band, or in any convenient band to which the RF signal is down-converted. This would be the normal situation with analog processing.

In order to move to a matched filter in the carrier band, we use the

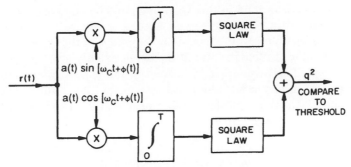

FIGURE 7.1   Quadrature receiver.

procedures of Sect. 3.4. In particular, the low-pass complex envelope relation Eq. (7.38) corresponds to the bandpass relation

$$y(t) = h(t)*r(t).$$

In our application, $h(t)$ is the real bandpass impulse response with complex envelope

$$\bar{h}(t) = 2a(T - t) \exp[-j\phi(T - t)].$$

That is,

$$h(t) = \text{Re}\{2a(T - t) \exp[j\omega_c t - j\phi(T - t)]\}$$
$$= 2a(T - t) \cos[\omega_c t - \phi(T - t)]. \qquad (7.39)$$

Then the quantity Eq. (7.38) can be realized by computing first the real signal

$$y(t)/2 = \{a(T - t) \cos[\omega_c t - \phi(T - t)]\}*r(t),$$

then computing its real envelope $|\bar{y}(t)|/2$, and finally sampling the real envelope at the end of the observation time to obtain $q = (1/2)|\bar{y}(T)|$. The detector Eq. (7.37) then appears as in Fig. 7.2. Since we are only interested in the real envelope of the filter output in Fig. 7.2, it is clear that a constant phase shift of the function Eq. (7.39) is irrelevant; that is, we could use $h(t)/2 = a(T - t) \cos[\omega_c(T - t) + \phi(T - t)]$. The system in Fig. 7.2 is called the *incoherent matched filter*. The term "incoherent" refers to the fact that we have not attempted to match to the carrier phase, only to the complex envelope of the signal.

FIGURE 7.2   Equivalent quadrature receiver for slowly varying signal envelope.

The procedures described in this section should be compared with the situation considered in Example 6.1 of Sect. 6.5. Figure 6.4 shows the output of a matched filter in the case that the phase of the input signal is known. We are able to "catch" the filter output exactly at its peak value when it is sampled. However, if the phase of the carrier is not known, we might sample at any point along the sinusoidal filter output. We are better off not to run the risk of sampling near a zero of the output and instead compute and sample the real envelope.

7.4 RECEIVER OPERATING CHARACTERISTICS

In this section we want to determine the performance of the detector

$$q > q_t,$$

where q is the detection statistic of Eq. (7.31). This applies to the problem Eq. (7.1) or Eq. (7.10) in the case $s_0(t) = 0$. The sought signal exists over a known interval $0 \le t \le T$ and is

$$s(t) = Aa(t) \cos[\omega_c t + \phi(t) + \theta], \qquad (7.40)$$

where the phase θ is unknown and $a(t)$ is normalized as in Eq. (7.24). The quantity $\tilde{r}(t)$ is the complex envelope of the received data $r(t)$ over the observation interval $0 \le t \le T$.

We will need the probability density of the test statistic q under the two hypotheses. These are $p_0(q)$ and

$$p_1(q) = \mathscr{E} p_1(q \mid \theta),$$

where the expectation is over the phase angle θ, which is assumed to have the uniform density. Writing

$$g = \frac{1}{2} \int_0^T a(t) \tilde{r}(t \mid \theta) \exp[-j\phi(t)] \, dt = x + jy, \qquad (7.41)$$

so that

$$q = |g| = (x^2 + y^2)^{1/2},$$

we seek the densities of x, y.

We have from Eq. (3.28) that

$$\tilde{r}(t) = [r(t) + j\hat{r}(t)] \exp(-j\omega_c t),$$

where $\hat{r}(t)$ is the Hilbert transform of $r(t)$. Since $r(t)$ is Gaussian under both hypotheses and the Hilbert transform is a linear operation on $r(t)$, the complex envelope $\tilde{r}(t)$ is a complex Gaussian random process. Then x, y are Gaussian random variables, and we need their means and variances to compute the receiver operating characteristic (ROC) curve. We have first

$$\mathcal{E}(x \mid H_0) = \mathcal{E}(y \mid H_0) = 0, \tag{7.42}$$

because the noise process $n(t)$ is zero mean. Also,

$$\mathcal{E}(x \mid H_1, \theta) = \frac{1}{2} \int_0^T a(t) \, \text{Re}\{\mathcal{E}[\tilde{r}(t) \mid H_1] \exp[-j\phi(t)]\} \, dt$$

$$= \frac{1}{2} \int_0^T a(t) \, \text{Re}\{\tilde{s}(t) \exp[-j\phi(t)]\} \, dt$$

$$= \frac{A}{2} \int_0^T a^2(t) \cos(\theta) \, dt = \left(\frac{AT}{2}\right) \cos(\theta), \tag{7.43}$$

where we use Eq. (7.24) and note $\tilde{s}(t) = Aa(t) \exp\{j[\phi(t) + \theta]\}$. Similarly,

$$\mathcal{E}(y \mid H_1, \theta) = (AT/2) \sin(\theta). \tag{7.44}$$

As to the variances of x, y, since the signal $s(t)$ is deterministic for given θ, which we are currently assuming, the variances are the same under both hypotheses and are just the variances resulting from Eq. (7.41) using $\tilde{r}(t) = \tilde{n}(t)$. Invoking the usual sampling plan, from Eqs. (7.13), (7.14), (7.41) we can write

$$\sigma^2 = \mathcal{E}(x^2) = \left(\frac{\delta t}{2}\right)^2 \sum_{k,l=1}^m a_k a_l \mathcal{E}[(n_{ik} \cos \phi_k + n_{qk} \sin \phi_k)$$

$$\times (n_{il} \cos \phi_l + n_{ql} \sin \phi_l)]$$

$$= \left(\frac{\delta t}{2}\right)^2 \sum_{k=1}^m a_k^2 [R_i(0) \cos^2 \phi_k + R_q(0) \sin^2 \phi_k]$$

$$= \left(\frac{\delta t}{2}\right)^2 N_0 B \sum_{k=1}^m a_k^2 = \left(\frac{N_0 B \delta t}{4}\right) \int_0^T a^2(t) \, dt = \frac{N_0 T}{4}. \quad *(7.45)$$

In this we used $\delta t = 1/B$. The same result holds for $\mathcal{E}(y^2)$. We then have

$$p_1(x, y \mid \theta) = (2\pi\sigma^2)^{-1} \exp\{(-1/2\sigma^2)[(x - \bar{x})^2 + (y - \bar{y})^2]\}, \tag{7.46}$$

where we have written $\bar{x}$, $\bar{y}$ for the means in Eqs. (7.43), (7.44). The density under the hypothesis H_0 is just Eq. (7.46) with the means set to zero. Now since $q = |g|$, we can write

$$g = x + jy = q[\cos(\beta) + j \sin(\beta)],$$

where β is some phase angle. Transforming the density Eq. (7.46) using the usual Eq. (1.33), we obtain

$$p_1(q, \beta \mid \theta) = q p_1(x, y)$$
$$= q(2\pi\sigma^2)^{-1}$$
$$\times \exp\{(-1/2\sigma^2)[q^2 + (\bar{x}^2 + \bar{y}^2) - 2q(\bar{x} \cos \beta + \bar{y} \sin \beta)]\}. \tag{7.47}$$

We now need to integrate the density Eq. (7.47) first over the random variable β to determine the density $p_1(q \mid \theta)$ and then over the carrier phase θ to determine $p_1(q)$, the density of the test statistic. The first integration involves

$$\int_{-\pi}^{\pi} \exp\left[\frac{q}{\sigma^2}(\bar{x}\cos\beta + \bar{y}\sin\beta)\right] d\beta = \int_{-\pi}^{\pi} \exp\left[\frac{qAT}{2\sigma^2}\cos(\theta - \beta)\right] d\beta.$$

Using Eq. (7.27) then yields

$$p_1(q \mid \theta) = (q/\sigma^2)\exp[-(q^2 + A^2T^2/4)/2\sigma^2]I_0(qAT/2\sigma^2), \qquad (7.48)$$

where again $I_0(\gamma)$ is the modified Bessel function of the first kind and order zero.

The density Eq. (7.48) does not involve the phase θ. Therefore

$$p_1(q) = \int_{-\pi}^{\pi} \frac{1}{2\pi} p_1(q \mid \theta)\, d\theta = p_1(q \mid \theta)$$

$$= (q/\sigma^2)\exp[-(q^2 + A^2T^2/4)/2\sigma^2]I_0(qAT/2\sigma^2). \qquad (7.49)$$

The density $p_0(q)$ results by setting $A = 0$ in Eq. (7.49). The result is

$$p_0(q) = (q/\sigma^2)\exp(-q^2/2\sigma^2). \qquad (7.50)$$

Equation (7.49) is the Rician density, studied in Chapter 4, and Eq. (7.50) is the Rayleigh density. The parameters of these densities are given by Eqs. (7.23), (7.45).

Using the statistic q of Eq. (7.35) and its density Eq. (7.50), we obtain the false alarm probability

$$P_{\mathrm{f}} = P(D_1 \mid H_0) = \int_{q_t}^{\infty} p_0(q)\, dq = \exp\left(\frac{-q_t^2}{2\sigma^2}\right), \qquad *(7.51)$$

where $\sigma^2 = N_0T/4$ and the detector is $q > q_t$. Solving this the other way, we have the required parameter value for a specified P_{f} as

$$q_t/\sigma = [-2\ln(P_{\mathrm{f}})]^{1/2}. \qquad (7.52)$$

The probability of detection is

$$P_d = P(D_1 \mid H_1) = \int_{q_t}^{\infty} p_1(q)\, dq$$

$$= \int_{q_t/\sigma}^{\infty} v\exp\left[-\left(\frac{v^2}{2} + \frac{A^2T^2}{8\sigma^2}\right)\right]I_0\left(\frac{ATv}{2\sigma}\right) dv$$

$$= Q(AT/2\sigma, q_t/\sigma), \qquad *(7.53)$$

where $Q(\alpha, \beta)$ is the Marcum Q-function, Eq. (4.68).

The probabilities Eqs. (7.51), (7.53) allow the construction of a ROC curve. For given P_{f}, Eq. (7.52) determines the required parameter q_t/σ.

With this, Eq. (7.53) yields the attained P_d for any value $AT/2\sigma$. For any given noise density N_0 and signal duration T, from Eq. (7.45) we can calculate σ, so that q_t/σ yields the detector threshold q_t. With Eq. (7.32) the parameter in Eqs. (7.53) is

$$AT/2\sigma = (A^2T/N_0)^{1/2} \cong (2E/N_0)^{1/2}. \qquad (7.54)$$

It is natural and conventional to choose the quantity Eq. (7.54) as the signal-to-noise ratio (SNR), so that

$$\alpha_{dB} = 20 \log(AT/2\sigma) \cong 10 \log(2E/N_0). \qquad *(7.55)$$

This has the virtue of being the same as the definition used for the problem of completely known signal in noise. The ROC curves for the two problems are of course different, even though the SNRs are the same.

Figure 7.3 is a plot of the ROC curve for this problem using the SNR Eq. (7.55) in conjunction with Eqs. (7.52), (7.53). For comparison, the

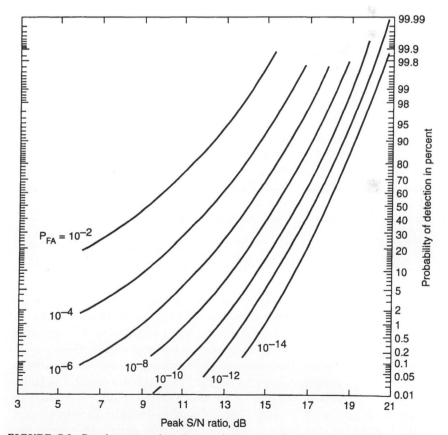

FIGURE 7.3 Receiver operating characteristic curves for incoherent detection (From "Radar Detection," by J. V. DiFranco and W. L. Rubin © 1980, Artech House, Norwood, MA.)

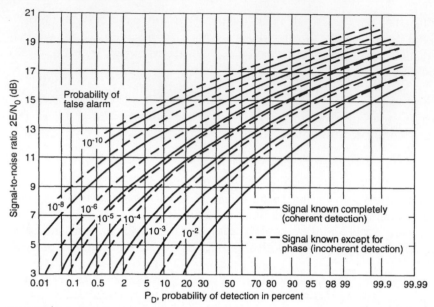

FIGURE 7.4 Receiver operating characteristic curves for coherent and incoherent detection.

ROC curve computed from Eq. (5.18) with the SNR Eq. (6.25) identical to Eq. (7.55) is overlaid in Fig. 7.4. (That is just Fig. 5.7 replotted.) Over regions of main interest, the SNRs for equal P_d, P_f in the two problems differ by only about 1 dB. That is the slight penalty paid in this problem for our lack of knowledge about carrier phase. In the case of detection decisions based on multiple observations, however, we will see in Chapter 8 that this difference has a major effect in practice.

7.5 SIGNALS WITH RANDOM PHASE AND AMPLITUDE

In the previous section, we considered the binary hypothesis-testing problem Eqs. (7.1), (7.2) in the case that the carrier phase θ was unknown but the signal complex envelope $A(t) \exp[j\phi(t)]$ was completely specified under both hypotheses. That is sometimes a realistic model. However, it is more common that the transmission or scattering medium has distorted the (presumably known) transmitted signal so that it is not precisely known at the receiver. At one extreme, we could simply consider $A(t)$, $\phi(t)$ to be random processes with specified probability models. In this section we will deal with a simpler case, yet one which is often adequate to the task. In particular, we will assume a signal of constant amplitude A, so that $A(t) = A$ in Eq. (7.2). We will assume no phase modulation, so that $\phi(t) = 0$. That is, we consider the problem

$$H_1: \quad r(t) = A \cos(\omega_c t + \theta) + n(t),$$
$$H_0: \quad r(t) = n(t). \tag{7.56}$$

The new element in this section is that we will now consider the envelope amplitude scale factor A to be a random variable as well as the carrier phase θ.

In communication problems, radar, and (especially) sonar, one physical factor that enters into the propagation path which leads to unknown carrier phase and unknown signal amplitude at the receiver is multipath propagation. We transmit a narrowband pulse which propagates along multiple paths to the receiver. This results in interference of those multiple signals at the receiver. The consequent coherent sum of sinusoids has a net amplitude and phase which depend on the details of the individual propagation paths and are not usually predictable. In seeking a model of the result, it is common to consider that enough paths exist that the central limit theorem comes into play. The received signal is then assumed to be a narrowband Gaussian process. As we saw in Sect. 4.4, such a process has a phase θ which is uniformly distributed and an amplitude A with the Rayleigh density:

$$p(A) = (A/A_0^2) \exp(-A^2/2A_0^2), \quad A \geq 0. \qquad *(7.57)$$

The density parameter is $A_0^2 = \mathscr{E}(A^2)/2$.

The model Eq. (7.57) also applies to scattering phenomena in which multiple scattering centers are simultaneously illuminated by an active source. The result is a received signal which is the coherent summation of many scattered signals from the elementary scattering elements of a larger object. This accounts for the common model of aircraft radar backscatter as a Gaussian process with Rayleigh amplitude and for the statistics of a coherent imaging system, such as a laser or synthetic aperture radar system. In communications, rain patches along the transmission path can lead to Rayleigh statistics for the received signal.

In this section, we will consider only the so-called slow Rayleigh fading case, in which the amplitude A is a random variable but has a fixed value over the entire observation interval $0 \leq t \leq T$. Slow fading is particularly applicable to communication problems, as the observation time available is only that of a single message character.

The detector for the problem Eq. (7.56) is just $q > q_t$, with q as in Eq. (7.31) and $a(t) = 1$, $\phi(t) = 0$. Alternatively, Eq. (7.37) can be used. Since q does not depend on the signal amplitude A, this detector is uniformly most powerful for the parameter A. That is, whatever the (unknown) value of A might be, we will use the same detector and we do not need to know A in order to compute q. However, the performance of that detector will depend on A. It is that performance which is depicted in the ROC curve of Fig. 7.3. It is the power function $P_d(A)$ of the test, parametrized by P_f. In this section, we will compute the expected value of the power, $\mathscr{E}P_d(A)$, in the case of the density Eq. (7.57).

Using Eq. (7.53) and the Rayleigh density Eq. (7.57), we have

$$P_d = \mathscr{E}P_d(A) = \int_0^\infty \left(\frac{A}{A_0^2}\right) \exp\left(\frac{-A^2}{2A_0^2}\right)$$

$$\times \int_{q_t/\sigma}^\infty v \exp\left[-\left(\frac{v^2}{2} + \frac{A^2T^2}{8\sigma^2}\right)\right] I_0\left(\frac{ATv}{2\sigma}\right) dv\, dA,$$

where as before $\sigma^2 = N_0T/4$. This integral turns out to have a closed-form expression [Gradshteyn and Ryzhik, 1965, #6.633.4], namely

$$P_d = \frac{N_0}{(A_0^2T + N_0)} \int_{q_t/\sigma}^\infty v \exp\left[-\frac{v^2}{2}\Big/\left(1 + \frac{A_0^2T}{N_0}\right)\right] dv$$

$$= \exp\{-2q_t^2/[T(A_0^2T + N_0)]\}. \qquad *(7.58)$$

For the problem Eq. (7.56) with specified value of A, in Eq. (7.54) we made the choice of SNR as

$$\alpha^2 = (AT/2\sigma)^2 = A^2T/N_0.$$

In the current problem, we correspondingly will take

$$\alpha^2 = (T/N_0)\mathscr{E}(A^2) = 2TA_0^2/N_0, \qquad (7.59)$$

using Eq. (4.57) for the mean square of the density Eq. (7.57). For the problem Eq. (7.56), we have the average signal energy as

$$E_{av} = \mathscr{E}(A^2) \int_0^T \cos^2(\omega_c t + \theta)\, dt \cong \frac{T}{2}\mathscr{E}(A^2) = A_0^2T.$$

Then the SNR Eq. (7.59) becomes

$$\alpha = (2E_{av}/N_0)^{1/2},$$
$$\alpha_{dB} = 10 \log_{10}(2E_{av}/N_0). \qquad *(7.60)$$

In terms of Eq. (7.60) as SNR, the ROC curve is described by Eq. (7.52), namely

$$q_t^2 = -(N_0T/2) \ln(P_f),$$

in conjunction with Eq. (7.58), which becomes

$$P_d = P_f^{(1+\alpha^2/2)^{-1}} = P_f^{(1+E_{av}/N_0)^{-1}}. \qquad *(7.61)$$

The form of this last expression would perhaps argue for the choice of SNR as E_{av}/N_0, rather than $2E_{av}/N_0$. Nonetheless, for uniformity with

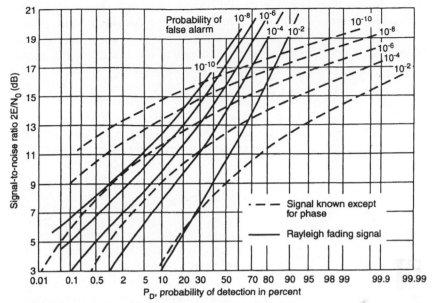

FIGURE 7.5 Detection performance for a Rayleigh fading signal and for a signal known except for phase.

our convention in other problems we will continue to use the latter. Figure 7.5 shows the relation Eq. (7.61) in terms of the SNR $2E_{av}/N_0$. We have superimposed the ROC curve Fig. 7.4 for the case of known value of A. At high SNR, there is a degradation in performance due to the Rayleigh fading amplitude. At low SNR, the converse is the case, a situation which we will discuss in the next section.

We should mention the following special case. Suppose that the problem we have been discussing is discretized by sampling the various waveforms at some time interval δt. It may be that the "slow" Rayleigh fading is so fast that the observation interval T is just the time δt of one sample. In that case, the received waveform has uniformly distributed phase and independent Rayleigh amplitude in every interval δt. That is, the quadrature components $x(t_i)$, $y(t_i)$ are independent Gaussian variables at each t_i. The envelope signal $\tilde{s}(t)$ can then be considered as arising from a signal $s(t)$ which is a narrowband Gaussian random process. The discrete samples are then processed as described in Sect. 6.8.

7.6 NONCOHERENT FREQUENCY SHIFT KEYING

In Sect. 5.8 we considered the problem of selecting among m hypotheses in the case that the criterion was minimum probability of error. The result was that, given data $r(t)$, we should choose that hypothesis which had maximum *a posteriori* probability. In Sect. 6.3 we considered the communications case of equal *a priori* probabilities of m characters, in which case the receiver is to choose the maximum of the likelihood functions:

$$j \Leftarrow \max_i p[r(t) \mid H_i].$$

With additive white Gaussian noise, and with the characters encoded as signals $s_i(t)$ of equal energies, Eq. (6.84) shows that the receiver is to choose the largest of the correlations between the signals and the received data:

$$j \Leftarrow \max_i \int_0^T s_i(t)r(t) \, dt. \tag{7.62}$$

That is, the receiver amounts to a bank of coherent matched filters. For the special case Eq. (6.85) of orthogonal signals of equal energy, Eq. (6.87) gives the attained error probability.

In a communication system, the encoding signals $s_i(t)$ will be narrowband signals modulating a carrier at frequency f_c. For the reasons we discussed in Sect. 7.2, the carrier phase is generally not known. In some systems it is estimated from the received signals and the estimate used in the matched filters Eq. (7.62). In this section, as in Sect. 7.2, we will take the point of view that the carrier phase is to be modeled as a uniform random variable and the error probability minimized on the average. For illustration, we will analyze the system using an m-ary orthogonal set of equal energy signals, as in Eq. (6.83):

$$\int_0^T s_i(t)s_j(t) \, dt = E\delta_{ij}, \tag{7.63}$$

where E is the signal energy.

Consider the signal parameters available for use in realizing an orthogonal encoding set. Phase is not in control in the incoherent case we are investigating here and is not available. Using different envelope waveforms for the different encoding signals leads to equipment complexity. On–off keying is suitable only for a binary system. One parameter that remains is frequency, and frequency shift keying is commonly used in an incoherent system. That is, we take the transmitted signals as

$$s_i(t) = A \, \cos[(\omega_c + \delta\omega_i)t + \theta_i], \qquad 0 \le t \le T, \tag{7.64}$$

where the frequency offsets $\delta\omega_i$ encode the character, and the phase angles θ_i are independent uniform random variables. We can arrange the set Eq. (7.64) to be orthogonal by considering the time cross-correlations of the signals $s_i(t)$. Where $\Delta\theta_{ij} = \theta_i - \theta_j$ and $\Delta f_{ij} = \delta f_i - \delta f_j$, these are

$$\int_0^T s_i(t)s_j(t)\ dt \cong (A^2T/2)\ \cos(\pi T\Delta f_{ij} + \Delta\theta_{ij})\ [\sin(\pi T\Delta f_{ij})]/(\pi T\Delta f_{ij}).$$

In order to obtain orthogonal signals, we select the frequency offsets such that

$$\delta f_i = k/T, \tag{7.65}$$

where k is an integer. The multiplier $A^2T/2$ is the signal energy.

As in Eq. (7.22), we need to compute $\mathscr{E}p_i[\bar{r}(t) \mid \theta_i]$ for the m signals $s_i(t)$ of Eq. (7.64). Comparing Eq. (7.23) with Eq. (7.64), the phase modulation is $\phi_i(t) = \delta\omega_i t$. The corresponding decision variables from Eq. (7.31) are

$$q_i = \frac{1}{2}\left|\int_0^T \bar{r}(t)\ \exp(-j\delta\omega_i t)\ dt\right| \cdot \tag{*(7.66)}$$

That is, the detection variables can be computed from the powers in the frequency bins of a Fourier analysis of the complex envelope of the received signal. As in Sect. 7.3, they can also be computed from the samples at time T of the real envelopes of the filter outputs as $q_i = \frac{1}{2}|\bar{y}_i(T)|$, where

$$y_i(t) = 2\int_0^t r(t')\ \cos[(\omega_c + \delta\omega_i)(t - t')]\ dt', \tag{7.67}$$

where we have dropped an irrelevant phase angle $\delta\omega_i T$ in the integrand. Corresponding to Eq. (7.33), the detector Eq. (7.62) becomes

$$j \Leftarrow \max_i I_0(2Aq_i/N_0) \quad \text{or} \quad \max_i(q_i) \quad \text{or} \quad \max_i(q_i^2). \tag{7.68}$$

The probability of error of the detector Eq. (7.68) is computed as was done in Sect. 6.7 in the case of known phase. As in Eq. (7.41) consider the complex random variable

$$g_i = \frac{1}{2}\int_0^T \bar{r}(t)\ \exp(-j\delta\omega_i t)\ dt = x_i + jy_i, \tag{7.69}$$

so that $q_i = |g_i|$. Then

$$\begin{aligned}
\mathscr{E}(g_i \mid H_j) &= \frac{1}{2}\int_0^T \mathscr{E}[\bar{r}(t) \mid H_j]\ \exp(-j\delta\omega_i t)\ dt \\
&= \frac{A}{2}\int_0^T \exp\{j[(\delta\omega_j - \delta\omega_i)t + \theta_j]\}\ dt \\
&= \left(\frac{AT}{2}\right)\ \exp(j\theta_i)\delta_{ij},
\end{aligned} \tag{7.70}$$

using the orthogonality condition Eq. (7.65).

The variance calculation uses Eq. (7.69) written for the noise complex envelope $\tilde{n}(t) = n_i(t) + jn_q(t)$. With the usual sampling plan leading to Eq. (7.45), with $\delta t = 1/B$, we have for example

$$
\begin{aligned}
\mathcal{E}(x_k x_l) &= \left(\frac{\delta t}{2}\right)^2 \sum_{i,j=1}^{m} \mathcal{E}\{[n_i(i/B)\cos(\delta\omega_k i/B) + n_q(i/B)\sin(\delta\omega_k i/B)] \\
&\qquad \times [n_i(j/B)\cos(\delta\omega_l j/b) + n_q(j/B)\sin(\delta\omega_l j/B)] \\
&= \left(\frac{\delta t}{2}\right)^2 \sum_{i=1}^{m} (N_0 B) \\
&\qquad \times [\cos(\delta\omega_k i/B)\cos(\delta\omega_l i/B) + \sin(\delta\omega_k i/B)\sin(\delta\omega_l i/B)] \\
&= \frac{N_0}{4} \int_0^T [\cos(\delta\omega_k t)\cos(\delta\omega_l t) + \sin(\delta\omega_k t)\sin(\delta\omega_l t)]\, dt \\
&= (N_0 T/4)\delta_{kl}, \qquad\qquad\qquad\qquad\qquad\qquad\qquad (7.71)
\end{aligned}
$$

where we used Eqs. (7.13), (7.14) and the orthogonality relation Eq. (7.65).

Since the noise $n(t)$ is assumed to be Gaussian, the decision variables x_i, y_i form a set of Gaussian variables. From Eq. (7.71) and the corresponding relations for $x_k y_l$, $y_k y_l$, they are independent from one channel to another. Therefore, the decision variables q_i are independent. As developed in Sect. 7.4, under hypothesis H_j the variable q_j is Rician with density Eq. (7.49) and the variables $q_{i\neq j}$ are Rayleigh, as in Eq. (7.50). The parameter in the densities is $\sigma^2 = \mathcal{E}[x^2 \mid r(t) = n(t)] = N_0 T/4$.

Following Sect. 6.7, the probability of error in this system can be determined in terms of the probability that, with symbol H_j sent, the decision variable $q_j \geq q_{i\neq j}$. That is,

$$
\begin{aligned}
1 - P_e &= \int_0^\infty P\{q_{i\neq j} < q_j \mid H_j\} p(q_j \mid H_j)\, dq_j \\
&= \int_0^\infty \left\{ \prod_{i\neq j} P(q_i < q_j \mid H_j) \right\} p(q_j \mid H_j)\, dq_j \\
&= \int_0^\infty \left(\frac{q_j}{\sigma^2}\right) \exp\left[-\left(q_j^2 + \frac{A^2 T^2}{4}\right) \Big/ 2\sigma^2\right] I_0\left(\frac{q_j A T}{2\sigma^2}\right) \\
&\qquad \times \left[\int_0^{q_j} \left(\frac{q}{\sigma^2}\right) \exp\left(\frac{-q^2}{2\sigma^2}\right) dq\right]^{m-1} dq_j.
\end{aligned}
$$

The inner integral in this is easily evaluated and the power expanded using the binomial series. The result is

$$1 - P_e = \sum_{k=0}^{m-1} (-1)^k \binom{m-1}{k} \exp\left(\frac{-A^2T^2}{8\sigma^2}\right)$$

$$\times \int_0^\infty \left(\frac{q_j}{\sigma^2}\right) \exp\left[\frac{-(k+1)q_j^2}{2\sigma^2}\right] I_0\left(\frac{q_j AT}{2\sigma^2}\right) dq_j$$

$$= \sum_{k=0}^{m-1} (-1)^k \binom{m-1}{k} (k+1)^{-1} \exp\left[\frac{-A^2T^2k}{8(k+1)\sigma^2}\right].$$

The parameter in the exponent can be written

$$A^2T^2/8\sigma^2 = A^2T/2N_0 = E/N_0,$$

where the energy of the symbol waveform is

$$E = \int_0^T A^2 \cos^2(\omega_c + \delta\omega_j)t \; dt = \frac{A^2T}{2}.$$

Then we have

$$P_e = 1 - \sum_{k=0}^{m-1} (-1)^k \binom{m-1}{k} (k+1)^{-1} \exp\left[\frac{-kE}{(k+1)N_0}\right]. \quad *(7.72)$$

This is plotted in Fig. 7.6 for various m, in terms of the SNR per bit:

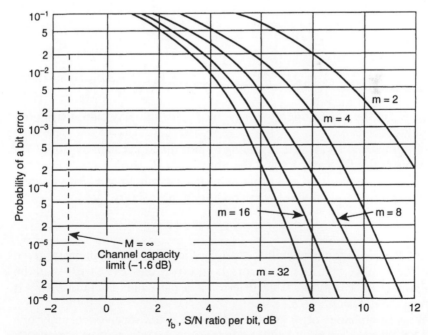

FIGURE 7.6 Probability of a bit error for noncoherent detection of orthogonal signals. (From J. G. Proakis, "Digital Communications," McGraw Hill, 1983, with permission.)

$$\alpha^2 = (E/N_0)/\log_2(m). \tag{7.73}$$

This is more commonly written as the energy per bit in ratio to the noise power spectral density: $\alpha^2 = E_b/N_0$. The character error probability P_e is converted to a per-bit basis as discussed in the following section. In the case of a binary system ($m = 2$), Eq. (7.72) becomes

$$P_e = (1/2) \exp(-E/2N_0) = (1/2) \exp(-A^2T/4N_0). \tag{7.74}$$

Random Amplitude—Rayleigh Fading

In a multipath environment, the amplitudes of the received signals Eq. (7.64) may differ from the amplitudes at the transmitter. Just as in Sect. 7.5, the received amplitudes are reasonably modeled with the Rayleigh density Eq. (7.57). We will again assume the case of slow fading, in which the amplitude A is constant over the duration T of the waveform but has a value which differs from pulse to pulse. Since the computation of the decision variables Eq. (7.69) does not require that the amplitude of the received signal be known, the detection strategy is again to choose as the transmitted symbol that for which the decision variable q_i is largest. The error probability will change, however. With the density Eq. (7.57), the average of the error probability Eq. (7.72) is

$$\mathcal{E}(P_e) = \int_0^\infty \left(\frac{A}{A_0^2}\right) \exp\left(\frac{-A^2}{2A_0^2}\right) P_e(A) \, dA$$

$$= 1 - \sum_{k=0}^{m-1} (-1)^k \binom{m-1}{k} \left[1 + k\left(\frac{E_{av}}{N_0} + 1\right)\right]^{-1}. \tag{7.75}$$

Here E_{av} is the signal energy E averaged over the Rayleigh distributed amplitudes A.

In normalized form, it is revealing to plot functions such as Eq. (7.75) in terms of the SNR per bit and the probability P_b of bit error. The latter can be determined from the probability P_s of symbol error if we assume equally probable symbols. From conditional probability, we have

$$P_b = P(b \mid s) \, P_s / P(s \mid b),$$

where $P(b \mid s)$ is the probability of any particular bit being in error, given that the symbol in which it occurs is in error, and $P(s \mid b) = 1$ is the probability of a symbol being in error if a particular bit is in error. It remains to determine $P(b \mid s)$.

With an alphabet of m symbols possible at the receiver, exactly one will be correct and $m - 1$ will be incorrect. Assume that m is even. (It is often a power of 2.) Let us suppose that each symbol is a bit sequence, and confine attention to the $m - 1$ incorrect symbols and to one particular bit position. Of the total m symbols, exactly $m/2$ will have a 0 bit in the position of interest and $m/2$ will have a 1 bit. Suppose the correct symbol has 1 in that position. Then of the $m - 1$ symbols in error, $(m/2) - 1$ will have 1 (correct) and $m/2$ will have 0 and thereby have a bit error. The same conclusion ($m/2$ wrong symbols with a bit error) holds if 0 is the bit in the correct symbol. With equally likely symbols, we then conclude that

$$P(b \mid s) = (m/2)/(m - 1),$$

so that

$$P_b = [(m/2)/(m - 1)]P_s.$$

The function Eq. (7.75) is plotted in Fig. 7.7 in terms of P_b for various values of m as a function of the SNR per bit $\alpha^2 = E_{av}/N_0 \log_2(m)$.

In the case of a binary system ($m = 2$), Eq. (7.75) becomes

$$\mathscr{E}(P_e) = 1/(2 + E_{av}/N_0). \tag{7.76}$$

Figure 7.8 shows the three binary cases Eqs. (6.35), (6.87), (7.76) on the same plot for comparison. The strong deleterious effect of fading is clear. In Chapter 8 we will discuss some techniques (diversity schemes) for mitigating the situation. In essence, the signal is transmitted simultane-

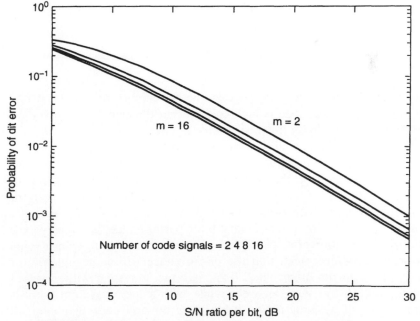

FIGURE 7.7 Performance of fading m-ary orthogonal incoherent system.

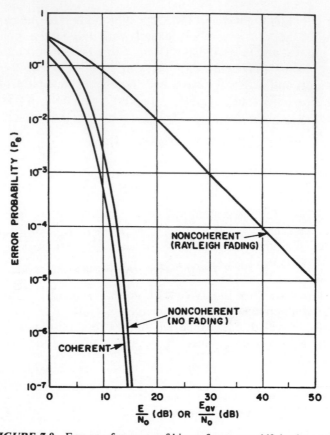

FIGURE 7.8   Error performance of binary frequency shift keying system.

ously over multiple channels, so that the probability that at least one channel is not in deep fade is high enough that the signal is received on at least one channel with low probability of error.

The effects of fading which are reflected in Eq. (7.75) are based on the assumption of a Rayleigh distribution for the received signal amplitude. If the SNR is rather low, the probability of error for a fading system with Rayleigh density and average energy E_{av} may actually be smaller than for a nonfading system with constant energy $E = E_{av}$. This is because the constant energy E may provide very poor performance, whereas with a statistical distribution of values E there is some chance of drawing a value E from the ensemble such that the performance will be acceptable for that symbol. On the other hand, if E_{av} is large, a system with constant $E = E_{av}$ will always perform well, while with a fading system there is some chance of drawing a low value of E for any particular symbol.

The exact situation depends on the distribution selected for the signal amplitude A.

7.7 SIGNALS WITH RANDOM FREQUENCY

Let us consider again the problem of detecting the presence of a narrow-band signal in white Gaussian noise. We will assume a known constant amplitude A for the received signal, over a specified interval of length T. The hypotheses are then:

$$H_1: \quad r(t) = A\cos(\omega t + \theta) + n(t),$$
$$H_0: \quad r(t) = n(t), \qquad 0 \le t \le T. \tag{7.77}$$

We assume that the frequency f is unknown but is located somewhere in a specified band of width B around a carrier frequency f_c. This would be the case, for example, if the signal to be detected were the return from a target in motion with unknown radial speed. In that case, the frequency f of the received signal would be the carrier f_c offset by the Doppler shift $f_D = 2Vf_c/c$, where V is the signed radial speed, which we assume to be algebraically much smaller than the speed of propagation c in the medium. (The factor 2 in the Doppler shift assumes an active system, such as a radar.)

In the case that we assume the carrier phase angle θ is unknown and is a uniformly distributed random variable, the likelihood ratio corresponding to problem Eq. (7.77), conditioned by the unknown Doppler frequency, is that in Eq. (7.30):

$$\lambda[\bar{r}(t) \mid \omega_D] = I_0(2Aq/N_0) \exp(-A^2T/2N_0). \tag{7.78}$$

Here we have, from Eq. (7.31),

$$q = \frac{1}{2} \left| \int_0^T \tilde{r}(t) \exp(-j\omega_D t) \, dt \right|. \tag{7.79}$$

In this we have treated the Doppler shift as a phase modulation $\phi(t) = \omega_D t$.

The detector for the problem Eq. (7.77) results by thresholding the expected value of the likelihood ratio Eq. (7.78), taken over some density assumed for the unknown Doppler frequency f_D. We take that to be some specified density $p(f_D)$ over a range $f_L \le f_D \le f_U$. Discretizing that frequency range into $M = (f_U - f_L)/\delta f$ bins of width δf centered at values f_i, we can write the expected likelihood ratio as

$$\mathscr{E}(\lambda) = \delta f \exp\left(\frac{-A^2T}{2N_0}\right) \sum_{i=1}^{M} p(f_i) I_0\left(\frac{2Aq_i}{N_0}\right), \tag{7.80}$$

where

$$q_i = \frac{1}{2}\left| \int_0^T \tilde{r}(t)\exp(-j\omega_i t)\,dt \right|$$

is obtained from the power in the ith bin of a Fourier analysis of the data complex envelope. The receiver Eq. (7.80) is diagrammed in Fig. 7.9.

In the special case of a uniform density assumed for the unknown Doppler shift and with a small signal-to-noise ratio so that A/N_0 is small, the approximation

$$I_0(x) \cong 1 + x^2/4$$

can be used to write the receiver Eq. (7.80) as

$$\mathscr{E}(\lambda) = \frac{1}{M}\exp\left(\frac{-A^2 T}{2N_0}\right)\sum_{i=1}^{M}\left[1 + \left(\frac{Aq_i}{N_0}\right)^2\right].$$

Absorbing constants into the threshold, this receiver may be written

$$\gamma = \sum_{i=1}^{M} q_i^2 > \gamma_t.$$

That is, the powers in the bins of the Fourier analysis are simply summed, and the signal declared present if the sum exceeds a threshold set by the desired false alarm probability.

Rayleigh Fading Amplitude

As in each previous case in this chapter, we will now assume that the signal amplitude A in the problem Eq. (7.77) is an unknown constant selected in accord with the Rayleigh density Eq. (7.57). This gives one more variable to average over in the likelihood ratio. From Eq. (7.80)

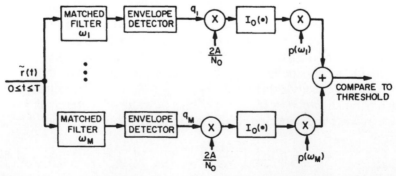

FIGURE 7.9 Optimum receiver for constant amplitude signal of unknown frequency.

we have

$$\mathscr{E}(\lambda) = \delta f \sum_{i=1}^{M} p(f_i) \int_0^\infty \left(\frac{A}{A_0^2}\right) \exp\left[-\left(\frac{A^2}{2A_0^2}\right)\left(1 + \frac{A_0^2 T}{N_0}\right)\right] I_0\left(\frac{2Aq_i}{N_0}\right) dA.$$

The integral has a simple closed form, and we obtain

$$\mathscr{E}(\lambda) = \delta f \left(1 + \frac{E_{av}}{N_0}\right)^{-1} \sum_{i=1}^{M} p(f_i) \exp\left[2\left(\frac{q_i^2 E_{av}}{N_0^2 T}\right) \Big/ \left(1 + \frac{E_{av}}{N_0}\right)\right],$$

where $E_{av} = A_0^2 T$ is the expected value of the signal energy. Assuming the uniform density for the unknown Doppler frequency and absorbing constants into the threshold, this receiver has the form

$$\gamma = \sum_{i=1}^{M} \exp\left[\frac{2q_i^2 E_{av}/N_0^2 T}{1 + E_{av}/N_0}\right] > \gamma_t. \qquad *(7.81)$$

In the case that the signal-to-noise ratio is reasonably high, the parameter q_i corresponding to the frequency bin with the signal will dominate the sum Eq. (7.81). In that case, the sum will be approximately equal to its largest term, and the detector Eq. (7.81) becomes

$$\gamma \cong \max_i \exp[2(q_i^2 E_{av}/N_0^2 T)/(1 + E_{av}/N_0)] > \gamma_t.$$

Absorbing constants into the threshold this becomes simply

$$\gamma' = \max_i q_i > \gamma_t'. \qquad (7.82)$$

This receiver is depicted in Fig. 7.10. The same argument can be made with respect to the receiver Eq. (7.80) and the monotonicity of the Bessel function used to obtain the receiver Eq. (7.82) in the case of known signal amplitude also.

The receiver Eq. (7.82) suggests that the value of the unknown Doppler shift can be determined from the number i of the variable q_i (if any) which exceeds the threshold γ_t'. That result comes about by considering the multiple-hypothesis problem:

$$H_i: \quad r(t) = A \cos(\omega_i t + \theta) + n(t), \qquad i = 1, M,$$
$$H_0: \quad r(t) = n(t).$$

From Sect. 5.8 the procedure for minimum probability of error is to choose that hypothesis for which the probability $P[\omega_i \mid \bar{r}(t)]$ is maximum. In the case of equal prior probabilities of the values ω_i, that corresponds to choosing the hypothesis for which $p[\bar{r}(t) \mid \omega_i]$ is maximum. If the data have random parameters, such as carrier phase or signal amplitude, the appropriate expected value is used. The form of receiver which results is just that of Fig. 7.10, with the additional feature that, if the threshold

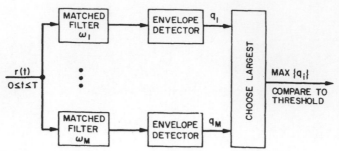

FIGURE 7.10   Receiver for detecting signal with one of M frequencies or noise only.

is exceeded, the value of the unknown frequency is taken to be that corresponding to the channel with largest q_i.

7.8 SIGNALS WITH RANDOM TIME OF ARRIVAL

In all of the preceding work, we have assumed that the signal, if present, occupied a known interval $0 \leq t \leq T$. In this section we assume that the signal, if present, occupies some interval of length T with unknown position on a larger segment of the time axis. That is the case, for example, with a radar system attempting to detect targets which might be present at any range. In this case, just as was done for unknown frequency in the preceding section, we can divide the full observation interval into some number of time segments each of length T, each of which starts at a time t_i. Each of these gives rise to a likelihood ratio conditioned by the (unknown) value of t_i which corresponds to a target. That is, assuming for example unknown carrier phase θ and specified signal amplitude A, as in Eq. (7.30) we compute

$$\lambda[\bar{r}(t) \mid t_i] = I_0(2Aq_i/N_0) \exp(-A^2 T/2N_0),$$

where now

$$q_i = \frac{1}{2} \left| \int_{t_i}^{t_i+T} \bar{r}(t) \, dt \right|. \tag{7.83}$$

The decision as to presence or absence of a target, regardless of range bin, is made by thresholding the averaged likelihood ratio. If we assume a uniform density for the bin times t_i, we then have the receiver

$$\gamma = \sum_{i=1}^{M} I_0 \left(\frac{2Aq_i}{N_0} \right) > \gamma_t.$$

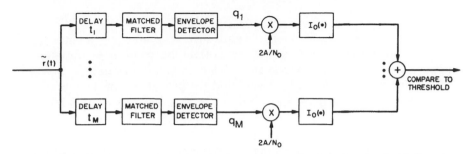

FIGURE 7.11 One way to implement optimum receiver for detecting signal with discrete arrival times.

This receiver is shown in Fig. 7.11. With the same approximations as discussed in the previous section with regard to unknown Doppler shift, both the presence and range of a target can be estimated by noting whether the maximum of the q_i exceeds a threshold and, if it does, which is the corresponding range bin $t_i \le t \le t_i + T$. In fact, since multiple targets may be present, we declare a target in every range bin for which the value q_i exceeds a threshold set by the false alarm rate which the system can tolerate. The same arguments apply in the case of random amplitude.

In the case of both unknown range and unknown Doppler shift, the two procedures of this section can be combined with those of the previous section. In short, in the case of uniform prior distributions of range and Doppler, we carry out a short-time Fourier analysis of the signal of length T corresponding to each range interval. A target is declared for each range–Doppler cell with power exceeding a threshold set by the false alarm rate. The receiver is shown in Fig. 7.12.

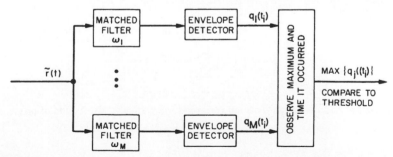

FIGURE 7.12 Optimum receiver for both detecting a signal and estimating its frequency and time of arrival.

Exercises

7.1   For the detection of a known signal such as $A \sin(\omega_c t + \theta)$ in white Gaussian noise, where θ is known, a correlation receiver is optimum. Suppose we believed that θ was zero, when in fact it was not.

(a)   For the correlation receiver, find the probability of detection as a function of θ and compare it to case where θ is correctly known. (Assume a radar-type problem where the null hypothesis is noise only.)

(b)   Also show that depending on the value of θ, the probability of detection may be less than the probability of false alarm.

7.2   For the incoherent matched filter of a sine wave (a matched filter followed by an envelope detector), show that the choice of phase for the filter is arbitrary.

7.3   For the hypotheses

$$H_1: \quad r(t) = A \sin(\omega_c t + \theta) + n(t), \qquad 0 \le t \le T$$
$$H_0: \quad r(t) = n(t), \qquad\qquad\qquad\quad 0 \le t \le T$$

assume that A and ω_c are constants, and $n(t)$ is white Gaussian noise of spectral density $N_0/2$. Assume that the phase is a random variable with density function $p(\theta) = [e^{\nu \cos \theta}/2\pi]I_0(\nu)$. Using the Neyman–Pearson criterion, show that the receiver can be implemented as shown in Fig. 7.13. Use the fact that

$$\int_0^{2\pi} \exp[(\nu + q_c) \cos \theta - q_s \sin \theta] \frac{d\theta}{2\pi} = I_0\{[(\nu + q_c)^2 + q_s^2]^{1/2}\}$$

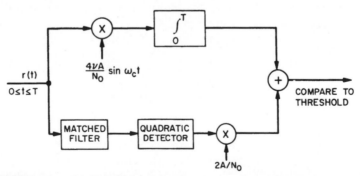

FIGURE 7.13   Neyman–Pearson receiver for a particular distribution of phase.

7.4   Consider the hypotheses

$$H_1: \quad r(t) = A \sin(\omega_c t + \theta) + n(t)$$
$$H_0: \quad r(t) = n(t)$$

where ω_c is a constant, θ is a random variable uniformly distributed $(0, 2\pi)$, and $n(t)$ is white Gaussian noise with spectral density $N_0/2$.

(a) Let A be a discrete random variable such that $P(A = 0) = 1 - p$ and $P(A = A_0) = p$. Determine the likelihood ratio using the Neyman–Pearson criterion. Can q be used as the test statistic?

(b) Show that the probability of detection is $P_D = (1 - p)P_f + pP_d(A_0)$ where P_{fa} is the probability of false alarm and $P_d(A_0)$ is the probability of detection for a constant amplitude signal of level A_0.

7.5 For the detection problem of the preceding exercise, assume that the amplitude has a distribution

$$p(A) = (1 - p)\,\delta(A) + p\,\frac{A}{A_0^2}\exp\left(-\frac{A^2}{2A_0^2}\right), \qquad A \geq 0,\ p \neq 0$$

(a) Determine the likelihood ratio using the Neyman–Pearson criterion. Can q be used as the test statistic?

(b) Show that the probability of detection is

$$P_d = (1 - p)P_f + pP_f^u$$

where

$$u = \frac{1}{1 + (TA_0^2/N_0)}$$

Note that as A_0 becomes arbitrary large, P_d approaches p (assuming low P_f).

7.6 For the detection problem of Exercise 7.4, assume that

$$p(A) = \sum_{i=1}^{M} p_i \delta(A - A_i)$$

Can q be used as the test statistic?

7.7 The probability of detection of a Rayleigh fading signal is given by Eq. (7.58). Rederive this result by considering the following problem. A signal

$$r(t) = a(t)\cos(\omega_c t + \phi(t)) + n(t), \qquad -\infty < t < \infty$$

is passed through a linear filter, $h(\tau) = \cos \omega_c \tau,\ 0 \leq \tau \leq T$, followed by an envelope detector. The signal $a(t)\cos(\omega_c t + \phi(t))$ is a Gaussian process, and $R_a(\tau)$ and $\phi(t)$ are substantially constant over any interval T. Make any reasonable narrowband assumption.

(a) Find the probability that a sample of the detector output is greater than a threshold η.

(b) This result is the same as Eq. (7.58). Explain why this is so.

7.8   Consider the detection problem

$$H_1: \quad r(t) = A \cos \omega_1 t + B \cos(\omega_2 t + \phi) + n(t), \qquad 0 \le t \le T$$
$$H_0: \quad r(t) = B \cos(\omega_2 t + \phi) + n(t), \qquad\qquad\quad 0 \le t \le T$$

where A, B, ω_1, ω_2 are known constants, $n(t)$ is white Gaussian, and ϕ is uniformly distributed over $(0, 2\pi)$. If

$$\int_0^T \cos \omega_1 t \cos \omega_2 t \, dt = \int_0^T \cos \omega_1 t \sin \omega_2 t \, dt = 0$$

show that the optimum receiver can use $\int_0^T r(t) \cos \omega_1 t \, dt$ as the test statistic. Discuss

7.9   Redo the preceding problem with the change

$$H_1: \quad r(t) = A \cos(\omega_1 t + \theta) + B \cos(\omega_2 t + \phi) + n(t)$$

where θ is uniformly distributed $(0, 2\pi)$ and is statistically independent of ϕ. Assume that

$$\int_0^T \cos(\omega_1 t + \theta) \cos(\omega_2 t + \phi) \, dt = 0$$

for all θ and ϕ. Show that the optimum receiver can use q as the test statistic, where

$$q^2 = \left[\int_0^T r(t) \cos \omega_1 t \, dt \right]^2 + \left[\int_0^T r(t) \sin \omega_1 t \, dt \right]^2$$

Discuss.

7.10   For the following series of problems, the hypotheses are of the form

$$H_1: \quad r(t) = A \cos(\omega_0 t + \theta) + [1 + m \cos(\omega_0 t + \phi)] s(t) + n(t)$$
$$H_0: \quad r(t) = n(t)$$

where $s(t)$ and $n(t)$ are bandlimited Gaussian processes, $|\omega| \le 2\pi B = \Omega$, with spectral density $S_0/2$ and $N_0/2$ respectively. Also, $\omega_0 \ll \Omega$ and $m \ll 1$. (The signal consists of an amplitude modulated noiselike signal and the modulating signal is also present as an additive term.) Assume the received signal is sampled at intervals $1/2B$ and the total number of samples is $2BT$. Assume these samples are statistically independent. Define V_n and $V_{sn}(i)$ as the variance of $r_i = r(t_i)$ under hypotheses H_0 and H_1 (for given θ and ϕ) respectively and assume that $\prod_{i=1}^{2BT} [V_n/V_{sn}(i)]^{1/2}$ is sensibly independent of ϕ.

   (a)   Show that the conditional likelihood ratio is monotonically related as

$$\lambda(\mathbf{r} \mid \theta, \phi) \sim \exp - \tfrac{1}{2} \sum \left\{ \frac{[r_i - A\cos(\omega_0 i/2B + \theta)]^2}{V_{sn}(i)} - \frac{r_i^2}{V_n} \right\}$$

(b) For $A = 0$ and ϕ known, show that the likelihood ratio receiver can be implemented as shown in Fig. 7.14. Because $m \ll 1$, assume that

$$\frac{1}{V_n} - \frac{1}{V_{sn}(i)} \simeq \alpha \left[\beta + \cos\left(\frac{\omega_0 i}{2B} + \phi \right) \right]$$

where

$$\alpha = \frac{2mS_0}{(S_0 + N_0)^2 B} \qquad \text{and} \qquad \beta = \left(\frac{S_0}{N_0} \right)\left(\frac{S_0 + N_0}{2mS_0} \right)$$

Discuss the receiver qualitatively.

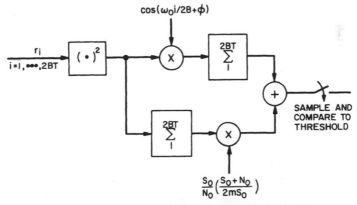

FIGURE 7.14 Maximum likelihood receiver ($A = 0$, Φ known).

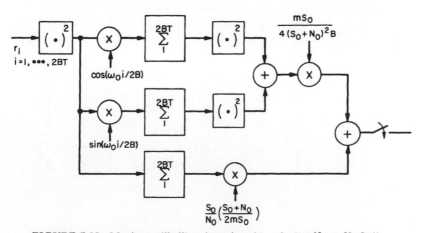

FIGURE 7.15 Maximum likelihood receiver ($A = 0$, ϕ uniform $[0, 2\pi)$).

(c)   Assume that $A = 0$, and that ϕ is uniformly distributed $(0, 2\pi)$. Show that the receiver may be implemented as shown in Fig. 7.15. Use the small signal assumption that $\ln I_0(v) \simeq v^2/4$. Discuss.

(d)   Assume that A is a constant not equal to zero, and that $\theta = \phi$ is a constant. Show that the receiver may be implemented as shown in Fig. 7.16. Hint: Use previous assumption for $1/V_n - 1/V_{sn}(i)$

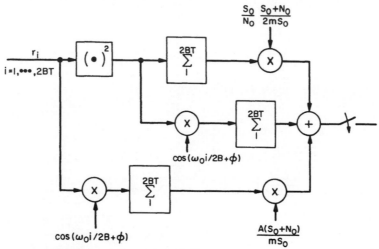

FIGURE 7.16   Maximum likelihood receiver ($\theta = \phi$ known).

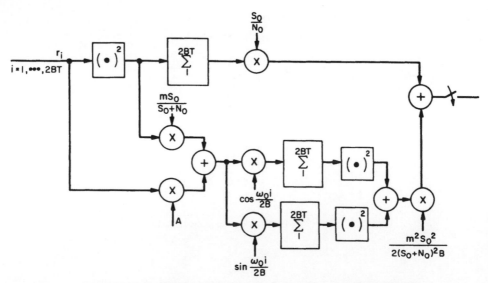

FIGURE 7.17   Maximum likelihood receiver ($\theta = \phi$ uniform $[0,2\pi)$).

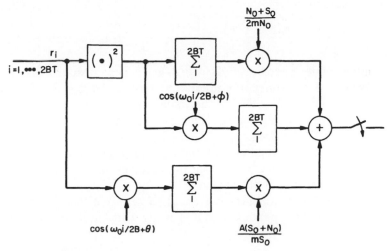

FIGURE 7.18   Maximum likelihood receiver (θ, ϕ known).

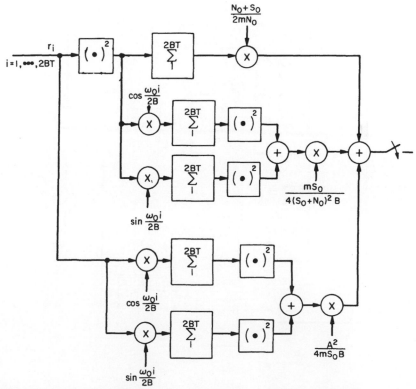

FIGURE 7.19   Maximum likelihood receiver (θ and ϕ independent and uniform $[0, 2\pi)$).

and assume

$$\sum \frac{r_i \cos(\omega_0 i/2B + \phi)}{V_{sn}(i)} \approx \frac{1}{(S_0 + N_0)B} \sum r_i \cos(\omega_0 i/2B + \phi)$$

and assume $\sum [\cos^2(\omega_0 i/2B + \phi)]/V_{sn}(i)$ is a constant. Discuss.

(e)   Assume A is a constant not equal to zero, and $\theta = \phi$ is uniformly distributed $(0, 2\pi)$. Using the usual small signal approximation for $I_0(\cdot)$, show that the receiver may be implemented as shown in Fig. 7.17. Discuss.

(f)   Assume A is a constant not equal to zero, and that θ and ϕ are known but unequal. Show that the receiver may be implemented as shown in Fig. 7.18. Discuss.

(g)   Assume A is a constant not equal to zero and that θ and ϕ are statistically independent and uniformly distributed $(0, 2\pi)$. Using

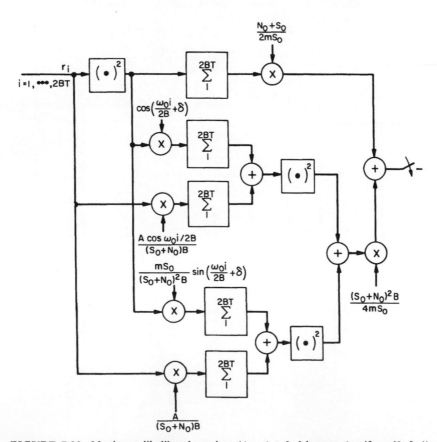

FIGURE 7.20   Maximum likelihood receiver ($\phi = \theta + \delta$, δ known, θ uniform $[0, 2\pi)$).

the small signal approximation for $I_0(\cdot)$, show that the receiver may be implemented as shown in Fig. 7.19.

(h)   Assume that A is a constant not equal to zero, that $\phi = \theta + \delta$, where δ is known, and that θ is uniformly distributed $(0, 2\pi)$. Using the small signal approximation for $I_0(\cdot)$, show that the receiver may be implemented as shown in Fig. 7.20. Discuss.

8 Multiple Pulse Detection of Signals

*I*n all of our discussions so far, it has been assumed that only one observation interval was available for use in deciding the presence or absence of a signal or in deciding which, if any, of a number of possible signals was present. In many situations, however, more than one observation is available. For example, a search radar beam with a width of 3° rotating at 10 rpm will illuminate a target for 0.05 seconds, during which time perhaps 50 pulses will be emitted. As another example, to defeat fading effects a communication system might be arranged to transmit the same information on multiple carrier frequencies or using multiple polarizations of the transmitting antenna. In such diversity transmission systems, multiple versions of the same transmitted signal are available for processing in the receiver.

In this chapter, we deal with detection strategies (receivers) which make decisions based on combined processing of all such multiple "looks" at the target or the transmitted signal. We assume that, in the case of radar, the target either is or is not present in every data interval and in the case of a communication system that the same signal is transmitted over every diversity channel. We further assume that the noise is independent from one data interval or channel to the next. In the case of radar, that is a realistic result of the fact that the noise is thermal in origin and is nominally white. (We will not discuss "clutter" noise, which results from echoes not of interest.) In the case of a diversity communication system, maximum benefit of the multiple channels results when the statistical properties of the channels are independent from one channel to another, so that such a system will be purposely arranged for that to be the case.

Using the assumption of independence of the noise from one pulse or channel to another, the detection problem amounts to a straightforward extension of the single-pulse procedures used in the preceding chapter. The performance resulting from multiple-pulse processing can be markedly superior to that of single pulse processing, especially in the case that the signal has unknown parameters, such as a random amplitude. In that case, if a particular data interval involves a signal with a low amplitude, there is a chance that the amplitude may be large in another interval. The receiver can then in effect concentrate attention on the favored interval in making its decision.

In considering various cases of multiple pulse problems, in this chapter we will follow the hierarchy of cases considered in Chapter 7, beginning with completely known signals and progressing through various levels of random parameters.

8.1 KNOWN SIGNALS

Suppose that the signals $s_i(t)$ whose presence or absence are to be decided, if present, are known in every detail in every observation interval, including carrier phase in the case of a bandpass signal. We emphasize that the signals $s_i(t)$ either are simultaneously present or none of them are present. Commonly the signals $s_i(t)$ are all the same and equal some $s(t)$, but it is easy to consider the slight generalization to a family of completely specified signals $s_i(t)$, $i = 1, M$, and we shall do so. The M intervals span known time segments of specified lengths T_i within which the signals begin at times τ_i, as in Fig. 8.1. For example, in a multipulse

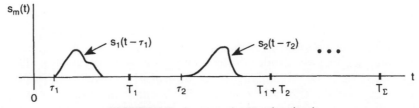

FIGURE 8.1   A composite M-pulse signal.

radar return, each interval begins at some specified range time delay after the transmission time of a pulse, and the signal within that interval spans a segment of time equal to the length of the transmitted pulse, in the case of a point target. The noise is assumed to be zero mean, additive, white, and Gaussian, with (two-sided) power spectral density $N_0/2$.

Under all these conditions, the problem can be viewed as the detection of a single known signal in white noise. That is, the M signals can be thought of as one signal

$$s_\Sigma(t) = \sum_{i=1}^{M} s_i(t - \tau_i), \qquad (8.1)$$

spanning a time interval of length $T_\Sigma = \Sigma\, T_i$. The noise waveforms $n_i(t)$ in the various intervals similarly can be thought of as together constituting a single noise waveform $n(t)$. Then the results of Sect. 6.2 apply directly. That is, the receiver is the matched filter Eq. (6.24), which now takes the form

$$S_\Sigma = \int_0^{T_\Sigma} s_\Sigma(t) r(t)\, dt$$

$$= \sum_{i=1}^{M} \int_0^{T_i} s_i(t) r_i(t)\, dt > S_{\Sigma t}, \qquad (8.2)$$

where T_Σ is the length of the composite signal $s_\Sigma(t)$ and $r_i(t)$ is the received signal in the ith interval. The receiver Eq. (8.2) is shown in Fig. 8.2 for

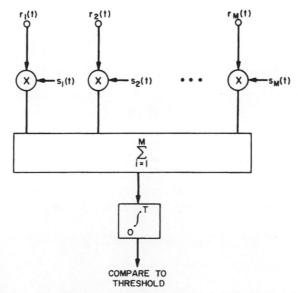

FIGURE 8.2   Correlation receiver for known signals in white Gaussian noise.

the common case that the observation intervals are all of the same length T. By interchanging the order of the summation and the integration in Fig. 8.2, to correspond directly to Eq. (8.2), the receiver might be realized more generally by a bank of matched filters whose outputs are sampled at the ends of the observation intervals and then summed.

The threshold for the receiver Eq. (8.2) might be set, for example, using the Neyman–Pearson criterion and Eq. (6.26):

$$P_f = \int_{\alpha S_{\Sigma t}/E_{\Sigma}}^{\infty} (2\pi)^{-1/2} \exp(-u^2/2)\, du. \tag{8.3}$$

Here α is the attained signal-to-noise ratio (SNR) of the receiver, which we take as

$$\alpha = (2E_{\Sigma}/N_0)^{1/2}, \tag{8.4}$$

and E_{Σ} is the energy of the composite signal Eq. (8.1):

$$E_{\Sigma} = \int_0^{T_{\Sigma}} s_{\Sigma}^2(t)\, dt.$$

In the usual case that the signals $s_i(t)$ are the same in each observation interval and equal a common $s(t)$, with the observation intervals having a common length T, the receiver Eq. (8.2) becomes

$$S = \sum_{i=1}^{M} \int_0^T s(t) r_i(t)\, dt, \tag{8.5}$$

where $r_i(t)$ is the received signal in the ith interval. The total energy is

$$E_{\Sigma} = \sum_{i=1}^{M} \int_0^T s^2(t)\, dt = ME,$$

where E is the energy of a single pulse $s(t)$. The (power) signal to noise ratio is then

$$\alpha^2 = 2ME/N_0, \qquad *(8.6)$$

which corresponds to a 3-dB increase in the single pulse SNR for every doubling of the number of pulses available.

Just as the false alarm performance of the multiple pulse receiver can be set by Eq. (8.3), which is Eq. (6.26) rewritten, the detection performance follows from Eq. (6.22). That is, for any value of the parameter U we have

$$P_f = \int_U^{\infty} (2\pi)^{-1/2} \exp\left(\frac{-u^2}{2}\right) du,$$

$$P_d = \int_{U-\alpha}^{\infty} (2\pi)^{-1/2} \exp\left(\frac{-u^2}{2}\right) du,$$

where α is the SNR Eq. (8.4) or its special case Eq. (8.6). The receiver operating characteristic of Fig. 5.7 applies, in which the SNR is taken as the value Eq. (8.6).

8.2 SIGNALS WITH UNKNOWN PHASE

Let us now consider the first loosening of the restrictions of the previous section. We shall suppose here that the signals $s_i(t)$ in the observation intervals are narrowband, with phase angles θ_i which are unknown to the receiver. This generalizes the case considered in Sect. 7.2 to multiple pulses in multiple observation intervals. As in Sect. 7.2, for compactness and generality we will take the signals in the form of their complex envelopes, assuming the processing of Sect. 3.6 has been carried out previously on the received data. This has the advantage that the formalism developed below will apply to any complex observables, such as the Fourier spectrum of the data.

Accordingly, from Eq. (7.23) we take the signal set as

$$\tilde{s}_i(t) = Aa(t) \exp\{j[\phi(t) + \theta_i]\}. \tag{8.7}$$

We assume the usual case that the observation intervals T_i are constant and equal T. Then the real envelope is assumed normalized such that

$$\frac{1}{T} \int_0^T a^2(t) \, dt = 1.$$

We take the phases θ_i to be uniformly distributed random variables on $[0, 2\pi)$. The scale factors A are for the time being assumed known and equal.

We might mention the trivial case that the phases θ_i in Eq. (8.7), although random, are all equal. This converts the problem to the single pulse case of Sect. 7.2, with the sought signal being the composite signal Eq. (8.1), with a single random phase angle for the composite. The sufficient statistic for the receiver is just the quantity q of Eq. (7.31), which becomes

$$q = \frac{1}{2} \left| \sum_{i=1}^M \int_0^T a(t)\tilde{r}_i(t) \exp[-j\phi(t)] \, dt \right|,$$

where the tilde indicates the complex envelope of the received data. The performance of the receiver is described by the ROC curve of Fig. 7.3 (dashed curves), using the SNR Eq. (7.55) as adjusted for the multiple pulse case:

$$\alpha_{dB} = 10 \log_{10}(2ME/N_0),$$

where ME is the energy of the M-pulse train of pulses each with energy E. Again note the 3-dB increase in SNR for each doubling of the number of pulses.

The case of more interest, however, is that in which the phase angles θ_i in Eq. (8.7) are independent random variables and therefore may differ in each observation interval. As developed in Chapter 5, for any of the usual detection criteria the strategy of the receiver is to threshold the likelihood ratio. For each pulse interval, from Eq. (7.30) the likelihood ratio is

$$\lambda[\tilde{r}_i(t)] = I_0(2Aq_i/N_0) \exp(-A^2T/2N_0), \tag{8.8}$$

where the sufficient statistic for the ith interval is

$$q_i = \frac{1}{2} \left| \int_0^T a(t)\tilde{r}_i(t) \exp[-j\phi(t)] \, dt \right|. \tag{8.9}$$

The amplitude A relates to the energy E of the signal on one interval through

$$E = A^2T/2, \tag{8.10}$$

and $I_0(x)$ is the modified Bessel function of the first kind and order zero.

Because the signal phase angles θ_i are independent of one another, the likelihood ratio for the M-interval problem is just the product of the likelihood ratios Eq. (8.8) which relate to the single interval problems. That is, letting $\mathbf{r}_i$ be the vector of data samples on interval i, we have

$$\lambda(\mathbf{r}) = \frac{p_1(\mathbf{r})}{p_0(\mathbf{r})} = \frac{p_1(\mathbf{r}_1, \ldots, \mathbf{r}_M)}{p_0(\mathbf{r}_1, \ldots, \mathbf{r}_M)}$$

$$= \frac{\prod p_1(\mathbf{r}_i)}{\prod p_0(\mathbf{r}_i)} = \prod_{i=1}^M \frac{p_1(\mathbf{r}_i)}{p_0(\mathbf{r}_i)} = \prod_{i=1}^M \lambda(\mathbf{r}_i).$$

Then from Eq. (8.8) we have

$$\lambda[\tilde{r}(t)] = \exp\left(\frac{-MA^2T}{2N_0}\right) \prod_{i=1}^M I_o\left(\frac{2Aq_i}{N_0}\right). \tag{8.11}$$

Converting to the log likelihood ratio, Eq. (8.11) shows that the multipulse receiver is

$$V = \sum_{i=1}^M \ln I_0\left(\frac{2Aq_i}{N_0}\right) > V_t, \tag{*8.12}$$

where V_t is a threshold which will depend on the detection criterion. Since the quantities q_i are first formed, involving "detection" of the envelope of the signal in each interval, and then combined before making a decision, the receiver Eq. (8.12) is said to include *postdetection integration*. As in Fig. 7.2, the receiver Eq. (8.12) can be implemented using carrier matched filters as shown in Fig. 8.3.

One of two approximations to the receiver Eq. (8.12) might be used. First, as we are often in the position of attempting to detect "small" signals, that is, those for which the SNR $E/N_0 = A^2T/2N_0$ is small, the approximations

$$I_0(x) \cong 1 + x^2/4,$$
$$\ln(1 + x) \cong x$$

can be used to yield the receiver

$$V = \sum_{i=1}^{M} q_i^2 > V_t \qquad \text{*(8.13)}$$

in which the constant A/N_0 has been absorbed into the threshold. [Note from Eq. (8.9) that q_i is of the order of AT, so that the argument of the Bessel function is of the order of the SNR. Even though the SNR is small, however, some particular sample value of the random variable q_i might be large. Nonetheless, the approximation is useful because such cases are rare.] The receiver Eq. (8.13) is the *small-signal receiver* for the problem at hand. It is a *quadratic receiver*, because the statistics q_i are squared before postdetection integration.

Although it is less useful as a data model, for ease of implementation one might use the asymptotic expansion of the Bessel function,

$$I_0(x) \sim \exp(x)/(2\pi x)^{1/2},$$

to determine a *large-signal receiver*:

$$\sum_{i=1}^{M} \left[\frac{2Aq_i}{N_0} - \frac{1}{2} \ln \left(\frac{4\pi Aq_i}{N_0} \right) \right] > V_t.$$

Dropping the small second term and absorbing constants into the threshold yields the receiver:

$$V = \sum_{i=1}^{M} q_i > V_t. \qquad \text{*(8.14)}$$

This is called a *linear receiver*, since the statistics q_i are not squared before postdetection integration.

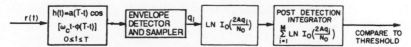

FIGURE 8.3   Receiver for incoherent detection of a train of M pulses.

8.3 PERFORMANCE OF THE QUADRATIC DETECTOR

The performance of the receiver Eq. (8.12) cannot be analyzed in terms of standard functions. However, the performance of the quadratic (small-signal) approximation Eq. (8.13) is easy to determine using results we have previously developed. The performance of the linear receiver Eq. (8.14), the large-signal approximation, has been calculated and tabulated. It turns out that these two receivers differ in their performances by an amount which is usually negligible. The per-pulse SNRs of the two receivers differ by less than a half dB for equivalent performance. It is usually assumed that these receivers bracket the possibilities and that the performance of the optimal receiver Eq. (8.12) is adequately described by that of either the linear or the quadratic receiver.

Let us first determine the performance (ROC curve) of the quadratic receiver Eq. (8.13). We need the density of the statistic q_i^2 in Eq. (8.9). From Eq. (7.49) we have

$$p_1(q_i) = (q_i/\sigma^2) \exp[-(q_i^2 + A^2T^2/4)/2\sigma^2]I_0(q_iAT/2\sigma^2), \qquad (8.15)$$

where the parameter $\sigma^2 = N_0T/4$ is the variance of either the in-phase (x) or quadrature (y) component of the noise output of the incoherent matched filter: $q^2 = x^2 + y^2$. From Eq. (4.65), Eq. (8.15) is the Rician density. The corresponding normalized density for $v = q_i/\sigma$ is that of Eq. (4.66) with parameter

$$\alpha^2 = (AT/2\sigma)^2 = A^2T/N_0 = 2E/N_0, \qquad (8.16)$$

the per-pulse power signal-to-noise ratio. That is,

$$p(v) = v \exp[-(v^2 + \alpha^2)/2]I_0(\alpha v). \qquad (8.17)$$

From Eq. (8.15) it is simple to find the density of the squared envelope $u = q_i^2$ as $p(u) = p(q_i)/(du/dq_i)$, taking account that q_i can only be nonnegative:

$$p(u_i) = (2\sigma^2)^{-1} \exp[-(u_i + A^2T^2/4)/2\sigma^2]I_0(u_i^{1/2}AT/2\sigma^2). \qquad (8.18)$$

From Eq. (4.81), this is seen to be the noncentral chi-squared density (with two degrees of freedom). The corresponding normalized density of $v = q_i^2/\sigma^2$ is

$$p(v) = \tfrac{1}{2} \exp[-(v + \alpha^2)/2]I_0(\alpha v^{1/2}), \qquad (8.19)$$

where again the noncentrality parameter of the density is the per-pulse SNR, $\alpha^2 = (AT/2\sigma)^2 = 2E/N_0$. The result Eq. (8.18) follows from Eq. (4.87) for $n = 2$, if we take account from Eqs. (8.7), (8.9) that $A_{1,2}$ in Eq. (4.84) are

$$A_1 = \left(\frac{A}{2}\right) \cos\theta \int_0^T a^2(t)\, dt, \qquad A_2 = \left(\frac{A}{2}\right) \sin\theta \int_0^T a^2(t)\, dt.$$

Then from Eq. (4.86) the parameter S in Eq. (4.87) is given by

$$S = A_1^2 + A_2^2 = A^2T^2/4,$$

leading to Eq. (8.18).

In the case of M pulses, the quadratic detector statistic Eq. (8.13) then has the noncentral chi-squared density with $2M$ degrees of freedom. Using the normalized receiver $v = V/\sigma^2$, from Eq. (4.89) we have

$$p_1(v) = \tfrac{1}{2}(v/\alpha^2)^{(M-1)/2} \exp[-(v + \alpha^2)/2]I_{M-1}(\alpha v^{1/2}), \qquad (8.20)$$

where the noncentrality parameter is the M-pulse SNR:

$$\alpha^2 = 2ME/N_0.$$

For the unnormalized receiver V itself, Eq. (8.20) yields

$$p_1(V) = (2\sigma^2)^{-1}(V/\sigma^2\alpha^2)^{(M-1)/2} \exp[-(V + \alpha^2\sigma^2)/2\sigma^2]I_{M-1}(\alpha V^{1/2}/\sigma).$$
$$*(8.21)$$

The density governing the false alarm probability of the quadratic receiver Eq. (8.13), after normalization by σ^2, is the density Eq. (8.20) with $\alpha = 0$. Taking account that, for small x, $I_n(x) \cong x^n/(2^n n!)$, there follows

$$p_0(v) = [2^M(M - 1)!]^{-1}v^{M-1} \exp(-v/2), \qquad (8.22)$$

which can also be obtained directly from Eq. (4.76) with $n = 2M$ degrees of freedom.

Using the densities Eqs. (8.20), (8.22) the ROC curve for the quadratic detector with postdetection integration can be found. From Eq. (8.22) for the normalized detector $v > v_t$ we have

$$P_f = \int_{v_t}^{\infty} p_0(v)\, dv = 1 - \int_0^{v_t} p_0(v)\, dv$$

$$= 1 - I(v_t/2M^{1/2}, M - 1), \qquad *(8.23)$$

where $I(u, p)$ is Pearson's form of the incomplete gamma function, Eq. (4.79), which has been tabulated. Pachares (1958) has tabulated the quantity Eq. (8.23) in terms of M and the detector threshold. Figure 4.11 shows a plot of Eq. (8.23), in which the ordinate is $v_t/2M$ and the abscissa is $100(1 - P_f)$. The curves are labeled by $2M$.

For $M \gg 1$, the expression Eq. (8.23) can be usefully approximated as (DiFranco and Rubin, 1980, p. 347)

$$P_f \cong (M/2\pi)^{1/2}(v_t/2M)^M[\exp(-v_t/2 + M)]/(v_t/2 - M + 1). \qquad (8.24)$$

The detection probability of the normalized receiver $v > v_t$ follows from Eq. (8.20) as

$$P_d = \int_{v_t}^{\infty} p_1(v)\, dv.$$

Making the change of variable $v = u^2$ in the integral yields

$$P_d = Q_M(\alpha, v_t^{1/2}), \qquad *(8.25)$$

where $\alpha^2 = 2ME/N_0$ is the M-pulse SNR and

$$Q_M(x, y) = \int_y^\infty u \left(\frac{u}{x}\right)^{M-1} \exp\left[\frac{-(u^2 + x^2)}{2}\right] I_{M-1}(xu) \, du \qquad (8.26)$$

is the generalized Marcum Q-function.

The detection probability can also be written in terms of the incomplete Toronto function, defined as:

$$T_u(m, n, r) = 2r^{n-m+1} \exp(-r^2) \int_0^u z^{m-n} \exp(-z^2) I_n(2rz) \, dz.$$

Then

$$P_d = 1 - T_{(v_t/2)^{1/2}}[2M - 1, M - 1, \alpha/\sqrt{2}].$$

For $M \gg 1$, a useful approximation for this case is (Helstrom, 1968, p. 224)

$$\Phi_C^{-1}(P_d) = \left[\frac{M}{\alpha^2 + M}\right]^{1/2} \left[\Phi_C^{-1}(P_f) - \frac{\alpha^2}{\sqrt{2M}}\right],$$

where $\alpha^2 = 2ME/N_0$ is the M-pulse SNR, and

$$\Phi_C(z) = \int_z^\infty (2\pi)^{-1/2} \exp\left(\frac{-u^2}{2}\right) du. \qquad (8.27)$$

8.4 GRAM–CHARLIER SERIES

The ROC curves for the M-pulse linear detector Eq. (8.14) must be computed numerically and are not available as standard tabulated functions. Robertson (1967) has done so, using a particular series expansion which we now wish to discuss. That is the Gram–Charlier series, which expresses an arbitrary density function as an infinite series whose leading term is a Gaussian density and whose higher-order terms are computed from the moments of the density being approximated. The series does not necessarily converge rapidly, if the density in question is not close to a Gaussian, but it has the advantages of allowing us to work in terms of moments, which are sometimes relatively easy to calculate, and also of linking the problem to the Hermite polynomials, which are standard functions.

Consider then some arbitrary probability density function $w(x)$, for which we assume that moments of all orders exist. The characteristic function is defined as

$$\Phi(\omega) = \int_{-\infty}^{\infty} w(x) \exp(j\omega x)\, dx = \mathcal{E}[\exp(j\omega x)]. \tag{8.28}$$

Letting $p = j\omega$, we obtain the moment generating function $\Phi(p)$ of the density. Expanding the exponential in a series and integrating term by term yields the moment generating function as

$$\Phi(p) = \sum_{n=0}^{\infty} (\mu_n/n!)p^n,$$

where $\mu_n = \mathcal{E}(x^n)$ is the nth moment of the density $w(x)$. [Hence the name: If we somehow obtain $\Phi(p)$, the moments of the density result by making a Taylor series expansion of it.] Then using the series

$$\ln(1 + z) = \sum_{m=1}^{\infty} (-1)^{m-1}\frac{z^m}{m}$$

yields the second moment function $\Psi(p)$ of the density as

$$\Psi(p) = \ln[\Phi(p)] = \sum_{m=1}^{\infty}\left[\frac{(-1)^{m-1}}{m}\right]\left[\sum_{n=1}^{\infty}\left(\frac{\mu_n}{n!}\right)p^n\right]^m$$

$$= \mu_1 p + \tfrac{1}{2}(\mu_2 - \mu_1^2)p^2 + \sum_{k=3}^{\infty}\left(\frac{\gamma_k}{k!}\right)p^k. \tag{8.29}$$

The *cumulants* (or *semiinvariants*) of a density are by definition the coefficients in the Taylor series of the second moment function $\Psi(p)$, so that the numbers γ_k are just the cumulants of the density $w(x)$. The leading terms involve the first- and second-order cumulants, which are the mean μ_1 and variance $\mu_2 - \mu_1^2 = \sigma^2$.

Letting the remainder term in Eq. (8.29) be $r(p)$, Eq. (8.29) yields the moment generating function as

$$\Phi(p) = \exp(\mu_1 p + \sigma^2 p^2/2)\exp[r(p)]. \tag{8.30}$$

Expanding the second exponential in a series in r yields

$$\exp[r(p)] = \sum_{m=0}^{\infty}\frac{r^m}{m!} = \sum_{m=0}^{\infty}\left(\frac{1}{m!}\right)\left[\sum_{k=3}^{\infty}\left(\frac{\gamma_k}{k!}\right)p^k\right]^m$$

$$= \sum_{n=0}^{\infty} c_n p^n, \tag{8.31}$$

where the c_n are combinations of the γ_n, which in turn from Eq. (8.29) are combinations of the moments μ_n of the original density $w(x)$. We will soon indicate a more convenient method of calculating them. The leading coefficients are $c_0 = 1$, $c_1 = c_2 = 0$.

Using Eq. (8.31) in Eq. (8.30) yields the characteristic function as

$$\Phi(\omega) = \exp\left(j\omega\mu_1 - \frac{\omega^2\sigma^2}{2}\right)\left[1 + \sum_{n=3}^{\infty} c_n(j\omega)^n\right]. \tag{8.32}$$

The leading term of this is the characteristic function of a Gaussian density with mean μ_1 and variance σ^2:

$$\exp(j\omega\mu_1 - \omega^2\sigma^2/2) \Leftrightarrow (2\pi\sigma^2)^{-1/2} \exp[-(x - \mu_1)^2/2\sigma^2].$$

Since multiplication by $j\omega$ in the frequency domain corresponds to differentiation in the x domain:

$$(j\omega)^n F(j\omega) \Leftrightarrow d^n f(t)/dt^n,$$

we have

$$(-j\omega)^n \Phi_w(j\omega) \Leftrightarrow d^n w(x)/dx^n, \tag{8.33}$$

where $w(x)$ is any density, and $\Phi_w(j\omega)$ is its characteristic function. The negative sign on the left arises because the characteristic function Eq. (8.28) is defined as the Fourier transform of the density, evaluated at $-\omega$.

Now let $\varphi(x)$ be the normalized Gaussian density:

$$\varphi(x) = (2\pi)^{-1/2} \exp(-x^2/2).$$

Applying Eq. (8.33) to Eq. (8.32) then results in

$$w(x) = (1/\sigma)\varphi[(x - \mu_1)/\sigma]$$

$$+ \frac{1}{\sigma}\sum_{n=3}^{\infty}\left[\frac{(-1)^n c_n}{\sigma^n}\right]\varphi^{(n)}\left(\frac{x - \mu_1}{\sigma}\right), \qquad *(8.34)$$

where as always the notation indicates that the derivative is $\varphi(u)$ differentiated with respect to u, with u then set to $(x - \mu_1)/\sigma$. This is the *Gram–Charlier series* for the density $w(x)$, expressing it as an infinite series of the Gaussian and its derivatives. Although the leading term is the Gaussian, the coefficients c_n do not necessarily decrease quickly with increasing n or even monotonically with increasing n. The series will be more or less useful depending on how well the density $w(x)$ is approximated by a Gaussian. It is in any event a useful form in which to express a numerical calculation, as we will see below in considering the ROC curve for the linear detector.

The coefficients c_n in the Gram–Charlier series Eq. (8.34) are conveniently calculated in terms of the Hermite polynomials. Let us write

$$a_n = (-1)^n c_n/\sigma^n,$$

and seek the coefficients a_n. It is convenient to consider the normalized density $w_0(y)$ with $y = (x - \mu_1)/\sigma$. Then Eq. (8.34) transforms to

$$w_0(y) = \sigma w(x) = \varphi(y) + \sum_{n=3}^{\infty} a_n\varphi^{(n)}(y). \tag{8.35}$$

Let us now consider explicitly the derivatives $\varphi^{(n)}(y)$. It happens that

$$\varphi^{(n)}(y) = (-1)^n \varphi(y) H_n(y), \tag{8.36}$$

where the $H_n(y)$ are the *Hermite polynomials*. These obey the recursion relation

$$H_{n+1}(y) = y H_n(y) - n H_{n-1}(y), \tag{8.37}$$

which is initialized with

$$H_0(y) = 1, \qquad H_1(y) = y. \tag{8.38}$$

They have the further property of being orthogonal to the Gaussian function and its derivatives:

$$\int_{-\infty}^{\infty} H_m(y) \varphi^{(n)}(y) \, dy = (-1)^n n! \delta_{mn}, \tag{8.39}$$

where δ_{mn} is the Kronecker delta.

Using the property Eq. (8.39) with Eq. (8.35) yields

$$\int_{-\infty}^{\infty} H_m(y) w_0(y) \, dy = \int_{-\infty}^{\infty} H_m(y) \varphi(y) \, dy$$

$$+ \sum_{n=3}^{\infty} a_n \int_{-\infty}^{\infty} H_m(y) \varphi^{(n)}(y) \, dy = (-1)^m m! a_m.$$

That is, the coefficients in the expansion Eq. (8.35) are given by:

$$a_n = \left[\frac{(-1)^n}{n!} \right] \int_{-\infty}^{\infty} H_n(y) w_0(y) \, dy. \qquad *(8.40)$$

Since the $H_n(y)$ are polynomials in y, Eq. (8.40) expresses the coefficients a_n as combinations of the moments of the normalized density $w_0(y)$. Let these latter be

$$\nu_n = \int_{-\infty}^{\infty} y^n w_0(y) \, dy,$$

with $\nu_0 = 1$, $\nu_1 = 0$, $\nu_2 = 1$, by the fact that the density is normalized. Using Eqs. (8.37), (8.38), it is found that

$$H_0(y) = 1,$$
$$H_1(y) = y,$$
$$H_2(y) = y^2 - 1,$$
$$H_3(y) = y^3 - 3y,$$
$$H_4(y) = y^4 - 6y^2 + 3,$$
$$H_5(y) = y^5 - 10y^2 + 15y,$$
$$H_6(y) = y^6 - 15y^4 + 45y^2 - 15,$$
$$H_7(y) = y^7 - 21y^5 + 105y^3 - 105y,$$

and so forth. Using these in Eq. (8.40) then yields the coefficients in the Gram–Charlier series Eq. (8.35) in terms of the moments of the normalized density as

$$a_0 = 1,$$
$$a_1 = 0,$$
$$a_2 = 0,$$
$$a_3 = -\nu_3/3!,$$
$$a_4 = (\nu_4 - 3)/4!,$$
$$a_5 = -(\nu_5 - 10\nu_3)/5!,$$
$$a_6 = (\nu_6 - 15\nu_4 + 30)/6!,$$
$$a_7 = -(\nu_7 - 21\nu_5 + 105\nu_3)/7!,$$
$$a_8 = (\nu_8 - 28\nu_6 + 210\nu_4 - 315)/8!,$$

and so forth.

Although the decrease as $1/n!$ in the coefficients a_n is encouraging, nonetheless for particular densities the moments ν_n may behave in such a way as to cause the series to converge rather slowly initially. When only a few terms of the series are to be used for calculation, it is sometimes recommended to use series constructed from coefficients of the following orders:

$$0, 3;$$
$$0, 3, 4, 6;$$
$$0, 3, 4, 6, 5, 7, 9;$$

and so forth. Such groupings lead to the family of approximations known as *Edgeworth series*.

8.5 PERFORMANCE OF THE LINEAR DETECTOR

There is no closed-form analysis of the performance of the multiple pulse linear detector Eq. (8.14). However, the results of the preceding section have been applied to calculation of the ROC curve for the linear detector Eq. (8.14), as we will summarize here.

The signal-plus-noise density for the corresponding normalized detector,

$$v = \frac{V}{\sigma} = \sum_{i=1}^{M} \frac{q_i}{\sigma} > \frac{V_t}{\sigma} = v_t, \tag{8.41}$$

with $\sigma^2 = N_0 T/4$, and for $M = 1$, is given by the Rician density Eq. (8.17),

$$p(v) = v \exp[-(v^2 + \alpha^2)/2]I_0(\alpha v), \tag{8.42}$$

where the parameter α is the single-pulse SNR:

$$\alpha = AT/2\sigma = A(T/N_0)^{1/2} = (2E/N_0)^{1/2}. \tag{8.43}$$

The moments of the density Eq. (8.42) are given in Eq. (4.67) and are known in terms of tabulated functions.

Considering the case of M pulses, there will be a density $p_M(v)$ resulting from M-fold convolution of the single-pulse density Eq. (8.42). From that, the probabilities of false alarm and detection could be determined as:

$$P_f = \int_{v_t}^{\infty} p_M(v \mid \alpha = 0) \, dv, \tag{8.44}$$

$$P_d(\alpha) = \int_{v_t}^{\infty} p_M(v \mid \alpha) \, dv. \tag{8.45}$$

From these the ROC curve could be constructed.

Robertson (1967) has carried out this formidable computation in the following way. In Eqs. (8.44), (8.45) a Gram–Charlier series Eq. (8.35) can be used to approximate the M-pulse density. As we developed earlier, the series coefficients a_n can be written in terms of the moments v_n of the M-pulse density. These in turn follow from expansion of the moment-generating function of the density:

$$\Phi_M(p) = \sum_{k=0}^{\infty} \left(\frac{v_n}{n!} \right) p^n. \tag{8.46}$$

Since the pulses are independent, the generating function $\Phi_M(p)$ is just the M-fold product of the generating function $\Phi_1(p)$ of the single-pulse case. That is,

$$\Phi_M(p) = \left[\sum_{i=0}^{\infty} \left(\frac{m_i}{i!} \right) p^i \right]^M, \tag{8.47}$$

where the m_i are the (known) moments of the density Eq. (8.42). Expanding the power in Eq. (8.47) and comparing powers of p with Eq. (8.46) then yields the moments v_n needed in the Gram–Charlier series. The ROC curve can then be calculated.

In Appendix 1 to this chapter we include a full set of Robertson's curves. A sample is shown in Fig. 8.4. The P_f scale markings in each decade are at values:

$$10, 5, 2, 1.$$

The signal-to-noise ratio values on the curves are $10 \log_{10}(E/N_0)$. The heavy dots on the right are at intervals of 0.2 dB in SNR.

An approximation to Robertson's curves has been given by Albersheim (1981). Particular values from the curves are given to within 0.2 dB by the relation

$$\alpha_{dB} = -5 \log_{10}(M) + [6.2 + 4.54/(M + 0.44)^{1/2}] \log_{10}(A + 0.12AB + 1.7B).$$

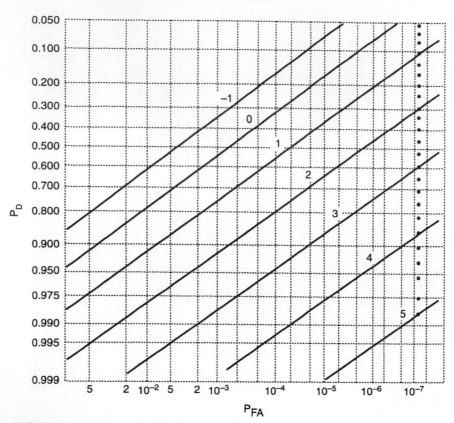

FIGURE 8.4   Receiver operating characteristics for detecting sine waves in white Gaussian noise (16 samples averaged). (From G. H. Robertson, BSTJ, **46**, April 1967, by permission of AT&T).

In this,

$$A = \ln(0.62/P_f),$$
$$B = \ln(P_d) - \ln(1 - P_d).$$

The accuracy of 0.2 dB holds over $1 \le M \le 8096$, $10^{-3} < P_f < 10^{-7}$, and $0.1 < P_d < 0.9$.

An alternative plotting of Robertson's curves is revealing. Figure 8.5 shows the behavior of two specific operating points in terms of required SNR as a function of number of pulses integrated. Beyond some number of pulses, depending on the operating point, the curves become nearly linear, with a slope such that the required SNR decreases 1.5 dB for every doubling of the number of pulses integrated. That behavior makes it easy to interpolate for M between any two of Robertson's curves, once the number of pulses is reasonably large.

Results of a final calculation by Robertson are shown in Fig. 8.6. That

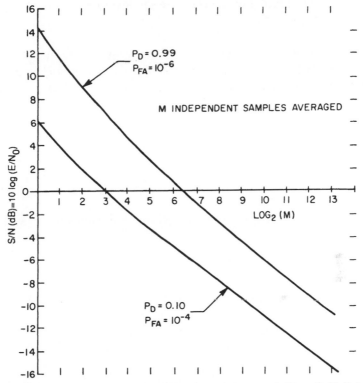

FIGURE 8.5 Performance related to number of samples averaged. (From G. H. Robertson, BSTJ, **46**, April 1967, by permission of AT&T.)

compares the SNR required by the linear and quadratic detectors, with the same number of pulses integrated, for three operating points. The result is the basis of the claim that the linear and quadratic detectors have very nearly equal performances.

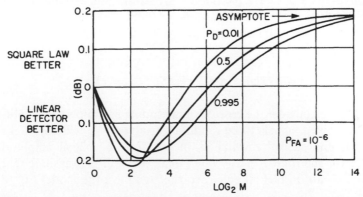

FIGURE 8.6 Comparison of square law and linear detectors for detection of M independent pulses. (From G. H. Robertson, BSTJ, **47**, March 1968, by permission of AT&T.)

8.6 THE CASE OF UNKNOWN PHASE AND KNOWN UNEQUAL AMPLITUDES

In the preceding sections we assumed not only that the amplitude A of each pulse was known but also that each pulse had the same amplitude. We now take the signals as in Eq. (8.7), but with unequal amplitudes:

$$\tilde{s}_i(t) = A_i a(t) \exp\{j[\phi(t) + \theta_i]\}. \tag{8.48}$$

The only effect is that in Eq. (8.8) the amplitude A becomes a different A_i for each pulse. The likelihood ratio Eq. (8.11) becomes

$$\lambda[\tilde{r}(t)] = \exp[-(\Sigma A_i^2)T/2N_0] \prod_{i=1}^{M} I_0(2A_i q_i/N_0), \tag{8.49}$$

and, absorbing the constant into the threshold, the receiver Eq. (8.12) becomes

$$V = \sum_{i=1}^{M} \ln I_0(2A_i q_i/N_0)] > V_t. \tag{8.50}$$

In the receiver combination, each matched filter output q_i is weighted by the corresponding signal amplitude A_i, so that signal replicas with larger power contribute more to the final decision. Absorbing constants into the threshold, the small-signal approximation to the receiver Eq. (8.50) is

$$V \cong \sum_{i=1}^{M} (A_i q_i)^2 > V_t. \tag{8.51}$$

The receiver Eq. (8.50) requires that the signal amplitudes be specified in advance and is different for each set of amplitudes. Therefore, the receiver is not uniformly most powerful for amplitude.

The performance of even the small-signal receiver Eq. (8.51) is not available in terms of standard functions. This is because each single pulse density involves a different parameter A_i, so that the convolution of the densities involves the individual parameter values. Fortunately, the model of known but unequal amplitudes is not particularly useful in applications.

8.7 UNKNOWN AMPLITUDE AND PHASE

We will now consider the case of signal fading and allow the amplitudes of the M received signals to be random variables. We will consider only the slow fading case, by which is meant that the signal amplitude is constant over the time T of any one pulse. We will first assume that the amplitudes A_i are in fact all equal, so that we have the model Eq. (8.7):

$$\tilde{s}_i(t) = Aa(t) \exp\{j[\phi(t) + \theta_i]\}. \tag{8.52}$$

We assume now, however, that the amplitude A is a random variable with some density $p(A)$. [This is called *scan-to-scan fading* in radar, because the fade condition stays the same over the many pulses placed on the target during a single scan of the rotating antenna. It is also known as the *Swerling-I model* or the *Swerling-III model,* depending on $p(A)$.] Later we will allow the M pulses to have different densities $p_i(A_i)$ (*pulse-to-pulse fading,* or the *Swerling-II* or *Swerling-IV model*). In this latter case we will assume that the amplitudes A_i are independent random variables. In every case the pulse phases θ_i are independent uniformly distributed random variables.

Scan-to-Scan Fluctuation

Let us assume first the case of scan-to-scan fading. We will treat the Swerling-I model, in which the amplitude A in Eq. (8.52) is a Rayleigh random variable, so that

$$p(A) = (A/A_0^2) \exp(-A^2/2A_0^2), \qquad A \geq 0. \tag{8.53}$$

Here the parameter is $A_0^2 = \mathcal{E}(A^2)/2$, as in Eq. (4.57). It is the average power of the carrier band signal

$$s_i(t) = Aa(t) \cos[\omega_c t + \phi(t) + \theta_i].$$

The Swerling-III model assumes the density

$$p(A) = (9A^3/2A_0^4) \exp(-3A^2/2A_0^2), \tag{8.54}$$

which models a single dominant scatterer together with multiple smaller scatters.

With the model Eq. (8.53), the M-pulse likelihood ratio Eq. (8.11) becomes a conditional likelihood ratio. The likelihood ratio for the problem of fading amplitude is then

$$\lambda[\bar{r}(t)] = \int_0^\infty (A/A_0^2) \exp(-A^2/2A_0^2)$$

$$\times \exp(-MA^2T/2N_0) \prod_{i=1}^M I_0(2Aq_i/N_0) \, dA. \qquad *(8.55)$$

This has no closed-form evaluation in terms of standard functions. However, we can consider the small-signal case $Aq_i \ll 1$. Then using $I_0(x) \cong 1 + x^2/4$, we have

$$\prod_{i=1}^M I_0(x_i) \cong 1 + (1/4) \sum_{i=1}^M x_i^2.$$

Using this in Eq. (8.55) yields

$$\lambda[\bar{r}(t)] \cong \int_0^\infty (A/A_0^2) \exp(-\gamma A^2) \, dA$$

$$+ \int_0^\infty (A^3/N_0^2 A_0^2) \exp(-\gamma A^2) \, dA \sum_{i=1}^M q_i^2,$$

where $\gamma = 1/2A_0^2 + MT/2N_0$. There results again the quadratic receiver

$$v = \sum_{i=1}^M q_i^2/\sigma^2 > v_t, \qquad\qquad *(8.56)$$

where again for convenience we normalize by $\sigma^2 = N_0 T/4$.

The analysis of the performance of the quadratic receiver Eq. (8.56) is complicated for the Swerling-I model under consideration. The situation with no signal present has already been analyzed, leading to Eq. (8.23) and the approximation Eq. (8.24). However, only after a long development (DiFranco and Rubin, 1980, p. 377) does it result that

$$P_d = 1 - I[v_t/2(M-1)^{1/2}, M-2]$$
$$+ (1 + 1/A_0^2)^{M-1} \exp[-v_t/2(1 + A_0^2)]$$
$$\times I\{v_t/[2(M-1)^{1/2}(1 + 1/A_0^2)], M-2\},$$

where $I(x, y)$ is the incomplete gamma function. For $MA_0^2 \gg 1$ and $P_f \ll 1$, a useful approximation is

$$P_d \cong (1 + 1/MA_0^2)^{M-1} \exp\{-v_t/[2(1 + MA_0^2)]\}. \qquad *(8.57)$$

An approximate inverse is (DiFranco and Rubin, 1980, p. 392)

$$A_0^2 = \Phi^{-1}(P_f)/[M^{1/2} \ln(1/P_d)], \qquad\qquad (8.58)$$

where $\Phi(z)$ is the Gaussian function Eq. (8.27). For specified P_f, P_d, and $M \geq 10$, this last is accurate to about 1 dB.

Pulse-to-Pulse Fluctuation

Let us now assume the fading model that the amplitudes of the pulses Eq. (8.52) are independent random variables with the common Rayleigh density:

$$p(A_i) = (A_i/A_0^2) \exp(-A_i^2/2A_0^2), \qquad A_i \geq 0. \qquad (8.59)$$

Here the average power of each pulse is the parameter $A_0^2 = \mathscr{E}(A_i^2)/2$, as in Eq. (4.57). The likelihood ratio Eq. (8.49) now becomes a conditional likelihood ratio, conditioned on the values of the A_i. The conditioning is removed by integration over the densities Eq. (8.59). Since the pulse amplitudes are independent, the multidimensional density of the M amplitudes factors, and we can carry out the integration of Eq. (8.49) as

$$\lambda(\bar{r}) = \prod_{i=1}^{M} \int_0^\infty \exp\left(\frac{-A_i^2 T}{2N_0}\right)$$

$$\times I_0(2A_i q_i/N_0)(A_i/A_0^2) \exp(-A_i^2/2A_0^2) \, dA_i. \tag{8.60}$$

As we noted in reaching Eq. (7.58), the integral in Eq. (8.60) has a closed-form expression [Gradshteyn and Ryzhik, 1965, #6.633.4] leading to

$$\lambda(\bar{r}) = \prod_{i=1}^{M} \frac{N_0}{N_0 + A_0^2 T} \exp\left[\frac{2A_0^2 q_i^2}{N_0(N_0 + A_0^2 T)}\right]. \tag{8.61}$$

Taking the log likelihood ratio and absorbing constants into the threshold, the receiver corresponding to Eq. (8.61) is just the quadratic receiver Eq. (8.13):

$$V = \sum_{i=1}^{M} q_i^2 > V_t. $$

The quadratic receiver is thus the optimum receiver for pulse-to-pulse fading signals, rather than an approximation to the optimum, as it is for the case of signals with specified amplitudes.

The performance of the quadratic receiver for the fading signal case can be analyzed in terms of the normalized receiver Eq. (8.56). For M pulses with any particular values of A_i, as in Eq. (8.20) the random variable v has the noncentral chi-squared density with $2M$ degrees of freedom as its conditional density:

$$p_1(v \mid A_1, \ldots, A_M) = \tfrac{1}{2}(v/\alpha^2)^{(M-1)/2}$$

$$\times \exp[-(v + \alpha^2)/2]I_{M-1}(\alpha v^{1/2}), \tag{8.62}$$

where the noncentrality parameter is now the M-pulse SNR:

$$\alpha^2 = \frac{2}{N_0} \sum_{i=1}^{M} E_i = \frac{T}{N_0} \sum_{i=1}^{M} A_i^2. \tag{8.63}$$

The false alarm probability for this problem of unknown signal amplitudes is just that of Eq. (8.23):

$$P_f = 1 - I(v_t/2M^{1/2}, M - 1). \tag{8.64}$$

The detection probability is

$$P_d = \int_{v_t}^\infty p_1(v) \, dv$$

$$= \int_{v_t}^\infty \int_0^\infty p_1(v \mid A_1, \ldots, A_M) p(\mathbf{A}) \, d\mathbf{A} \, dv = \mathscr{E} P_d(\mathbf{A}),$$

where the expectation is over the vector of random variables A_i and $P_d(\mathbf{A})$ is the detection probability for fixed $\mathbf{A}$. As seen in Eq. (8.62), the conditional detection probability depends on the signal amplitude vector $\mathbf{A}$ only through the SNR Eq. (8.63). We can therefore calculate

$$P_d = \mathscr{E} P_d(\alpha^2). \tag{8.65}$$

To that end we need the density of α^2.

Because A_i has the Rayleigh density Eq. (8.59), the normalized variable $u_i = A_i/A_0$ has the density

$$p(u_i) = A_0 p(A_i) = u_i \exp(-u_i^2/2),$$

which is the normalized Rayleigh density Eq. (4.56). Then $z = u_i^2$ has the exponential density:

$$p(z_i) = p(u_i)/2u_i = \tfrac{1}{2} \exp(-z_i/2),$$

taking account that u_i must be positive. This is also the chi-squared density with two degrees of freedom. Accordingly, since the A_i are independent, the variable

$$y = \sum_{i=1}^{M} \left(\frac{A_i}{A_0}\right)^2 \tag{8.66}$$

has the chi-squared density with $2M$ degrees of freedom:

$$p(y) = [2^M(M-1)!]^{-1} y^{M-1} \exp(-y/2). \tag{8.67}$$

Since the variable y of Eq. (8.66) is proportional to the SNR α^2, we can compute Eq. (8.65) in terms of $p(y)$. Using Eq. (8.67) in Eq. (8.65) then yields

$$P_d = \int_0^\infty p(y) \int_{v_t}^\infty \frac{1}{2}\left(\frac{v}{\varepsilon y}\right)^{(M-1)/2}$$

$$\times \exp[-(v + \varepsilon y)/2] I_{M-1}[(\varepsilon y v)^{1/2}] \, dv \, dy,$$

where $\varepsilon = A_0^2 T/N_0 = \mathscr{E}(A_i^2) T/2N_0 = \mathscr{E}(E_i)/N_0 = E_{av}/N_0$ is the ratio of the average per-pulse energy to the noise power spectral density. The integral over y has a closed-form expression [Gradshteyn and Ryzhik, 1965, #6.643.2] leading to

$$P_d = \int_{v_t/2(1+\varepsilon)}^\infty [(M-1)!]^{-1} z^{M-1} \exp(-z) \, dz.$$

Integrating by parts, this results in

$$P_d = \exp\left[\frac{-v_t}{2(1+\varepsilon)}\right] \sum_{n=0}^{M-1} \left[\frac{v_t}{2(1+\varepsilon)}\right]^n / n!. \qquad *(8.68)$$

Although the form Eq. (8.68) is attractive, for large M it is more easily computed as Pearson's form of the incomplete gamma function Eq. (4.79):

$$I(u, p) = [1/\Gamma(p + 1)] \int_0^{u(p+1)^{1/2}} z^p \exp(-z)\, dz,$$

so that Eq. (8.68) becomes

$$P_d = 1 - I[v_t/2(1 + \varepsilon)\sqrt{M}, M - 1].$$

The expression Eq. (8.68) is therefore one minus the distribution of the chi-squared density. It is plotted in Fig. 4.11. Equation (8.68) together with Eq. (8.64) can be used to compute the ROC curve in terms of the average single-pulse SNR $\varepsilon = \mathscr{E}(E_i)/N_0$. In Fig. 8.7 we show an example for $M = 16$, and in Appendix 2 to this chapter we show a set of curves for M ranging from 1 to 1024.

8.8 DIVERSITY RECEPTION

In Fig. 7.8 we have indicated the very severe detrimental effect of fading on detection of a single transmitted pulse. This is because there is some chance that the particular sample of pulse amplitude at hand may

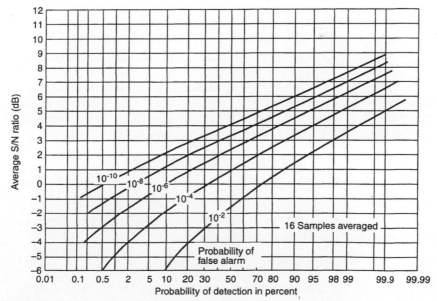

FIGURE 8.7 Detection performance for slow Rayleigh fading signals (16 samples averaged).

be very much below the average. A *diversity system* defeats this tendency. In such a system, more than one channel is used to send the same piece of information, or more than one "look" at a cell to be interrogated for a target is made available. The diversity may be in time, frequency, polarization, or space. The channels or looks are designed so that the properties of each are, as far as possible, independent of the properties of the other channels or looks. Then it becomes unlikely that all channels will simultaneously be in deep fade. If only one channel is not in fade, the message gets through. A much improved probability of target detection or probability of bit error is the result.

Usually the term "diversity" is reserved for a communication system which makes multiple transmissions of the same information over different channels. The diversity can be realized by successive transmission of the same signal in different time intervals, or by transmission on different carrier frequencies, or by reception by multiple receivers (space diversity) as might be done in a tropospheric scatter communication system. In this section we consider only a single, and rather limited, situation in order to draw on the results developed earlier in the chapter. Specifically, we consider the frequency shift keying (FSK) communication system described in Sect. 7.6, with the use of multiple carriers to provide the diversity. We assume a likelihood ratio receiver.

Diversity Improvement for FSK

Let us consider incoherent binary frequency shift keying in the presence of Rayleigh amplitude fading, using a diversity system of order M. There is assumed to be a carrier f_c, with the signals 0, 1 encoded as respectively $f_c + \delta f_0, f_c + \delta f_1$. The particular signal in question is sent over M channels with independent and unknown amplitudes and phases and with independent additive white Gaussian noises. We assume that the received signals in the channels $r_i(t)$, $i = 1, M$, have been converted to complex envelopes before processing. This might be done, for example, as

$$\tilde{r}(t) = x(t) + jy(t) = Aa(t) \exp[j\phi(t) + \theta]$$
$$= [2r(t) \cos(\omega_c t)]_{\text{LP}} - j[2r(t) \sin(\omega_c t)]_{\text{LP}},$$

where LP indicates low-pass filtering of the quantity in brackets.

Under hypothesis H_1, symbol 1 and frequency shift δf_1 were sent, so that the signals in the M channels are

$$\tilde{r}_i(t) = A_i a(t) \exp[j(\delta f_1 t + \theta_i)] + n_i(t).$$

We will assume that the amplitude A_i of each channel is Rayleigh distributed with common average power $\mathscr{E} A_i^2 = A_0^2/2$ over the observation interval $[0, T]$ and that the phase θ_i is uniform over $[0, 2\pi)$. The noise $n_i(t)$ is zero-mean white Gaussian with two-sided power spectral density $N_0/2$. The amplitude and phase in each channel are independent of those in every other channel. The noise $n_i(t)$ is independent of the noise waveform in every other channel. Under the alternative hypothesis H_0, the same expression holds, with the same values A_i, θ_i and the same noise waveforms $n_i(t)$ but with frequency offset δf_0.

As the received signals from all channels are available and we want to process them together, we need the joint likelihood ratio of the signals $\tilde{r}_i(t)$, $i = 1, M$, under each hypothesis H_1, H_0. Since the channel signals are independent, the joint likelihood function of the channels taken together is the product of the M individual likelihood functions. Equation (8.61) is the likelihood ratio corresponding to the test of the M channels with one signal present against the alternative of no signal present. That is,

$$\lambda_j(\mathbf{r}) = p(\mathbf{r} \mid H_j)/p_0(\mathbf{r}),$$

where $p(\mathbf{r} \mid H_j)$ is the probability density of the M-channel data ensemble given that one of the signals, 0 or 1, is present, and $p_0(\mathbf{r})$ is the M-channel density with noise only present. Then $p(\mathbf{r} \mid H_j) = \lambda_j(\mathbf{r})p_0(\mathbf{r})$.

Therefore, using Eq. (8.61), the M-channel likelihood ratio test for deciding between H_0 and H_1 is

$$\lambda(\mathbf{r}) = p(\mathbf{r} \mid H_1)/p(\mathbf{r} \mid H_0) = \lambda_1(\mathbf{r})/\lambda_0(\mathbf{r}) > 1,$$

where we have chosen the threshold corresponding to the criterion of minimum probability of error. That is,

$$\lambda(\mathbf{r}) = \frac{\prod_{i=1}^{M} \dfrac{N_0}{N_0 + TA_0^2} \exp\left[\dfrac{2A_0^2 q_{i1}^2}{N_0(N_0 + TA_0^2)}\right]}{\prod_{i=1}^{M} \dfrac{N_0}{N_0 + TA_0^2} \exp\left[\dfrac{2A_0^2 q_{i0}^2}{N_0(N_0 + TA_0^2)}\right]} > 1.$$

Converting to the log likelihood ratio and canceling constants, the receiver is

$$\ln[\lambda(\mathbf{r})] > 0,$$

$$\sum_{i=1}^{M} q_{i1}^2 > \sum_{i=1}^{M} q_{i0}^2. \tag{8.69}$$

Here, as in Eq. (8.9),

$$q_{i1} = \frac{1}{2}\left| \int_0^T \tilde{r}_i(t)a(t)\exp[-j(\delta f_1)t]\,dt \right|,$$

taking account that for FSK the known phase function is $\phi_1(t) = (\delta f_1)t$. The statistic q_{i0} is computed in the same way, using δf_0. With the usual choice $a(t) = 1$, the receiver statistics are the powers in the bins of a

Fourier analysis of the received signal envelope corresponding to the two frequency offsets used. The extension to an m-ary system amounts to examining more bins and choosing the largest power.

Performance of the Diversity Receiver

The performance of the receiver Eq. (8.69) is conveniently found in terms of the normalized receiver

$$Q_1 = \sum_{i=1}^{M} \frac{q_{i1}^2}{\sigma^2} > \sum_{i=1}^{M} \frac{q_{i0}^2}{\sigma^2} = Q_0, \tag{8.70}$$

where $\sigma^2 = N_0 T/4$. The analysis is the same as that in Sect. 7.6, except that here we consider only a binary system, so that $m = 2$, while on the other hand the decision variables Eq. (8.70) are based on M pulses, rather than a single pulse. We continue to assume that the frequency offsets are such that Q_0, Q_1 are independent.

We make an error if symbol H_1 is sent, but nonetheless there results $Q_0 > Q_1$. The probability of that happening depends on the collection of single-pulse SNRs $\alpha_i^2 = (A_i T/2\sigma)^2 = 2E_i/N_0$ and on the number of pulses M. In this fading case the SNRs α_i are also random variables, with the amplitudes A_i having a common Rayleigh density Eq. (8.59).

We first compute a conditional error probability as

$$P_e(\mathbf{A}) = \Pr(Q_0 > Q_1 | A_1, \ldots, A_M)$$

$$= \int_0^\infty \left[\int_{Q_1}^\infty p(Q_0) \, dQ_0 \right] p(Q_1 | \mathbf{A}) \, dQ_1. \tag{8.71}$$

In this, $p(Q_0)$ is the normalized chi-squared density with $2M$ degrees of freedom, as in Eq. (8.22), as by assumption the signal corresponding to H_1 was sent, and the matched filters for H_0 are orthogonal to that signal. Integrating Eq. (8.22) by parts as in Eq. (8.68) yields

$$\int_{Q_1}^\infty p(Q_0) \, dQ_0 = \exp\left(\frac{-Q_1}{2}\right) \sum_{n=0}^{M-1} \frac{Q_1^n}{(n! 2^n)}. \tag{8.72}$$

Because Q_1 results from the filters matched to the signal, Q_1 has the normalized noncentral chi-squared density with $2M$ degrees of freedom, Eq. (8.62), where the noncentrality parameter is as in Eq. (8.63). Since the density Eq. (8.62) involves the pulse amplitudes A_i only through the noncentrality parameter, i.e., the M-pulse SNR α, the error probability depends only on that quantity.

To determine the average probability of error over the distribution of fading pulse amplitudes, the conditional error probability $P_e(\alpha)$ must be averaged over the SNRs α. The density of α is given by Eq. (8.67), using the relation

$$\alpha^2 = TA_0^2 y/N_0,$$

so that

$$p(\alpha^2) = [1/\varepsilon 2^M (M-1)!](\alpha^2/\varepsilon)^{M-1} \exp(-\alpha^2/2\varepsilon), \qquad (8.73)$$

where, as before, ε is the average per-pulse SNR, defined (as usual in communication problems) as

$$\varepsilon = A_0^2 T/N_0 = \mathscr{E}(A_i^2)T/2N_0 = \mathscr{E}(E_i)/N_0, \qquad (8.74)$$

where E_i is the energy of the ith pulse.

We now use the density Eq. (8.73) to average over the conditional error probability Eq. (8.71). Since the term Eq. (8.72) does not involve the SNR, using Eq. (8.62) we can first carry out the portion of the integration involving the SNR as

$$\int_0^\infty p(\alpha^2)p(Q_1 \mid \alpha^2)\, d\alpha^2 = \exp(-Q_1/2)Q_1^{(M-1)/2}[\varepsilon^M(M-1)!2^{M+1}]^{-1}$$

$$\times \int_0^\infty \alpha^{M-1} \exp\left[\frac{-\alpha^2(1+\varepsilon)}{2\varepsilon}\right] I_{M-1}(\alpha Q_1^{1/2})\, d\alpha^2$$

$$= Q_1^{M-1} \exp\left[\frac{-Q_1}{2(\varepsilon+1)}\right] [2^M(M-1)!(\varepsilon+1)^M]^{-1},$$

using [Gradshteyn and Ryzhyk, 1965, #6.643.2].

Using this last with Eq. (8.72) to complete the integration in Eq. (8.71), we finally have

$$P_e = [2^M(M-1)!(\varepsilon+1)^M]^{-1}$$

$$\times \sum_{n=0}^{M-1} (n!2^n)^{-1} \int_0^\infty Q_1^{M+n-1} \exp\left\{-\left(\frac{Q_1}{2}\right)\left[\frac{(\varepsilon+2)}{(\varepsilon+1)}\right]\right\} dQ_1$$

$$= (\varepsilon+2)^{-M} \sum_{n=0}^{M-1} \binom{M+n-1}{n}\left(\frac{\varepsilon+1}{\varepsilon+2}\right)^n, \qquad (8.75)$$

using [Gradshteyn and Ryzhik, 1965, #3.381.4].

The error performance Eq. (8.75) for several orders of diversity is shown in Fig. 8.8. Also shown is the performance for a binary noncoherent FSK system without fading and with fading, from Fig. 7.8. A diversity level of only $M = 4$ nearly recovers the performance degradation of the nondiversity system caused by fading.

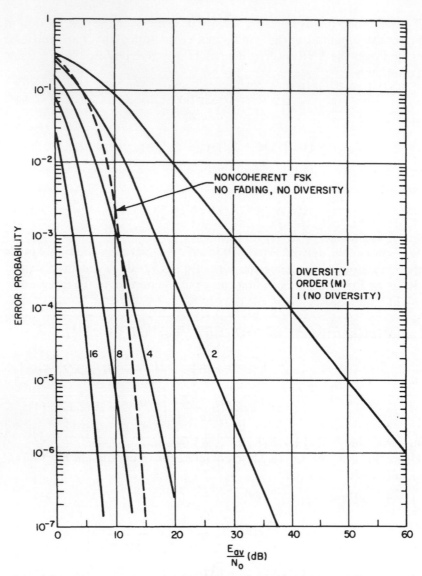

FIGURE 8.8   Diversity performance for frequency shift keying and slow Rayleigh fading. The average signal energy in each channel is E_{av} and $N_0/2$ is the two-sided noise power spectral density.

APPENDIX 1

We reproduce here the full set of Robertson's curves for performance of the M-pulse linear detector Eq. (8.41). This is the incoherent detector (random phase) with known equal amplitudes for each pulse. The signal-to-noise ratio is $10 \log_{10}(E/N_0)$, where E is the energy of each of the M pulses. The heavy dots on the right are at intervals of 0.2 dB in SNR. Figures A1.1 through A1.14 show receiver operating characteristics for detecting sine waves in white Gaussian noise (xxx samples averaged).

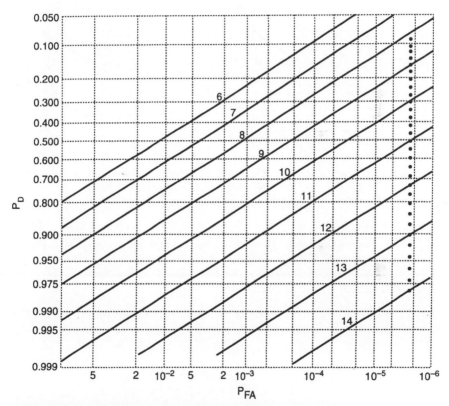

FIGURE A1.1 Receiver operating characteristic for detecting a single sine wave pulse in white Gaussian noise. [From G. H. Robertson, Operating characteristics for a linear detector of CW signals in narrow-band gaussian noise, *Bell Syst. Tech. J.* **46**, No. 4, 755–774 (1967); copyright 1967, AT&T, reprinted by permission.]

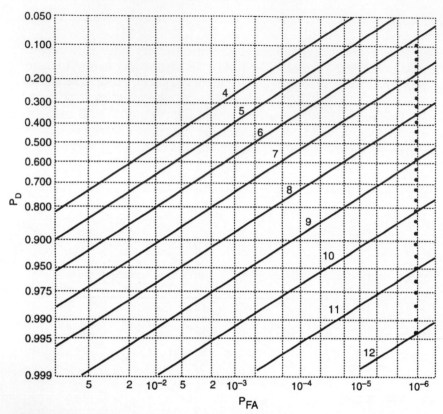

FIGURE A1.2   Receiver operating characteristic for detecting sine waves in white Gaussian noise (2 samples averaged). [From G. H. Robertson, Operating characteristics for a linear detector of CW signals in narrow-band gaussian noise, *Bell Syst. Tech. J.* **46,** No. 4, 755–774 (1967); copyright 1967, AT&T, reprinted by permission.]

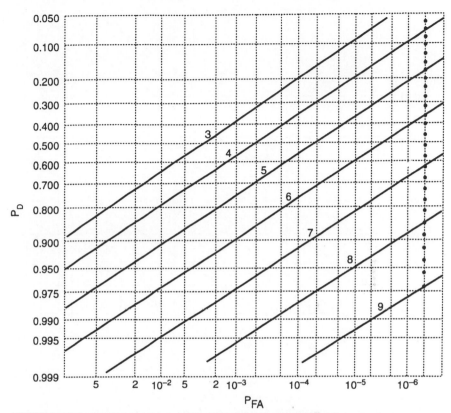

FIGURE A1.3 Receiver operating characteristics for detecting sine waves in white Gaussian noise (4 samples averaged). [From G. H. Robertson, Operating characteristics for a linear detector of CW signals in narrow-band gaussian noise, *Bell Syst. Tech. J.* **46,** No. 4, 755–774 (1967); copyright 1967, AT&T, reprinted by permission.]

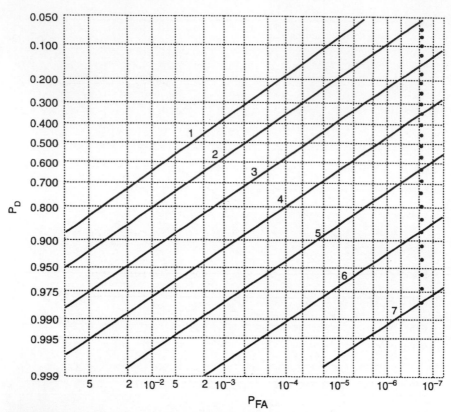

FIGURE A1.4 Receiver operating characteristics for detecting sine waves in white Gaussian noise (8 samples averaged). [From G. H. Robertson, Operating characteristics for a linear detector of CW signals in narrow-band gaussian noise, *Bell Syst. Tech. J.* **46**, No. 4, 755–774 (1967); copyright 1967, AT&T, reprinted by permission.]

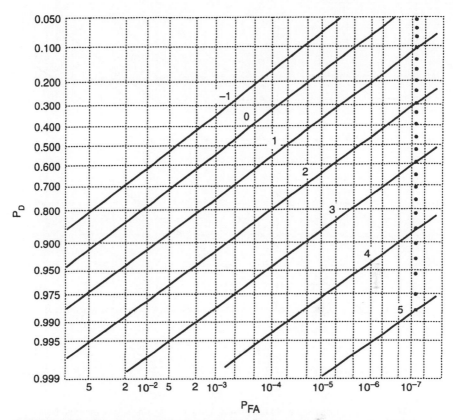

FIGURE A1.5 Receiver operating characteristics for detecting sine waves in white Gaussian noise (16 samples averaged). [From G. H. Robertson, Operating characteristics for a linear detector of CW signals in narrow-band gaussian noise, *Bell Syst. Tech. J.* **46**, No. 4, 755–774 (1967); copyright 1967, AT&T, reprinted by permission.]

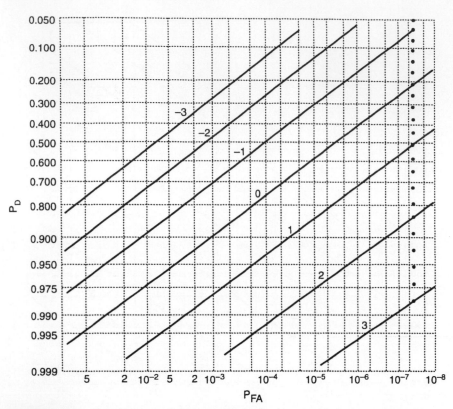

FIGURE A1.6   Receiver operating characteristics for detecting sine waves in white Gaussian noise (32 samples averaged). [From G. H. Robertson, Operating characteristics for a linear detector of CW signals in narrow-band gaussian noise, *Bell Syst. Tech. J.* **46**, No. 4, 755–774 (1967); copyright 1967, AT&T, reprinted by permission.]

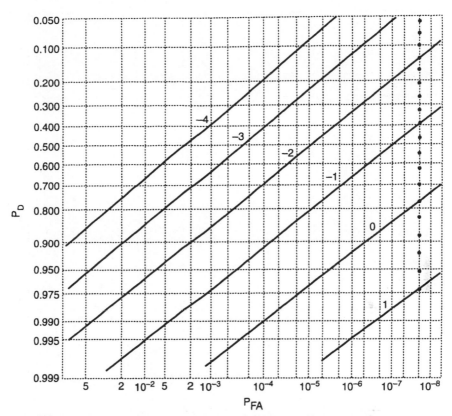

FIGURE A1.7   Receiver operating characteristics for detecting sine waves in white Gaussian noise (64 samples averaged). [From G. H. Robertson, Operating characteristics for a linear detector of CW signals in narrow-band gaussian noise, *Bell Syst. Tech. J.* **46,** No. 4, 755–774 (1967); copyright 1967, AT&T, reprinted by permission.]

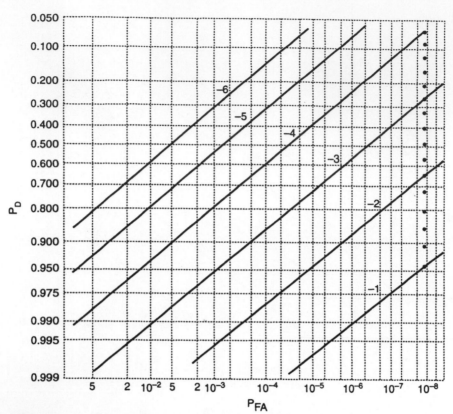

FIGURE A1.8 Receiver operating characteristics for detecting sine waves in white Gauss-ian noise (128 samples averaged). [From G. H. Robertson, Operating characteristics for a linear detector of CW signals in narrow-band gaussian noise, *Bell Syst. Tech. J.* **46**, No. 4, 755–774 (1967); copyright 1967, AT&T, reprinted by permission.]

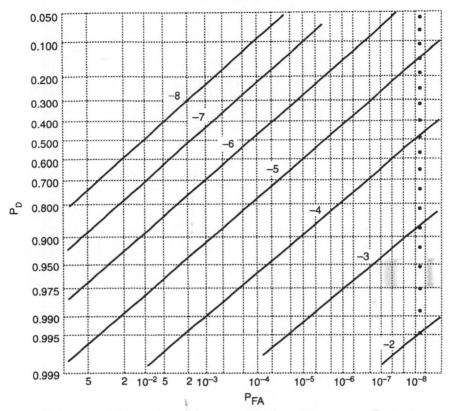

FIGURE A1.9   Receiver operating characteristics for detecting sine waves in white Gaussian noise (256 samples averaged). [From G. H. Robertson, Operating characteristics for a linear detector of CW signals in narrow-band gaussian noise, *Bell Syst. Tech. J.* **46**, No. 4, 755–774 (1967); copyright 1967, AT&T, reprinted by permission.]

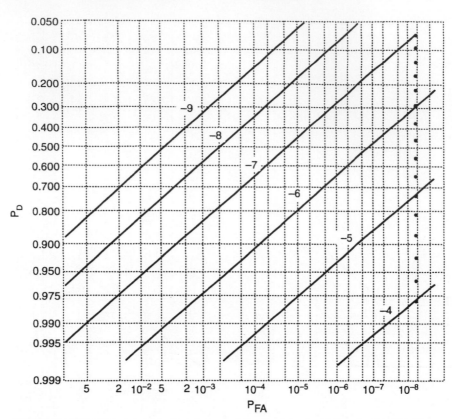

FIGURE A1.10   Receiver operating characteristics for detecting sine waves in white Gaussian noise (512 samples averaged). [From G. H. Robertson, Operating characteristics for a linear detector of CW signals in narrow-band gaussian noise, *Bell Syst. Tech. J.* **46,** No. 4, 755–774 (1967); copyright 1967, AT&T, reprinted by permission.]

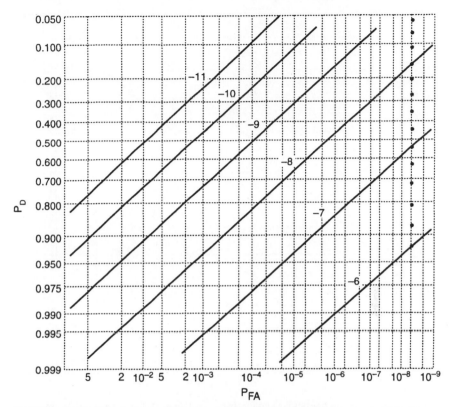

FIGURE A1.11   Receiver operating characteristics for detecting sine waves in white Gaussian noise (1024 samples averaged). [From G. H. Robertson, Operating characteristics for a linear detector of CW signals in narrow-band gaussian noise, *Bell Syst. Tech. J.* **46,** No. 4, 755–774 (1967); copyright 1967, AT&T, reprinted by permission.]

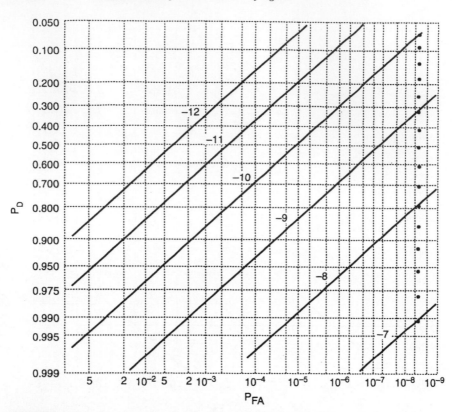

FIGURE A1.12   Receiver operating characteristics for detecting sine waves in white Gaussian noise (2048 samples averaged). [From G. H. Robertson, Operating characteristics for a linear detector of CW signals in narrow-band gaussian noise, *Bell Syst. Tech. J.* **46**, No. 4, 755–774 (1967); copyright 1967, AT&T, reprinted by permission.]

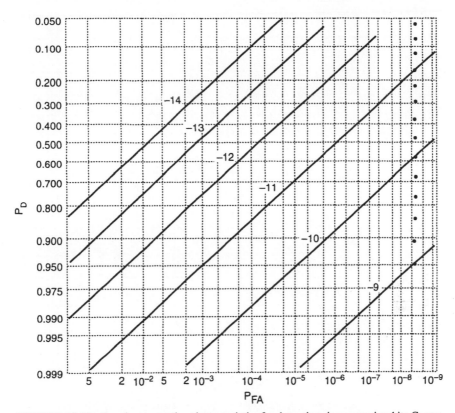

FIGURE A1.13 Receiver operating characteristics for detecting sine waves in white Gaussian noise (4096 samples averaged). [From G. H. Robertson, Operating characteristics for a linear detector of CW signals in narrow-band gaussian noise, *Bell Syst. Tech. J.* **46**, No. 4, 755–774 (1967); copyright 1967, AT&T, reprinted by permission.]

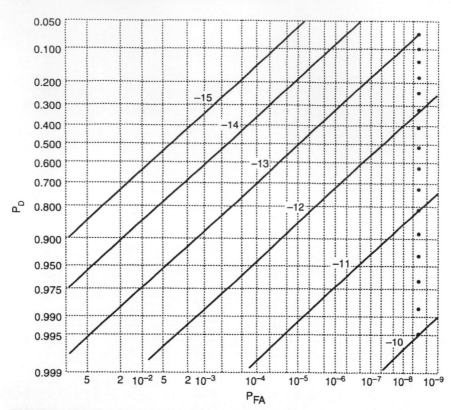

FIGURE A1.14   Receiver operating characteristics for detecting sine waves in white Gaussian noise (8192 samples averaged). [From G. H. Robertson, Operating characteristics for a linear detector of CW signals in narrow-band gaussian noise, *Bell Syst. Tech. J.* **46,** No. 4, 755–774 (1967); copyright 1967, AT&T, reprinted by permission.]

APPENDIX 2

We reproduce here a set of performance curves for the M-pulse quadratic detector Eq. (8.56) in the case of amplitudes which are independently distributed Rayleigh random variables with uniformly distributed independent phases. The signal-to-noise ratio is $10 \log_{10}(E_{av}/N_0)$, where E_{av} is the average energy of each pulse. Figures A2.1 through A2.11 show the detection performance for slow Rayleigh fading signals (xxx samples averaged).

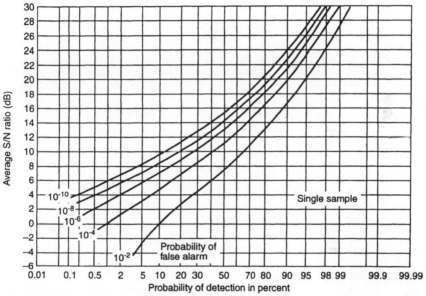

FIGURE A2.1 Detection performance for a slow Rayleigh fading signal (single sample).

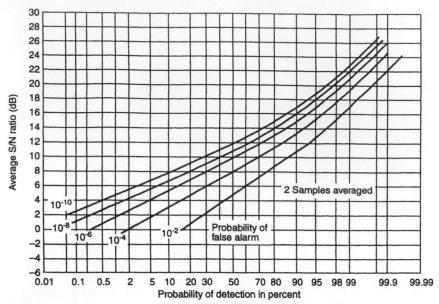

FIGURE A2.2 Detection performance for slow Rayleigh fading signals (2 samples averaged).

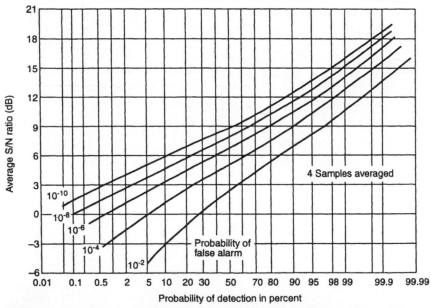

FIGURE A2.3 Detection performance for slow Rayleigh fading signals (4 samples averaged).

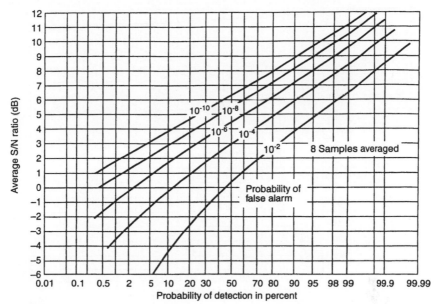

FIGURE A2.4 Detection performance for slow Rayleigh fading signals (8 samples averaged).

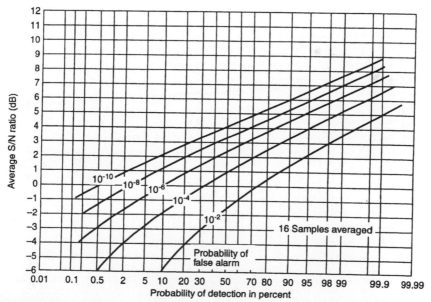

FIGURE A2.5 Detection performance for slow Rayleigh fading signals (16 samples averaged).

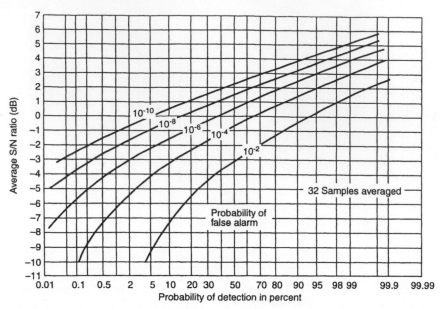

FIGURE A2.6 Detection performance for slow Rayleigh fading signals (32 samples averaged).

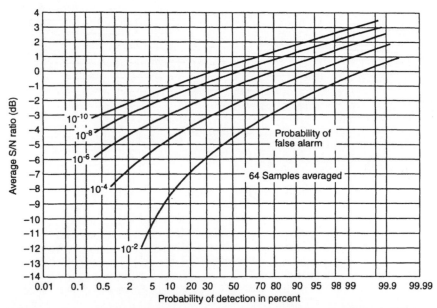

FIGURE A2.7 Detection performance for slow Rayleigh fading signals (64 samples averaged).

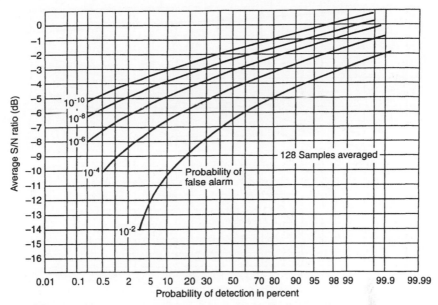

FIGURE A2.8 Detection performance for slow Rayleigh fading signals (128 samples averaged).

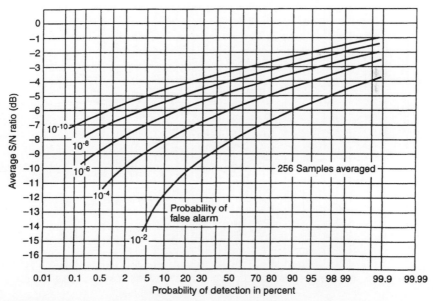

FIGURE A2.9 Detection performance for slow Rayleigh fading signals (256 samples averaged).

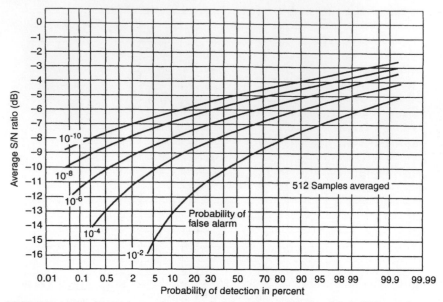

FIGURE A2.10   Detection performance for slow Rayleigh fading signals (512 samples averaged).

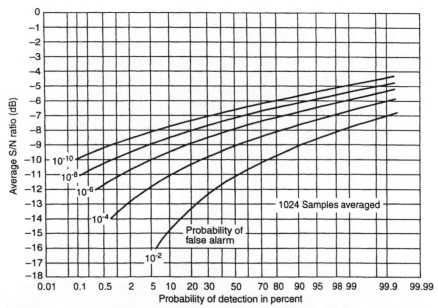

FIGURE A2.11   Detection performance for slow Rayleigh fading signals (1024 samples averaged).

Exercises

8.1 Find an expression to approximate the Rayleigh density function

$$p(x) = xe^{-x^2/2}, \quad x \geq 0$$

using two terms of the Gram–Charlier series. Would you expect this to be a reasonable approximation?

8.2 Determine the coefficient a_9 in Eq. (8.40) in terms of the central moments denoted ν_n.

8.3 The probability of detection for the case of M pulses having uniform phase and Rayleigh amplitude distribution is given in Sec. 8.7. Rederive that result by considering the following problem. (See Exercise 7.7 for similar results in the single-pulse case.)

A signal $r(t) = a(t)\cos(\omega_c t + \phi(t)) + n(t)$, $-\infty < t < \infty$, is passed through a linear filter, $h(\tau) = \cos\omega_c\tau$, $0 \leq \tau \leq T$ followed by a quadratic detector and a constant multiplier $(1/\sigma_T^2)$. The signal $a(t)\cos(\omega_c t + \phi(t))$ is a Gaussian process, and $R_a(\tau)$ and $\phi(t)$ are substantially constant over any interval T. Assume that M statistically independent samples at the output of the detector are summed. Verify that the probability that this sum exceeds a threshold is as given in Eq. (8.68).

8.4 For the coherent combining problem as posed in Fig. 8.9, assume $n_1(t)$ and $n_2(t)$ are statistically independent, and are zero mean with variances N_1 and N_2 respectively.

(a)   Find the value of K which will maximize the signal-to-noise ratio of the output. What is the maximum value of signal-to-noise ratio?

(b)   Suppose that K is chosen so that the signals are combined on a basis of equal noise powers. That is, $K^2 N_1 = N_2$. Find the resulting signal-to-noise ratio. Under what condition is the resulting signal-to-noise ratio greater than the maximum of either S_1/N_1 or S_2/N_2?

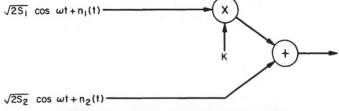

FIGURE 8.9   Coherent addition of two signals.

8.5 For the incoherent combining problem posed in Fig. 8.10, $n_1(t)$ and $n_2(t)$ are statistically independent, zero mean, with variances N_1 and N_2 respectively. Define the deflection signal-to-noise ratio of the output as

$$\text{DSNR} \triangleq \frac{E_{sn}\{z\} - E_n\{z\}}{\sigma_n\{z\}}$$

where $E_{sn}\{z\}$ and $E_n\{z\}$ are expected values with signal plus noise and noise only, respectively, and $\sigma_n\{z\}$ is the standard deviation of z with noise only present.

(a)   Show that the maximum value of DSNR is

$$\frac{S_2}{N_2}\left(\frac{S_1}{N_1} + \frac{S_2}{N_2}\right)\bigg/\left[\left(\frac{S_1}{N_1}\right)^2 + \left(\frac{S_2}{N_2}\right)^2\right]^{1/2}$$

Consider the ratio of DSNR to the maximum input signal-to-noise ratio. What is its maximum value?

(b)   Combine the signals on an equal power basis, that is $GN_1 = N_2$. Under what conditions is the DSNR greater than the maximum input signal-to-noise ratio?

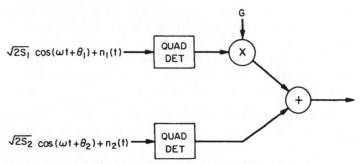

FIGURE 8.10   Incoherent addition of two signals.

8.6   For the receiver shown in Fig. 8.11, denote the output as

$$Q = \sum_{i=1}^{M} q_i$$

Assume detection of constant-amplitude sine waves in white Gaussian noise. For a given performance level, as $M \to \infty$ the signal-to-noise ratio becomes small.

(a)   Show that the asymptotic form (M becoming large and signal-to-noise ratio approaching zero) of the deflection signal-to-noise ratio is

$$\text{DSNR} \triangleq \frac{E_1\{Q\} - E_0\{Q\}}{[V_0\{Q\}]^{1/2}} \approx \left(\frac{\pi M}{4 - \pi}\right)^{1/2}\frac{\alpha^2}{4}$$

where $\alpha^2/2 = A^2/2\sigma_T^2$, $\sigma_T^2 = N_0 T/4$, $E_i\{Q\}$ is the average value of Q for hypothesis i, and $V_0\{Q\}$ is the variance of Q for the noise-only case.

(b) Show that the asymptotic form of the detection probability is

$$P_D \approx \int_\xi^\infty \frac{1}{(2\pi)^{1/2}} e^{-u^2/2} \, du$$

where

$$\xi = [V_T - (\pi/2)^{1/2}\sigma_T(\alpha^2/4)M]/\gamma^{1/2}$$

where

$$\gamma^{1/2} = \left(\frac{4-\pi}{2}\right)^{1/2} M^{1/2}\sigma_T$$

and V_T is the detection threshold. (Assume that the variance of Q for signal plus noise is the same as for noise alone.)

(c) Then show that as the number of pulses is doubled, the signal-to-noise ratio required to maintain the same detection probability is reduced by $2^{1/2}$.

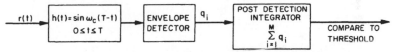

FIGURE 8.11 Approximation to optimum receiver for high signal-to-noise ratio.

8.7 Repeat the preceding problem for the receiver which uses the quadratic detector, Fig. 8.12. Specifically, show the asymptotic results

(a) $DSNR = M^{1/2}\alpha^2/2$

(b) $P_D \approx \int_\xi^\infty (2\pi)^{1/2} e^{-u^2/2} \, du$ where

$$\xi = \frac{V_Q - M\sigma_T^2\alpha^2}{2\sigma_T^2 M^{1/2}}$$

and V_Q is the threshold for $\sum_{i=1}^M q_i^2$.

(c) As M is doubled, the signal-to-noise ratio required to maintain the same performance is reduced by $2^{1/2}$.

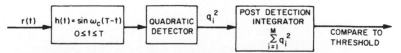

FIGURE 8.12 Approximation to optimum receiver for low signal-to-noise ratio.

8.8 Consider the receiver shown in Fig. 8.13 with a step function nonlinearity. The nonlinearity can be expressed mathematically as $z = u(q - q_0)$

where $u(x)$ is 1 for $x \geq 0$, 0 for $x < 0$. The resulting receiver performs what is called binary integration.

Assume that q_0 is chosen such that the probability of q_0 being exceeded due to noise only is small. Denote this by p_f. Denote the probability of q_0 being exceeded by signal plus noise as p_d. Assume that H_1 is chosen if $L \geq I$.

(a) What are the expressions for the probabilities of false alarm and detection in terms of p_d, p_f, I, and M?

(b) Assume that with a signal-to-noise ratio of 6 dB, q_0 is chosen such that $p_d = \frac{1}{2}$ and $p_f = 10^{-2}$. Assume $M = 4$. For $I = 1$, 2, and 3, what is the quantitative degradation in detectability, in terms of signal-to-noise ratio, incurred by using the nonlinearity?

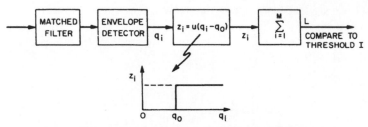

FIGURE 8.13 A receiver using binary integration.

8.9 Consider the detection problem for which the hypotheses are

$$H_1: \quad r_i(t) = A_i \sin(\omega_c t + \theta_i) + n_i(t), \qquad i = 1, \ldots, M, \quad 0 \leq t \leq T$$
$$H_0: \quad r_i(t) = n_i(t), \qquad\qquad\qquad\qquad i = 1, \ldots, M, \quad 0 \leq t \leq T$$

The $n_i(t)$ are white Gaussian noise processes with spectral density $N_0/2$, and $n_i(t)$ is uncorrelated with $n_j(t)$, $i \neq j$. The phases θ_i are uniformly distributed $(0, 2\pi)$, and θ_i and θ_j are uncorrelated for $i \neq j$.

(a) Assume that A_i is a discrete random variable such that

$$P(A_i = 0) = 1 - p, \qquad P(A_i = A_0) = p$$

Determine the likelihood ratio. What is the asymptotic form of the likelihood ratio as A_0 approaches zero?

8.10 For the preceding problem assume that the probability density function for A_i is

$$p(A_i) = (1 - p)\,\delta(A_i) + p\frac{A_i}{A_0^2}\exp\left(-\frac{A_i^2}{2A_0^2}\right)$$

Determine the likelihood ratio, and its form as A_0 approaches zero.

9 Detection of Signals in Colored Gaussian Noise

*I*n nearly all of the previous developments, it has been assumed that the signal, if present, occurred with additive white noise, usually Gaussian. That the noise is white means that successive time samples are uncorrelated, or independent in the Gaussian case. That allowed the likelihood function of the received data to be written as the product of the likelihood functions of each time sample. The log likelihood ratios were then simple sums over time, which in the limit of infinite bandwidth corresponded to integrals. Thereby all our processors amounted to one kind or another of filtering on the data signals.

The procedure in its simplest form was illustrated in Sect. 5.7. Suppose we deal with a signal $s(t)$ and data

$$r(t) = s(t) + n(t) \tag{9.1}$$

and that we observe n uniformly spaced samples $r(i\,\delta t)$ over some interval $T = n\,\delta t$. If the noise is white, with two-sided constant power spectral density $N_0/2$ over some band $|f| \leq B$, the noise autocorrelation function is

339

$$R_n(\tau) = \int_{-B}^{B} \frac{N_0}{2} \exp(j2\pi f\tau) \, df$$

$$= (N_0 B)[\sin(2\pi B\tau)]/(2\pi B\tau). \tag{9.2}$$

If we sample at the Nyquist rate, $n_i = n(i \, \delta t) = n(i/2B)$, we have

$$\mathscr{E}n_i n_j = R(|i - j|/2B) = (N_0 B)\delta_{ij}, \tag{9.3}$$

where δ_{ij} is the Kronecker delta. The noise samples are then independent, since they are Gaussian, and the (perhaps conditional) likelihood ratio of the data Eq. (9.1) can be written

$$\lambda(\mathbf{r}) = \prod_{i=1}^{n} \lambda(r_i),$$

where $\mathbf{r}$ is the vector of data samples. The corresponding log likelihood ratio is

$$\ln[\lambda(\mathbf{r})] = \sum_{i=1}^{n} \ln[\lambda(r_i)] = \sum_{i=1}^{n} \ln\{\lambda[r(i \, \delta t)]\}. \tag{9.4}$$

In Sect. 6.1 we passed to the limit as the noise bandwidth $B \Rightarrow \infty$, so that $\delta t \Rightarrow 0$ and the sums such as Eq. (9.4) passed over to integrals over $0 \leq t \leq T$. We then proceeded with all the subsequent developments.

In this chapter, we remove the restriction that the noise must have a white spectrum. The procedure will be to prefilter the data in such a way that the noise on the filtered data is white. Then the detectors we have studied can be applied. The composite system of whitening filter and detector with white noise is the detector for the case of colored noise.

9.1 MATRIX FORMULATION

In Sect. 6.1 we considered the problem of detecting a signal vector $\mathbf{s}(t)$ in vector additive zero-mean Gaussian noise:

$$\mathbf{r}(t) = \mathbf{s}(t) + \mathbf{n}(t). \tag{9.5}$$

Letting $\mathbf{N} = \mathscr{E}(\mathbf{n}\mathbf{n}^T)$ be the covariance matrix of the noise, where the superscript T means vector transpose, as in Eq. (6.4) the (conditional, if the signal has random parameters) log likelihood ratio for the problem is

$$\lambda(\mathbf{r}) = \mathbf{s}^T\mathbf{N}^{-1}\mathbf{r} - \tfrac{1}{2}\mathbf{s}^T\mathbf{N}^{-1}\mathbf{s}. \tag{9.6}$$

We assume the problem is not singular, which means that we assume the noise covariance matrix $\mathbf{N}$ is nonsingular, so that its inverse exists.

In Sect. 6.6 we proceeded in another way. We sought a nonsingular (and therefore square) matrix $\mathbf{H}$ such that the noise, when filtered by $\mathbf{H}$, was white. That is, we let

$$w = Hn, \tag{9.7}$$

and sought **H** such that

$$\mathcal{E}(ww^T) = H\mathcal{E}(nn^T)H^T = HNH^T = I, \tag{9.8}$$

where **I** is the unit matrix. Letting $A = H^{-1}$, Eq. (9.8) becomes

$$AA^T = N. \tag{9.9}$$

In Sect. 6.6 we considered solution of the problem Eq. (9.9). One solution is provided by the *singular value decomposition* of the matrix **N**. In general, any $m \times n$ matrix **N**, singular or not, square or rectangular (with $m \geq n$, say), can be written as

$$N = U\Sigma V^T, \tag{9.10}$$

where **U** is an $m \times m$ orthogonal matrix (orthonormal column vectors), **V** is an $n \times n$ orthogonal matrix, and Σ is an $m \times n$ diagonal matrix, with a number of nonzero elements equal the rank of **N**. The singular value decomposition is accurately computed by standard computer codes. It is the case that the columns of **U** are the eigenvectors of NN^T, the columns of **V** are the eigenvectors of N^TN, and the entries along the diagonal of Σ are the correspondingly ordered nonnegative square roots of the eigenvalues of N^TN, which are also the eigenvalues of NN^T. The numbers Σ_i are the *singular values* of the matrix **N**.

In application to a nonsingular positive definite noise covariance matrix **N**, we have $NN^T = N^2 = N^TN$, so that $U = V$ and $\Sigma = \Lambda^2$, where Λ is the diagonal matrix of the eigenvalues of **N**. Then Eq. (9.10) becomes

$$N = U\Lambda^2U^T. \tag{9.11}$$

Comparing this with Eq. (9.9), we have $A = U\Lambda$, so that the noise whitening transformation is

$$H = \Lambda^{-1}U^{-1} = \Lambda^{-1}U^T. \tag{9.12}$$

This can be compared with the earlier transformation Eq. (6.76):

$$H = \Lambda^{-1/2}T^T,$$

where **T** is the matrix whose columns are the eigenvectors of **N** and Λ is the diagonal matrix of the corresponding eigenvalues. We see $T = U\Lambda^{-1/2}$, and the singular value decomposition is an alternative way to do the computation in Sect. 6.6.

Alternatively, as we pointed out in Sect. 6.6, we could require that the filter **H** be physically realizable. That is, interpreting the elements r_i of **r** as ordered time samples, we require that the jth element of **w** depend only on $r_{i \le j}$. Correspondingly, $\mathbf{H}$ is to be lower triangular, or to have only zeros above the diagonal. Then $\mathbf{A} = \mathbf{H}^{-1}$ is also to be lower triangular. The *LDU decomposition* of an arbitrary $m \times n$ matrix $\mathbf{N}$ can be applied to the problem. Let $\mathbf{N}$ be of row rank r. That is, let there be r of the m rows of $\mathbf{N}$ which form linearly independent vectors. Then standard computer codes will determine unique matrices such that

$$\mathbf{N} = \mathbf{LDU}, \tag{9.13}$$

where the $m \times r$ matrix $\mathbf{L}$ is lower triangular with units on the diagonal, $\mathbf{U}$ is upper triangular and $r \times n$ with units on the diagonal, and $\mathbf{D}$ is an $r \times r$ nonsingular diagonal matrix.

In our case of a symmetric nonsingular positive definite noise covariance matrix $\mathbf{N}$, Eq. (9.13) becomes

$$\mathbf{N} = \mathbf{LDL}^{\mathsf{T}},$$

where the elements of $\mathbf{D}$ are positive numbers. Then taking

$$\mathbf{A} = \mathbf{LD}^{1/2}$$

in Eq. (9.9) shows that the whitening transformation can be taken as

$$\mathbf{H} = \mathbf{A}^{-1} = \mathbf{D}^{-1/2}\mathbf{L}^{-1}. \tag{*9.14}$$

Since $\mathbf{L}$ is triangular, it is easy to invert. This use of the LDU decomposition is an alternative way to compute the transformation Eq. (6.81).

9.2 DISCRETE SPECTRAL FACTORIZATION

The procedures we have described are general. However, they are not useful if the number of time samples to be processed is large. In effect, they yield batch-processing whitening algorithms, whereas we usually need a sequential algorithm (filter) to apply to an ongoing data stream. There exists a general procedure for factoring the power spectrum of the noise to be whitened in a way which indicates what filter should be used. In a later section we will discuss the method in the continuous time domain. As a prelude, in this section we will develop the procedure in overview in discrete time.

We need some results about discrete linear filters. Suppose that a sequence of time samples x_i is used as input to a digital filter with coefficients h_i. The output sequence is the convolution

$$y_i = \sum_k h_{i-k} x_k = \sum_k h_k x_{i-k}. \tag{9.15}$$

If the input is a noise sequence, the cross-correlation sequence of the output with the input is

$$r_{xy}(m) = \mathscr{E}(x_{i+m}y_i) = \sum_k h_k \mathscr{E}(x_{i+m}x_{i-k})$$
$$= \sum_k h_k r_x(m+k) = \sum_k h_{-k} r_x(m-k) = h(-m) * r_x(m).$$

Using the fact that the transform of a discrete convolution is the product of the transforms of the convolved sequences, the discrete cross-power spectrum of the input with the output is then

$$S_{xy}(z) = \sum_m r_x(m)z^{-m} = H\left(\frac{1}{z}\right)S_x(z), \qquad (9.16)$$

since

$$\sum_m h(-m)z^{-m} = \sum_m h_m z^m = H\left(\frac{1}{z}\right).$$

Again using Eq. (9.15), the autocorrelation sequence of the output is

$$r_y(m) = \mathscr{E}(y_{i+m}y_i) = \sum_k h_k \mathscr{E}(x_{i+m-k}y_i) = \sum_k h_k r_{xy}(m-k).$$

Then the output power spectrum is

$$S_y(z) = H(z)S_{xy}(z) = H(z)H(1/z)S_x(z). \qquad (9.17)$$

If we wish the output sequence to be white with variance σ_w^2, we want $r_y(i) = \sigma_w^2 \delta_{i0}$. That is, we want $S_y(z) = \sigma_w^2$. Using this in Eq. (9.17) then yields the equation to be solved for the whitening filter $H(z)$:

$$H(z)H(1/z)S_x(z) = \sigma_w^2. \qquad (9.18)$$

We will require that $H(z)$ be a stable filter, so that all the poles of $H(z)$ lie inside the unit circle of the z plane. The poles of $H(1/z)$ then lie outside the unit circle.

In the case that the spectrum $S_x(z)$ of the sequence x_i to be whitened is a rational fraction, there is a relatively simple way to solve Eq. (9.18) for the stable whitening filter $H(z)$. (It is the discrete version of the procedure discussed in Sect. 6.5.3.) We will seek a certain factorization of the given spectrum. To that end, consider a sequence of (stationary zero mean) real noise samples x_i with correlation sequence $r_k = \mathscr{E}(x_{i+k}x_i)$. The correlation sequence is even in time:

$$r_{-k} = \mathscr{E}(x_{i-k}x_i) = \mathscr{E}(x_i x_{i+k}) = \mathscr{E}(x_{i+k}x_i) = r_k.$$

Hence the discrete power spectrum of the noise is of the form

$$S(z) = \sum_{k=-\infty}^{\infty} r_k z^{-k}$$

$$= r_0 + \sum_{k=1}^{\infty} r_k(z^k + z^{-k}). \qquad (9.19)$$

By Eq. (9.19), a power spectrum must be unchanged if z is replaced by $1/z$. Therefore any numerator or denominator factor $z - z_0$ of a rational $S_x(z)$ must be accompanied by a factor $(1/z) - z_0$. The roots of the numerator or denominator must occur in pairs $z_0 = a \exp(j\phi)$, $1/z_0 = (1/a) \exp(-j\phi)$. Since $S_x(z)$ is real, the roots must also occur in conjugate pairs in the z plane. Taking these together, the roots of $S_x(z)$ must have inverse symmetry around the unit circle.

Using the fact that the numerator and denominator roots of a rational spectrum $S_x(z)$ occur in pairs relative to the unit circle, we can write a given rational spectrum in the form

$$S_x(z) = [A(z)A(1/z)]/[B(z)B(1/z)]$$
$$= G(z)G(1/z), \qquad\qquad *(9.20)$$

where $G(z) = A(z)/B(z)$. If we can find $G(z)$ by factoring $S_x(z)$, then from Eq. (9.18) the solution for the whitening filter $H(z)$, for the usual choice $\sigma_w^2 = 1$, can be taken as

$$H(z) = 1/G(z). \qquad\qquad *(9.21)$$

Since we want the filter to be stable, the poles of $H(z)$ should be inside the unit circle, so we should ensure that the roots of $A(z)$ are inside the unit circle. There is an arbitrariness in the choice of $B(z)$, which can be removed by requiring that $H(z)$ also be minimum phase. That is, we will choose the zeros of $H(z)$, which is to say the roots of $B(z)$, also to be inside the unit circle. An example will illustrate the procedure.

EXAMPLE 9.1   Suppose the noise sequence to be whitened has spectrum

$$S_x(z) = \frac{-z^6 + 5z^5 - 9z^4 + 82z^3 - 9z^2 + 5z - 1}{z^6 + 3z^5 + 3z^4 + 22z^3 + 3z^2 + 3z + 1}.$$

[Note that we are not free to pick an arbitrary function for $S_x(z)$. Just as a power spectral density function $S(f)$ must be positive at every frequency f, since it is after all power, a discrete spectrum $S(z)$ must be positive everywhere on the unit circle.] Factoring the numerator and denominator of this $S_x(z)$ yields numerator roots $-0.0336 \pm j0.2664$, $-0.4664 \pm j3.6943$, 0.1716, 5.8284, and denominator roots $0.0747 \pm j0.4122$, $0.4253 \pm j2.3487$, -0.2679, -3.7321. Collecting together those of these roots which lie inside the unit circle, we form tentatively

$$G_1(z) = \frac{z^3 - 0.1044z^2 + 0.0606z - 0.0124}{z^3 + 0.1185z^2 + 0.1355z + 0.0470}.$$

From this we obtain

$$G_1(1/z) = \frac{-.0124z^3 + 0.0606z^2 - 0.1044z + 1}{0.0470z^3 + 0.1355z^2 + 0.1185z + 1}.$$

Then we have

$$G_1(z)G_1(1/z) = S_x(z)/3.80,$$

so that the factorization Eq. (9.20) is obtained, except for a constant. We then take $G(z) = (3.80)^{1/2}G_1(z)$, to finally obtain the whitening filter as

$$H(z) = 1/G(z) = \frac{0.513z^3 + 0.0608z^2 + 0.0695z + 0.0241}{z^3 - 0.1044z^2 + 0.0606z - 0.0124}.$$

With this, the sequence x_i is whitened by the algorithm

$$w_i = 0.1044w_{i-1} - 0.0606w_{i-2} + 0.0124w_{i-3}$$
$$+ 0.513x_i + 0.0608x_{i-1} + 0.0695x_{i-2} + 0.0241x_{i-3}. \quad (9.22)$$

It is important to note that, by assuming a rational spectrum $S_x(z)$, we have implied that the input sequence x_i begins in the infinite past, so that any initialization effects are ignored in the algorithm Eq. (9.22).

9.3 CONTINUOUS TIME SPECTRAL FACTORIZATION

Let us illustrate again the procedure discussed in Sect. 6.5.3. In analogy with the developments of the preceding section, a continuous time real random process $x(t)$ with rational power spectral density $S_x(f)$ can be whitened by a filter found by factoring the power spectrum. Instead of using the unit circle to sort the poles and zeros of the spectrum, we use the j axis. Again, by assuming a rational power spectrum, we imply that the signal $x(t)$ extends into the infinite past, and we ignore any initialization effects.

Because $x(t)$ is real, its autocorrelation function is even. The power spectrum, being the Fourier transform of the autocorrelation function, is therefore a real function of f. Letting $s = j\omega = j2\pi f$, it follows that the spectrum is an even function of s. Hence replacing s by $-s$ must leave the spectrum unchanged. Therefore any factor $s + s_0$ of the numerator or denominator has a companion $-s + s_0$. We can therefore factor the spectrum as

$$S_x(s) = G(s)G(-s),$$

where we can choose $G(s)$ to be a stable minimum phase transfer function by assigning to it the left plane poles and zeros of $S_x(s)$.

Now consider the filter $H(s) = 1/G(s)$, which is also stable and minimum phase. With the process $x(t)$ as input, by Eq. (2.65) the power spectrum of the output is

$$|H(j\omega)|^2 S_x(f) = H(j\omega)H(-j\omega)S_x(f)$$
$$= S_x(f)/G(j\omega)G(-j\omega) = 1,$$

so that the output noise process is white. An example will illustrate the procedure.

EXAMPLE 9.2   Let a real random process $x(t)$ have power spectral density

$$S_x(f) = \frac{11\omega^4 - 44.25\omega^2 + 85}{2\omega^6 - 13\omega^4 + 21.125\omega^2 + 36.125}.$$

We rewrite this as

$$S_x(s) = \frac{11s^4 + 44.25s^2 + 85}{-2s^6 - 13s^4 - 21.125s^2 + 36.125}.$$

The numerator roots are $\pm 0.6199 \pm j1.5478$. The denominator roots are $\pm 0.5 \pm j2$, ± 1. Collecting together the left plane roots we use a trial

$$G_1(s) = \frac{s^2 + 1.2398s + 2.7800}{(s^2 + s + 4.25)(s + 1)}.$$

Then we find

$$G_1(s)G_1(-s) = (2/11)S_x(s).$$

We therefore determine the whitening filter as

$$H(s) = 1/G(s) = (2/11)^{1/2}/G_1(s)$$

$$= \frac{0.4264s^3 + 0.8528s^2 + 2.2386s + 1.8122}{s^2 + 1.2398s + 2.7800}$$

$$= 0.4264s + 0.3241 + \frac{0.6513s + 0.9111}{s^2 + 1.2398s + 2.7800}.$$

Since the noise is to be whitened, it is not surprising that the filter involves differentiation of the input process.

9.4 FINITE OBSERVATION TIME AND THE KARHUNEN–LOÈVE EXPANSION

In the procedures of the preceding two sections, we assumed implicitly that the process to be whitened had been going on for so long that any initial transients of the whitening filter had died away. That assumes in effect an infinite observation time of the signal to be processed. In this section we will begin an exact treatment of the case of finite observation interval $0 \le t \le T$. The matter could be viewed as a return to the case of the Sect. 9.1, which considered only a finite number of data samples, but with a formulation convenient for a continuous time stationary process.

The procedure to be used amounts to solving the Wiener–Hopf equation, Eq. (6.58), for the case of finite observation time.

The key to the method is to make a transition from the process $x(t)$ or its samples as observables to new observables computed from $x(t)$ by a filtering algorithm. The new observables are defined in such a way that the corresponding noise is white (the observables have equal variances), even though the noise associated with the original data is colored. This is done with the finiteness of the observation interval taken into account. If the filter is invertible, any detection problem can be worked equally well in terms of the original data or in terms of the filtered data.

As a preliminary, let us introduce the notion of least square error expansion of a (possibly complex) finite energy signal $x(t)$ on a set of (possibly complex) functions $\phi_i(t)$, the basis functions of the approximation, over the interval $0 \le t \le T$. We seek the coefficients x_i in the approximation

$$x(t) \cong x_n(t) = \sum_{i=1}^{n} x_i \phi_i(t). \tag{9.23}$$

The coefficients are to be such that the integrated squared error in the approximation is minimum:

$$x_i \Leftarrow \min_{x_i} \int_0^T |x(t) - x_n(t)|^2 \, dt. \tag{9.24}$$

Let us assume that the basis functions have been orthonormalized, perhaps using the Gram–Schmidt procedure of Sect. 6.6. That is, we assume that the functions $\phi_i(t)$ are such that

$$\int_0^T \phi_i^*(t)\phi_j(t) \, dt = \delta_{ij}, \tag{9.25}$$

where δ_{ij} is the Kronecker delta.

A necessary condition for the coefficients x_i results by using Eq. (9.23) in Eq. (9.24) and setting the derivative of the result equal to zero for each variable x_i. In doing this, we can use the convenient shorthand (discussed in Sect. 11.2) of considering a variable and its conjugate to be formally independent in the process of extremizing a real function of complex parameters. Letting the integrated squared error in Eq. (9.24) be ε, using Eq. (9.23) in Eq. (9.24) we have

$$\varepsilon = \int_0^T [x(t) - x_n(t)][x(t) - \sum_{i=1}^{n} x_i \phi_i(t)]^* \, dt,$$

$$\frac{\partial \varepsilon}{\partial x_j^*} = \int_0^T [x(t) - x_n(t)][-\phi_j^*(t)] \, dt = 0.$$

Substituting Eq. (9.23) into this last and using Eq. (9.25) results in

$$x_j = \int_0^T \phi_j^*(t)x(t) \, dt. \tag{9.26}$$

This is the continuous time version of the dot product between a "vector" $x(t)$ and a set of orthonormal unit vectors $\phi_j(t)$.

Although in practice it is not of much interest, for theoretical developments it is convenient to assume that the given orthonormal set of functions $\phi_i(t)$ has infinitely many members: $i = 1, \infty$. Then we can investigate the question of whether or not the integrated squared error ε of Eq. (9.24) tends to zero as $n \Rightarrow \infty$. If it does, the set $\phi_i(t)$ is called *complete*. We shall assume that is the case. Furthermore, if the integrated squared error in Eq. (9.24) is zero, then the function $x(t)$ and its approximation can differ at most at isolated points. With that understanding of what we mean by equality, we shall henceforth without comment write simply

$$x(t) = \sum_{i=1}^{\infty} x_i\phi_i(t), \qquad 0 \le t \le T, \tag{9.27}$$

where the expansion coefficients are those specified by Eq. (9.26).

Now suppose that $x(t)$ is a (possibly complex) colored noise process. We want to find conditions under which the coefficients x_i are uncorrelated random variables. Such an expansion Eq. (9.27) with uncorrelated coefficients is called the *Karhunen–Loève expansion* of the process $x(t)$. Since the mean can always be removed separately, we will assume that $x(t)$ is a zero-mean process. Initially we will allow the process to be nonstationary, so that the correlation function is $\mathcal{E}x(t)x^*(t') = R(t, t')$.

Using the expression Eq. (9.26) for the coefficients, let us investigate their correlation. We have

$$\mathcal{E}x_i x_j^* = \int_0^T \int_0^T \mathcal{E}[x(t)x^*(t')]\phi_i^*(t)\phi_j(t') \, dt \, dt'$$
$$= \int_0^T \int_0^T \phi_i^*(t)R(t, t')\phi_j(t') \, dt \, dt'. \tag{9.28}$$

[This is the form that a quadratic form takes when the "vectors" are complex time functions $\phi_i(t)$.] Now suppose that we succeed in finding a set of orthonormal functions such that

$$\int_0^T R(t, t')\phi_j(t') \, dt' = \lambda_j\phi_j(t), \qquad 0 \le t \le T. \qquad *(9.29)$$

Using Eq. (9.29) in Eq. (9.28) then yields

$$\mathcal{E}x_i x_j^* = \lambda_j \int_0^T \phi_i^*(t)\phi_j(t) \, dt = \lambda_j\delta_{ij}. \tag{9.30}$$

Thereby the coefficients x_i in the expansion of the process $x(t)$ on this particular basis $\phi_i(t)$ become uncorrelated, with variances λ_j. Our problem is solved, provided we can find an orthonormal set of functions satisfying Eq. (9.29).

Equation (9.29) is a homogeneous integral equation. The functions $\phi_j(t)$ are eigenfunctions of the kernel $R(t, t')$ of the equation, and the numbers λ_j are the corresponding eigenvalues. The problem is much studied. We will enumerate some properties of the solutions, all of which have corresponding statements in the form of the finite matrix eigenvector problem discussed in Sect. 6.6. That a kernel of Eq. (9.29) is *symmetric* means

$$R(t, t') = R^*(t', t). \tag{9.31}$$

In particular, because

$$R(t, t') = \mathscr{E}\, x(t)x^*(t'),$$

Eq. (9.31) holds for the correlation function of a random process, and we will henceforth assume the property Eq. (9.31).

For a symmetric kernel Eq. (9.31), which is also square integrable, i.e., for which

$$\int_0^T \int_0^T |R(t, t')|^2\, dt\, dt' < \infty,$$

the principal result is that there exist at least one number λ and function $\phi(t)$ for which Eq. (9.29) holds true. That is, eigenvalues and eigenfunctions exist for a symmetric square integrable kernel. For such a kernel, whatever eigenvalues and eigenfunctions exist satisfy some strong properties, which are easy to demonstrate.

We first verify that solutions of Eq. (9.29) are indeed orthonormal, almost automatically. Assume Eq. (9.29) holds true, and form

$$\lambda_j \int_0^T \phi_i^*(t)\phi_j(t)\, dt$$

$$= \int_0^T \int_0^T \phi_i^*(t)R(t, t')\phi_j(t')\, dt'\, dt.$$

Now conjugate this, interchange the arbitrary indices i, j, and interchange the dummy variables t, t'. The result is

$$\lambda_i^* \int_0^T \phi_j(t')\phi_i^*(t')\, dt'$$

$$= \int_0^T \phi_j(t') \int_0^T R^*(t', t)\phi_i^*(t)\, dt\, dt'$$

$$= \int_0^T \int_0^T \phi_i^*(t)R(t, t')\phi_j(t')\, dt'\, dt,$$

where we used the symmetry property Eq. (9.31). Subtracting this last equation from the previous equation yields

$$(\lambda_j - \lambda_i^*) \int_0^T \phi_i^*(t)\phi_j(t) \, dt = 0. \tag{9.32}$$

From Eq. (9.32), taken for $i = j$, it follows that $\lambda_i = \lambda_i^*$, since the integral is positive. That is, the eigenvalues λ_i of the symmetric kernel $R(t, t')$ are real. We can therefore drop the conjugate notation on the eigenvalues.

Furthermore, from Eq. (9.32) taken for $i \neq j$, it follows that the eigenfunctions $\phi_i(t)$ belonging to numerically different eigenvalues are orthogonal. Since Eq. (9.29) is homogeneous, the orthogonal solutions can be normalized to unit energy to produce an orthonormal set.

However, it is not necessarily true that all the eigenvalues of Eq. (9.29) are numerically distinct. Suppose then that there are two linearly independent solutions $\phi_1(t)$, $\phi_2(t)$ with the same eigenvalue λ. Then any linear combination of the two functions is also an eigenfunction, with eigenvalue λ. This follows by forming $\phi(t) = a_1\phi_1(t) + a_2\phi_2(t)$ and noting that it satisfies Eq. (9.29) with value λ. The process extends to m linearly independent eigenfunctions with the same numerical eigenvalue. In that case, the m functions can be orthonormalized using the Gram–Schmidt procedure of Sect. 6.6 to produce m orthonormal eigenfunctions with eigenvalue λ, each of which is still orthogonal to the eigenfunctions with eigenvalues numerically not equal λ.

In addition to the symmetry property Eq. (9.31), a correlation function has the further property of being *nonnegative definite*. In the context of the operator on the left of Eq. (9.29), this means that, for arbitrary $f(t)$,

$$I = \int_0^T \int_0^T f^*(t)R(t, t')f(t') \, dt \, dt' \geq 0 \tag{9.33}$$

for all $f(t) \neq 0$. That this is true follows from substituting the definition $R(t, t') = \mathscr{E}[x(t)x^*(t')]$ to obtain

$$I = \mathscr{E} \int_0^T \int_0^T f^*(t)x(t)x^*(t')f(t') \, dt \, dt'$$

$$= \mathscr{E} \left| \int_0^T f^*(t)x(t) \, dt \right|^2 \geq 0. \tag{9.34}$$

Using the property Eq. (9.33), it is easy to show that all the eigenvalues λ_i of the kernel are nonnegative. From Eq. (9.29) we have

$$\int_0^T \int_0^T \phi_j^*(t)R(t, t')\phi_j(t) \, dt \, dt' = \lambda_j \int_0^T \phi_j^*(t)\phi_j(t) \, dt = \lambda_j \geq 0,$$

the last following from Eq. (9.33).

Except in singular cases, a random process $x(t)$ satisfies Eq. (9.34) in the stronger sense that equality cannot occur as long as $f(t)$ is not identically zero. That is, Eq. (9.33) cannot be satisfied nontrivially with equality. Such a kernel $R(t, t')$ is *positive definite,* rather than only nonnegative definite. In that case, the very strong property is true that there is in fact a countable infinity of eigenvalues λ_i which satisfy Eq. (9.29), that the corresponding eigenfunctions $\phi_i(t)$ belonging to numerically equal eigenvalues are linearly independent, and that the set of all eigenfunctions forms a complete set. The proofs are straightforward but lengthy, and appear for example in Courant and Hilbert, 1953. The conclusion is that solutions exist of the integral equation Eq. (9.29) which serve the purpose of the Karhunen–Loève expansion Eq. (9.27).

A further property of a nonnegative definite kernel is (*Mercer's theorem*) that it has an expansion in terms of its eigenvalues and eigenfunctions as:

$$R(t, t') = \sum_{i=1}^{\infty} \lambda_i \phi_i(t) \phi_i^*(t'). \qquad *(9.35)$$

The inverse kernel $R^{-1}(t, t')$ is defined by

$$\int_0^T R^{-1}(t, t') R(t', \tau) \, dt' = \delta(t - \tau), \qquad 0 \le t, \tau \le T. \qquad (9.36)$$

It has the expansion

$$R^{-1}(t, t') = \sum_{i=1}^{\infty} \left(\frac{1}{\lambda_i}\right) \phi_i(t) \phi_i^*(t'). \qquad *(9.37)$$

Note from this that the inverse of a symmetric kernel is symmetric:

$$R^{-1*}(t', t) = \sum_{i=1}^{\infty} \left(\frac{1}{\lambda_i}\right) \phi_i^*(t') \phi_i(t) = R^{-1}(t, t'), \qquad (9.38)$$

where we used the fact that the eigenvalues are real.

We have now argued that linearly independent solutions of Eq. (9.29) exist for a nonsingular correlation function and that they form a complete orthonormal set of functions. Therefore the coefficients in the expansion Eq. (9.27) are uncorrelated. We shall soon consider how to find them in the common case that the noise is stationary with power spectral density which is a rational fraction. First we will discuss the detector which will result.

9.5 DETECTION OF KNOWN SIGNALS WITH FINITE OBSERVATION TIME

Let us consider again the binary hypothesis-testing problem with completely known signals in zero-mean (not necessarily stationary) Gaussian noise:

$$H_1: \quad r(t) = s_1(t) + n(t),$$
$$H_0: \quad r(t) = s_0(t) + n(t), \qquad 0 \le t \le T. \tag{9.39}$$

Since we do not assume that the noise is white, as the data upon which to base the decision we will take the coefficients r_i in the expansion Eq. (9.27) using the Karhunen–Loève functions $\phi_i(t)$ determined from Eq. (9.29). That is, for the given noise covariance function $R(t, t')$, we solve Eq. (9.29) for the eigenvalues λ_i and eigenfunctions $\phi_i(t)$ and then take as data the numbers

$$r_i = \int_0^T \phi_i^*(t) r(t)\ dt. \qquad *(9.40)$$

Since the noise is Gaussian and the signals are known, the coefficients r_i, being determined by linear operations on a Gaussian process, are Gaussian random variables. By construction of the Karhunen–Loève basis functions, they are independent. To determine the joint densities needed for solving the detection problem, we then need only the means and variances of the coefficients Eq. (9.40) under the two hypotheses. We have at once

$$\mathscr{E}_1(r_i) = \int_0^T \phi_i^*(t) s_1(t)\ dt = s_{1i},$$

$$\mathscr{E}_0(r_i) = \int_0^T \phi_i^*(t) s_0(t)\ dt = s_{0i}.$$

From Eq. (9.30), the variances of the coefficients r_i are the eigenvalues λ_i. For compactness, we will proceed with the assumption that the signals are real. Then so also are the correlation function and its eigenvalues and eigenfunctions. In Chapter 11 we will indicate how to handle complex signals (complex envelopes) with a convenient notation. Where conjugates appear in the following equations, the result is true for complex envelope signals as well as the real signals assumed in the derivation.

For a finite number N of real coefficients r_i in Eq. (9.40), we have the multidimensional densities

$$p_{1,0}(\mathbf{r}) = \prod_{i=1}^N (2\pi\lambda_i)^{-1/2} \exp\left[\frac{-(r_i - s_{1,0i})^2}{2\lambda_i} \right]. \tag{9.41}$$

We want to take the limit as $N \Rightarrow \infty$ in Eq. (9.41) to obtain filters for use in the optimal decision processing. Letting $s(t)$ be either of $s_1(t)$, $s_0(t)$, the exponential factors in Eq. (9.41) involve the term

$$A = \sum_{i=1}^{\infty} \left(\frac{-1}{2\lambda_i}\right) [r_i^2 - (2r_i - s_i)s_i].$$

(9.42)

For the first part of this, we have

$$\sum_{i=1}^{\infty} \frac{r_i^2}{\lambda_i} = \sum_{i=1}^{\infty} \left(\frac{1}{\lambda_i}\right) \left[\int_0^T \phi_i(t)r(t) \, dt\right]^2$$

$$= \int_0^T \int_0^T r(t) \sum_{i=1}^{\infty} \left[\frac{\phi_i(t)\phi_i(t')}{\lambda_i}\right] r(t') \, dt \, dt'$$

$$= \int_0^T \int_0^T r(t)R^{-1}(t, t')r(t') \, dt \, dt'.$$

(9.43)

In this last we have used Eq. (9.37) for the inverse kernel.

The second term in Eq. (9.42) can be manipulated as follows. We have

$$\sum_{i=1}^{\infty} \frac{(2r_i - s_i)s_i}{\lambda_i}$$

$$= \sum_{i=1}^{\infty} \left(\frac{s_i}{\lambda_i}\right) \int_0^T \phi_i(t)[2r(t) - s(t)] \, dt$$

$$= \int_0^T [2r(t) - s(t)] \left[\sum_{i=1}^{\infty} \frac{s_i\phi_i(t)}{\lambda_i}\right] dt$$

$$= \int_0^T [2r(t) - s(t)]h(t) \, dt,$$

(9.44)

where the function $h(t)$ is so defined.

For the function $h(t)$ in Eq. (9.44) we have the property

$$\int_0^T R(t, t')h(t') \, dt' = \sum_{i=1}^{\infty} \left(\frac{s_i}{\lambda_i}\right) \int_0^T R(t, t')\phi_i(t') \, dt'$$

$$= \sum_{i=1}^{\infty} \left(\frac{s_i}{\lambda_i}\right) \lambda_i\phi_i(t) = \sum_{i=1}^{\infty} s_i\phi_i(t) = s(t). \quad *(9.45)$$

In this we used the definition Eq. (9.29) for the eigenfunctions. Using Eq. (9.45) we can further identify $h(t)$ as follows. We have

$$\int_0^T R^{-1}(t, t')s(t') \, dt'$$

$$= \int_0^T \int_0^T R^{-1}(t, t')R(t', t'')h(t'') \, dt'' \, dt'$$

$$= \int_0^T \delta(t - t'') \, h(t'') \, dt'' = h(t), \qquad *(9.46)$$

using the definition Eq. (9.36) for the inverse kernel.

Using Eq. (9.46) in Eq. (9.44) and substituting the result together with Eq. (9.43) into Eq. (9.41) yields the probability density of the infinite number of data expansion coefficients in the form

$$p[r(t)] = C \exp \left[\left(\frac{-1}{2} \right) \int_0^T \int_0^T [r(t) - s(t)] \right.$$

$$\left. \times R^{-1}(t, t')[r(t') - s(t')] \, dt \, dt' \right]. \qquad (9.47)$$

Here the constant provides the necessary normalization.

Using Eq. (9.47) we can at once write down the receiver for the problem Eq. (9.39). The log likelihood ratio is

$$G = \ln p_1[r(t)] - \ln p_0[r(t)]$$

$$= -\frac{1}{2} \int_0^T \int_0^T [r(t) - s_1(t)]R^{-1}(t, t')[r(t') - s_1(t')] \, dt \, dt'$$

$$+ \frac{1}{2} \int_0^T \int_0^T [r(t) - s_0(t)]R^{-1}(t, t')[r(t') - s_0(t')] \, dt \, dt'.$$

Thresholding this yields the receiver

$$G = \text{Re} \int_0^T \int_0^T [s_1(t) - s_0(t)]^* \, R^{-1}(t, t')r(t') \, dt \, dt'$$

$$- \frac{1}{2} \left[\int_0^T \int_0^T s_1^*(t)R^{-1}(t, t')s_1(t') \, dt \, dt' \right.$$

$$\left. - \int_0^T \int_0^T s_0^*(t)R^{-1}(t, t')s_0(t') \, dt \, dt' \right] > G_t. \qquad (9.48)$$

The expression Eq. (9.48) is correct as written for complex envelopes, although the derivation assumes real signals.

Using the fact Eq. (9.38) that the inverse kernel is symmetric, Eq. (9.46) yields

$$\int_0^T s_{1,0}^*(t) R^{-1}(t, t') \, dt$$

$$= \int_0^T s_{1,0}^*(t) R^{-1*}(t', t) \, dt = h_{1,0}^*(t'). \tag{9.49}$$

With this the detector Eq. (9.48) becomes

$$G = \mathrm{Re} \int_0^T h_1^*(t)[r(t) - \tfrac{1}{2} s_1(t)] \, dt$$

$$- \mathrm{Re} \int_0^T h_0^*(t)[r(t) - \tfrac{1}{2} s_0(t)] \, dt > G_0. \tag{9.50}$$

By redefining the threshold, the receiver Eq. (9.50) can also be written as

$$G' = \mathrm{Re} \int_0^T [h_1(t) - h_0(t)]^* r(t) \, dt > G_0'. \tag{*9.51}$$

This can be implemented as a correlation receiver or matched filter as in Fig. 9.1. The filter functions must be determined from the inverse kernel

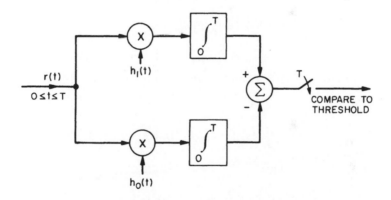

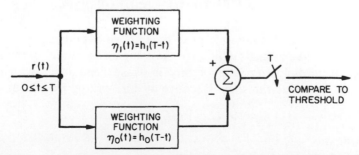

FIGURE 9.1 Equivalent receivers for detecting known signals in nonwhite Gaussian noise.

$R^{-1}(t, t')$ as in Eq. (9.46). The inverse kernel itself must be found from Eq. (9.36). That is not an easy task. We will give a method below for the special case of stationary noise with a rational power spectral. For the case of stationary noise with white spectrum $R(\tau) = (N_0/2)\delta(\tau)$, the functions $h_1(t)$, $h_0(t)$ are just the two signals themselves, scaled by a factor $2/N_0$ which does not affect the receiver.

9.6 RECEIVER PERFORMANCE

The receiver Eq. (9.50) dictates the choice of hypothesis H_1 if $G > G_0$. To determine the performance of this receiver, in terms of a receiver operating characteristic (ROC) curve, we require the probability density of the test statistic G under the two hypotheses. Since G is a linear operation on the Gaussian data waveform, the mean and variance suffice to determine the matter.

Because the noise in the problem Eq. (9.39) is zero mean, after some rearrangement, from the receiver Eq. (9.48) we have

$$\mathscr{E}_1(G) = -\mathscr{E}_0(G) = \frac{1}{2}\int_0^T \int_0^T [s_1(t) - s_0(t)]^*$$

$$\times R^{-1}(t, t')[s_1(t') - s_0(t')]\, dt\, dt'. \qquad (9.52)$$

Using Eq. (9.46), Eq. (9.52) can be written as

$$\mathscr{E}_1(G) = -\mathscr{E}_0(G)$$

$$= \frac{1}{2}\int_0^T [s_1(t) - s_0(t)]^*[h_1(t) - h_0(t)]\, dt. \qquad (9.53)$$

From Eq. (9.50) we have

$$G = \operatorname{Re}\int_0^T [h_1(t) - h_0(t)]^* r(t)\, dt$$

$$- \frac{1}{2}\operatorname{Re}\int_0^T [h_1^*(t)s_1(t) - h_0^*(t)s_0(t)]\, dt.$$

We want to compute the variance of G for the two hypotheses $i = 0, 1$. In the case of complex signals which are the complex envelopes of random processes, the cross-covariance of any in-phase component with any quadrature component vanishes. Because of this, writing Eq. (9.50) as

$$G = \operatorname{Re}\tilde{G} = \operatorname{Re}(G + j\hat{G}),$$

we can compute

$$\mathscr{E}(G^2) = \frac{1}{2}\mathscr{E}|\bar{G}|^2.$$

The application of this to Eq. (9.50) leads to lengthy expressions, however, so that for the moment we will assume real signals. Then, remembering that $R(t, t')$ is the noise correlation, Eq. (9.50) yields

$$\mathscr{E}_i(G^2) = \int_0^T \int_0^T [h_1(t) - h_0(t)][R(t, t') + s_i(t)s_i(t')]$$

$$\times [h_1(t') - h_0(t')] \, dt \, dt'$$

$$- \int_0^T [h_1(t) - h_0(t)]s_i(t) \, dt$$

$$\times \int_0^T [s_1(t)h_1(t) - s_0(t)h_0(t)] \, dt$$

$$+ \frac{1}{4}\left[\int_0^T [s_1(t)h_1(t) - s_0(t)h_0(t)] \, dt\right]^2$$

$$= \int_0^T \int_0^T [h_1(t) - h_0(t)]R(t, t')[h_1(t') - h_0(t')] \, dt \, dt'$$

$$+ \left[\int_0^T [h_1(t) - h_0(t)]s_i(t) \, dt\right.$$

$$\left. - \frac{1}{2}\int_0^T [s_1(t)h_1(t) - s_0(t)h_0(t)] \, dt\right]^2. \qquad (9.54)$$

In simplifying this last expression, we need the following fact deducible from Eqs. (9.45), (9.46):

$$\int_0^T h_1(t)s_0(t) \, dt = \int_0^T \int_0^T h_1(t)R(t, t')h_0(t') \, dt \, dt'$$

$$= \int_0^T \int_0^T \int_0^T R^{-1}(t, t'')s_1(t'')R(t, t')h_0(t') \, dt \, dt' \, dt''$$

$$= \int_0^T \int_0^T \delta(t' - t'')s_1(t'')h_0(t') \, dt' \, dt''$$

$$= \int_0^T h_0(t')s_1(t') \, dt'.$$

In this we have used the fact that $R(t, t')$ is symmetric. From this,

$$\int_0^T h_1(t)s_0(t) \, dt = \frac{1}{2}\int_0^T [h_1(t)s_0(t) + h_0(t)s_1(t)] \, dt.$$

Using this last we have

$$\int_0^T [h_1(t) - h_0(t)]s_i(t) \, dt$$

$$= (-1)^i \int_0^T \{-h_i(t)s_i(t) + [h_1(t)s_0(t) + h_0(t)s_1(t)]/2\} \, dt.$$

Then Eq. (9.54) appears as

$$\mathcal{E}_i(G^2) = \int_0^T \int_0^T [h_1(t) - h_0(t)]R(t, t')[h_1(t') - h_0(t')] \, dt \, dt'$$

$$+ [\mathcal{E}_i(G)]^2,$$

so that, using Eqs. (9.45), (9.46), the variance of G is

$$\sigma_G^2 = \int_0^T \int_0^T [h_1(t) - h_0(t)]^*R(t, t')[h_1(t') - h_0(t')] \, dt \, dt'$$

$$= \int_0^T [h_1(t) - h_0(t)]^*[s_1(t) - s_0(t)] \, dt$$

$$= \int_0^T \int_0^T [s_1(t) - s_0(t)]^*R^{-1}(t, t')[s_1(t') - s_0(t')] \, dt \, dt'. \quad *(9.55)$$

It is to be noted that

$$\sigma_G^2 = 2 |\mathcal{E}(G)|.$$

With these, the probability densities of the detector statistic G follow:

$$p_{0,1}(G) = (2\pi\sigma_G^2)^{-1/2} \exp[-(G \pm \sigma_G^2/2)/2\sigma_G^2].$$

For the detector which selects H_1 whenever

$$G > G_t,$$

we have

$$P_f = \int_{[G_t + \sigma_G^2/2]/\sigma_G}^{\infty} (2\pi)^{-1/2} \exp\left(\frac{-u^2}{2}\right) \, du,$$

$$P_d = \int_{[G_t + \sigma_G^2/2]/\sigma_G - \sigma_G}^{\infty} (2\pi)^{-1/2} \exp\left(\frac{-u^2}{2}\right) \, du.$$

That is, the problem is described by the standard ROC curve Fig. 5.7 for Gaussian signal in noise, with the SNR

$$\alpha = \sigma_G,$$
$$\alpha_{dB} = 20 \log(\sigma_G).$$

In the case of a communication system, in which we want minimum probability of error with equal prior probabilities of the two signals, we set

$$P(D_1 \mid H_0) = P_f = P(D_0 \mid H_1) = 1 - P_d,$$

and obtain $G_t = 0$ and an average probability of error

$$P_e = \tfrac{1}{2}[P(D_1 \mid H_0) + P(D_0 \mid H_1)]$$

$$= \int_{\sigma_G/2}^{\infty} (2\pi)^{-1/2} \exp\left(\frac{-u^2}{2}\right) du. \qquad *(9.56)$$

The ROC curve is that of Fig. 6.2 with $\rho = 0$ and $E_{av}/N_0 = (\sigma_G/2)^2$. The central difficulty in using these expressions is solving the integral equation Eq. (9.45) to determine the necessary filter functions $h_0(t)$, $h_1(t)$. We will present a method for doing that in the case of a stationary noise process with rational power spectral density in Sect. 9.9.

9.7 OPTIMUM SIGNAL WAVEFORM

In Sect. 9.4 the homogeneous integral equation Eq. (9.29) involving the noise kernel $R(t, t')$ arose in the course of defining the Karhunen–Loève expansion of a signal. The same equation arises in the context of determining the optimum signal waveforms $s_0(t)$, $s_1(t)$ for a binary communication system in the presence of nonwhite noise with finite observation interval.

Equation (9.56) expresses the error probability of such a system in terms of the single parameter σ_G^2 in Eq. (9.55). The larger σ_G, the smaller is the error probability P_e. Hence it is reasonable to seek the signals $s_0(t)$, $s_1(t)$ in Eq. (9.55) to maximize σ_G for given noise covariance $R(t, t')$. From Eq. (9.55) it is clear that a trivial solution is to increase the energy of the signals. In order to obtain a more useful formulation, we will require that the parameter σ_G be maximized subject to the constraint that the average energy of the two signals is some specified value E:

$$\int_0^T [|s_0(t)|^2 + |s_1(t)|^2] \, dt = 2E. \qquad (9.57)$$

We now extremize the variance Eq. (9.55) with the constraint Eq. (9.57) adjoined with a Lagrange multiplier μ^*. We consider the augmented "cost" function

$$Q = \int_0^T \int_0^T [s_1(t) - s_0(t)]^* R^{-1}(t, t')[s_1(t') - s_0(t')] \, dt \, dt'$$

$$- 2\mu^* \int_0^T [|s_0(t)|^2 + |s_1(t)|^2] \, dt. \qquad (9.58)$$

This is a real functional of complex functions $s_0(t)$, $s_1(t)$, in general, if we allow complex envelopes. As we will discuss in Sect. 11.2, the correct extremizing condition results by setting equal to zero the variations of Q with respect to the conjugate functions $s_0^*(t)$, $s_1^*(t)$, while formally considering $s_0(t)$, $s_1(t)$ to be constant.

The result of doing that is

$$\int_0^T \int_0^T [s_1(t) - s_0(t)]^* R^{-1}(t, t')[\delta s_1(t') - \delta s_0(t')] \, dt \, dt'$$

$$= 2\mu^* \int_0^T [s_1^*(t')\delta s_1(t') + s_0^*(t')\delta s_0(t')] \, dt'.$$

This must hold true for every pair of functions $\delta s_1(t)$, $\delta s_0(t)$. Hence we must have

$$\int_0^T [s_1(t) - s_0(t)]^* R^{-1}(t, t') \, dt$$

$$= 2\mu^*(-1)^{i+1}s_i^*(t'), \qquad 0 \le t' \le T, \qquad i = 0, 1. \tag{9.59}$$

Since the left sides of the two equations Eq. (9.59) are equal, we must have

$$s_1(t) = -s_0(t). \tag{9.60}$$

That is, the optimal binary system uses antipodal waveforms.

The condition on the optimal waveforms themselves can be found by substituting Eq. (9.60) into the necessary condition Eq. (9.59). The result for $s_1(t)$ is

$$\int_0^T s_1^*(t)R^{-1}(t, t') \, dt = \mu^* s_1^*(t'), \qquad 0 \le t' \le T. \tag{9.61}$$

Multiplying this through by $R(t', t'')$ and integrating over t', using Eq. (9.36), yields

$$\int_0^T s_1^*(t)\delta(t - t'') \, dt = \mu^* \int_0^T R(t', t'')s_1^*(t') \, dt',$$

that is,

$$\int_0^T R(t', t)s_1^*(t') \, dt' = \left(\frac{1}{\mu^*}\right)s_1^*(t),$$

$$\int_0^T R(t, t')s_1(t') \, dt = \left(\frac{1}{\mu}\right)s_1(t), \tag{9.62}$$

where in this last we again use the fact that the covariance function is symmetric.

Equation (9.62) is a homogeneous integral equation for the sought signal $s_1(t)$. In general it has countably infinitely many solutions, the eigenvalues and eigenfunctions of the covariance kernel. We recall that the eigenvalues μ are real, since the kernel is symmetric. Substituting the condition Eq. (9.61) into Eq. (9.55) yields the attained variance σ_G^2 for any of these solutions. Again taking into account Eq. (9.60), we have

$$\sigma_G^2 = 4 \int_0^T \int_0^T s_1^*(t) R^{-1}(t, t') s_1(t') \, dt \, dt'$$

$$= 4\mu \int_0^T |s_1(t)|^2 \, dt = 4\mu E.$$

Since E is fixed, we should use the solution μ which is largest. That is, $1/\mu$ should be the smallest eigenvalue of the covariance kernel, and $s_1(t)$ should be the corresponding eigenfunction, with $s_0(t) = -s_1(t)$. This puts the signals into parts of the vector space where the noise is weakest.

As a special case we can consider $R(t, t') = (N_0/2)\delta(t - t')$, that is, white noise with power spectral density $N_0/2$. Equation (9.62) becomes

$$(N_0/2)s_1(t) = (1/\mu)s_1(t).$$

Any function $s_1(t)$ is a solution, and all the eigenvalues are equal: $\mu = 2/N_0$. Provided we choose an antipodal system $s_0(t) = -s_1(t)$, every waveform $s_1(t)$ leads to the same error probability. The noise intensity is omnidirectional in the signal space.

9.8 INTEGRAL EQUATIONS

We have encountered two types of integral equations in various contexts. In the first case we have the homogeneous integral equation of the continuous time eigenproblem:

$$\int_0^T R(t, t')\phi(t') \, dt' = \lambda\phi(t), \qquad 0 \le t \le T. \tag{9.63}$$

We have also dealt with the inhomogeneous equation

$$\int_0^T R(t, t')q(t') \, dt' = s(t), \qquad 0 \le t \le T. \tag{9.64}$$

This last is the *Fredholm integral equation of the first kind*. A second type of inhomogeneous equation results if the kernel is of the mixed form

$$R(t, t') = (N_0/2)\,\delta(t - t') + C(t, t'),$$

where the kernel $C(t, t')$ corresponds to nonwhite noise. Using this in Eq. (9.64) leads to

$$\int_0^T C(t, t')q(t')\, dt' + (N_0/2)q(t) = s(t), \qquad 0 \le t \le T. \qquad (9.65)$$

This is the *Fredholm equation of the second kind*. It blends the homogeneous equation Eq. (9.63) and the Fredholm equation of first kind, Eq. (9.64).

If the eigenvalues and eigenfunctions of the kernel of the homogeneous equation Eq. (9.63) have been found, the Fredholm equation Eq. (9.64) can be solved in terms of them. We continue to assume that the kernel is symmetric and positive definite, so that the eigenvalues are positive and the eigenfunctions form a complete set.

If a solution $q(t)$ with integrable square exists for Eq. (9.64), it then has the expansion

$$q(t) = \sum_{i=1}^{\infty} q_i \phi_i(t), \qquad 0 \le t \le T,$$

in terms of the eigenfunctions of the kernel in question. The expansion coefficients are

$$q_i = \int_0^T \phi_i^*(t)q(t)\, dt.$$

Similarly, the known function on the right of Eq. (9.64) may be expanded as

$$s(t) = \sum_{i=1}^{\infty} s_i \phi_i(t),$$

$$s_i = \int_0^T \phi_i^*(t)s(t)\, dt.$$

Substituting these expansions into Eq. (9.64) and equating coefficients of $\phi_i(t)$ yields the solution

$$q_i = s_i/\lambda_i,$$

$$q(t) = \sum_{i=1}^{\infty} \left(\frac{s_i}{\lambda_i}\right)\phi_i(t).$$

A solution exists provided

$$\sum_{i=1}^{\infty} \frac{|s_i|^2}{\lambda_i^2} < \infty.$$

It is not generally possible to compute this formal solution. We now describe a procedure which is sometimes adequate for applications and which always yields a solution. The method applies to all three types of equations, Eqns. (9.63), (9.64), (9.65).

9.9 INTEGRAL EQUATION SOLUTION FOR A RATIONAL POWER SPECTRUM

Usually, solving the homogeneous integral equation Eq. (9.29) for the eigenfunctions $\phi_i(t)$ needed in the Karhunen–Loève expansion Eq. (9.27) is difficult. However, in one case there is a straightforward method. That is the case that the power spectrum of a real process $x(t)$ to be expanded over $0 \le t \le T$ is a rational fraction. Because any spectrum may be approximated arbitrarily closely by a rational fraction of sufficiently high order, in principle the problem is solved. Since we deal with the power spectrum, the implication is that the random process $x(t)$ is stationary, so that the kernel of the integral equation Eq. (9.29) is $R(t, t') = R(t - t') = R(\tau)$.

For generality, let us consider the integral equation

$$\int_0^T R(t - t')q(t') \, dt' = f(t), \qquad 0 \le t \le T. \tag{9.66}$$

The covariance function $R(\tau)$ of a real process is an even function of its argument, so that its Fourier transform, the power spectrum, is an even function of frequency. That is, as we discussed in Sect. 9.3, we have

$$S(f) = \mathscr{F}[R(\tau)] = N(-\omega^2)/D(-\omega^2). \tag{9.67}$$

We now assume that $N(-\omega^2)$ is a polynomial of degree $2m$ in ω and that $D(-\omega^2)$ is a polynomial of degree $2n$ in ω. Since the process $x(t)$ is colored noise, it has finite variance. Therefore $S(f)$ must be integrable, and hence $m < n$.

The method involves converting the integral equation Eq. (9.66) to a differential equation. If the limits $[0, T]$ of the region of interest were instead $(-\infty, \infty)$, Eq. (9.66) would be a convolution. Fourier transformation would then convert the equation to

$$S(f)Q(j\omega) = N(-\omega^2)Q(j\omega)/D(-\omega^2) = F(j\omega),$$
$$N(-\omega^2)Q(j\omega) = D(-\omega^2)F(j\omega). \tag{9.68}$$

Now recall that, for the Laplace transform, for a function vanishing except over $[0, T]$ we have

$$\mathscr{L}\{df/dt\} = \int_{0+}^{T-} (df/dt) \exp(-st) \, dt$$
$$= sF(s) - f(0+) + f(T-) \exp(-sT). \tag{9.69}$$

Since all our functions vanish along with all their derivatives at $\pm\infty$, over the infinite interval we can ignore the end-point conditions in this last and make the correspondence

$$s^2 = -\omega^2 \Leftrightarrow d^2(\ )/dt^2.$$

Now write

$$D(-\omega^2) = \sum_{k=0}^{n} a_k(-\omega^2)^k,$$

$$N(-\omega^2) = \sum_{k=0}^{m} b_k(-\omega^2)^k.$$

From Eq. (9.68) we then have a differential equation to be solved for $q(t)$ given $f(t)$:

$$\sum_{k=0}^{m} b_k \frac{d^{2k}}{dt^{2k}} q(t) = \sum_{k=0}^{n} a_k \frac{d^{2k}}{dt^{2k}} f(t). \tag{9.70}$$

Since we have ignored all the impulses arising in iterations of Eq. (9.69) for the actual range $0 \le t \le T$ of interest, the solution of Eq. (9.70) does not solve the integral Eq. (9.66). However, with an adjustment the solution of Eq. (9.70), calculated using $f(t)$ over the given range $[0, T]$ while ignoring boundary conditions, can be made to fit the case of Eq. (9.66).

It is straightforward but lengthy to show (Davenport and Root, 1958) that the necessary modification of the solution $q_0(t)$ of Eq. (9.70) is to add impulse functions and derivatives of impulses. Specifically, it is necessary to take

$$q(t) = q_0(t) + \sum_{i=0}^{2(n-m-1)} [a_i\delta^{(i)}(t - 0+) + b_i\delta^{(i)}(t - T + 0+)]. \tag{9.71}$$

Here the notation indicates that the impulses occur just inside the interval $[0, T]$. The function $q_0(t)$ should include both the homogeneous solution and the particular integral of the differential equation Eq. (9.70). Since we have assumed $m < n$, the upper limit on the sum is nonnegative.

EXAMPLE 9.3 Consider the nonhomogeneous integral equation

$$\int_0^T \alpha \exp(-\beta|t - t'|)q(t')\, dt' = f(t), \tag{9.72}$$

where $f(t)$ is some specified function. The kernel is $R(\tau) = \alpha \exp(-\beta|\tau|)$, and its spectrum (everywhere positive, as it must be) is

$$S(f) = 2\alpha\beta/(\omega^2 + \beta^2). \tag{9.73}$$

(Usually we will be given the spectrum and will need to determine the kernel from that.)

The differential equation Eq. (9.70) is

$$2\alpha\beta q(t) = -f''(t) + \beta^2 f(t).$$

The homogeneous solution is found with $f(t) = 0$ and is $q_h(t) = 0$. The particular integral is trivially

$$q_p(t) = [-f''(t) + \beta^2 f(t)]/(2\alpha\beta).$$

Since $m = 0$, $n = 1$, Eq. (9.71) specifies that the solution should be taken as

$$q(t) = [-f''(t) + \beta^2 f(t)]/(2\alpha\beta) + a_1\delta(t) + a_2\delta(t - T), \quad (9.74)$$

where a_1, a_2 are constants to be determined.

To determine the constants, the solution Eq. (9.74) is substituted into the original equation, Eq. (9.72), and it is at this point that we need to know the form of the kernel. Integrating by parts twice and remembering that the impulses occur just inside the interval of integration, the result is

$$\int_0^T \alpha \exp(-\beta|t - t'|)q(t') \, dt'$$

$$= \frac{1}{2\beta}\int_0^t \exp[-\beta(t - t')][-f''(t') + \beta^2 f(t')] \, dt'$$

$$+ \frac{1}{2\beta}\int_t^T \exp[\beta(t - t')][-f''(t') + \beta^2 f(t')] \, dt'$$

$$+ \alpha\{a_1 \exp(-\beta t) + a_2 \exp[\beta(t - T)]\}$$

$$= (1/2\beta)\{-f'(t) + \beta f(t) + [f'(0) - \beta f(0)] \exp(-\beta t)\}$$

$$+ (1/2\beta)\{f'(t) + \beta f(t) + [-f'(T) - \beta f(T)] \exp[\beta(t - T)]\}$$

$$+ \alpha a_1 \exp(-\beta t) + \alpha a_2 \exp[\beta(t - T)]$$

$$= f(t) + (1/2\beta) \exp(-\beta t)[f'(0) - \beta f(0) + 2\alpha\beta a_1]$$

$$+ (1/2\beta) \exp[\beta(t - T)][-f'(T) - \beta f(T) + 2\alpha\beta a_2] = f(t).$$

This is to be satisfied as an identity, so that we require

$$a_1 = [-f'(0) + \beta f(0)]/(2\alpha\beta),$$
$$a_2 = [f'(T) + \beta f(T)]/(2\alpha\beta).$$

Using these in Eq. (9.74), the solution to Eq. (9.72) becomes

$$q(t) = (1/2\alpha\beta)\{-f''(t) + \beta^2 f(t) + [-f'(0) + \beta f(0)]\delta(t)$$
$$+ [f'(T) + \beta f(T)]\delta(t - T)\}, \quad 0 \le t \le T.$$

EXAMPLE 9.4 In the preceding example, the necessary constants were introduced by means of impulses at the boundaries of the interval of interest. In this example, additional necessary constants enter through the homogeneous solution of the differential equation Eq. (9.70).

The equation to be solved is

$$\frac{1}{2}\int_0^1 \exp(-|t - t'|) \cos[2(t - t')]q(t') \, dt'$$

$$= \exp(-t), \quad 0 \le t \le 1. \quad (9.75)$$

The transform of the kernel, that is, the power spectrum of the noise of interest, is

$$S(f) = \mathcal{F}\{\exp(-|\tau|)\cos(2\tau)\} = \frac{\omega^2 + 5}{\omega^4 - 6\omega^2 + 25}.$$

The differential equation to be solved, Eq. (9.70), is then

$$-q''(t) + 5q(t) = f^{(iv)}(t) + 6f''(t) + 25f(t),$$

where

$$f(t) = \exp(-t).$$

That is,

$$-q''(t) + 5q(t) = 32 \exp(-t).$$

The complete solution (homogeneous solution plus particular integral) is

$$q_0(t) = 8\exp(-t) + A_1 \exp(\sqrt{5}t) + A_2 \exp(-\sqrt{5}t). \tag{9.76}$$

Since $m = 2$, $n = 1$, Eq. (9.71) prescribes an impulse at each end of the interval, so that we take the solution of Eq. (9.75) as

$$q(t) = q_0(t) + a_1\delta(t) + a_2\delta(t-1), \tag{9.77}$$

where $q_0(t)$ is as in Eq. (9.76). Substituting Eq. (9.77) into Eq. (9.75) yields the following requirements on the constants:

$$-0.111803A_1 + 0.111803A_2 + 3.91201a_1 = 0,$$
$$2.76393A_1 + 14.4721A_2 = 0,$$
$$1.04609A_1 - 0.0119493A_2 + 2.98458a_2 = 0.367879,$$
$$1.69260A_1 + 0.00738508A_2 = -0.367879.$$

The final solution is

$$q(t) = -0.2175 \exp(\sqrt{5}t) + 0.04154 \exp(-\sqrt{5}t) + 8\exp(-t)$$
$$-0.0074040\delta(t) + 0.1997\delta(t-1).$$

EXAMPLE 9.5   Now consider the homogeneous integral equation corresponding to Example 9.3:

$$\int_0^T \alpha \exp(-\beta|t - t'|)q(t')\,dt' = \lambda q(t), \qquad 0 \le t \le T. \tag{9.78}$$

The procedure is the same as before, with the difference that the solution will not need any impulse functions to deal with boundary conditions. Clearly if $q(t')$ on the left of Eq. (9.78) contains an impulse, it will not appear in the value of the integral and will not be present on the left to match its occurrence on the right. Furthermore, the differential equation Eq. (9.70) is homogeneous, since $f(t) \equiv q(t)$.

The covariance kernel in Eq. (9.78) has the spectrum Eq. (9.73). Hence we take the solution $q(t)$ of Eq. (9.78) as the homogeneous solution of the differential equation

$$2\alpha\beta q(t) = -\lambda f''(t) + \lambda\beta^2 f(t) = -\lambda q''(t) + \lambda\beta^2 q(t),$$
$$q''(t) + \beta(2\alpha/\lambda - \beta)q(t) = 0. \qquad (9.79)$$

The roots of the characteristic equation are $\pm\gamma$, where

$$\gamma = (\beta^2 - 2\alpha\beta/\lambda)^{1/2}, \qquad (9.80)$$

and the solution is

$$q(t) = A\,\exp(\gamma t) + B\,\exp(-\gamma t). \qquad (9.81)$$

Substituting Eq. (9.81) into Eq. (9.78), integrating, and equating coefficients of $\exp(\gamma_1 t)$, $\exp(\gamma_2 t)$ leads to two equations for the two unknown constants A, B:

$$A(\beta - \gamma) + B(\beta + \gamma) = 0,$$
$$A(\beta + \gamma)\exp(\gamma T) + B(\beta - \gamma)\exp(-\gamma T) = 0.$$

For a solution to exist, the determinant must vanish. This yields an equation for the eigenvalue γ:

$$\exp(\gamma T) = (\beta - \gamma)/(\beta + \gamma). \qquad (9.82)$$

For real γ, the left side of this is always greater than unity and the right side is always less than unity. No real solutions for γ exist.

Accordingly, we take Eq. (9.80) in the form $\gamma = j\omega$, where

$$\omega = (2\alpha\beta/\lambda - \beta^2)^{1/2}. \qquad (9.83)$$

Now Eq. (9.82) reads

$$\cos(\omega T) + j\,\sin(\omega T) = (\beta - j\omega)^2/(\beta^2 + \omega^2).$$

Equating real and imaginary parts, this yields the two equations

$$\cos(\omega T) = \lambda(\beta^2 - \omega^2)/2\alpha\beta,$$
$$\sin(\omega T) = -\lambda\omega/\alpha.$$

Formally eliminating λ between these yields the single equation in ω:

$$\tan(\omega T) = 2\beta\omega/(\omega^2 - \beta^2). \qquad (9.84)$$

Now recall the identity in x:

$$\tan(2x) = \frac{2\sin(x)\cos(x)}{\cos^2(x) - \sin^2(x)}.$$

If we make the correspondences

$$\beta = \sin(x),$$
$$\omega = \cos(x),$$

then with $x = \omega T/2$ Eq. (9.84) yields

$$\tan(x) = \beta/\omega = \tan(\omega T/2),$$
$$\tan(\omega T/2) = (\beta T/2)/(\omega T/2). \tag{9.85}$$

This expresses the intersection of the tangent curve on the left with the hyperbola on the right, as sketched in solid in Fig. 9.2. There are infinitely many solutions, due to the multivalued inverse of the tangent function.

Alternatively, we can make the correspondences

$$\omega = \sin(x),$$
$$\beta = \cos(x).$$

With $x = -\omega T/2$, we than have

$$\tan(x) = \omega/\beta = -\tan(\omega T/2),$$
$$-1/\tan(\omega T/2) = (\beta T/2)/(\omega T/2). \tag{9.86}$$

This expresses the intersection of the negative inverse of the tangent curve in Eq. (9.85) with the same hyperbola. Again there are countably infinitely many solutions, and they interleave the solutions of Eq. (9.85).

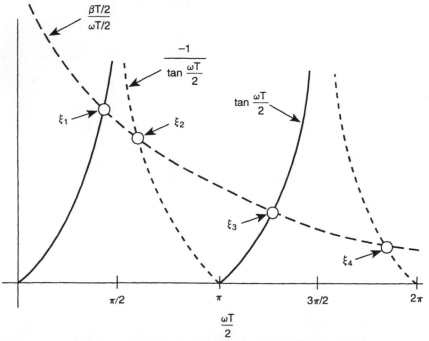

FIGURE 9.2   Solutions ξ_k relating to eigenvalues of Example 9.5.

With $\xi_k = \omega_k T/2$, the combined set of solutions can be expressed as

$$\xi_k \tan(\xi_k) = \beta T/2,$$
$$\xi_k \cot(\xi_k) = -\beta T/2. \tag{9.87}$$

From Eq. (9.83), the solutions ξ_k yield the eigenvalues of the original problem Eq. (9.78) as

$$\lambda_k = 2\alpha\beta/(\omega_k^2 + \beta^2). \tag{9.88}$$

With complex roots now being identified, Eq. (9.81) is conveniently written as

$$q(t) = a \cos(\omega t) + b \sin(\omega t).$$

Again this is substituted into the original Eq. (9.78) and conditions are set up on the constants a, b in order that the equation hold true. There result

$$\beta a - \omega b = 0,$$
$$[-\beta \cos(\omega T) + \omega \sin(\omega T)]a + [-\beta \sin(\omega T) - \omega \cos(\omega T)]b = 0. \tag{9.89}$$

Now we are assured that solutions exist, so long as we choose ω to be one of the solutions $\omega_k = 2\xi_k/T$ of Eq. (9.87).

Proceeding to solve for a, b, substituting the first of Eq. (9.89) into the second, we require

$$[-\beta \cos(\omega T) + \omega \sin(\omega T)]a + (\beta/\omega)[-\beta \sin(\omega T) - \omega \cos(\omega T)]a = 0,$$

that is,

$$[\tan(\omega T) - 2\beta\omega/(\omega^2 - \beta^2)]a = 0.$$

But from Eq. (9.84), this is satisfied identically for any of the solutions ω_k of Eq. (9.87). Hence corresponding to the values ω_k and associated λ_k of Eq. (9.88) we obtain the eigenfunctions

$$q_k(t) = a_k[\cos(\omega_k t) + (\beta/\omega_k) \sin(\omega_k t)],$$

for arbitrary a_k. The constants a_k can be taken as

$$\frac{1}{a_k^2} = \int_0^T \left[\cos(\omega_k t) + \left(\frac{\beta}{\omega_k}\right) \sin(\omega_k t) \right]^2 dt$$

in order to obtain an orthonormal set over $[0, T]$.

EXAMPLE 9.6 Suppose the noise has a white component and that the covariance function is

$$R(\tau) = \exp(-|\tau|) + 2\delta(\tau).$$

With this covariance kernel, the integral equation Eq. (9.66) is a Fredholm equation of second kind:

$$\int_0^T \exp(-|t - t'|)q(t') \, dt' + 2q(t) = \exp(-2t), \qquad 0 \le t \le T,$$

where we have specified a particular forcing function $f(t)$ on the right.

First observe that the solution $q(t)$ cannot contain any impulsive components, because they will disappear after integration of the first term, and there are none present on the right which could match them in the second term on the left.

The spectrum of the kernel $R(\tau)$ of the integral equation is

$$\frac{1}{s + 1} + \frac{1}{-s + 1} + 2 = \frac{-2s^2 + 4}{-s^2 + 1},$$

so that the differential equation Eq. (9.70) is

$$-2q''(t) + 4q(t) = -f'' + f = -3 \exp(-2t).$$

The characteristic equation has roots $\pm \sqrt{2}$, so that the general solution is

$$q(t) = A \exp(\sqrt{2}t) + B \exp(-\sqrt{2}t) + \tfrac{3}{4}\exp(-2t).$$

Substituting this into the integral equation yields the conditions
$$(1 - \sqrt{2})A + (1 + \sqrt{2})B = -\tfrac{3}{4},$$
$$A(1 + \sqrt{2}) \exp[(-1 + \sqrt{2})T] + B(1 - \sqrt{2}) \exp[(-1 - \sqrt{2})T]$$
$$= \tfrac{1}{4}\exp(-3T).$$

These have a solution for every $T > 0$. For example, for $T = 1$ we obtain

$$q(t) = 0.0003 \exp(\sqrt{2}t) - 0.311 \exp(-\sqrt{2}t) + 0.75 \exp(-2t).$$

9.10 DETECTION OF SIGNALS WITH UNKNOWN PHASE

We will now apply the foregoing results to the problem of Sect. 7.2, the detection of narrowband signals with random carrier phase. Whereas earlier we assumed the noise was white, here we will allow the noise to be colored. Then time samples taken at the Nyquist rate are correlated, and we are not able to calculate the necessary likelihood functions as products of likelihood functions for single time samples. Here we will use the coefficients of the Karhunen–Loève signal decompositions as independent samples.

In this section we will deal explicitly with the complex envelopes of the signals and noises in question. Consider then a real narrowband signal

$$r(t) = a(t) \cos[\omega_c t + \gamma(t) + \theta] + n(t), \qquad 0 \le t \le T. \tag{9.90}$$

The amplitude and phase functions $a(t)$, $\gamma(t)$ are completely known, the phase θ is a uniform random variable, and the noise $n(t)$ is a nonwhite zero-mean Gaussian random process. We first consider a fixed value of θ and develop the conditional likelihood functions. The nuisance parameter θ is then integrated out to yield the unconditioned likelihood functions and the likelihood ratio detector.

Using any of the techniques of Chapter 3, the complex envelope of the signal Eq. (9.90) is determined:

$$\tilde{r}(t) = a(t) \exp[j\gamma(t)] \exp(j\theta) + \tilde{n}(t)$$
$$= \tilde{A}(t) \exp(j\theta) + \tilde{n}(t), \tag{9.91}$$

where $\tilde{A}(t) = a(t) \exp[j\gamma(t)]$ is known. Writing $\tilde{n}(t) = x(t) + jy(t)$, the covariance function of the complex envelope is

$$\text{Cov}[\tilde{r}(t)] = R_{\tilde{n}}(\tau) = \mathscr{E}\{[x(t) + jy(t)][x(t - \tau) - jy(t - \tau)]\}$$
$$= [R_x(\tau) + R_y(\tau)] + j[R_{yx}(\tau) - R_{xy}(\tau)].$$

From Eq. (3.75) we have

$$R_x(\tau) = R_y(\tau). \tag{9.92}$$

Using Eq. (3.83) we also have

$$R_{yx}(\tau) = R_{xy}(-\tau) = -R_{xy}(\tau). \tag{9.93}$$

Therefore we have

$$R_{\tilde{n}}(\tau) = 2[R_x(\tau) - jR_{xy}(\tau)]. \tag{9.94}$$

Now consider that

$$n(t) = \text{Re}\{[x(t) + jy(t)] \exp(j\omega_c t)\}$$
$$= x(t) \cos(\omega_c t) - y(t) \sin(\omega_c t).$$

Then it is easy to show that

$$R_n(\tau) = \mathscr{E}[n(t)n(t - \tau)]$$
$$= R_x(\tau) \cos(\omega_c \tau) + R_{xy}(\tau) \sin(\omega_c \tau)$$
$$= \text{Re}\{[R_x(\tau) - jR_{xy}(\tau)] \exp(j\omega_c \tau)\}.$$

That is, the complex envelope of the noise correlation function $R_n(\tau)$ is

$$\tilde{R}_n(\tau) = R_x(\tau) - jR_{xy}(\tau) = \tfrac{1}{2}R_{\tilde{n}}(\tau). \tag{9.95}$$

The complex envelope Eq. (9.95) is a symmetric and positive definite function. For the first, we have

$$\tilde{R}_n(-\tau) = R_x(-\tau) - jR_{xy}(-\tau)$$
$$= R_x(\tau) + jR_{xy}(\tau) = \tilde{R}_n^*(\tau).$$

For the second, from Eq. (9.95) we have

$$\tilde{R}_n(\tau) = \tilde{R}_n(t - t') = \tfrac{1}{2} R_{\tilde{n}}(t - t'),$$

so that, for any function $f(t) \neq 0$,

$$\int_0^T \int_0^T f^*(t)\tilde{R}_n(t - t')f(t')\, dt\, dt' = \frac{1}{2}\mathscr{E}\left|\int_0^T f^*(t)\tilde{n}(t)\, dt\right|^2 > 0.$$

In writing the inequality as strict, we assume there is no degeneracy in the noise process.

Since $\tilde{R}_n(\tau)$ is symmetric and positive definite, we can generate a complete orthonormal set of functions $\phi_i(t)$ from the eigenproblem:

$$\int_0^T \tilde{R}_n(t - t')\phi_i(t')\, dt' = \lambda_i\phi_i(t), \qquad 0 \le t \le T. \tag{9.96}$$

We then use those eigenfunctions in the Karhunen–Loève procedure to obtain an expansion of the complex envelope of the received signal, with uncorrelated coefficients. The noise coefficients are

$$n_i = \alpha_i + j\beta_i = \int_0^T \phi_i^*(t)\tilde{n}(t)\, dt. \tag{9.97}$$

We need the covariance properties of the real and imaginary parts α_i, β_i. With zero mean noise we have

$$\mathscr{E}(n_i n_j^*) = \int_0^T \int_0^T \phi_i^*(t)\phi_j(t')\mathscr{E}[\tilde{n}(t)\tilde{n}^*(t')]\, dt\, dt'$$

$$= 2\int_0^T \int_0^T \phi_i^*(t)\tilde{R}_n(t - t')\phi_j(t')\, dt\, dt'$$

$$= 2\lambda_j \int_0^T \phi_i^*(t)\phi_j(t)\, dt = 2\lambda_j\delta_{ij}, \tag{9.98}$$

where we used Eqs. (9.95), (9.96). Further,

$$\mathscr{E}(n_i n_j) = \int_0^T \int_0^T \phi_i^*(t)\phi_j^*(t')\mathscr{E}[\tilde{n}(t)\tilde{n}(t')]\, dt\, dt' = 0. \tag{9.99}$$

In this last we have used the fact that

$$\mathscr{E}[\tilde{n}(t)\tilde{n}(t')] = \mathscr{E}[x(t) + jy(t)][x(t') + jy(t')]$$
$$= [R_x(\tau) - R_y(\tau)] + j[R_{xy}(\tau) + R_{yx}(\tau)] = 0,$$

using Eqs. (9.92), (9.93).

From Eqs. (9.98), (9.99) we now have

$$\mathscr{E}(n_i n_j^*) = \mathrm{Cov}(\alpha_i + j\beta_i)(\alpha_j - j\beta_j)$$
$$= \mathrm{Cov}(\alpha_i\alpha_j + \beta_i\beta_j) - j\,\mathrm{Cov}(\alpha_i\beta_j - \alpha_j\beta_i) = 2\lambda_i\delta_{ij},$$

$$\mathcal{E}(n_i n_j) = \text{Cov}(\alpha_i + j\beta_i)(\alpha_j + j\beta_j)$$
$$= \text{Cov}(\alpha_i \alpha_j - \beta_i \beta_j) + j \text{ Cov}(\alpha_i \beta_j + \alpha_j \beta_i) = 0.$$

Taking into account that λ_i is real, since the kernel of the integral equation Eq. (9.96) is symmetric, from these last we have

$$\text{Cov}(\alpha_i \alpha_j) = \text{Cov}(\beta_i \beta_j),$$
$$\text{Cov}(\alpha_i \alpha_j) + \text{Cov}(\beta_i \beta_j) = 2 \text{ Cov}(\alpha_i \alpha_j) = 2 \text{ Cov}(\beta_i \beta_j) = 2\lambda_i \delta_{ij},$$
$$\text{Cov}(\alpha_i \alpha_j) = \text{Cov}(\beta_i \beta_j) = \lambda_i \delta_{ij}. \tag{9.100}$$

Furthermore,

$$\text{Cov}(\alpha_i \beta_j) = 0. \tag{9.101}$$

These are also the covariances of the data coefficients. The means of the coefficients follow from Eqns. (9.91), (9.97). We have

$$\mathcal{E}(r_i) = \int_0^T \phi_i^*(t) \mathcal{E}\tilde{r}(t) \, dt$$

$$= \exp(j\theta) \int_0^T \phi_i^*(t) \tilde{A}(t) \, dt = a_i \exp(j\theta), \tag{9.102}$$

where the a_i are the expansion coefficients of the signal complex envelope $\tilde{A}(t)$ on the basis $\phi_i(t)$. From this, for the data coefficients

$$\mathcal{E}(\alpha_i) = \text{Re}[a_i \exp(j\theta)],$$
$$\mathcal{E}(\beta_i) = \text{Im}[a_i \exp(j\theta)]. \tag{9.103}$$

Since we assume the noise is Gaussian, so also is the received signal for fixed carrier phase θ. Since the coefficients α_i, β_i are linear functionals of the signal, they are also Gaussian. With the means Eq. (9.103) and covariance properties Eqs. (9.100), (9.101) taken into account, we can now write the likelihood function for the signal Eq. (9.90) in terms of the vector **r** of real and imaginary parts α_i, β_i of the expansion coefficients r_i of the complex envelope. We have

$$p(\mathbf{r} \mid \theta) = \prod_{i=1}^{\infty} (2\pi\lambda_i)^{-1}$$

$$\times \exp\left(\frac{-1}{2\lambda_i}\right) \{[\alpha_i - \text{Re } a_i \exp(j\theta)]^2 + [\beta_i - \text{Im } a_i \exp(j\theta)]^2\}$$

$$= \prod_{i=1}^{\infty} (2\pi\lambda_i)^{-1} \exp\left[\frac{-|r_i - a_i \exp(j\theta)|^2}{2\lambda_i}\right]$$

$$= C \exp \sum_{i=1}^{\infty} \left[\frac{-|r_i - a_i \exp(j\theta)|^2}{2\lambda_i}\right], \tag{9.104}$$

where the constant C is of no interest.

Let us now define an amplitude q and phase η by

$$q \exp(j\eta) = \sum_{i=1}^{\infty} \left(\frac{r_i a_i^*}{\lambda_i} \right)$$

$$= \int_0^T \tilde{r}(t) \sum_{i=1}^{\infty} \left[\frac{a_i^* \phi_i^*(t)}{\lambda_i} \right] dt$$

$$= \int_0^T \tilde{h}^*(t) \tilde{r}(t) \, dt. \tag{9.105}$$

Here we used Eq. (9.97) written for the data and defined $\tilde{h}(t)$. This function can be further defined as follows. From the definition of $\tilde{h}(t)$ in Eq. (9.105) we have

$$\int_0^T \tilde{R}_n(t - t') \tilde{h}(t') \, dt'$$

$$= \sum_{i=1}^{\infty} \left(\frac{a_i}{\lambda_i} \right) \int \tilde{R}_n(t - t') \phi_i(t') \, dt'$$

$$= \sum_{i=1}^{\infty} a_i \phi_i(t) = \tilde{A}(t). \tag{9.106}$$

Here we used Eqs. (9.96), (9.102). That is, the filter $\tilde{h}(t)$ is obtained as the solution of an inhomogeneous integral equation using the noise covariance kernel and the known signal complex envelope.

In terms of q, η of Eq. (9.105), we have

$$p(\mathbf{r} \mid \theta) = C \exp \sum_{i=1}^{\infty} \frac{|r_i|^2 + |a_i|^2}{-2\lambda_i}$$

$$\times \exp \sum_{i=1}^{\infty} \mathrm{Re} \left[\left(\frac{r_i a_i^*}{\lambda_i} \right) \exp(-j\theta) \right]$$

$$= C \exp[q \cos(\eta - \theta)] \exp \sum_{i=1}^{\infty} \frac{|r_i|^2 + |a_i|^2}{-2\lambda_i}. \tag{9.107}$$

We now calculate the unconditioned likelihood function by integrating over the uniform random variable θ. We have

$$\frac{1}{2\pi} \int_{-\pi}^{\pi} \exp[q \cos(\eta - \theta)] \, d\theta$$

$$= \frac{1}{2\pi} \int_{-\pi}^{\pi} \exp[q \cos(\theta)] \, d\theta = I_0(q),$$

where $I_0(q)$ is the modified Bessel function of the first kind and order zero. Using this we obtain

$$p(\mathbf{r}) = CI_0(q) \exp \sum_{i=1}^{\infty} \frac{|r_i|^2 + |a_i|^2}{-2\lambda_i}. \qquad (9.108)$$

For a binary hypothesis problem, the log likelihood ratio corresponding to Eq. (9.108) is

$$\ln[\lambda(\mathbf{r})] = \ln[p_1(\mathbf{r})] - \ln[p_0(\mathbf{r})]$$

$$= \ln I_0(q_1) - \ln I_0(q_0) - \sum_{i=1}^{\infty} \frac{|a_{i1}|^2}{2\lambda_i} + \sum_{i=1}^{\infty} \frac{|a_{i0}|^2}{2\lambda_i}.$$

From Eqs. (9.102), (9.105), (9.106) the last terms can be simplified as

$$\sum_{i=1}^{\infty} \frac{|a_i|^2}{\lambda_i} = \sum_{i=1}^{\infty} \left(\frac{a_i^*}{\lambda_i} \right) \int_0^T \phi_i^*(t) \tilde{A}(t) \, dt$$

$$= \int_0^T \tilde{h}^*(t) \tilde{A}(t) \, dt$$

$$= \int_0^T \int_0^T \tilde{h}^*(t) \tilde{R}_n(t - t') \tilde{h}(t') \, dt \, dt' = d^2, \qquad (9.109)$$

where we so define d^2. With this the likelihood ratio becomes

$$\ln[\lambda(\mathbf{r})] = \ln I_0(q_1) - \ln I_0(q_0) - (d_1^2 - d_0^2)/2. \qquad *(9.110)$$

From Eq. (9.97) written for the data we also have

$$\sum_{i=1}^{\infty} \frac{|r_i|^2}{\lambda_i} = \int_0^T \int_0^T \tilde{r}^*(t) \left[\sum_{i=1}^{\infty} \frac{\phi_i(t)\phi_i^*(t')}{\lambda_i} \right] \tilde{r}(t') \, dt \, dt'$$

$$= \int_0^T \int_0^T \tilde{r}^*(t) \tilde{R}_n^{-1}(t - t') \tilde{r}(t') \, dt \, dt'. \qquad (9.111)$$

The last equality uses Mercer's theorem in the form Eq. (9.37) and the fact, from Eq. (9.96), that the eigenfunctions $\phi_i(t)$ are generated from the kernel $R_n(t, t')$. Furthermore, from Eq. (9.106) we can write

$$\int_0^T \tilde{R}_n^{-1}(t'', t) \tilde{A}(t) \, dt$$

$$= \int_0^T \int_0^T \tilde{R}_n^{-1}(t'', t) \tilde{R}_n(t, t') \tilde{h}(t') \, dt \, dt' = \tilde{h}(t''),$$

using the definition Eq. (9.36) of the inverse kernel. With this, Eq. (9.109) becomes

$$\sum_{i=1}^{\infty} \frac{|a_i|^2}{\lambda_i} = \int_0^T \int_0^T \tilde{A}^*(t) \tilde{R}_n^{-1}(t, t') \tilde{A}(t') \, dt \, dt'. \qquad (9.112)$$

Using Eqs. (9.111), (9.112) in Eq. (9.108) then yields an alternative form of the density Eq. (9.108) in terms of the signal envelopes rather than their expansion coefficients.

The Binary Communications Case

Equation (9.110) can be applied to the case

$$H_i: \quad r(t) = a_i(t) \cos[\omega_i t + \gamma_i(t) + \theta_i], \qquad i = 0, 1, \qquad (9.113)$$

where $a_i(t)$ and $\gamma_i(t)$ are given functions, ω_i are specified frequencies, and θ_i are independent uniform random variables. We can write the signals under the two hypotheses as

$$r(t) = a_i(t) \cos[\omega_c t + (\omega_i - \omega_c)t + \gamma_i(t) + \theta_i],$$

where ω_c is some frequency intermediate to ω_0, ω_1. We assume that the difference $|\omega_0 - \omega_1|$ is small enough that the terms $\omega_i - \omega_c$ can be absorbed into the complex envelopes under the two hypotheses. That is, we take the complex envelopes of the received signal under the two hypotheses as

$$\tilde{A}_0(t) = a_0(t) \exp\{j[\gamma_0(t) + (\omega_0 - \omega_c)t + \theta_0]\},$$
$$\tilde{A}_1(t) = a_1(t) \exp\{j[\gamma_1(t) + (\omega_1 - \omega_c)t + \theta_1]\}.$$

The log likelihood ratio for this problem is just Eq. (9.110). There q_0, q_1 are the magnitudes of the envelopes resulting from the processing Eq. (9.105) using filters derived from the complex envelopes $\tilde{A}_0(t)$, $\tilde{A}_1(t)$ as in Eq. (9.106). The constant is determined from

$$d_i^2 = \int_0^T \tilde{h}_i^*(t) \tilde{A}_i(t) \, dt.$$

The Radar Case

This is the special case of Eq. (9.113) in which $a_0(t) = 0$. Then $\tilde{A}_0(t) = 0$, and the likelihood ratio detector resulting from Eq. (9.110) is simply

$$\ln[I_0(q)] > t,$$

where t is a threshold set by the false alarm probability. Since the Bessel function is monotonic, this is just

$$q > q_t, \qquad (9.114)$$

where q is the magnitude of the envelope of the output of the correlation receiver Eq. (9.105) using a weighting function derived from the signal complex envelope and the complex envelope of the noise covariance by solving the integral equation Eq. (9.106). The receiver can be implemented as a matched filter, using the impulse response $\tilde{h}^*(-t)$ and sampling the output at $t = 0$.

The detector Eq. (9.114) is the same as that for the white noise case, analyzed in Sect. 7.4. The only difference is the way in which the statistic q is computed. Its performance is easy to find from Sect. 7.4. Let us first write

$$q \exp(j\eta) = q_r + jq_i.$$

Since the data $\tilde{r}(t)$ is Gaussian, from Eq. (9.105) q is a Rayleigh variable and q_r and q_i are Gaussian random variables. We seek their variances. Since the noise is zero mean, we have

$$\mathscr{E}_0(q_r + jq_i) = 0,$$

$$\mathscr{E}_1(q_r + jq_i) = \int_0^T \tilde{h}^*(t)\tilde{A}(t) \, dt \, \exp(j\theta)$$

$$= d^2[\cos(\theta) + j\sin(\theta)], \tag{9.115}$$

where d^2 is as in Eq. (9.109).

Further, note from Eqs. (9.95), (9.105), (9.109) that

$$\mathscr{E}_0|q_r + jq_i|^2 = \mathscr{E}_0(q_r^2) + \mathscr{E}_0(q_i^2) = \mathscr{E}(q^2)$$

$$= \int_0^T \int_0^T \tilde{h}^*(t)R_{\tilde{n}}(t - t')\tilde{h}(t') \, dt \, dt'$$

$$= 2 \int_0^T \int_0^T \tilde{h}^*(t)\tilde{R}_n(t - t')\tilde{h}(t') \, dt \, dt' = 2d^2.$$

Also

$$\mathscr{E}_0(q_r + jq_i)^2 = \mathscr{E}_0(q_r^2) - \mathscr{E}_0(q_i^2) + 2j\mathscr{E}_0(q_r q_i)$$

$$= \int_0^T \int_0^T \tilde{h}^*(t)\mathscr{E}_0[\tilde{n}(t)\tilde{n}(t')]\tilde{h}^*(t') \, dt \, dt' = 0.$$

The last follows from

$$\mathscr{E}_0[\tilde{n}(t)\tilde{n}(t')] = \mathscr{E}[x_n(t) + jy_n(t)][x_n(t') + jy_n(t')]$$
$$= R_x(\tau) - R_y(\tau) + j[R_{xy}(\tau) + R_{yx}(\tau)] = 0,$$

using Eqs. (9.92), (9.93). Then we have finally

$$\mathscr{E}_0(q_r^2) = \mathscr{E}_0(q_i^2) = d^2,$$
$$\mathscr{E}_0(q_r q_i) = 0. \tag{9.116}$$

With these, it follows that the distribution of the test statistic q under hypothesis H_0 is that of the Rayleigh density Eq. (4.55):

$$p_0(q) = (q/d^2) \exp(-q^2/2d^2).$$

The false alarm probability of the detector Eq. (9.114) is then

$$P_f = \int_{q_t}^{\infty} \left(\frac{q}{d^2}\right) \exp\left(\frac{-q^2}{2d^2}\right) dq = \exp\left(\frac{-q_t^2}{2d^2}\right).$$

In the case of signal present, we compare Eq. (9.115) with Eq. (7.43), and Eq. (9.116) with Eq. (7.45), and make the correspondences

$$AT/2 = d^2,$$
$$\sigma^2 = d^2.$$

Then Eq. (7.49) shows the density of the test statistic q under hypothesis H_1:

$$p_1(q) = (q/d^2) \exp[-(q^2 + d^4)/2d^2]I_0(q).$$

This is the Rician density. As in Eq. (7.53) the corresponding detection probability is

$$P_d = \int_{q_t}^{\infty} \left(\frac{q}{d^2}\right) \exp\left[\frac{-(q^2 + d^4)}{2d^2}\right] I_0(q) \, dq$$
$$= Q(d, q_t/d),$$

where $Q(\alpha, \beta)$ is the Marcum Q-function. The signal-to-noise ratio for the problem is that in Eq. (7.55):

$$\alpha_{dB} = 20 \log(d).$$

The ROC curve is that of Fig. 7.3.

Exercises

9.1 Assume λ_k are the eigenvalues associated with the kernel $R_m(\tau)$. Recall $R_m(t_1, t_2) \triangleq E\{m(t_1)m(t_2)\}$. Show that

(a)
$$\int_0^T R_m(t_1, t_1) \, dt_1 = \sum_k \lambda_k$$

(b)
$$\int_0^T \int_0^T R_m^2(t_1, t_2) \, dt_1 \, dt_2 = \sum_k \lambda_k^2$$

9.2 Consider the signal

$$r(t) = m(t) + n(t), \qquad 0 \le t \le T$$

where $m(t)$ is a zero-mean colored Gaussian noise process with correlation function $R_m(\tau)$, and $n(t)$ is a zero-mean white Gaussian noise process with autocorrelation function $N_0/2 \, \delta(\tau)$.

(a)   Suppose we expand $r(t)$ in a series

$$r(t) = \sum_{k=0}^{K} r_k\psi_k(t) = \sum_{k=0}^{K} (m_k + n_k)\psi_k(t)$$

where $r_k = \int_0^T r(t)\psi_k(t) \, dt$. (Similar definitions apply for m_k and n_k.) Show that the coefficients r_k will be statistically independent if the $\psi_k(t)$ are chosen to be the eigenfunctions of $R_m(\tau)$.

(b) Determine the covariance of the coefficients.
(c) Using a finite set of coefficients, what is the likelihood function of $r(t)$?

9.3 Find the likelihood function of the preceding problem by first determining the conditional density function of the r_k given the coefficients m_k.

9.4 Consider the detection problem having the hypotheses

$$H_1: \quad r(t) = m(t) + n(t), \qquad 0 \leq t \leq T$$
$$H_0: \quad r(t) = n(t), \qquad\qquad 0 \leq t \leq T$$

where $m(t)$ and $n(t)$ are the same as given in Exercise 9.2.
(a) Show that we may use as the statistic

$$\gamma_T = \sum_{k=1}^{K} \frac{\lambda_k r_k^2}{2\lambda_k + N_0}$$

where λ_k are the eigenvalues associated with the kernel $R_m(\tau)$.
(b) Determine the mean and variance of γ_T for each hypothesis.

9.5 In the preceding problem, neglect any convergence problems and allow K to approach ∞ so that

$$\gamma_T = \sum_{k=1}^{\infty} \frac{\lambda_k r_k^2}{2\lambda_k + N_0}$$

Show that this may be put in the integral form

$$\gamma_T = \int_0^T \int_0^T r(t) r(v) h(t, v) \, dt \, dv$$

where $h(t, v)$ is the solution to the integral equation

$$h(t, v) + \frac{2}{N_0} \int_0^T R_m(v - u) h(t, u) \, du = \frac{1}{N_0} R_m(t, v), \qquad 0 \leq v, t \leq T$$

9.6 Consider the detection of a known signal in a mixture of white and colored noise. That is,

$$H_1: \quad r(t) = s(t) + m(t) + n(t), \qquad 0 \leq t \leq T$$
$$H_0: \quad r(t) = m(t) + n(t), \qquad\qquad 0 \leq t \leq T$$

where $m(t)$ and $n(t)$ are described in Exercise 9.2. Show that we may use as the test statistic

$$\int_0^T [r(t) - s(t)] h(t) \, dt$$

where $h(t)$ is the solution to the integral equation

$$h(t) + \frac{2}{N_0} \int_0^T h(\tau)R_m(t - \tau) \, d\tau = \frac{1}{N_0} s(t), \qquad 0 \le t \le T$$

9.7 In Sect. 9.7, the optimum signal for a binary communication system was derived. Discuss qualitatively the significance of the solution.

9.8 Consider the detection of a known signal in colored Gaussian noise

$$H_1: \quad r(t) = s(t) + n(t), \qquad 0 \le t \le T$$
$$H_0: \quad r(t) = n(t), \qquad\qquad 0 \le t \le T$$

Denote the noise correlation function as $R_n(\tau)$. Using a Neyman–Pearson criterion, find the signal $s(t)$ which will maximize the detection probability.

9.9 Consider the integral equation for $h(t)$

$$\int_0^T h(\tau)R_n(t - \tau) \, d\tau = g(t), \qquad 0 \le t \le T$$

where $R_n(\tau)$ and $g(t)$ are known. Assume that $S_n(\omega) = \omega_1^2/(\omega^2 + \omega_1^2)$ or $R_n(\tau) = \frac{1}{2}\omega_1 \, e^{-\omega_1|\tau|}$. Show that

$$h(t) = g(t) - \frac{1}{\omega_1^2}\frac{d^2}{dt^2}g(t)$$

will be a solution to the integral equation only if

$$\frac{g'(0)}{\omega_1} - g(0) = 0 \qquad \text{and} \qquad \frac{g'(T)}{\omega_1} + g(T) = 0$$

where $g'(0)$ and $g'(T)$ are the derivatives of $g(t)$ evaluated at $t = 0$, T respectively.

9.10 If the boundary values of the signal in the preceding problem are not satisfied, what is the total solution to the integral equation?

9.11 If the kernel to the Fredholm integral equation of the first kind is $R_n(\tau) = \alpha\delta(\tau) + \frac{1}{2}\beta\omega_1 \, e^{-\omega_1|\tau|}$,
  (a) What is the differential equation to be solved?
  (b) What is the form of the homogeneous solution?
  (c) If $g(t) = 1 - \cos \omega_0 t$, what is the form of the particular solution?

9.12 Assume a kernel of the form

$$R_n(\tau) = \begin{cases} 1 - \dfrac{|\tau|}{L}, & |\tau| \le L \\[2mm] 0, & \text{otherwise} \end{cases}$$

What differential equation is associated with determining the eigenfunctions of the integral equation

$$\int_0^T f(\tau) R_n(t - \tau) \, d\tau = \lambda f(t), \qquad 0 \le t \le T$$

where $T < L$?

10 Estimation of Signal Parameters

*I*n all of the discussions so far, we have been concerned with making a decision. A signal is or is not present in additive noise. The frequency of a carrier is one or the other of two possible values over an observation interval. Exactly one symbol of an *m*-ary alphabet is sent, and we must decide which one. No matter how we go about making such decisions, we will sometimes make errors. Under various cost criteria, we have determined algorithms for processing data into decision variables (sufficient statistics) whose values indicated the "best" decision to make.

In this chapter, we will consider the closely related problem of *estimation*. We would like to know the numerical value of some quantity. We have available measurements of related quantities, and from them we want to make the best guess about the value of the quantity of interest. For example, a moving target imparts a Doppler shift to the carrier frequency of a radar signal. The target radial speed is determined by that shift. If we know a target is present, and if we want to measure its speed, we have the problem of estimating the frequency of the carrier of the radar return, given a noisy version of the received signal.

In another situation, we may want to build a model of some physical process, given data of some kind about the process. We may be able to use the known physics of the process to determine a model having some parameters. We then need to estimate the values of those parameters from some amount of data. In other cases we may only be able to do curve fitting, in which case we seek a model with some parameters which is broad enough to replicate, and perhaps predict, the process. Again we need to estimate the values of the parameters to complete the model.

In all these cases, we will be more or less successful in building good models or accurately estimating parameter values. Just as in the case of detection algorithms, our estimation algorithms will need to be based on some criteria of goodness. Those will depend on how much information about the process being observed we have available and on how much we know about the relation between the quantities of interest and the available data.

In this chapter, we will present the standard procedures for building "optimal" estimation algorithms for various situations commonly encountered. In a general statement of the problem, we have available a vector $\mathbf{y}$ of measured data numbers y_i, and we want to know the value of a vector $\mathbf{x}$ of parameters x_i. Either or both of these might be a vector of random variables, and the relationship between them can be structured in various ways. Three versions of the problem can be distinguished. The vector $\mathbf{x}$ might constitute some parameters of the density function of the data $\mathbf{y}$, and we want to estimate $\mathbf{x}$ given some values of $\mathbf{y}$. The parameters are viewed as nonrandom constants with unknown values. In other situations, we may want to model a deterministic data source. In that case, the vector $\mathbf{x}$ might be the parameters of some model which we propose, which we want to estimate to get the best fit to some collection of values of the noisy data vector $\mathbf{y}$. Finally, in the case that most closely parallels the detection schemes we considered in earlier chapters, and which we consider first below, $\mathbf{x}$ and $\mathbf{y}$ might be random variables whose interrelationship is modeled by a given joint density.

10.1 BAYES ESTIMATION

Consider a vector $\mathbf{y}$ of random variables whose values can be observed by measurement. The vector $\mathbf{y}$ is related in some way to a vector $\mathbf{x}$ of parameters whose values we would like to know. In this section, we will suppose that $\mathbf{x}$ is also a random vector, and we will take as the given relationship between them the joint probability density $p(\mathbf{x}, \mathbf{y})$. We will seek an algorithm

$$\hat{\mathbf{x}}(\mathbf{y}) \cong \mathbf{x} \qquad (10.1)$$

which will indicate how to calculate an approximation $\hat{\mathbf{x}}$ of $\mathbf{x}$ from a given numerical value of the data vector $\mathbf{y}$.

No matter what algorithm Eq. (10.1) we use, sometimes we will make errors. In any particular trial of the experiment underlying the random variables, the algorithm will result in an error

$$\boldsymbol{\varepsilon} = \mathbf{x} - \hat{\mathbf{x}}(\mathbf{y}) \qquad (10.2)$$

which is also a random variable. In making that error, we will incur some cost which we take as a specified scalar function $C(\boldsymbol{\varepsilon})$. In the procedure known as *Bayes estimation*, we require that the algorithm Eq. (10.1) be such that the average cost over all cases of the underlying experiment be minimum:

$$\hat{\mathbf{x}}(\mathbf{y}) \Leftarrow \min_{\hat{\mathbf{x}}} \mathscr{E} \, C(\boldsymbol{\varepsilon}), \qquad (10.3)$$

here $\mathscr{E}(\ )$ is the expectation operator.

The problem Eq. (10.3) is a problem of functional minimization. We seek the form of the function $\hat{\mathbf{x}}(\mathbf{y})$ which minimizes the expected cost over the joint ensemble of random variables $\mathbf{x}, \mathbf{y}$. the first step in solution of the problem is to replace the problem of selecting a minimizing function $\hat{\mathbf{x}}(\mathbf{y})$ by the ordinary calculus problem of selecting a minimizing value $\hat{\mathbf{x}}$. We do that by writing the quantity to be minimized as

$$\mathscr{E}C(\boldsymbol{\varepsilon}) = \mathscr{E}C[\mathbf{x}, \hat{\mathbf{x}}(\mathbf{y})] = \int_{-\infty}^{\infty} \int_{-\infty}^{\infty} C[\mathbf{x}, \hat{\mathbf{x}}(\mathbf{y})] p(\mathbf{x}, \mathbf{y}) \, d\mathbf{x} \, d\mathbf{y}$$

$$= \int_{-\infty}^{\infty} \left[\int_{-\infty}^{\infty} C[\mathbf{x}, \hat{\mathbf{x}}(\mathbf{y})] p(\mathbf{x} \mid \mathbf{y}) \, d\mathbf{x} \right] p(\mathbf{y}) \, d\mathbf{y}. \qquad *(10.4)$$

Since the function $p(\mathbf{y})$ is nonnegative, this average cost will be minimized if, for every fixed value of $\mathbf{y}$, we select $\hat{\mathbf{x}}(\mathbf{y})$ to be such that the inner integral of Eq. (10.4) is minimum. That is, the functional minimization problem Eq. (10.4) is solved if we solve the ordinary calculus problem of selecting the number $\hat{\mathbf{x}}$ such that

$$\hat{\mathbf{x}} \Leftarrow \min_{\hat{\mathbf{x}}} \mathscr{E}[C(\mathbf{x}, \hat{\mathbf{x}}) \mid \mathbf{y}], \qquad *(10.5)$$

where the expectation is the conditional mean over $\mathbf{x}$ for given $\mathbf{y}$.

EXAMPLE 10.1   A scalar random variable (RV) x has the exponential density $p_x(x) = \exp(-x)U(x)$, where $U(x)$ is the unit step function. We observe the value of x in additive noise n, also with an exponential density: $p_n(n) = 2\exp(-2n)U(n)$. That is, we observe $y = x + n$. The noise is independent of the RV x. As the error function we take the quadratic: $C(x, \hat{x}) = (x - \hat{x})^2$. We seek the Bayes estimate Eq. (10.3) of x, given the observation y.

We need the conditional density $p(x \mid y)$. We have

$$p_{y|x}(y \mid x) = p_n(y - x) = 2 \exp[-2(y - x)]U(y - x).$$

Then we seek to minimize (over $\hat{x}$)

$$\mathscr{E}(C \mid y) = \frac{1}{p_y(y)} \int_0^\infty (x - \hat{x})^2 p_{y|x}(y \mid x)p_x(x) \, dx.$$

Setting the derivative of this with respect to $\hat{x}$ equal to zero yields

$$\int_0^y (x - \hat{x}) \exp(x) \, dx = 0,$$
$$\hat{x} = y/[1 - \exp(-y)] - 1.$$

EXAMPLE 10.2   Some number N of observations are made of a Gaussian random variable x, with mean m_x, in additive independent zero mean Gaussian noise n: $y_i = x + n_i$, $i = 1, N$. The variances are respectively σ_x^2, σ_n^2. We have

$$p_{x|y}(x \mid \mathbf{y}) = [1/p_y(\mathbf{y})](2\pi \sigma_n^2)^{-N/2}$$

$$\times \exp\left[-\sum_{i=1}^N \frac{(y_i - x)^2}{2\sigma_n^2} \right](2\pi \sigma_x^2)^{-1/2} \exp\left[\frac{-(x - m_x)^2}{2\sigma_x^2} \right].$$

Using this and setting equal to zero the derivative with respect to $\hat{x}$ of $\mathscr{E}[(x - \hat{x})^2 \mid \mathbf{y}]$ yields

$$\hat{x} = \left[\frac{1}{\sigma_n^2/N} + \frac{1}{\sigma_x^2} \right]^{-1} \left[\frac{1}{\sigma_n^2/N} \left(\frac{1}{N} \right) \sum_{i=1}^N y_i + \frac{m_x}{\sigma_x^2} \right]. \tag{10.6}$$

That is, as an estimate of x we take a mixture of the sample mean of the data and the ensemble mean m_x of the quantity being estimated, weighted in proportion to the inverses of their variances. If the variance σ_x is much smaller than the noise variance σ_n, adjusted for the number of samples, then the mixture leans toward taking the ensemble mean as estimate and puts less weight on the noisy data. On the other hand, if σ_x is large, the data are emphasized in the estimate. In the case of one data sample, Eq. (10.6) becomes simply

$$\hat{x} = (y/\sigma_n^2 + m_x/\sigma_x^2)/(1/\sigma_n^2 + 1/\sigma_x^2). \tag{10.7}$$

That is, we mix the *a priori* estimate m_x and the *a posteriori* estimate y of x in inverse proportion to their variances.

10.2 THE CONDITIONAL MEAN AS BAYES ESTIMATOR

The conditional mean $\hat{\mathbf{x}}(\mathbf{y}) = \mathscr{E}(\mathbf{x} \mid \mathbf{y})$ occupies a special place as an estimate of a random variable $\mathbf{x}$ in terms of a second random variable $\mathbf{y}$. First, regardless of the densities involved, it is the Bayes estimator if the cost function is quadratic, and, second, for a broad class of cost functions it is the Bayes estimator under mild restrictions on the densities.

To see the first of these statements, suppose that the cost of estimation errors is a quadratic function:

$$C(\boldsymbol{\varepsilon}) = (\mathbf{x} - \hat{\mathbf{x}})^{\mathrm{T}} A (\mathbf{x} - \hat{\mathbf{x}}), \tag{10.8}$$

where A is any symmetric positive definite matrix and the superscript T indicates transpose. (We require that A be positive definite in order that a nonzero error always have a cost greater than zero.) The Bayes estimator results by minimizing the conditional mean cost Eq. (10.5), which also results in a minimum of the unconditioned cost Eq. (10.4). It should be noted that, in the case Eq. (10.8) with $A = I$, the unit matrix, we thereby minimize the variance of the error, defined as the sum of the variances of the components of the error vector in the multidimensional case. The resulting estimator is therefore often called the minimum variance estimator.

The extrema of the conditional cost Eq. (10.5) result from

$$0 = \left(\frac{d}{d\hat{\mathbf{x}}}\right) \int_{-\infty}^{\infty} (\mathbf{x} - \hat{\mathbf{x}})^{\mathrm{T}} A(\mathbf{x} - \hat{\mathbf{x}}) p(\mathbf{x} \mid \mathbf{y}) \, dx$$

$$= 2 \int_{-\infty}^{\infty} (\mathbf{x} - \hat{\mathbf{x}})^{\mathrm{T}} A \, p(\mathbf{x} \mid \mathbf{y}) \, dx,$$

$$\hat{\mathbf{x}}^{\mathrm{T}} A \int_{-\infty}^{\infty} p(\mathbf{x} \mid \mathbf{y}) \, dx = \int_{-\infty}^{\infty} \mathbf{x}^{\mathrm{T}} p(\mathbf{x} \mid \mathbf{y}) \, dx \, A,$$

$$\hat{\mathbf{x}} = \int_{-\infty}^{\infty} \mathbf{x} \, p(\mathbf{x} \mid \mathbf{y}) \, dx = \mathscr{E}(\mathbf{x} \mid \mathbf{y}). \tag{10.9}$$

In the last step we used the fact that A, being positive definite, has an inverse. That the extremum in Eq. (10.9) is a minimum follows from the second derivative:

$$\left(\frac{d}{d\hat{\mathbf{x}}}\right) \int_{-\infty}^{\infty} (\mathbf{x} - \hat{\mathbf{x}})^{\mathrm{T}} A p(\mathbf{x} \mid \mathbf{y}) \, dx = -A \int_{-\infty}^{\infty} p(\mathbf{x} \mid \mathbf{y}) \, dx = -A.$$

This matrix is negative definite, so that the extremum is a minimum rather than a maximum or a saddle point of the cost surface.

EXAMPLE 10.3 A Gaussian random vector $\mathbf{x}$ is to be estimated in terms of a Gaussian random vector $\mathbf{y}$. The cost is quadratic, as in Eq. (10.8).

The Bayes estimator, minimizing the average cost as in Eq. (10.5), is the conditional mean Eq. (10.9), which we now seek.

Using Eq. (4.4) for the multidimensional Gaussian density, and where K is an irrelevant constant, ignoring for the moment the means of the variables we have

$$\ln[p(\mathbf{x} \mid \mathbf{y})] = \ln[p(\mathbf{x}, \mathbf{y})] - \ln[p(\mathbf{y})]$$

$$= (\mathbf{x}^\mathrm{T} \, \mathbf{y}^\mathrm{T}) \begin{bmatrix} R_{xx} & R_{xy} \\ R_{yx} & R_{yy} \end{bmatrix}^{-1} \begin{bmatrix} \mathbf{x} \\ \mathbf{y} \end{bmatrix} - \mathbf{y}^\mathrm{T} R_{yy}^{-1} \mathbf{y} + K,$$

where the entries in the partitioned matrix are the variance and covariance matrices of the indicated random vectors. The inverse of a nonsingular partitioned matrix is

$$\begin{bmatrix} A & B \\ C & D \end{bmatrix}^{-1}$$

$$= \begin{bmatrix} (A - BD^{-1}C)^{-1} & -(A - BD^{-1}C)^{-1}BD^{-1} \\ -(D - CA^{-1}B)^{-1}CA^{-1} & (D - CA^{-1}B)^{-1} \end{bmatrix}. \quad (10.10)$$

In this, the matrices A and D are to be square, perhaps of different dimensions, while B and C are conformable. In the common case of a symmetric matrix, so that A and D are symmetric, and $C = B^\mathrm{T}$, we have necessarily that the inverse is also symmetric, so that we have the identity

$$(D - B^\mathrm{T} A^{-1}B)^{-1}B^\mathrm{T}A^{-1} = [(A - BD^{-1}B^\mathrm{T})^{-1}BD^{-1}]^\mathrm{T}$$
$$= D^{-1}B^\mathrm{T}(A - BD^{-1}B^\mathrm{T})^{-1}. \quad (10.11)$$

Using Eq. (10.10) and taking into account that the covariance matrices are related by $R_{yx} = R_{xy}^\mathrm{T}$, so that Eq. (10.11) applies, the conditional density becomes

$$\ln[p(\mathbf{x} \mid \mathbf{y})] = (\mathbf{x} - R_{xy}R_{yy}^{-1}\mathbf{y})^\mathrm{T}$$
$$\times (R_{xx} - R_{xy}R_{yy}^{-1}R_{xy}^\mathrm{T})^{-1}(\mathbf{x} - R_{xy}R_{yy}^{-1}\mathbf{y}) + f(\mathbf{y}),$$

where $f(\mathbf{y})$ is an irrelevant function of the conditioning variables. From this last follow the important relations for a multidimensional Gaussian that

$$\mathscr{E}(\mathbf{x} \mid \mathbf{y}) = \mathbf{x}_0 + (C_{xy}C_{yy}^{-1})(\mathbf{y} - \mathbf{y}_0), \qquad *(10.12)$$

$$\mathrm{Var}(\mathbf{x} \mid \mathbf{y}) = C_{xx} - C_{xy}C_{yy}^{-1}C_{xy}^\mathrm{T}. \qquad *(10.13)$$

In arriving at Eq. (10.12) we have replaced $\mathbf{x}$ and $\mathbf{y}$ by $\mathbf{x} - \mathbf{x}_0$, where $\mathbf{x}_0 = \mathscr{E}(\mathbf{x})$, and $\mathbf{y} - \mathbf{y}_0$, and noted that trivially $\mathscr{E}(\mathbf{x}_0 \mid \mathbf{y}) = \mathbf{x}_0$.

It is noteworthy that, in this case of a quadratic cost, the Bayes estimator $\hat{\mathbf{x}} = \mathscr{E}(\mathbf{x} \mid \mathbf{y})$ of Eq. (10.12) is a linear function of the conditioning variables $\mathbf{y}$. That is a sufficiently strong advantage that one should have good reason before selecting a cost function other than quadratic in a Bayesian estimation problem.

The result Eq. (10.12) yields the estimator of Example 10.2. Letting $\mathbf{1}$ be the column vector of N ones, we have:

$$
\begin{aligned}
\mathbf{x} &= x, \\
\mathbf{x}_0 &= m_x, \\
C_{xx} &= \sigma_x^2, \\
\mathbf{y} &= x\mathbf{1} + \mathbf{n}, \\
\mathbf{y}_0 &= \mathbf{1}m_x, \\
C_{yy} &= \mathscr{E}[(x - m_x)\mathbf{1} + \mathbf{n}][(x - m_x)\mathbf{1} + \mathbf{n}]^{\mathrm{T}} \\
&= \sigma_x^2 \mathbf{1}\mathbf{1}^{\mathrm{T}} + \sigma_n^2 I, \\
C_{xy} &= \mathscr{E}[(x - m_x)][(x - m_x)\mathbf{1} + \mathbf{n}]^{\mathrm{T}} = \sigma_x^2 \mathbf{1}^{\mathrm{T}}.
\end{aligned}
$$

Equation (10.12) then reads

$$
\hat{x} = m_x + \sigma_x^2 \mathbf{1}^{\mathrm{T}} (\sigma_n^2 I + \sigma_x^2 \mathbf{1}\mathbf{1}^{\mathrm{T}})^{-1} (\mathbf{y} - \mathbf{1}\, m_x)
$$

$$
= m_x + \left[\frac{\sigma_x^2}{(\sigma_n^2 + N\sigma_x^2)} \right] \left[\sum_{i=1}^{N} y_i - m_x \right],
$$

which agrees with Eq. (10.6). In this we have used $\mathbf{1}\mathbf{1} = N$ and the ubiquitous identity, valid for any B, C, and nonsingular A:

$$
(A + BC^{\mathrm{T}})^{-1} = A^{-1} - A^{-1}B(I + C^{\mathrm{T}}A^{-1}B)^{-1}C^{\mathrm{T}}A^{-1}. \tag{10.14}
$$

Equation (10.9) indicates that the conditional mean is the Bayes estimator for the quadratic cost function Eq. (10.8), for any positive definite weighting matrix A. The conditional mean is also the Bayes estimator for broader classes of situations. Suppose that the scalar cost function of the error vector $\boldsymbol{\varepsilon} = \hat{\mathbf{x}} - \mathbf{x}$ is symmetric: $C(\boldsymbol{\varepsilon}) = C(-\boldsymbol{\varepsilon})$. Further suppose that the error function is convex:

$$
C[(\boldsymbol{\varepsilon}_1 + \boldsymbol{\varepsilon}_2)/2] \leq [C(\boldsymbol{\varepsilon}_1) + C(\boldsymbol{\varepsilon}_2)]/2,
$$

for any vectors $\boldsymbol{\varepsilon}_1$, $\boldsymbol{\varepsilon}_2$. That is, the surface C evaluated at the midpoint of the line between any two vectors $\boldsymbol{\varepsilon}_1$, $\boldsymbol{\varepsilon}_2$ lies below the mean of the surface C evaluated at the two vectors. [Figure 10.1 shows a convex function $C(\boldsymbol{\varepsilon})$ in the one dimensional case.] Finally, assume the conditional density $p(\mathbf{x} \mid \mathbf{y})$ is symmetric about its mean:

$$
p(\Delta \mathbf{x} \mid \mathbf{y}) = p(-\Delta \mathbf{x} \mid \mathbf{y}), \qquad \Delta \mathbf{x} = \mathbf{x} - \mathscr{E}(\mathbf{x} \mid \mathbf{y}).
$$

Now consider any estimate $\hat{\mathbf{x}}$ of $\mathbf{x}$, and the conditional mean estimate $\hat{\mathbf{x}}_{cm} = \mathscr{E}(\mathbf{x} \mid \mathbf{y})$. Letting $\mathbf{u} = \mathbf{x} - \hat{\mathbf{x}}_{cm}$, the conditional Bayes cost for the

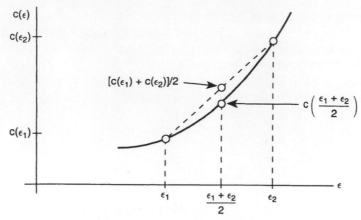

FIGURE 10.1   A convex cost function.

arbitrary estimator is

$$\mathscr{E}[C(\mathbf{x} - \hat{\mathbf{x}}) \mid \mathbf{y}] = \mathscr{E}[C(\mathbf{u} + \hat{\mathbf{x}}_{cm} - \hat{\mathbf{x}}) \mid \mathbf{y}]$$

$$= \int_{-\infty}^{\infty} C(\mathbf{u} + \hat{\mathbf{x}}_{cm} - \hat{\mathbf{x}}) p(\mathbf{u} \mid \mathbf{y}) \, d\mathbf{u}$$

$$= \int_{-\infty}^{\infty} C(-\mathbf{u} + \hat{\mathbf{x}}_{cm} - \hat{\mathbf{x}}) p(\mathbf{u} \mid \mathbf{y}) \, d\mathbf{u},$$

using the symmetry of the conditional density. Using the symmetry of the cost function, this is further

$$\mathscr{E}[C(\mathbf{x} - \hat{\mathbf{x}}) \mid \mathbf{y}] = \int_{-\infty}^{\infty} C(\mathbf{u} - \hat{\mathbf{x}}_{cm} + \hat{\mathbf{x}}) p(\mathbf{u} \mid \mathbf{y}) \, d\mathbf{u}.$$

Adding these two forms for the mean cost and using the convexity of the cost function, we have

$$\mathscr{E}[C(\mathbf{x} - \hat{\mathbf{x}}) \mid \mathbf{y}] = \int_{-\infty}^{\infty} \tfrac{1}{2} [C(\mathbf{u} + \hat{\mathbf{x}}_{cm} - \hat{\mathbf{x}}) + C(\mathbf{u} - \hat{\mathbf{x}}_{cm} + \hat{\mathbf{x}})] p(\mathbf{u} \mid \mathbf{y}) \, d\mathbf{u}$$

$$\geq \int_{-\infty}^{\infty} C[\tfrac{1}{2}(\mathbf{u} + \hat{\mathbf{x}}_{cm} - \hat{\mathbf{x}}) + \tfrac{1}{2}(\mathbf{u} - \hat{\mathbf{x}}_{cm} + \hat{\mathbf{x}})] p(\mathbf{u} \mid \mathbf{y}) \, d\mathbf{u}$$

$$= \int_{-\infty}^{\infty} C(\mathbf{u}) p(\mathbf{u} \mid \mathbf{y}) \, d\mathbf{u} = \mathscr{E}[C(\mathbf{x} - \hat{\mathbf{x}}_{cm}) \mid \mathbf{y}].$$

That is, the cost of any other estimator is at least as large as the cost of the conditional mean as estimator. Hence the Bayes estimator is the conditional mean.

One important cost function is not convex—the uniform norm (Fig. 10.2), and the above proof is not applicable. However, if the convexity

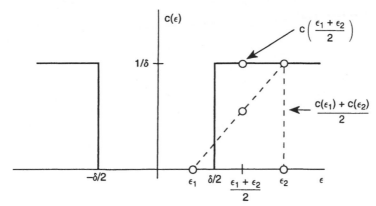

FIGURE 10.2 Uniform cost function, illustrating nonconvexity.

condition is replaced with the condition

$$\lim_{u \Rightarrow \infty} C(\mathbf{u}) p(\mathbf{u} \mid \mathbf{y}) = 0,$$

then the result can also be obtained for the uniform norm.

10.3 MAXIMUM A POSTERIORI ESTIMATION

Bayes estimation, discussed in the preceding sections, requires (or allows) us to specify a cost of errors $C(\varepsilon)$, which is then minimized on average. If a cost function is not in question, another type of estimator is often used, the *maximum a posteriori* (MAP) estimator. As the name indicates, this estimator results by choosing the most probable value of the quantity being estimated, given the data which are at hand:

$$\hat{\mathbf{x}} \Leftarrow \max_{x} p(\mathbf{x} \mid \mathbf{y}). \tag{10.15}$$

Since $p(\mathbf{x} \mid \mathbf{y}) = p(\mathbf{y} \mid \mathbf{x})p(\mathbf{x})/p(\mathbf{y})$ and $p(\mathbf{y})$ does not involve $\mathbf{x}$, the MAP estimator also results as

$$\hat{\mathbf{x}} \Leftarrow \max_{x} p(\mathbf{y} \mid \mathbf{x})p(\mathbf{x}). \tag{10.16}$$

In the case of Gaussian $\mathbf{x}$, $\mathbf{y}$, the *a posteriori* density in Eq. (10.15) is also Gaussian, and its maximum is its mean. That is, in the Gaussian case the MAP estimator is the conditional mean $\mathscr{E}(\mathbf{x} \mid \mathbf{y})$ and also minimizes the Bayes cost under a variety of conditions, as detailed in the previous section. In any case, the MAP estimator results from Bayes estimation using the limiting form of the uniform cost function, Fig. 10.2, for small δ. In this case the conditional mean of the cost is

$$\mathscr{E}[C(\hat{\mathbf{x}} - \mathbf{x}) \mid \mathbf{y}] = \int_{-\infty}^{\infty} C(\hat{\mathbf{x}} - \mathbf{x}) \, p(\mathbf{x} \mid \mathbf{y}) \, d\mathbf{x}$$

$$= \frac{1}{\delta}\left[\left[\int_{-\infty}^{\hat{x}-\delta/2} + \int_{\hat{x}+\delta/2}^{\infty}\right] p(\mathbf{x} \mid \mathbf{y}) \, d\mathbf{x}\right]$$

$$= \frac{1}{\delta}\left[1 - \int_{\hat{x}-\delta/2}^{\hat{x}+\delta/2} p(\mathbf{x} \mid \mathbf{y}) \, d\mathbf{x}\right]$$

$$\cong (1/\delta)[1 - \delta\, p(\hat{\mathbf{x}} \mid \mathbf{y})] = 1/\delta - p(\hat{\mathbf{x}} \mid \mathbf{y}).$$

For any fixed small δ, this is minimized by picking $\hat{\mathbf{x}}$ to be that value of $\mathbf{x}$ that maximizes $p(\mathbf{x} \mid \mathbf{y})$, in accord with Eq. (10.15).

Considering Eq. (10.16), suppose that $p(\mathbf{x})$ is a broadly spreading function with respect to $p(\mathbf{y} \mid \mathbf{x})$. Then the maximum of the product Eq. (10.16) will nearly coincide with the maximum of the factor $p(\mathbf{y} \mid \mathbf{x})$. The MAP estimate will be very nearly

$$\hat{\mathbf{x}} \Leftarrow \max_{\mathbf{x}} p(\mathbf{y} \mid \mathbf{x}). \tag{10.17}$$

This is an important estimator in its own right, which we will consider in detail below: the *maximum likelihood estimator* of $\mathbf{x}$. It does not involve the *a priori* density $p(\mathbf{x})$, which may be difficult to assign. In fact, the assumption of broad $p(\mathbf{x})$ is equivalent to an admission of no prior reason to pick one value of $\mathbf{x}$ over another.

The maximum likelihood estimate Eq. (10.17) is used when no cost function is relevant, and when no prior statistics are available, or else in a case in which the Bayes estimate is difficult to compute. Although the maximum likelihood estimate itself may be difficult to compute, to specify it we only need to know how the data $\mathbf{y}$ arise from a specified signal $\mathbf{x}$. That information is often available.

EXAMPLE 10.4   We make N independent observations of a zero-mean Gaussian parameter x in zero-mean Gaussian noise: $y_i = x + n_i$. We have

$$p(\mathbf{y} \mid x)p(x) = \prod_{i=1}^{N} (2\pi\sigma_n^2)^{-1/2} \exp\left[\frac{-(y_i - x)^2}{2\sigma_n^2}\right]$$

$$\times (2\pi\sigma_x^2)^{-1/2} \exp(-x^2/2\sigma_x^2).$$

The MAP estimator results by maximizing the log of the density:

$$\hat{x} \Leftarrow \frac{\partial}{\partial x}\left[\frac{1}{2\sigma_n^2}\sum_{i=1}^{N} (y_i - x)^2 + \frac{x^2}{2\sigma_x^2}\right] = 0.$$

There is only one solution, which is

$$\hat{x} = \left(1 + \frac{\sigma_n^2}{N\sigma_x^2}\right)^{-1} \frac{1}{N} \sum_{i=1}^{N} y_i. \tag{10.18}$$

This is the Bayes estimate Eq. (10.6), as expected. All the densities involved are Gaussian, so the Bayes estimate is the conditional mean. Furthermore, the conditional density is Gaussian, so the conditional mean is also the mode. Hence it is the MAP estimator.

10.4 MAXIMUM LIKELIHOOD ESTIMATION

The Bayes estimator optimizes some cost criterion, perhaps the variance of the estimator. The maximum *a posteriori* estimator is that value of the sought parameter with the greatest probability of having been in effect to produce the data at hand. In contrast, the *maximum likelihood estimator* Eq. (10.17) makes no claim to optimality. Its virtue is that it requires no *a priori* information, and it may be easy to compute. We simply form the *likelihood function* of the data, $p(\mathbf{y} \mid \mathbf{x})$, and maximize it.

EXAMPLE 10.5 Events occur randomly in time at an average interval of μ time units. The waiting time T between events is often modeled as an exponential random variable: $p(T) = (1/\mu) \exp(-T/\mu)$. We observe N values of T and seek the maximum likelihood estimate of μ. We have

$$p(T_1, \ldots, T_N \mid \mu) = (1/\mu)^N \exp[-(1/\mu) \sum T_i].$$

Setting the derivative of this with respect to μ equal to zero yields the reasonable estimator

$$\hat{\mu} = \frac{1}{N} \sum_{i=1}^{N} T_i.$$

EXAMPLE 10.6 A classic problem is estimation of the mean and/or the variance of a Gaussian density given a sequence of N values $\mathbf{y}_i$ drawn independently from the population. We have

$$p(\mathbf{y} \mid m, \sigma^2) = (2\pi\sigma^2)^{-N/2} \exp\left[-\sum_{i=1}^{N} \frac{(y_i - m)^2}{2\sigma^2}\right].$$

If σ^2 is known, we obtain the maximum likelihood estimate of m by

$$0 = \frac{\partial}{\partial m} \sum_{i=1}^{N} (y_i - m)^2,$$

which yields the sample mean:

$$\hat{m} = \frac{1}{N} \sum_{i=1}^{N} y_i. \tag{10.19}$$

Since the variance σ^2 does not appear in this, we obtain the same estimator whether or not the variance is known. Similarly, if m is known, the maximum likelihood estimate of σ^2 follows from $\partial p(\mathbf{y} \mid m, \sigma^2)/\partial \sigma^2 = 0$ and is easily seen to be the sample variance given the true ensemble mean:

$$\hat{\sigma}^2 = \frac{1}{N} \sum_{i=1}^{N} (y_i - m)^2. \tag{10.20}$$

In the usual case that neither the true mean m nor the true variance σ^2 is known, the density $p(\mathbf{y} \mid m, \sigma^2)$ is maximized jointly over the two variables m, σ^2. The result is just the estimator $\hat{\sigma}^2$ with $m = \hat{m}$.

10.5 PROPERTIES OF ESTIMATORS

In the preceding four sections, we have discussed the formation of the standard best estimators of parameters about which we have some information through measurements of random variables whose distributions involve the parameters of interest. Which estimator to use depends on the purpose at hand and more likely on the amount of information we have, or are willing to assume, about the problem. Regardless of which estimator we choose, it is important to characterize how well the estimator will do over some class of data. In this section we introduce the common measures of quality of a parameter estimator and discuss the measures for the specific estimators introduced in the preceding sections.

First, it is important to understand that an estimator $\hat{x}(\mathbf{y})$ of a parameter $\mathbf{x}$, based on data $\mathbf{y}$, is a random variable, since the data set $\mathbf{y}$ is random. Each particular measurement situation will produce a particular data set $\mathbf{y}$ and a particular value of the estimate $\hat{x}(\mathbf{y})$. That estimate will be a more or less accurate approximation to the true value $\mathbf{x}$ of the parameter of interest which was in effect at the time the data were produced. We will be interested in judging how good an estimate $\hat{x}$ is of $\mathbf{x}$ over the full ensemble of random data values and over the ensemble of values of $\mathbf{x}$ in the case that $\mathbf{x}$ is also a random variable, rather than an unknown but constant parameter.

In the case that the true parameter value $\mathbf{x}$ is a random variable, we can regard the computed value $\hat{x}$ as a particular realization of that random variable. The density $p(\hat{x} \mid \mathbf{x})$ is called the *sampling distribution* of the random variable $\mathbf{x}$. We will be interested in estimators $\hat{x}$ whose sampling distributions have good properties. We now proceed to discuss the properties of sampling distributions which are commonly introduced.

The first desirable quality of any estimate $\hat{x}(y)$ of x is that, on average, we should get the right answer. The *bias* of a particular estimate $\hat{x}$ of x is defined as

$$b(x, y) = \hat{x}(y) - x, \qquad (10.21)$$

and we would like it to be small over the ensemble of situations encountered. If the quantity x being estimated is an unknown constant, and if

$$\mathscr{E}[b(x, y)] = \mathscr{E}[\hat{x}(y)] - x = 0, \qquad (10.22)$$

then the estimator (not the estimate) is said to be *unbiased*. If x is modeled as a random variable, then an estimator $\hat{x}$ is said to be *conditionally unbiased* if

$$\mathscr{E}[b(x, y) \mid x] = \mathscr{E}[\hat{x}(y) \mid x] - x = 0. \qquad (10.23)$$

[Equation (10.23) is the requirement that the mean of the sampling distribution of the estimator $\hat{x}$ be the true value x of the parameter in the particular trial at hand.] If

$$\mathscr{E}[b(x, y)] = \mathscr{E}[\hat{x}(y)] - \mathscr{E}(x) = 0, \qquad (10.24)$$

the estimator is (*unconditionally*) unbiased. If the appropriate unbiasedness condition Eq. (10.22), (10.23), (10.24) is not satisfied, then the estimator is said to be *biased* and the bias is correspondingly

$$b = \mathscr{E}[b(x, y)] = \mathscr{E}[\hat{x}(y)] - x,$$
$$b = \mathscr{E}[b(x, y) \mid x] = \mathscr{E}[\hat{x}(y) \mid x] - x,$$
$$b = \mathscr{E}[b(x, y)] = \mathscr{E}[\hat{x}(y)] - \mathscr{E}(x).$$

If the bias b is a known constant, it can simply be subtracted from the estimator x. However, that is not the usual situation.

With an unbiased estimator, we are at least assured that, for a particular trial of a random experiment or over an ensemble of trials, the conditional or unconditional mean respectively of the values $\hat{x}(y)$ obtained will be the true value of the parameter x we are trying to estimate or its mean. However, that is not much comfort if the scatter of the estimator $\hat{x}(y)$ about its mean is large. In such a case, for any particular sample y of data we may be unlucky and compute a value $\hat{x}(y)$ which is far from its mean. This consideration leads to the second property which we require of a good estimator—*consistency*.

Suppose the estimator $\hat{x}(y)$ of a scalar x depends on n data numbers y_i; that is, suppose that y is an n-vector. Accordingly, we will write the estimator as $\hat{x}_n(y)$. Such an estimator is called *consistent* (or sometimes *convergent*) provided that, for any given small number ε,

$$\lim_{n \Rightarrow \infty} P(|\hat{x}_n(y) - x| > \varepsilon) = 0. \qquad *(10.25)$$

That is, we are assured that, if we take enough data, on any particular trial of the random experiment the value $\hat{x}(\mathbf{y})$ will have zero probability of being farther than any small amount from the true value x we are trying to estimate. The difficulty comes in judging how much data is "enough" to make the probability Eq. (10.25) adequately small.

In the case of a vector parameter $\mathbf{x}$, Eq. (10.25) is taken with a vector $\boldsymbol{\varepsilon}$ each of whose components is arbitrarily small, and the absolute value expression is taken componentwise.

For any random variable z, the Chebyshev inequality states that, for any arbitrary $\varepsilon > 0$,

$$P\{|z - \mathcal{E}(z)| > \varepsilon\} < [\text{Var}(z)]/\varepsilon^2. \tag{10.26}$$

In the case of a conditionally unbiased estimator $\hat{x}_n(\mathbf{y})$, the expectation and variance in Eq. (10.26) can be taken as conditional on the true parameter value x, and Eq. (10.26) reads

$$P\{|\hat{x}_n(\mathbf{y}) - x| > \varepsilon\} < [\text{Var}(\hat{x}_n \mid x)]/\varepsilon^2. \qquad *(10.27)$$

If then the variance of the sampling distribution $\text{Var}[\hat{x}_n(\mathbf{y}) \mid x] \Rightarrow 0$ as $n \Rightarrow \infty$, we obtain Eq. (10.25). Accordingly, a conditionally unbiased estimator with a sampling variance that tends to zero is consistent.

EXAMPLE 10.7   Consider again the estimator Eq. (10.6) of Example 10.2. There we process N samples $y_i = x + n_i$ of a parameter in zero-mean Gaussian noise. The estimator is

$$\hat{x} = \left[1 + \frac{\sigma_n^2}{N\sigma_x^2}\right]^{-1} \left\{\frac{1}{N}\sum_{i=1}^{N} y_i + \frac{m_x \sigma_n^2}{N\sigma_x^2}\right\}.$$

This estimator has a conditional mean

$$\mathcal{E}(\hat{x} \mid x) = \left[1 + \frac{\sigma_n^2}{N\sigma_x^2}\right]^{-1} \left(x + \frac{m_x \sigma_n^2}{N\sigma_x^2}\right) \neq x$$

and is therefore conditionally biased. However,

$$\lim_{N \Rightarrow \infty} \mathcal{E}(\hat{x} \mid x) = x,$$

so that the estimator is *asymptotically* conditionally unbiased. Further, since

$$\mathcal{E}(\hat{x}) = \mathcal{E}[\mathcal{E}(\hat{x} \mid x)]$$
$$= [1 + \sigma_n^2/N\sigma_x^2]^{-1}(m_x + m_x \sigma_n^2/N\sigma_x^2) = m_x,$$

the estimator is (unconditionally) unbiased for all N.

The conditional variance of the estimator is

$$\mathcal{E}\{[\hat{x} - \mathcal{E}(\hat{x} \mid x)]^2 \mid x\} = (\sigma_n^2/N)[1 + \sigma_n^2/N\sigma_x^2]^{-1},$$

which tends to zero with increasing N. The estimator is therefore consistent. We are assured that, if we take enough data samples, the computed value of the estimator will equal the true value x of the random variable in that trial of the experiment with probability one.

Another property which an estimator may have is that of *sufficiency*, in the sense of the sufficient statistics that we discussed earlier. A sufficient estimator is one such that no other estimator can yield more information about the quantity being estimated. In order that an estimator $\hat{x}(y)$ be sufficient, it is necessary and sufficient that the likelihood function of the data y and the parameter x being estimated factor as

$$p(\mathbf{y} \mid \mathbf{x}) = p[\hat{\mathbf{x}}(\mathbf{y}) \mid \mathbf{x}] f(\mathbf{y}),$$

where $f(\mathbf{y})$ is any function of the data alone [Stuart and Ord, 1991]. Any function of a sufficient estimator is also sufficient, so that there is some flexibility in choosing the particular form of estimator.

EXAMPLE 10.8 Consider again Example 10.6, in which the maximum likelihood estimator of the mean of a Gaussian process was found to be the sample mean:

$$\hat{m} = \frac{1}{N} \sum_{i=1}^{N} y_i,$$

with the y_i being independent samples of the process. Since the y_i are Gaussian, so also is $\hat{m}$. Its mean is

$$\mathscr{E}(\hat{m}) = \mathscr{E}(y_i) = m,$$

so that the estimator is unbiased. The variance is

$$\mathscr{E}(\hat{m} - m)^2 = \sigma^2/N,$$

so that the estimator is consistent. We then have

$$p(\hat{m} \mid m) = (2\pi\sigma^2/N)^{-1/2} \exp[-N(\hat{m} - m)^2/2\sigma^2).$$

We now seek to factor this quantity from the likelihood function

$$p(\mathbf{y} \mid m) = \prod_{i=1}^{N} (2\pi\sigma^2)^{-1/2} \exp\left[\frac{-(y_i - m)^2}{2\sigma^2}\right].$$

We are successful, obtaining

$$p(\mathbf{y} \mid m) = p(\hat{m} \mid m)N^{-1/2}(2\pi\sigma^2)^{-(N-1)/2} \exp[-(\Sigma\, y_i^2 - N\hat{m}^2)/2\sigma^2].$$

Since the function on the right involves only the data y_i and not the mean m, the estimator is sufficient. No other estimator can tell us more about the ensemble mean m, given the data samples y_i.

10.6 THE CRAMÉR–RAO BOUND

As we indicated in discussing consistent estimators, the conditional variance of an estimator is one of its important characteristics. We would like our (scalar, say) estimators to be conditionally unbiased, so that (also assuming scalar data)

$$\mathscr{E}[\hat{x}(y) \mid x] = x,$$

and to have small conditional variance

$$\text{Var}(\hat{x} \mid x) = \mathscr{E}[(\hat{x} - x)^2 \mid x].$$

Thereby in any particular trial of the experiment, the value $\hat{x}(y)$ is likely to be near the true value of the random variable x. In Sect. 10.2 we pointed out that the conditional mean $\mathscr{E}(x \mid y)$ minimizes the error variance over the ensemble of values x. We now investigate the error variance over the data ensemble for any particular value of x.

The *Cramér–Rao bound* (sometimes called the *minimum variance bound*) expresses the minimum conditional error variance of any estimator $\hat{x}(y)$ of x in terms of the conditional density $p(y \mid x)$ of the data. Since this latter is usually not difficult to compute in practical applications, the bound gives a convenient standard against which to judge the performance of any other estimator which we may wish to use. We will proceed generally and assume that the estimator $\hat{x}$ may be conditionally biased, with a bias

$$b(x) = \mathscr{E}(\hat{x} \mid x) - x$$

which might depend on the true value of the parameter.

The bound in question is easy to derive in this scalar case [Sage and Melsa, 1971]. Consider first the definition of the conditional bias:

$$b(x) = \mathscr{E}\{[\hat{x}(y) - x] \mid x\} = \int_{-\infty}^{\infty} [\hat{x}(y) - x]p(y \mid x)\, dy.$$

Differentiating both sides of this with respect to x yields

$$\int_{-\infty}^{\infty} [\hat{x}(y) - x]\left[\frac{dp(y \mid x)}{dx}\right] dy = \int_{-\infty}^{\infty} p(y \mid x)\, dy + \frac{db(x)}{dx} = 1 + \frac{db(x)}{dx}.$$

Now use the identity

$$(d/dx) \ln p(y \mid x) = [dp(y \mid x)/dx]/p(y \mid x)$$

to write the above as

$$\int_{-\infty}^{\infty} [\hat{x}(y) - x]\left[\left(\frac{d}{dx}\right) \ln p(y \mid x)\right]p(y \mid x)\, dy = 1 + \frac{db(x)}{dx}. \qquad (10.28)$$

Now recall the Schwartz inequality, valid for arbitrary functions $f(t)$, $g(t)$, that

$$\left| \int_{-\infty}^{\infty} f^*(t)g(t)\,dt \right| \le \left[\int_{-\infty}^{\infty} |f(t)|^2\,dt \int_{-\infty}^{\infty} |g(t)|^2\,dt \right]^{1/2}. \qquad (10.29)$$

Now in this take

$$f(y) = [p(y \mid x)]^{1/2}[\hat{x}(y) - x],$$
$$g(y) = [p(y \mid x)]^{1/2}(d/dx) \ln p(y \mid x).$$

Using Eq. (10.28), the result is

$$|1 + db(x)/dx| \le \{\mathcal{E}[\hat{x}(y) - x]^2 \mid x\}^{1/2} \{\mathcal{E}[(d/dx) \ln p(y \mid x)]^2 \mid x^{1/2}\},$$

which can finally be put in the form

$$\mathcal{E}\{[\hat{x}(y) - x]^2 \mid x\}$$
$$\ge [1 + db(x)/dx]^2/\mathcal{E}\{[(d/dx) \ln p(y \mid x)]^2 \mid x\}. \qquad *(10.30)$$

This is the *Cramér–Rao lower bound* on the conditional error variance of the estimator $\hat{x}(y)$. The denominator of Eq. (10.30) is the one-dimensional case of the *Fisher information matrix* and is related to the idea of the amount of information about x which is carried by the data y.

The bound Eq. (10.30) has a companion for the case in which we know the density $p(x)$ of the variable being estimated, so that, for example, we can compute the Bayes estimate. In that case, removing the conditioning by x in the above derivation, and noting that the conditioned bias $b(x)$ is replaced by the constant unconditioned bias $b = \mathcal{E}(\hat{x} - x)$, the unconditioned Cramér–Rao bound is obtained as

$$\mathcal{E}\{[\hat{x}(y) - x]^2\} \ge 1 / \mathcal{E}\{[(d/dx) \ln p(y \mid x)]^2\}. \qquad (10.31)$$

Since equality holds in the Schwartz inequality Eq. (10.29) only when $f(t) = Kg(t)$, with K a constant, we know that the variance bound Eq. (10.30) is attained if and only if

$$(d/dx) \ln p(y \mid x) = K(x)[\hat{x}(y) - x]. \qquad (10.32)$$

The denominator in the bound Eq. (10.30) can be put in an alternative form by noting that, on the one hand,

$$\mathcal{E}\left\{ \left(\frac{d}{dx}\right) \ln p(y \mid x)]^2 \mid x \right\} = \int_{-\infty}^{\infty} \left[\frac{dp(y \mid x)}{dx} \right]^2 p^{-1}(y \mid x)\,dy,$$

while

$$\mathcal{E}\{(d/dx)^2 \ln p(y \mid x)\} = \int_{-\infty}^{\infty} \left\{ \frac{d^2 p(y \mid x)}{dx^2} - \left[\frac{dp(y \mid x)}{dx} \right]^2 p^{-1}(y \mid x) \right\} dy$$

$$= -\int_{-\infty}^{\infty} \left[\frac{dp(y \mid x)}{dx} \right]^2 p^{-1}(y \mid x)\, dy.$$

The last step follows because

$$\int_{-\infty}^{\infty} \left[\frac{d^2 p(y \mid x)}{dx^2} \right] dy = \left(\frac{d}{dx} \right)^2 \int_{-\infty}^{\infty} p(y \mid x)\, dy = \left(\frac{d}{dx} \right)^2 (1) = 0.$$

With this, we have

$$\mathcal{E}\{[(d/dx) \ln p(y \mid x)]^2 \mid x\} = - \mathcal{E}\{(d/dx)^2 \ln p(y \mid x) \mid x\},$$

so that the bound Eq. (10.30) appears as

$$\mathcal{E}\{[\hat{x}(y) - x]^2 \mid x\}$$
$$\geq - [1 + db(x)/dx]^2 / \mathcal{E}\{(d/dx)^2 \ln p(y \mid x) \mid x\}. \quad *(10.33)$$

The form corresponding to Eq. (10.31) is

$$\mathcal{E}\{[\hat{x}(y) - x]^2\} \geq - 1/\mathcal{E}\{(d/dx)^2 \ln p(y \mid x)\}. \qquad (10.34)$$

In the case of a vector estimator $\hat{\mathbf{x}}(\mathbf{y})$ of a vector, with bias $\mathbf{b}(\mathbf{x}) = \mathcal{E}\{[\hat{\mathbf{x}}(\mathbf{y}) - \mathbf{x}] \mid \mathbf{x}\}$, the Cramér–Rao bound corresponding to Eq. (10.30) is

$$\text{Cov}\{\hat{\mathbf{x}}(\mathbf{y})\} = \mathcal{E}\{\hat{\mathbf{x}}(\mathbf{y}) - \mathcal{E}[\hat{\mathbf{x}}(\mathbf{y})]\}\{\hat{\mathbf{x}}(\mathbf{y}) - \mathcal{E}[\hat{\mathbf{x}}(\mathbf{y})]\}^T$$
$$\geq [I + \partial \mathbf{b}(\mathbf{x})/\partial \mathbf{x}]^T \, \mathcal{F}^{-1}(\mathbf{x})[I + \partial \mathbf{b}(\mathbf{x})/\partial \mathbf{x}]. \quad *(10.35)$$

Here the expectations are conditioned on $\mathbf{x}$, and the inequality is in the sense that the difference matrix of the two sides is nonnegative definite. The matrix $\mathcal{F}(\mathbf{x})$ is the *Fisher information matrix* for the information about $\mathbf{x}$ carried by $\mathbf{y}$:

$$\mathcal{F}(\mathbf{x}) = \mathcal{E}\{\partial \ln [p(\mathbf{y} \mid \mathbf{x})]/\partial \mathbf{x}\}^T \{\partial \ln[p(\mathbf{y} \mid \mathbf{x})]/\partial \mathbf{x}\}$$
$$= -\mathcal{E} \, \partial^2 [\ln p(\mathbf{y} \mid \mathbf{x})]/\partial \mathbf{x}^2. \qquad *(10.36)$$

Here again the expectation is conditioned by $\mathbf{x}$. Recall that the derivative of a scalar with respect to a vector is defined as a row vector, in our notation. The second derivative of a scalar with respect to a vector is then the matrix of elements $\partial^2 p/\partial x_i \, \partial x_j$ in the case Eq. (10.36).

An estimator is called *efficient* if it is (conditionally) unbiased and has a variance which attains the Cramér–Rao bound. It therefore has the smallest possible variance which an unbiased estimator can have. Such an estimator need not exist. It if does not, we will be interested in finding the unbiased estimator which has the minimum possible variance for the problem at hand.

EXAMPLE 10.9   Consider again Examples 10.2 and 10.7, in which the value of a random variable x is estimated based on N samples $y_i = x + n_i$. As in Eq. (10.6), the Bayes estimator is

$$\hat{x} = \left(1 + \frac{\sigma_n^2}{N\sigma_x^2}\right)^{-1} \left\{\frac{1}{N}\sum_{i=1}^{N} y_i + \frac{m_x \sigma_n^2}{N\sigma_x^2}\right\}.$$

This is (conditionally) biased, with

$$b(x) = \mathscr{E}[(\hat{x} - x) \mid x] = (1 + \sigma_n^2/N\sigma_x^2)^{-1}(x + m_x\sigma_n^2/N\sigma_x^2) - x.$$

Then

$$1 + db(x)/dx = (1 + \sigma_n^2/N\sigma_x^2)^{-1}.$$

Furthermore,

$$\ln p(y \mid x) = \left(\frac{-N}{2}\right)\ln(2\pi\sigma_n^2) - \frac{1}{2\sigma_n^2}\sum_{i=1}^{N}(y_i - x)^2,$$

so that

$$d^2[\ln p(y \mid x)]/dx^2 = -N/\sigma_n^2.$$

The bound Eq. (10.33) then becomes

$$\mathscr{E}\{[\hat{x}(y) - x]^2 \mid x\} \geq (\sigma_n^2/N)(1 + \sigma_n^2/N\sigma_x^2)^{-2}.$$

On the other hand, the left side of this can easily be calculated directly from the expression for $\hat{x}(y)$ as being just the right side. That is, the Bayes estimator for this particular situation satisfies the Cramér–Rao bound with equality and is (conditionally) efficient for all N.

The maximum likelihood estimator occupies a special place among estimators in that, if a conditionally efficient estimator exists, it must be the maximum likelihood estimator for the problem at hand. To see this, recall from Eq. (10.32) that it is necessary and sufficient for an estimator to be efficient that it satisfy

$$d[\ln p(y \mid x)]/dx = K(x)(\hat{x} - x).$$

This says that, if an efficient estimator exists, it must be a root of the expression on the left of this last. On the other hand, the maximum likelihood estimator $\hat{x}_{ml}(y)$ results by maximizing $p(y \mid x)$. To that end, consider the equation

$$dp(y \mid x)/dx = 0,$$

which is equivalent to

$$d[\ln p(y \mid x)]/dx = p^{-1}(y \mid x) dp(y \mid x)/dx = 0.$$

If an efficient estimator exists, the expression on the left of this last has a root (the efficient estimator), hence this equation has at least one solution. It turns out that it is a unique and maximizing solution, so that it is also the maximum likelihood estimator. That is, if an efficient estimator exists for a problem, we can find it by finding the maximum likelihood estimator. The converse is not true, however. That is, a maximum likelihood estimator need not be efficient, nor indeed even unbiased. (The maximum likelihood estimator could occur, for example, at an end point of the domain of definition of $p(y \mid x)$.) The following example illustrates on the other hand that it may be both.

EXAMPLE 10.10   Consider again Example 10.5, in which we observed independent values T_i of the waiting time T in some system. The waiting time T has the exponential density $p(T) = (1/\mu) \exp(-T/\mu)$, with (conditional) mean μ and variance μ^2. We want to estimate the parameter μ, which is the average waiting time. The maximum likelihood estimator is the sample mean

$$\hat{\mu} = \frac{1}{N} \sum_{i=1}^{N} T_i.$$

This is unbiased, since $\mathscr{E}(\hat{\mu}) = \mu$, and has variance μ^2/N. The conditional density $p(y \mid x)$ is

$$p(T_1, \ldots, T_N \mid \mu) = \mu^{-N} \exp[-(1/\mu) \Sigma T_i)],$$

so that the bound Eq. (10.33) is

$$\text{Var}(\hat{\mu} \mid \mu) \geq -1/\mathscr{E}\left\{\left[\frac{N}{\mu^2} - \left(\frac{2}{\mu^3}\right) \sum_{i=1}^{N} T_i\right] \mid \mu\right\} = \frac{\mu^2}{N}.$$

On the other hand, we have directly that

$$\text{Var}(\hat{\mu} \mid \mu) = (1/N^2)\mathscr{E}\{[\Sigma(T_i - \mu)]^2 \mid \mu\}$$
$$= (1/N) \text{Var}(T_i \mid \mu) = \mu^2/N.$$

The estimator is therefore efficient for any N.

The following example illustrates that a maximum likelihood estimator need not be unbiased, and therefore need not be efficient, but it is asymptotically efficient.

EXAMPLE 10.11   Consider again Example 10.6, and the maximum likelihood estimator of the variance of a Gaussian density, based on N independent samples, in the case that the ensemble mean is not known:

$$\hat{\sigma}^2 = \frac{1}{N} \sum_{i=1}^{N} (y_i - \hat{m})^2$$

$$= \frac{1}{N} \sum_{i=1}^{N} [(y_i - m) - \frac{1}{N} \sum_{j=1}^{N} (y_j - m)]^2.$$

Taking account that the quantities $y_i - m$ are zero mean and independent, we have

$$\mathcal{E}(\hat{\sigma}^2) = \sigma^2(1 - 1/N).$$

The estimator is biased, but asymptotically unbiased. It is tedious but straightforward to verify that the estimator error variance is

$$\text{Var}(\hat{\sigma}^2) = 2\sigma^4/N.$$

Let us now evaluate the Cramér–Rao bound for this example. The parameter and bias vectors are

$$\mathbf{x} = \begin{bmatrix} m \\ \sigma^2 \end{bmatrix},$$

$$\mathbf{b}(\mathbf{x}) = \begin{bmatrix} 0 \\ -\sigma^2/N \end{bmatrix}.$$

Taking into account that

$$p(\mathbf{y} \mid \mathbf{x}) = (2\pi\sigma^2)^{-N/2} \exp\left[\frac{-1}{2\sigma^2} \sum_{i=1}^{N} (y_i - m)^2\right],$$

the Fisher information matrix is

$$\mathcal{F} = \mathcal{E} \begin{bmatrix} N/\sigma^2 & \Sigma(y_i - m)/\sigma^4 \\ \Sigma(y_i - m)/\sigma^4 & -N/2\sigma^4 + \Sigma(y_i - m)^2/\sigma^6 \end{bmatrix}$$

$$= \begin{bmatrix} N/\sigma^2 & 0 \\ 0 & N/2\sigma^4 \end{bmatrix}.$$

Then the Cramér–Rao bound appears as

$$\text{Cov}\left\{ \begin{bmatrix} \hat{m} \\ \hat{\sigma}^2 \end{bmatrix} \right\} \geq \begin{bmatrix} 1 & 0 \\ 0 & 1 - 1/N \end{bmatrix}^T \begin{bmatrix} \sigma^2/N & 0 \\ 0 & 2\sigma^4/N \end{bmatrix} \begin{bmatrix} 1 & 0 \\ 0 & 1 - 1/N \end{bmatrix}$$

$$= \begin{bmatrix} \sigma^2/N & 0 \\ 0 & 2\sigma^4(1 - 1/N)^2/N \end{bmatrix}.$$

In particular, this shows that

$$\text{Var}(\hat{\sigma}^2) \geq (2\sigma^4/N)(1 - 1/N)^2.$$

Comparing the bound with the actual variance, we see that the estimator is not efficient for arbitrary N. It is, however, asymptotically efficient. Among asymptotically efficient estimators we would prefer the one that approaches the Cramér–Rao bound most closely for finite N.

In the example just above, the matrix in the Cramér–Rao bound is diagonal. More generally, the Cramér–Rao bound supplies a matrix M such that the error covariance matrix C satisfies

$$C \geq M,$$

in the sense that the difference $C - M$ is nonnegative definite:

$$\mathbf{a}'(C - M)\mathbf{a} \geq 0,$$
$$\mathbf{a}'C\mathbf{a} \geq \mathbf{a}'M\mathbf{a},$$

for any vector $\mathbf{a}$. Taking $\mathbf{a}$ as zero everywhere except unity in the ith place, then we can write

$$\text{Var}[\hat{x}_i(\mathbf{y})] \geq M_{ii}.$$

In the case of an unbiased estimator, we have

$$M = \mathscr{F}^{-1}.$$

If the Fisher matrix $\mathscr{F}$ is diagonal, as in the preceding example, we then have

$$M_{ii} = 1/\mathscr{F}_{ii} = -1/\mathscr{E}\{\partial^2 \ln[p(\mathbf{y} \mid \mathbf{x})]/\partial x_i^2\}.$$

More generally, the bound on the error variance of the ith component of the unknown vector will involve all the mixed second partial derivatives of the density.

10.7 PARAMETERS OF SIGNALS IN ADDITIVE WHITE NOISE

Let us now consider the problem of estimating a vector $\mathbf{x}$ of constant but unknown parameters of a signal $s_r(t \mid \mathbf{x})$ which is available in additive zero-mean white Gaussian noise $n(t)$ over an observation interval of length T:

$$r(t) = s_r(t \mid \mathbf{x}) + n(t), \qquad t_0 \leq t \leq t_0 + T. \qquad (10.37)$$

We will consider in turn various of the standard parameter estimation problems in radar and sonar. These are estimation of the amplitude of a signal, assumed to be present, as well as estimation of its phase, frequency, or time of arrival, or various combinations of these. We will consider both low-pass signals and bandpass signals in conjunction with a high-frequency carrier.

For various of these parameters, taken alone or in combinations, we will determine the maximum likelihood estimate and the Cramér–Rao lower bound on the variance of any unbiased estimator. Our results will mostly assume the signal-to-noise ratio is relatively high, so that various practical approximations will be useful.

Development of both the maximum likelihood estimator and the Cramér–Rao bound begins with the likelihood function for the signal Eq. (10.37) of interest. We will assume that the signal is known to be present over some particular time interval of length T, which we will take as the interval $0 \leq t \leq T$. Introducing a sampling plan, and then letting the sampling interval $\delta t \Rightarrow 0$, leads to the likelihood function in the form Eq. (6.36). Since we assume the parameters in the signal of Eq. (10.37) are deterministic constants, we have the mean received waveform as

$$\bar{r}(t) = \mathscr{E}[r(t)] = s_r(t \mid \mathbf{x}),$$

so that the likelihood function from Eq. (6.36) is

$$p(r \mid \mathbf{x}) = \alpha \exp \left\{ -\left(\frac{1}{N_0}\right) \int_0^T [r(t) - s_r(t \mid \mathbf{x})]^2 \, dt \right\}. \qquad (10.38)$$

Here α is a constant of no interest, and the noise is assumed to be Gaussian, and white with two-sided power spectral density $N_0/2$.

The maximum likelihood estimates result by setting to zero the quantities

$$\frac{\partial [\ln p(r \mid \mathbf{x})]}{\partial x_i} = \frac{2}{N_0} \int_0^T [r(t) - s_r(t \mid \mathbf{x})] \left[\frac{\partial s_r(t \mid \mathbf{x})}{\partial x_i} \right] dt \qquad *(10.39)$$

and solving the resulting equations for the components x_i of the parameter vector $\mathbf{x}$.

In Eq. (10.39) we can take account that the function $r(t) - s_r(t \mid \mathbf{x})$ is just $n(t)$ and that

$$\mathscr{E}[n(t)n(t')] = R_n(t - t') = (N_0/2)\delta(t - t').$$

Then the Fisher information matrix Eq. (10.36) has elements

$$\mathscr{F}_{ij} = \left(\frac{2}{N_0}\right)^2 \int_0^T \int_0^T \left(\frac{N_0}{2}\right) \delta(t - t')$$

$$\times [\partial s_r(t \mid \mathbf{x})/\partial x_i][\partial s_r(t' \mid \mathbf{x})/\partial s_j] \, dt \, dt'$$

$$= \frac{2}{N_0} \int_0^T \left[\frac{\partial s_r(t \mid \mathbf{x})}{\partial x_i} \right] \left[\frac{\partial s_r(t \mid \mathbf{x})}{\partial s_j} \right] dt. \qquad *(10.40)$$

A signal may well involve one or more unknown parameters of no interest in a particular problem at hand. Such *nuisance parameters* can be suppressed by assuming them to be random variables with some density function. The likelihood function of the parameters of interest is then taken as the expected value of the full likelihood function, with the expectation taken over the density of the nuisance parameters.

The most common such situation is that of an information waveform modulating a high-frequency carrier. Let us take the noiseless received signal as

$$s_r(t) = Aa(t) \cos[\omega_c t + \phi(t) + \theta_c].$$

The received signal then has a complex envelope

$$\tilde{r}(t) = A\tilde{s}_{r0}(t) \exp(j\theta_c) + \tilde{n}(t), \tag{10.41}$$

where

$$\tilde{s}_{r0}(t) = a(t) \exp[j\phi(t)].$$

The normalized envelope $\tilde{s}_{r0}(t)$ may contain unknown parameters such as a Doppler shift in the carrier of the received waveform or arrival time. The carrier phase θ_c is a nuisance parameter, for which we will take the uniform density. The resulting likelihood function for the complex envelope $\tilde{r}(t)$, averaged over carrier phase θ_c, is given in Eq. (7.28):

$$\mathscr{E}p(\tilde{r}) = \alpha I_0(2Aq/N_0)$$

$$\times \exp\left[-\left(\frac{1}{2N_0}\right) \int_0^T [|\tilde{r}(t)|^2 + A^2|\tilde{s}_{r0}(t)|^2] \, dt\right]. \tag{10.42}$$

The function $I_0(x)$ is the modified Bessel function of the first kind and order zero, and the statistic in its argument is

$$q = \frac{1}{2}\left|\int_0^T \tilde{r}(t)\tilde{s}_{r0}^*(t) \, dt\right|. \tag{10.43}$$

The maximum likelihood estimates of the remaining parameters (in addition to carrier phase) are computed from the density Eq. (10.42). We will consider the case of high signal-to-noise ratio. That is the situation in which maximum likelihood estimation is most commonly useful. There is also considerable analytical simplification. Accordingly, in the exponential term of Eq. (10.42) we assume from Eq. (10.41) that

$$\tilde{r}(t) \cong A\tilde{s}_{r0}(t) \exp(j\theta_c), \tag{10.44}$$

and that the large-argument approximation applies for the Bessel function:

$$\ln I_0(x) \cong x.$$

Then Eq. (10.42) becomes

$$\ln \mathscr{E}p(\tilde{r}) \cong \ln(\alpha) + 2Aq/N_0 - \frac{A^2}{N_0} \int_0^T |\tilde{s}_{r0}(t)|^2 \, dt. \tag{10.45}$$

This is the form we will use below in various special cases.

10.8 ESTIMATION OF SPECIFIC PARAMETERS

We now consider estimation of the common radar or sonar signal parameters of interest. At first we assume a low-pass signal with only one parameter unknown. In that case, the equation to be solved for the maximum likelihood estimate is Eq. (10.39):

$$\int_0^T [r(t) - s_r(t \mid x)] \left[\frac{\partial s_r(t \mid x)}{\partial x} \right] dt = 0, \tag{10.46}$$

where $s_r(t \mid x)$ is the noiseless received signal. The Cramér–Rao bound for the variance of any unbiased estimator of the parameter is given by Eq. (10.35), with the bias $b = 0$, in conjunction with Eq. (10.40):

$$\text{Var}(\hat{x} - x) \geq \left[\frac{2}{N_0} \int_0^T \left[\frac{\partial s_r(t \mid x)}{\partial x} \right]^2 dt \right]^{-1}. \tag{10.47}$$

Estimation of Amplitude

Let us take the noiseless received signal as $s_r(t \mid A) = As_t(t)$, where $s_t(t)$ is the known transmitted waveform and the signal fills a known interval $0 \leq t \leq T$. Equation (10.46) at once yields the maximum likelihood estimate of the signal amplitude as

$$\hat{A} = \frac{1}{E_t} \int_0^T r(t)s_t(t) \, dt, \tag{10.48}$$

where

$$E_t = \int_0^T s_t^2(t) \, dt.$$

Then the estimate $\hat{A}$ is just the scaled output of a correlator receiver matched to the signal $s_t(t)$. This can equally well be implemented using a matched filter, as discussed earlier.

Taking into account that

$$r(t) = As_t(t) + n(t),$$

and that the noise is zero mean, the mean of the estimator Eq. (10.48) is

$$\mathscr{E}(\hat{A}) = A,$$

so that the estimator is unbiased. Furthermore,

$$\mathscr{E}(\hat{A}^2) = \frac{1}{E_t^2} \left[\int_0^T \int_0^T \mathscr{E}\{[As_t(t) + n(t)] \right.$$

$$\left. \times [As_t(t') + n(t')]\}s_t(t)s_t(t') \, dt \, dt' \right]$$

$$= A^2 + N_0/2E_t.$$

The variance of the estimator is then

$$\text{Var}(\hat{A}) = N_0/2E_t.$$

On the other hand, Eq. (10.47) yields the Cramér–Rao bound as

$$\text{Var}(\hat{A}) \geq N_0/2E_t.$$

The estimator Eq. (10.48) is therefore efficient, as well as unbiased.

Estimation of Phase

Now consider a signal with known amplitude and frequency, but one in which the phase is unknown and also of interest:

$$s_r(t) = A \cos(\omega_c t + \theta), \qquad 0 \leq t \leq T.$$

Equation (10.39) for the maximum likelihood estimate of the phase is

$$\int_0^T [r(t) - A \cos(\omega_c t + \theta)] \sin(\omega_c t + \theta) \, dt = 0.$$

This could be solved as it stands for the estimate of θ, but let us assume the case that $\omega_c T \gg 1$, that is, $T \gg 1/\omega_c$. Then we retain a reasonable approximation to the integral if we change T by an amount at most π/ω_c to obtain $T = k\pi/\omega_c$, for some (large) integer k. That is, we take $\omega_c T = k\pi$. Then the equation to be solved is

$$\int_0^T r(t) \sin(\omega_c t + \theta) \, dt = \frac{A}{2} \int_0^T \sin[2(\omega_c t + \theta)] \, dt = 0, \qquad (10.49)$$

which is to say

$$\cos(\theta) \int_0^T r(t) \sin(\omega_c t) \, dt = -\sin(\theta) \int_0^T r(t) \cos(\omega_c t) \, dt,$$

$$\tan(\theta) = \frac{-\displaystyle\int_0^T r(t) \sin(\omega_c t) \, dt}{\displaystyle\int_0^T r(t) \cos(\omega_c t) \, dt}, \qquad (10.50)$$

from which the estimate $\hat{\theta}$ follows.

One situation in which the carrier phase is of interest is that of the phase-locked loop. In this circuit, sketched in Fig. 10.3, the phase of an oscillator is to be locked to the phase of a sinusoid in additive noise. The circuit operates in effect to solve Eq. (10.50) by selecting the angle θ such that Eq. (10.49) holds true. To see this, suppose that the input signal $r(t)$ in Fig. 10.3 is noiseless, so that

$$r(t) = A \cos(\omega_c t + \theta).$$

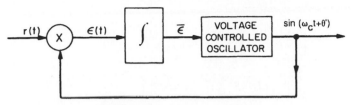

FIGURE 10.3 Phase-locked loop.

Then the output of the multiplier is

$$\varepsilon(t) = (A/2)[\sin(2\omega_c t + \theta + \theta') - \sin(\theta - \theta')].$$

The averager eliminates the double frequency term of this and leaves a voltage proportional to $\sin(\theta - \theta')$ which adjusts the oscillator phase in a direction to drive θ' toward θ as long as the two angles are not equal.

Estimation of Time of Arrival

In Chapter 7 we discussed the problem of detecting a signal whose time of arrival was not known. The process amounted to setting up a number of time "bins," each of length T equal to the length of the signal. A matched filter was operated over each bin, and any filter output which exceeded the threshold was taken as indication of a signal present in that time bin. In practice, the output of a matched filter is observed in running time and threshold crossings are taken as indicators of targets present.

In this section, we assume that we have already determined the presence of a signal in some interval $0 \leq t \leq T$, and we now want to refine its arrival time to a precision considerably finer than T. To do so as a practical matter, it is necessary to assume that the signal-to-noise ratio is reasonably high.

We will first consider a low-pass signal, in which there is no phase uncertainty and only an unknown arrival time. Then we will consider the bandpass case, in which phase is also unknown but has been integrated out as a nuisance parameter.

Low-Pass Signal

The signal $s_r(t \mid \tau)$, with arrival time τ being the unknown parameter, can be written $s_r(t - \tau)$. We assume that we know the signal exists in some interval $0 \leq t \leq T$ and that $\tau \cong 0$. We want to compute the maximum likelihood estimate of τ. Placing the signal in the middle of the observation interval, Eq. (10.39) for the maximum likelihood estimate is

$$\int_{-T/2}^{T/2} [r(t) - s_r(t - \tau)] \left[\frac{\partial s_r(t - \tau)}{\partial \tau} \right] dt = 0.$$

The second term of this is

$$-\int_{-T/2}^{T/2} s_r(t - \tau) \left[\frac{\partial s_r(t - \tau)}{\partial \tau} \right] dt$$

$$= \int_{-T/2-\tau}^{T/2-\tau} s_r(u) \left[\frac{\partial s_r(u)}{\partial u} \right] du = \frac{1}{2} \int_{-T/2-\tau}^{T/2-\tau} \left[\frac{\partial s_r^2(u)}{\partial u} \right] du$$

$$= \frac{1}{2}[s_r^2(T/2 - \tau) - s_r^2(-T/2 - \tau)] \cong \frac{1}{2}[s_r^2(T/2) - s_r^2(-T/2)] = 0.$$

In this last we assume that the signal occupies some part of the interior of the observation interval. Alternatively, it suffices to assume that the signal is symmetric, as we will do below.

Since only the waveform of the received signal enters the likelihood equation, the maximum likelihood estimate $\hat{\tau}$ of the delay τ is that value of τ for which

$$\int_{-T/2}^{T/2} r(t) \left[\frac{\partial s_t(t - \tau)}{\partial \tau} \right] dt = 0. \tag{10.51}$$

This processor then amounts to correlating the received signal with a signal of known waveform and adjusting the delay τ until the filter output is zero. In practice, in a tracking radar for example, the procedure can be implemented by assuming the signal to be symmetric around its midpoint, as in Fig. 10.4. The derivative is then antisymmetric and can be approximated as the bipolar gate function in the figure. The gate function is centered at the delay estimate if the received voltage integrates to zero over the gates, so that Eq. (10.51) holds true. From one radar pulse to the next the gate placement is adjusted by a feedback loop to keep the integrator output nulled, which maintains the center of the gates on the maximum likelihood estimate of the target position. Within the assumptions made, the estimate is unbiased.

The Cramér–Rao bound for the variance of any unbiased estimate $\hat{\tau}$ of a time delay τ is given by Eq. (10.47), which in this case becomes

$$\text{Var}(\hat{\tau} - \tau) \geq \left[\frac{2}{N_0} \int_{-T/2}^{T/2} \left[\frac{\partial s_r(t - \tau)}{\partial \tau} \right]^2 dt \right]^{-1}$$

$$= \left[\frac{2}{N_0} \int_{-T/2}^{T/2} \left[\frac{\partial s_r(t)}{\partial t} \right]^2 dt \right]^{-1}. \tag{10.52}$$

In the last step we assumed that $\tau \cong 0$. Using Parseval's theorem, the bound can be written in terms of the Fourier transform $(j\omega)S_r(j\omega)$ of $\partial s_r(t)/\partial t$ as

$$\text{Var}(\hat{\tau} - \tau) \geq \left[\frac{2}{N_0} \int_{-\infty}^{\infty} \omega^2 |S_r(j\omega)|^2 df \right]^{-1}. \tag{10.53}$$

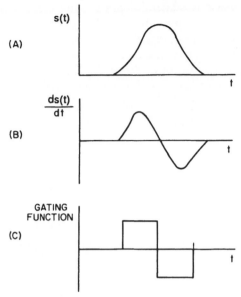

(A) s(t)

(B) $\dfrac{ds(t)}{dt}$

(C) GATING FUNCTION

FIGURE 10.4 An example of a video pulse (a) and its derivative (b). The gating function (c) may be used to approximate the derivative.

Let us now note that the received signal energy is

$$E_r = \int_{-T/2}^{T/2} s_r^2(t)\ dt = \int_{-\infty}^{\infty} |S_r(j\omega)|^2\ df.$$

Then we introduce a measure of the signal (radian) bandwidth β, defined by

$$\beta^2 = \frac{1}{E_r} \int_{-\infty}^{\infty} \omega^2 |S_r(j\omega)|^2\ df, \qquad \qquad *(10.54)$$

which is the second moment of the spectrum about $f = 0$. In terms of this bandwidth measure, the Cramér–Rao bound Eq. (10.53) becomes

$$\mathrm{Var}(\hat{\tau} - \tau) \ge [(2E_r/N_0)\beta^2]^{-1}. \qquad \qquad *(10.55)$$

This uncertainty in arrival time translates to an uncertainty in range in a tracking radar.

Narrowband Signal

We now want to estimate the arrival time τ of a narrowband signal which is known to lie in some interval which we will take to be $-T/2 \le t \le T/2$. That is, the signal arrival time, the center of the envelope of duration T, is approximately in the middle of some interval around $t = 0$, so the arrival time τ is small. The maximum likelihood estimate $\hat{\tau}$ of τ is that value of τ which maximizes the likelihood function Eq. (10.42), for which the carrier phase has been integrated out as a nuisance parameter. Since we assume now that only the arrival time is unknown, the exponential term in Eq. (10.42), involving the received energy, is constant. Then the likelihood function is maximized if we maximize the statistic q, since the Bessel function is monotonic. From Eq. (10.43), assuming again that the received signal envelope is approximately that of the transmitted signal, we finally want to choose τ to maximize

$$
2q(\tau) = \left| \int_{-T/2}^{T/2} \tilde{r}(t)\tilde{s}_{t0}^{*}(t)\, dt \right|
$$

$$
= \left| \int_{-T/2}^{T/2} \tilde{r}(t)\tilde{s}_{t}^{*}(t - \tau)\, dt \right|. \tag{10.56}
$$

In this, the range bin of interest is taken centered on the time origin, and $\tilde{r}(t)$ is the received envelope over that bin. The normalized received signal is the transmitted signal:

$$
\tilde{s}_{t}(t) = a(t)\, \exp[j\phi(t)],
$$

delayed by τ.

As discussed in Sect. 3.6, Eq. (10.56) can be implemented as follows. The received signal $r(t)$ is multiplied with sine and cosine terms at the carrier frequency in a quadrature detector. The result is passed through the matched filter corresponding to the transmitted envelope $\tilde{s}_{t}(t)$. That filter, being low pass, also removes the double frequency terms resulting from the quadrature detection process. The envelope of the result is determined and observed over the running time variable of the matched filter output, which is τ. That time at which the filter output envelope peaks is the approximate maximum likelihood estimate of signal delay.

We now want to determine the Cramér–Rao lower bound on the error variance of any unbiased estimator of the delay τ. We will write the received signal envelope as

$$
\tilde{r}(t) = A\tilde{s}_{t}(t - \tau) + \tilde{n}(t),
$$

where $\tilde{s}_{t}(t)$ is the transmitted waveform. In the large-signal approximation Eq. (10.45), we again assume the last term is approximately independent of τ. Then the log likelihood function is

$$
\ln p(\tilde{r} \mid \tau) \cong K(A) + 2Aq/N_0
$$

$$
= K(A) + \frac{A}{N_0} \left| \int_{-T/2}^{T/2} [A\tilde{s}_{t}(t - \tau) + \tilde{n}(t)]\tilde{s}_{t}^{*}(t - \tau)\, dt \right|.
$$

Because of the large signal-to-noise ratio we will drop the noise term in this. Further assuming $\tau \approx 0$ in the second factor, we have

$$\ln p(\tilde{r} \mid \tau) \cong K(A)$$

$$+ \frac{A}{N_0} \left| \int_{-T/2}^{T/2} A\tilde{s}_t(t - \tau)\tilde{s}_t^*(t) \, dt \right|.$$

Let us now introduce the complex *ambiguity function* of the transmitted signal, taken as

$$\chi(\tau, \omega) = \frac{1}{2E_t} \int_{-T/2}^{T/2} \tilde{s}_t(t - \tau)\tilde{s}_t^*(t) \exp(-j\omega\tau) \, dt. \qquad *(10.57)$$

(Other definitions exist, differing from this by a phase angle.) The normalization is defined such that $\chi(0) = 1$, which means that

$$E_t = \frac{1}{2} \int_{-T/2}^{T/2} |\tilde{s}_t(t)|^2 \, dt$$

is the energy of the transmitted signal.

In terms of the ambiguity function Eq. (10.57) we have

$$\ln p(\tilde{r} \mid \tau) \cong K(A) + (2A^2E_t/N_0)|\chi(\tau, 0)|. \qquad (10.58)$$

Then the Cramér–Rao bound Eq. (10.31) becomes

$$\text{Var}(\hat{\tau} - \tau) \geq [-(2E_r/N_0) \, \partial^2|\chi(\tau)|/\partial\tau^2]^{-1}, \qquad (10.59)$$

where we write $\chi(\tau)$ for $\chi(\tau, 0)$. Here we used the bound in the equivalent form indicated by Eq. (10.36) and wrote $E_r = A^2E_t$ for the energy of the noiseless received signal.

Let us examine the bound Eq. (10.59) further. Writing

$$|\chi(\tau)| = [\chi(\tau)\chi^*(\tau)]^{1/2},$$

we have

$$d|\chi(\tau)|/d\tau = [\chi(\tau)\chi'^*(\tau) + \chi'(\tau)\chi^*(\tau)]/2|\chi(\tau)|,$$
$$d^2|\chi(\tau)|/d\tau^2 = \text{Re}[\chi(\tau)\chi''^*(\tau) + |\chi'(\tau)|^2]/|\chi(\tau)|$$
$$- \{\text{Re}[\chi(\tau)\chi'(\tau)]\}^2/|\chi(\tau)|^3.$$

Evaluating this last at the approximate delay $\tau = 0$, remembering that $\chi(0) = 1$, yields

$$d^2|\chi(\tau)|/d\tau^2 \bigg|_{\tau=0} = \text{Re}[\chi''(0)] + \{\text{Im}[\chi'(0)]\}^2. \qquad (10.60)$$

From Eq. (10.57), we have further that

$$\chi'(\tau) = \left(\frac{-1}{2E_t}\right) \int_{-T/2}^{T/2} \tilde{s}_t'(t - \tau)\tilde{s}_t^*(t) \, dt,$$

$$\chi''(\tau) = \frac{1}{2E_t} \int_{-T/2}^{T/2} \tilde{s}_t''(t - \tau)\tilde{s}_t^*(t) \, dt,$$

so that

$$\chi'(0) = \left(\frac{-1}{2E_t}\right) \int_{-T/2}^{T/2} \tilde{s}_t'(t) \tilde{s}_t^*(t) \, dt,$$

$$\chi''(0) = \frac{1}{2E_t} \int_{-T/2}^{T/2} \tilde{s}_t''(t) \, \tilde{s}_t^*(t) \, dt$$

$$= \left(\frac{-1}{2E_t}\right) \int_{-T/2}^{T/2} |\tilde{s}'(t)|^2 \, dt.$$

In the last we integrated by parts and assumed that the transmitted signal envelope is continuous at the ends of the interval of integration and therefore vanishes there.

Now let the Fourier transform of the transmitted signal envelope $\tilde{s}_t(t)$ be $S_t(f)$. Then we have these last quantities as

$$\chi'(0) = \left(\frac{-j}{2E_t}\right) \int_{-\infty}^{\infty} \omega |S_t(f)|^2 \, df,$$

$$\chi''(0) = \left(\frac{-1}{2E_t}\right) \int_{-\infty}^{\infty} \omega^2 |S_t(f)|^2 \, df.$$

Now define a (radian) bandwidth measure β^2 for the envelope of the transmitted signal in terms of the second moment of the spectrum around the mean frequency:

$$\beta^2 = \frac{1}{2E_t} \int_{-\infty}^{\infty} \omega^2 |S_t(f)|^2 \, df$$

$$- \left[\frac{1}{2E_t} \int_{-\infty}^{\infty} \omega |S_t(f)|^2 \, df\right]^2. \qquad \text{*(10.61)}$$

Then finally, using Eq. (10.60), the bound Eq. (10.59) becomes

$$\text{Var}(\hat{\tau} - \tau) \geq 1/[(2E_r/N_0)\beta^2]. \qquad \text{*(10.62)}$$

Estimation of Frequency

Let us suppose now that the amplitude and time of arrival of the received signal are known. We seek an estimate of the frequency of the received signal. Presumably this is different from the frequency ω_c of the transmitted signal due to (most commonly) a Doppler shift ω_D. We suppose again that the transmitted signal is

$$s_t(t) = a(t) \cos[\omega_c t + \phi(t)],$$

where we allow for both amplitude and phase modulation. The received signal is then

$$s_r(t) = Aa(t) \cos[(\omega_c + \omega_D)(t - \tau) + \phi(t - \tau) + \theta_c],$$

with additive zero-mean white Gaussian noise with two-sided power spectral density $N_0/2$. Take the known delay τ as the origin of time for the received signal. Then the complex envelope of the received signal, taken with respect to the known carrier frequency, is

$$\tilde{r}(t) = \tilde{s}_r(t) + \tilde{n}(t) = A\tilde{s}_t(t) \exp(j\omega_D t) \exp(j\theta_c) + \tilde{n}(t).$$

If we treat the carrier phase as a nuisance parameter with the uniform density, the likelihood function of the received envelope is again that given by Eq. (10.42). From Eq. (10.43) we have

$$q = \frac{1}{2} \left| \int_0^T \tilde{r}(t)\tilde{s}_t^*(t) \exp(-j\omega_D t) \, dt \right|. \qquad *(10.63)$$

Again taking account that the Bessel function is monotonic and that the amplitude A is assumed known, so that the exponential term in Eq. (10.42) is known, the maximum likelihood estimate of the Doppler frequency is the value which maximizes Eq. (10.63). This amounts to passing the received signal through a bank of matched filters using the transmitted modulation with a range of Doppler shifts and selecting the Doppler frequency which results in maximum envelope of the output. The scheme is sketched in Fig. 10.5. It could be realized with either analog elements operating on signals at the carrier frequency or digitally by directly implementing Eq. (10.63) after determining the received signal complex envelope.

We now want to determine the Cramér–Rao bound on an unbiased estimate of the Doppler shift. This will indicate the minimum useful spacing of the Doppler filters in Fig. 10.5. We again assume a high signal-to-noise ratio so that we can drop the noise terms. We also assume that the Doppler shift is small. Then we have Eq. (10.45) as

$$\ln[p(\tilde{r} \mid \omega_D)] \cong K(A) + 2Aq/N_0$$

$$= K(A) + \frac{A}{N_0} \left| \int_0^T [A\tilde{s}_t(t) \exp(j\omega_D t) \right.$$

$$\left. + \tilde{n}(t)]\tilde{s}_t^*(t) \exp(-j\omega_D t) \, dt \right|$$

$$\approx K(A) + \frac{A}{N_0} \left| \int_0^T |s_t(t)|^2 \exp(j\omega_D t) \, dt \right|$$

$$= K(A) + (2A^2 E_t/N_0)|\chi(0, -\omega_D)|.$$

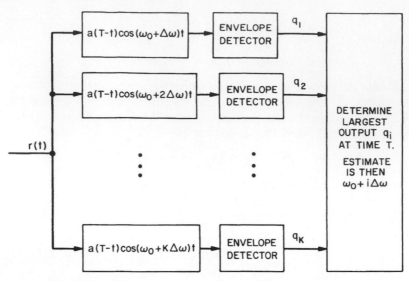

FIGURE 10.5   Receiver for estimating frequency with known arrival time.

The complex ambiguity function is again taken in the form Eq. (10.57). We will write

$$\chi(-\omega) = \chi(0,-\omega) = \frac{1}{2E_t} \int_0^T |\bar{s}_t(t)|^2 \exp(j\omega t)\, dt. \tag{10.64}$$

The normalization is again defined such that $\chi(0) = 1$, so that E_t is the energy of the transmitted signal and the energy of the received signal is $E_r = A^2 E_t$.

The Cramér–Rao bound Eq. (10.31), in the form Eq. (10.36), is

$$\text{Var}(\hat{\omega} - \omega) \geq [-(2E_r/N_0)\, \partial^2 |\chi(-\omega)|/\partial\omega^2]^{-1}. \tag{10.65}$$

The derivative is evaluated at the true value of the Doppler frequency ω, which we assume is zero. Similar to the above developments relative to delay τ, we have

$$d^2|\chi(-\omega)|/d\omega^2 = \text{Re}[\chi(-\omega)\chi''^*(-\omega) + |\chi'(-\omega)|^2]/|\chi(-\omega)|$$
$$- \{\text{Re}[\chi(-\omega)\chi'(-\omega)]\}^2/|\chi(-\omega)|^3,$$

where the prime indicates differentiation with respect to ω. Evaluating this at $\omega = 0$ yields

$$d^2|\chi(-\omega)|/d\omega^2\Big|_{\omega=0} = \text{Re}[\chi''(0)] + \{\text{Im}[\chi'(0)]\}^2. \tag{10.66}$$

From Eq. (10.64), we have further that

$$\chi'(-\omega) = \frac{j}{2E_t} \int_0^T t|\bar{s}_t(t)|^2 \exp(j\omega t) \, dt,$$

$$\chi''(-\omega) = \left(\frac{-1}{2E_t}\right) \int_0^T t^2|\bar{s}_t(t)|^2 \exp(j\omega t) \, dt,$$

so that

$$\chi'(0) = \frac{j}{2E_t} \int_0^T t|\bar{s}_t(t)|^2 \, dt,$$

$$\chi''(0) = \left(\frac{-1}{2E_t}\right) \int_0^T t^2|\bar{s}_t(t)|^2 \, dt.$$

Now define a time duration measure t_d^2 for the envelope of the transmitted signal in terms of the second moment around the mean time:

$$t_d^2 = \frac{1}{2E_t} \int_0^T t^2|\bar{s}_t(t)|^2 \, dt$$

$$- \left[\frac{1}{2E_t} \int_0^T t|\bar{s}_t(t)|^2 \, dt\right]^2. \qquad *(10.67)$$

Then finally, using Eq. (10.66), the bound Eq. (10.65) becomes

$$\text{Var}(\hat{\omega} - \omega) \geq 1/[(2E_r/N_0)t_d^2]. \qquad *(10.68)$$

10.9 SIMULTANEOUS ESTIMATION OF TIME AND FREQUENCY

In the previous section we developed the system of Eq. (10.56) for estimation of the arrival time of the envelope of a bandpass signal. The complex envelope of the received signal is correlated with the transmitted signal envelope and the arrival time taken as the time of peak correlation. The usual implementation uses a filter with response matched to the transmitted signal. In the case of known arrival time but unknown Doppler, the filter bank of Fig. 10.5 is used. If neither Doppler nor arrival time is known, the joint maximum likelihood estimate is obtained approximately by observing the outputs of the filters in Fig. 10.5 over "running" time. If a particular filter output peaks at some time, we take the arrival time as the time of the peak and the Doppler as the frequency of the particular filter in question.

In the framework of this chapter so far, it is assumed that the signal is known to be present. In a later section, we will discuss an approximate procedure for detection of signals with unknown parameters which amounts to declaring a signal present, with the indicated parameters, if the output of some filter of the filter bank Fig. 10.5 crosses an appropriate threshold.

In this section, we want to determine the Cramér–Rao lower bound for the variance of an unbiased joint estimator of arrival time and Doppler frequency. We will continue to treat carrier phase as a nuisance parameter. Accordingly, for a transmitted signal $s_t(t)$ with energy E_t, we take the received signal as

$$s_r(t) = As_t(t - \tau) \cos[(\omega_c + \omega_D)(t - \tau) + \theta_c].$$

The signal is available over a region $-T/2 \le t \le T/2$ centered on the delay τ, which we assume is small with respect to the origin at the received pulse center. The Doppler shift ω_D is small with respect to the bandwidth of the signal. We can absorb the phase angle $(\omega_c + \omega_D)\tau$ into the nuisance angle θ_c. The complex envelope of the received signal is then taken as

$$\tilde{s}_r(t) = A\tilde{s}_t(t - \tau) \exp(j\omega_D t). \tag{10.69}$$

We continue to assume the high signal-to-noise ratio case. Since the parameter ω_D enters the received signal as in Eq. (10.69), it cancels in the integral terms of Eq. (10.45), and we again have

$$\ln p(\tilde{r} \mid \tau, \omega_D) \cong K(A) + 2Aq/N_0,$$

where in this case, using Eq. (10.69),

$$2q = \left| \int_{-T/2}^{T/2} \tilde{r}(t)\tilde{s}_t^*(t - \tau) \exp(-j\omega_D t) \, dt \right|.$$

This can be implemented as a bank of Doppler filters as in Fig. 10.5. The envelopes are observed in running time to determine the delay τ which maximizes the likelihood function.

As we have done before, in determining the Cramér–Rao bound we will approximate this as

$$2q \cong \left| \int_{-T/2}^{T/2} A\tilde{s}_t(t - \tau) \exp(j\omega_D t) \right.$$

$$\left. \times \tilde{s}_t^*(t) \, dt \right|$$

$$= 2AE_t|\chi(\tau, -\omega_D)|. \tag{10.70}$$

In this we used the two-dimensional ambiguity function of the transmitted signal as in Eq. (10.57).

Now the Fisher information matrix Eq. (10.36) is of dimension two. The diagonal elements are just those found in Eqs. (10.62), (10.68):

$$\mathcal{F}_{11} = -(2E_r/N_0)\beta^2,$$
$$\mathcal{F}_{22} = -(2E_r/N_0)t_d^2.$$

The off-diagonal elements are given by

$$\mathcal{F}_{12} = (2A/N_0)\, \partial^2 q/\partial\tau\, \partial\omega = (2A^2 E_t/N_0)\, \partial^2|\chi(\tau, -\omega)|/\partial\tau\, \partial\omega$$
$$= (2E_r/N_0)\, \partial^2|\chi(\tau, -\omega)|/\partial\tau\, \partial\omega,$$

evaluated at $\tau = 0$, $\omega = 0$. Following the techniques of the previous section, at the point in question we find

$$\partial^2|\chi(\tau, -\omega)|/\partial\tau\, \partial\omega = \mathrm{Re}(\chi_{\tau\omega} + \chi_\tau\chi_\omega^*) - [\mathrm{Re}(\chi_\omega)][\mathrm{Re}(\chi_\tau)],$$

where the subscripts indicate differentiation and $\chi = \chi(\tau, -\omega)$. Evaluating the derivatives at zero from the above we have

$$\chi_\tau = \frac{-1}{2E_t}\int_{-T/2}^{T/2} \tilde{s}'(t)\tilde{s}_t^*(t)\, dt$$

$$= \frac{-j}{2E_t}\int_{-\infty}^{\infty} \omega|S_t(f)|^2\, df = -j\overline{\omega}, \tag{10.71}$$

$$\chi_\omega = \frac{j}{2E_t}\int_{-T/2}^{T/2} t|\tilde{s}_t(t)|^2\, dt = j\overline{t}, \tag{10.72}$$

while the definition in Eq. (10.57) yields

$$\chi_{\tau\omega} = \frac{-j}{2E_t}\int_{-T/2}^{T/2} t\tilde{s}'(t)\tilde{s}_t^*(t)\, dt.$$

In these, we define $\overline{t}$ and $\overline{\omega}$ as a mean time location and frequency center of the envelope of the transmitted signal and its spectrum, respectively.

With these last we have

$$\mathcal{F}_{12} = \mathcal{F}_{21}$$

$$= \frac{2E_r}{N_0}\left\{ \mathrm{Re}\left[\frac{-j}{2E_t}\int_{-T/2}^{T/2} t\tilde{s}_t'(t)\tilde{s}_t^*(t)\, dt \right] - \overline{\omega}\,\overline{t} \right\}.$$

This also has an interpretation in terms of the average time and frequency properties of the transmitted signal envelope. We have

$$\tilde{s}_t'(t) = [a'(t) + ja(t)\phi'(t)]\, \exp[j\phi(t)],$$

so that

$$\mathcal{F}_{12} = \left(\frac{2E_r}{N_0}\right)\left\{ \mathrm{Re}\left[\left(\frac{-j}{2E_t}\right)\int_{-T/2}^{T/2} ta(t)[a'(t) \right.\right.$$

$$\left.\left. + ja(t)\phi'(t)]\, dt \right] - \overline{\omega}\,\overline{t} \right\}$$

$$= \left(\frac{2E_r}{N_0}\right)\left\{ \left(\frac{1}{2E_t}\right)\int_{-T/2}^{T/2} ta^2(t)\phi'(t)\, dt - \overline{\omega}\,\overline{t} \right\}.$$

The derivative $\phi'(t)$ is the instantaneous frequency of the waveform, so that the integral in this last is an average of the time–frequency product of the transmitted signal, weighted by the signal envelope power. We accordingly write

$$\frac{1}{2E_t} \int_{-T/2}^{T/2} ta^2(t)\phi'(t) \, dt = \overline{\omega t} \qquad *(10.73)$$

as the average frequency–time product of the transmitted signal envelope.

We now have the full information matrix available, so that we can write the Cramér–Rao bound as

$$\mathrm{Cov}(\hat{\tau} - \tau, \hat{\omega} - \omega) \geq -\mathscr{F}^{-1}$$

$$= (\mathscr{F}_{11}\mathscr{F}_{22} - \mathscr{F}_{12}^2)^{-1} \begin{bmatrix} -\mathscr{F}_{22} & \mathscr{F}_{12} \\ \mathscr{F}_{12} & -\mathscr{F}_{11} \end{bmatrix}$$

$$= [(2E_r/N_0)(\beta^2 t_d^2 - (\overline{\omega t} - \overline{\omega}\,\overline{t})^2]^{-1}$$

$$\times \begin{bmatrix} t_d^2 & \overline{\omega t} - \overline{\omega}\,\overline{t} \\ \overline{\omega t} - \overline{\omega}\,\overline{t} & \beta^2 \end{bmatrix}. \qquad (10.74)$$

It is often the case that the signal is such that the mean time and frequency are near the center of the pulse and band, which we have taken to be the origins of time and frequency. In any event, by redefining the carrier frequency and time origin if necessary, we can always arrange that $\overline{\omega}$ and $\overline{t}$ are zero, and we have

$$\mathrm{Cov}(\hat{\tau} - \tau, \hat{\omega} - \omega)$$

$$\geq \left[\left(\frac{2E_r}{N_0}\right)(\beta^2 t_d^2 - (\overline{\omega t})^2]^{-1} \begin{bmatrix} t_d^2 & \overline{\omega t} \\ \overline{\omega t} & \beta^2 \end{bmatrix} \right]. \qquad *(10.75)$$

From this last, the error variances of the estimates of τ and ω can be read off as the diagonal elements. In the case that $\overline{\omega t} = 0$, these reduce to the earlier results in the case that only one parameter is unknown. The off-diagonal elements of the covariance matrix express the degradation in the estimate of time if frequency is not known, and conversely.

From Eq. (10.75), for the special case that $\overline{\omega t} = 0$, we have again the minimum error variance bounds for any unbiased estimator of time of arrival and frequency:

$$\sigma_\tau^2 \geq 1/[(2E_r/N_0)\beta^2],$$
$$\sigma_\omega^2 \geq 1/[(2E_r/N_0)t_d^2].$$

As a result, we have

$$\sigma_\tau \sigma_\omega \geq 1/[(2E_r/N_0)\beta t_d]. \qquad *(10.76)$$

That is, we can trade off accuracy in the estimate of τ against accuracy in the estimate of ω, or vice versa, but only to an extent governed by the received signal-to-noise ratio and the bandwidth–time product of the transmitted signal. We would like to have high signal-to-noise ratio and large bandwidth–time product to achieve a small lower bound on the variance product.

An Uncertainty Relation

As Eq. (10.76) shows, large bandwidth–time products are desirable in a transmitted signal. Seldom are we interested in small bandwidth–time products. Nonetheless, there exists a classical lower bound on the product βt_d which we are in a position to derive easily. There is no theoretical upper bound, although technology imposes certain practical constraints.

Suppose that the time and frequency origins are placed such that $\overline{\omega} = 0$, $\overline{t} = 0$, with these being the mean frequency and time of the signal as in Eqs. (10.71), (10.72). (These are different from the frequency and time extents β, t_d, which are the quantities of main interest.) Then for the product of time extent and bandwidth, as defined by Eqs. (10.61), (10.67), we have

$$\beta^2 t_d^2 = \left(\frac{1}{2E_t}\right)^2 \int_{-\infty}^{\infty} \omega^2 |\tilde{S}_t(f)|^2 \, df \int_{-T/2}^{T/2} t^2 |\tilde{s}_t(t)|^2 \, dt. \qquad (10.77)$$

Using in turn Parseval's relation and integration by parts, we have

$$\int_{-\infty}^{\infty} \omega^2 |\tilde{S}_t(f)|^2 \, df = -\int_{-T/2}^{T/2} \tilde{s}_t''(t) \tilde{s}_t^*(t) \, dt$$

$$= \int_{-T/2}^{T/2} |\tilde{s}_t'(t)|^2 \, dt.$$

Then we have

$$\beta^2 t_d^2 = \left(\frac{1}{2E_t}\right)^2 \int_{-T/2}^{T/2} |\tilde{s}_t'(t)|^2 \, dt \int_{-T/2}^{T/2} t^2 |\tilde{s}_t(t)|^2 \, dt$$

$$\geq \left(\frac{1}{2E_t}\right)^2 \left| \int_{-T/2}^{T/2} t\tilde{s}_t'(t) \tilde{s}_t^*(t) \, dt \right|^2.$$

The inequality follows from the Schwartz inequality. Then finally we have the bound

$$\beta t_d \geq \frac{1}{2E_t} \left| \int_{-T/2}^{T/2} t\tilde{s}_t'(t) \tilde{s}_t^*(t) \, dt \right|. \qquad (10.78)$$

The transmitted signal energy E_t appears explicitly on the right in order to cancel the energy implied by any amplitude scale factor in the signal

$\tilde{s}_t(t)$. The bandwidth–time product must be independent of scale factor. The relation Eq. (10.78) is sometimes called the uncertainty relation for signals. It states that duration cannot be reduced arbitrarily without increasing bandwidth, and vice versa. It is a consequence of the definitions of duration and bandwidth given earlier. A similar bound exists for other definitions. They are all a consequence of the definition of bandwidth through the Fourier transform of the time signal.

The bound Eq. (10.78) is different in meaning from the bound Eq. (10.76). The former states that bandwidth must be traded off against duration. If we want to transmit narrow pulses, we must provide adequately large bandwidth. On the other hand, Eq. (10.76) states that we can estimate both arrival time and Doppler shift to arbitrarily good accuracy, provided only that we provide enough signal-to-noise ratio or bandwidth–time product.

Let us write the signal in Eq. (10.78) explicitly in terms of the amplitude and phase modulation as

$$\tilde{s}_t(t) = A a(t) \exp[j\phi(t)].$$

Then the bound Eq. (10.78) becomes

$$(\beta t_d)^2 \geq \left(\frac{1}{2E_t}\right)^2 \left| \int_{-T/2}^{T/2} t a(t)[a'(t) + j a(t)\phi'(t)] \, dt \right|^2$$

$$= (\overline{\omega t})^2 + \left(\frac{1}{2E_t}\right)^2 \left[\int_{-T/2}^{T/2} t a(t) a'(t) \, dt \right]^2$$

$$= (\overline{\omega t})^2 + \tfrac{1}{4}. \qquad \qquad *(10.79)$$

In this we used the definition Eq. (10.73) for the first integral and integration by parts for the second. In the special case that the mean of the product ωt is zero, Eq. (10.79) becomes the standard result that

$$\beta t_d \geq \tfrac{1}{2}. \qquad \qquad *(10.80)$$

10.10 ESTIMATION IN NONWHITE GAUSSIAN NOISE

In the case of nonwhite (but still Gaussian) noise, the maximum likelihood estimates of parameters of interest and the corresponding Cramér–Rao lower bounds are obtained in just the ways we have detailed, with the difference that, in effect, the computations are preceded by a noise-whitening filter. In this section we show how the filter is obtained and where it enters into the final estimates and bounds. We consider in detail only the case of estimating a single unknown parameter.

In the case of a low-pass signal (not a complex envelope), the appropriate log likelihood function is that in Eq. (9.47):

$$\ln p[r(t)\,|\,x] = \ln(C) - \frac{1}{2} \int_0^T \int_0^T [r(t) - s(t)]$$

$$\times R_n^{-1}(t, t')[r(t') - s(t')]\, dt\, dt'. \quad (10.81)$$

Here C is a normalization constant, $r(t)$ is the received signal, $s(t)$ is the received signal without noise, x is some parameter of $s(t)$, and $R_n(t, t')$ is the correlation function of the zero-mean additive Gaussian noise $n(t)$. The signal exists over some interval $0 \le t \le T$. The inverse kernel R_n^{-1} is by definition the solution of Eq. (9.36):

$$\int_0^T R_n^{-1}(t, t')R_n(t', \tau)\, dt' = \delta(t - \tau), \qquad 0 \le t, \tau \le T. \quad (10.82)$$

The maximum likelihood estimates are those parameter values which maximize the log likelihood function (10.81). We seek the derivative of Eq. (10.81) with respect to the parameter of interest. Taking into account that the inverse kernel is symmetric, as in Eq. (9.38),

$$R_n^{-1}(t', t) = R_n^{-1}(t, t'), \quad (10.83)$$

we have

$$\frac{\partial \ln p(r)}{\partial x} = \int_0^T \int_0^T \left[\frac{\partial s(t')}{\partial x} \right] R_n^{-1}(t, t')$$

$$\times [r(t) - s(t)]\, dt\, dt'. \quad (10.84)$$

Now define the filter $h(t)$, essentially the noise whitener, as in Eq. (9.45):

$$\int_0^T R_n(t, t')h(t')\, dt' = s(t), \qquad 0 \le t \le T. \quad (10.85)$$

The solution is that given in Eq. (9.46):

$$h(t) = \int_0^T R_n^{-1}(t, t')s(t')\, dt', \qquad 0 \le t \le T. \quad (10.86)$$

Note that $h(t)$ is a function of whatever parameters are of interest in the signal $s(t)$. Now differentiating Eq. (10.86) with respect to the parameter x, substituting the result in (10.84), and setting the result equal to zero yields the equation to be solved for the maximum likelihood parameter estimate:

$$\frac{\partial[\ln p(r)]}{\partial x} = \int_0^T \frac{\partial h(t)}{\partial x}[r(t) - s(t)]\, dt = 0. \quad (10.87)$$

In the case of multiple parameters to be estimated, there is one equation Eq. (10.87) for each parameter, involving $\partial h(t, \mathbf{x})/\partial x_i$, and the set must be solved jointly for the estimates $\hat{x}_i$.

The Cramér–Rao lower bound for the variance of any unbiased estimator of the parameter x follows easily. In the case of Eq. (10.81), we can use Eq. (10.87) to find the elements of the Fisher information matrix Eq. (10.36) as

$$
\begin{aligned}
\mathcal{F}_{ij} &= \mathscr{E}\left\{ \frac{\partial[\ln p(r)]}{\partial x_i} \frac{\partial[\ln p(r)]}{\partial x_j} \right\} \\
&= \int_0^T \int_0^T \frac{\partial h(t)}{\partial x_i} \frac{\partial h(t')}{\partial x_j} R_n(t, t') \, dt \, dt' \\
&= \int_0^T \frac{\partial h(t)}{\partial x_i} \frac{\partial s(t)}{\partial x_j} \, dt.
\end{aligned}
\tag{10.88}
$$

In this we replaced $r(t) - s(t)$ by the noise $n(t)$, inserted the noise covariance $R_n(t, t')$ in place of $\mathscr{E} n(t)n(t')$, and used Eq. (10.85) differentiated once.

If the filter is not explicitly needed, Eq. (10.86) can be used in Eq. (10.88) to write

$$
\mathcal{F}_{ij} = \int_0^T \int_0^T \frac{\partial s(t)}{\partial x_i} R_n^{-1}(t, t') \frac{\partial s(t')}{\partial x_j} \, dt \, dt'.
\tag{*10.89}
$$

In either case, the Cramér–Rao bound is then

$$
\mathrm{Cov}(\hat{\mathbf{x}} - \mathbf{x}) \geq \mathcal{F}^{-1},
$$

in the sense that the difference matrix is nonnegative definite.

As an example, consider the case of estimating the amplitude of a signal received in noise, with no other unknown parameters. That is, we estimate A in

$$
r(t) = As_0(t) + n(t),
$$

where $s_0(t)$ is entirely known. The hard work is solving Eq. (10.85). If the noise is stationary, this is

$$
\int_0^T R_n(t - t')h(t') \, dt' = As_0(t), \qquad 0 \leq t \leq T.
$$

Clearly $h(t) = Ag(t)$, where $g(t)$ is the solution of

$$
\int_0^T R_n(t - t')g(t') \, dt' = s_0(t), \qquad 0 \leq t \leq T.
$$

Suppose that we have found a solution $g(t)$. Then we have a solution $h(t) = Ag(t)$, and also $\partial h/\partial A = g(t)$. Then the estimate of A is found by solving Eq. (10.87), which is

$$\int_0^T g(t)[r(t) - As_0(t)] \, dt = 0.$$

The solution is

$$\hat{A} = \frac{\displaystyle\int_0^T g(t)r(t) \, dt}{\displaystyle\int_0^T g(t)s_0(t) \, dt}. \qquad (10.90)$$

In this, the special form of the correlating waveform $g(t)$ takes into account that the noise is not white, and the observation interval is finite.

The Cramér–Rao bound follows from Eq. (10.89), which for this example is

$$\mathscr{F}_{11} = \int_0^T \int_0^T R_n^{-1}(t - t')s_0(t)s_0(t') \, dt \, dt'. \qquad (10.91)$$

Signals with Random Phase

Let us now consider parameter estimation in nonwhite Gaussian noise for signals consisting of amplitude and phase modulation on a carrier. If the carrier phase is of interest, then the preceding procedures apply. However, in the usual case that carrier phase is a nuisance parameter, it can be integrated away by assuming it is uniformly distributed. Parameter estimation is conveniently carried out using the complex envelope.

The received signal can again be written

$$r(t) = Aa(t) \cos[\omega_c t + \phi(t) + \theta_c] + n(t), \qquad 0 \le t \le T.$$

The corresponding relation among complex envelopes is as in Eq. (10.44):

$$\begin{aligned} \tilde{r}(t) &= Aa(t) \exp[j\phi(t)] \exp(j\theta_c) + \tilde{n}(t) \\ &= A\tilde{s}_{r0}(t) \exp(j\theta_c) + \tilde{n}(t). \end{aligned} \qquad (10.92)$$

The log likelihood ratio now results by using Eqs. (9.111), (9.112) in Eq. (9.108), taken with $\tilde{A}(t) = A\tilde{s}_{r0}(t)$ from Eq. (9.91). That is,

$$\begin{aligned} \ln p(\tilde{r}) = \ln C &+ \ln[I_0(q)] \\ &- \frac{1}{2} \int_0^T \int_0^T [\tilde{r}^*(t)\tilde{R}^{-1}(t, t')\tilde{r}(t') \\ &\qquad + A^2 \tilde{s}_{r0}^*(t)\tilde{R}^{-1}(t, t')\tilde{s}_{r0}(t')] \, dt \, dt'. \end{aligned} \qquad (10.93)$$

In this, as in Eq. (9.105),

$$q = \left| \int_0^T \tilde{h}^*(t)\tilde{r}(t) \, dt \right|,$$

where $h(t)$ is a complex filter response defined by Eq. (9.106):

$$\int_0^T \tilde{R}(t, t')\tilde{h}(t')\, dt' = A\tilde{s}_{r0}(t), \qquad 0 \le t \le T.$$

It is convenient here to write

$$\tilde{h}(t) = A\tilde{g}(t),$$

where $g(t)$ is defined by

$$\int_0^T \tilde{R}(t, t')\tilde{g}(t')\, dt' = \tilde{s}_{r0}(t), \qquad 0 \le t \le T. \tag{10.94}$$

Then

$$q = A \left| \int_0^T \tilde{g}^*(t)\tilde{r}(t)\, dt \right|. \tag{10.95}$$

To avoid complexity, and again because it is the usual case for which maximum likelihood estimation succeeds, we will assume the signal-to-noise ratio is large enough that we can approximate the log Bessel function in Eq. (10.93) by its argument. We will also absorb the first integral expression in Eq. (10.93) into the constant, since it does not explicitly involve any signal parameters and we will be concerned only with derivatives of the likelihood function. Then the log likelihood function Eq. (10.93) becomes approximately

$$\ln p(\tilde{r}) = \ln(C) + q$$
$$- \frac{1}{2}A^2 \int_0^T \int_0^T \tilde{s}_{r0}^*(t)\tilde{R}^{-1}(t, t')\tilde{s}_{r0}(t')]\, dt\, dt'$$
$$= \ln(C) + q - \tfrac{1}{2}A^2 w(\mathbf{x}_0). \tag{10.96}$$

Here we have written $\mathbf{x}_0$ for whatever unknown parameters (such as Doppler shift) are present in the normalized received envelope $\tilde{s}_{r0}(t)$. That is, $\mathbf{x} = [A, \mathbf{x}_0]$. The function $w(\mathbf{x}_0)$ is defined by the last equality in Eq. (10.96).

The equations to be solved for the maximum likelihood estimates of the parameters result by setting the derivatives of Eq. (10.96) equal to zero. If (as usual) the amplitude A is unknown, using Eq. (10.95) the first of these is

$$\frac{\partial q}{\partial A} = \left| \int_0^T \tilde{g}^*(t)\tilde{r}(t)\, dt \right| = Aw(\mathbf{x}_0).$$

The integral is a function of the parameter vector $\mathbf{x}_0$ because the function $\tilde{s}_{r0}(t)$ in Eq. (10.94) involves those parameters, and hence so does the filter $\tilde{g}(t)$. In any event, this equation can be solved for A to yield the maximizing value $\hat{A}(\mathbf{x}_0)$ in terms of the other parameters (if any). Then substituting in Eq. (10.96) yields an expression to be maximized over $\mathbf{x}_0$:

$$v(\mathbf{x}_0) = q(\hat{A}) - \hat{A}^2 w(\mathbf{x}_0)/2 = \hat{A}^2 w(\mathbf{x}_0) - \hat{A}^2 w(\mathbf{x}_0)/2 = \hat{A}^2 w(\mathbf{x}_0)/2$$

$$= \left[\frac{1}{2w(\mathbf{x}_0)}\right] \left|\int_0^T \tilde{g}^*(t)\tilde{r}(t)\,dt\right|^2.$$

Because the filter $\tilde{g}(t)$ involves the parameters $\mathbf{x}_0$, these equations to be solved for the estimates of $\mathbf{x}_0$ can be formidable. However, sometimes those parameters do not enter into the function $w(\mathbf{x}_0)$ defined in Eq. (10.96). Then the quantity $v(\mathbf{x}_0)$ can be maximized by setting up a bank of filters $\tilde{g}(t \mid \mathbf{x}_0)$ and observing which filter maximizes the quantity q in Eq. (10.95). The usual such situation is that in which the unknown parameters are time delay or Doppler shift. We then recover the filter bank of Fig. 10.5, but in which the filters are not matched directly to the signal envelope but are the filters $\tilde{g}(t)$ specified by Eq. (10.94).

The filters $\tilde{g}(t)$ can be specified formally in terms of the inverse noise covariance kernel by using Eq. (9.36) to write Eq. (10.94) as

$$\int_0^T \tilde{R}^{-1}(t, t')\tilde{s}_{r0}(t')\,dt'$$

$$= \int_0^T \int_0^T \tilde{R}^{-1}(t, t')\tilde{R}(t', t'')\tilde{g}(t'')\,dt'\,dt''$$

$$= \int_0^T \delta(t - t'')\tilde{g}(t'')\,dt'' = \tilde{g}(t), \qquad 0 \le t \le T. \qquad (10.97)$$

We will follow Helstrom (1968) in discussing the Cramér–Rao bound for this case of estimation in nonwhite Gaussian noise. Suppose that, rather than integrating out the carrier phase θ_c, we try to estimate it, along with whatever other parameters are unknown in the received signal. That is, we now consider the density of the received envelope Eq. (10.92) directly. That is given in Eq. (9.107). We again use Eqs. (9.111), (9.112) and absorb the expression Eq. (9.111) into the constant, since we will be concerned only with maximization of the log likelihood function. We obtain

$$\ln p(\tilde{r}) = \ln(C) + q\cos(\eta - \theta_c)$$

$$- \frac{A^2}{2} \int_0^T \int_0^T \tilde{s}_{r0}^*(t)\tilde{R}^{-1}(t, t')\tilde{s}_{r0}(t')]\,dt\,dt'. \qquad (10.98)$$

Here θ_c is the carrier phase and the angle η is defined by Eq. (9.105):

$$q\exp(j\eta) = A\int_0^T \tilde{g}^*(t)\tilde{r}(t)\,dt.$$

We now seek the maximum likelihood estimate of carrier phase θ_c. The appropriate equation is

$$0 = \partial[\ln p(\tilde{r})]/\partial\theta_c = q \sin(\eta - \theta_c).$$

The solution of this last is just $\theta_c = \eta$. Using that in Eq. (10.98) yields just Eq. (10.96). That is, the estimates of amplitude and any other parameters are the same, whether phase is estimated or not, in this case of high signal-to-noise ratio. Therefore we can find the Cramér–Rao bounds for the parameters other than carrier phase by assuming that phase is estimated, even though we do not actually carry out the estimation.

Now we want the density of the received signal envelope function $\tilde{r}(t)$ without averaging out the carrier phase as a nuisance parameter. That is given by Eq. (9.104). In that we can use the definitions of the expansion coefficients of the received envelope and the noiseless signal envelope to write

$$\ln p(\tilde{r}) = \ln(C) - \frac{1}{2}\int_0^T \int_0^T [\tilde{r}(t) - A\tilde{s}_{r0}(t)\exp(j\theta_c)]^*$$

$$\times \tilde{R}_n^{-1}(t, t')[\tilde{r}(t') - A\tilde{s}_{r0}(t')\exp(j\theta_c)] \, dt \, dt'.$$

In this we used Mercer's theorem Eq. (9.37) for the inverse kernel. The unnormalized noiseless signal envelope is

$$\tilde{s}(t) = A\tilde{s}_{r0}(t)\exp(j\theta).$$

This contains all the parameters **x** of the received envelope.

The Cramér–Rao bound involves the quantities

$$\partial[\ln p(\tilde{r})]/\partial x_i = -\frac{1}{2}\int_0^T \int_0^T [\tilde{r}(t) - \tilde{s}(t)]^*$$

$$\times \tilde{R}_n^{-1}(t, t')\left[\frac{-\partial\tilde{s}(t')}{\partial x_i}\right] dt \, dt'$$

$$-\frac{1}{2}\int_0^T \int_0^T \left[\frac{-\partial\tilde{s}(t)}{\partial x_i}\right]^* \tilde{R}_n^{-1}(t, t')$$

$$\times [\tilde{r}(t') - \tilde{s}(t')] \, dt \, dt'$$

$$= \text{Re}\int_0^T \int_0^T [\tilde{r}(t) - \tilde{s}(t)]^* \tilde{R}_n^{-1}(t, t')\frac{\partial\tilde{s}(t')}{\partial x_i} \, dt \, dt'$$

$$= \text{Re}\int_0^T [\tilde{r}(t) - \tilde{s}(t)]^* \frac{\partial\tilde{G}(t)}{\partial x_i} \, dt. \qquad (10.99)$$

In this last we used the property of the symmetric noise inverse kernel that $R^{-1}(t', t) = R^{-1*}(t, t')$. We also used Eq. (10.97) and defined

$$\tilde{G}(t) = A\tilde{g}(t)\exp(j\theta_c).$$

Now write the integral in Eq. (10.99) as $c_i = a_i + jb_i$, so that

$$\partial[\ln p(\tilde{r})]/\partial x_i = \text{Re}(c_i) = a_i.$$

The elements of the Fisher information matrix are then

$$\mathscr{F}_{ij} = \mathscr{E}(a_i a_j).$$

This can be written in another form as follows. The quantity $\tilde{r}(t) - \tilde{s}(t)$ is just the noise envelope $\tilde{n}(t)$, with covariance function

$$R_{\tilde{n}}(t, t') = 2\tilde{R}_n(t, t'),$$

using Eq. (9.95). That is, the complex covariance of the noise envelope $\tilde{n}(t)$ is just twice the envelope of the real covariance function of the noise $n(t)$. Then we can write the cross-covariance of the integral in Eq. (10.99), taken for two different parameters, as

$$\mathscr{E}(c_i c_j^*) = 2\int_0^T \int_0^T \tilde{R}_n(t, t') \left[\frac{\partial \tilde{G}(t)}{\partial x_j}\right]^*$$

$$\times [\partial \tilde{G}(t')/\partial x_i]\, dt\, dt'$$

$$= 2\int_0^T \left[\frac{\partial \tilde{G}(t)}{\partial x_j}\right]^* \frac{\partial \tilde{s}(t)}{\partial x_i}\, dt$$

$$= \mathscr{E}(a_i a_j + b_i b_j) + j\mathscr{E}(a_j b_i - a_i b_j). \qquad (10.100)$$

using Eq. (10.94).

Now if we write

$$\tilde{n}(t) = n_x(t) + jn_y(t),$$
$$\partial G(t)/\partial x_i = u_i(t) + jv_i(t),$$

and similarly for $\partial G(t)/\partial x_j$, we have

$$a_i = \int_0^T [n_x(t)u_i(t) + n_y(t)v_i(t)]\, dt,$$

$$b_i = \int_0^T [n_x(t)v_i(t) - n_y(t)u_i(t)]\, dt.$$

Then using Eqs. (9.92), (9.93) it is easy to see that

$$\mathscr{E}(a_i a_j) = \mathscr{E}(b_i b_j).$$

Then from Eq. (10.100) the elements of the Fisher matrix are

$$\mathcal{F}_{ij} = \mathcal{E}(a_i a_j) = \frac{1}{2}\mathcal{E}(a_i a_j + b_i b_j)$$

$$= \text{Re} \int_0^T \left[\frac{\partial \tilde{s}(t)}{\partial x_i}\right]\left[\frac{\partial \tilde{G}(t)}{\partial x_j}\right]^* dt. \qquad (10.101)$$

10.11 GENERALIZED LIKELIHOOD RATIO DETECTION

In this section we return to the subject of signal detection, rather than estimation of signal parameters. However, we now assume that the signal we want to detect contains unknown parameters:

$$r(t) = s(t \mid \mathbf{x}) + n(t).$$

Using the procedure of Chapter 7 we could postulate some probability density for the unknown parameter vector and, using that density, compute the average likelihood ratio:

$$\lambda(r) = [\mathcal{E}_x p_1(r \mid \mathbf{x})]/p_0(r).$$

In Chapter 7 we considered carrier phase with uniform density, received signal amplitude with Rayleigh distribution, Doppler frequency shift with any specified density, and arrival time with uniform density. The uniform density was selected because it is noncommittal about the parameter in question and is the most pessimistic assumption.

In this section we consider the common procedure of first using the data at hand to make maximum likelihood estimates of the unknown parameters. Those estimates are then used in the model of the signal we are trying to detect, and the detector is built as if those estimated values were in fact the correct values. That is, we will do detection using the *generalized likelihood ratio*:

$$p_1(r \mid \hat{\mathbf{x}}_{ml})/p_0(r) > \lambda_0,$$

where

$$\hat{\mathbf{x}}_{ml} \Leftarrow \max_x p_1(r \mid \mathbf{x}).$$

For simplicity, we assume there are no unknown parameters involved in the null hypothesis, but the generalization is easy. In our case, the maximum likelihood estimates of the parameters result by maximizing the conditional likelihood ratio

$$\lambda(r \mid \mathbf{x}) = p_1(r \mid \mathbf{x})/p_0(r). \qquad (10.102)$$

We will consider in turn the standard cases of Chapter 7 appropriate to the detection of signals on a carrier. The procedure makes no claims in general to optimality of the solution in the joint problem of estimation and detection, but the method is common and often successful. It is called *generalized likelihood ratio detection*.

We will not treat any form of *adaptive detector*. That subject is vast and has its own literature. We note only that another way to proceed is to use past data for estimation of the signal parameters, which are then applied to detection at the current time. The resulting detector changes its characteristics as more data accumulate. Recursive implementations are usual.

To proceed, assume the received signal is as usual

$$r(t) = Aa(t) \cos[\omega_c t + \phi(t) + \theta] + n(t), \qquad 0 \le t \le T. \tag{10.103}$$

The phase modulation $\phi(t)$ we will take as that of the transmitted signal. For simplicity, we will assume white Gaussian noise with two-sided power spectral density $N_0/2$. From Eq. (6.36), it is easy to see that the likelihood ratio Eq. (10.102) in this case corresponds to

$$\ln[\lambda(r)] = \frac{2A}{N_0} \int_0^T r(t) a(t) \cos[\omega_c t + \phi(t) + \theta] \, dt$$

$$- \frac{A^2}{2N_0} \int_0^T a^2(t) \, dt. \tag{10.104}$$

Then the receiver based on thresholding this conditional likelihood ratio is

$$\int_0^T r(t) a(t) \cos[\omega_c t + \phi(t) + \theta] \, dt > \gamma. \tag{10.105}$$

We consider in turn the signal parameters which might not be known in this receiver.

Unknown Phase

The general equation for the maximum likelihood estimates of parameters of the signal Eq. (10.103) is that given by Eq. (10.39). In the case of phase as the only unknown, the estimate is given by

$$\int_0^T \{r(t) - Aa(t) \cos[\omega_c t + \phi(t) + \hat{\theta}]\}$$

$$\times a(t) \sin[\omega_c t + \theta(t) + \hat{\theta}] \, dt$$

$$= A \int_0^T r(t) a(t) \sin[\omega_c t + \phi(t) + \hat{\theta}] \, dt = 0. \tag{10.106}$$

The second equality follows because the double frequency term integrates to approximately zero. The solution is an extension of Eq. (10.50):

$$\tan(\hat{\theta}) = \frac{-\int_0^T r(t)a(t) \sin[\omega_c t + \phi(t)] \, dt}{\int_0^T r(t)a(t) \cos[\omega_c t + \phi(t)] \, dt}. \tag{10.107}$$

The corresponding estimate $\hat{\theta}$ would then be inserted into Eq. (10.105) to implement the receiver. The procedure is shown in Fig. 10.6a, with a delay inserted to indicate that the receiver cannot begin to operate until the parameter estimate has been computed.

The receiver Eq. (10.105) can be expanded as

$$\cos(\hat{\theta}) \int_0^T r(t)a(t) \cos[\omega_c t + \phi(t)] \, dt$$

$$- \sin(\hat{\theta}) \int_0^T r(t)a(t) \sin[\omega_c t + \phi(t) \, dt > \gamma. \tag{10.108}$$

Let us define two new variables q, α by

$$q \cos(\alpha) = \int_0^T r(t)a(t) \cos[\omega_c t + \phi(t)] \, dt,$$

$$q \sin(\alpha) = \int_0^T r(t)a(t) \sin[\omega_c t + \phi(t)] \, dt.$$

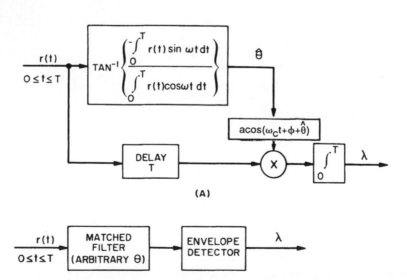

(A)

(B)

FIGURE 10.6 (a) Maximum likelihood receiver for unknown phase and (b) equivalent receiver (noncoherent matched filter).

From Eq. (10.107) we see that $\tan(\alpha) = -\tan(\hat{\theta})$. Then Eq. (10.108) becomes

$$q \cos^2(\hat{\theta}) + q \sin^2(\hat{\theta}) = q > \gamma.$$

That is, the receiver can be implemented as the incoherent matched filter of Eq. (7.37) and diagrammed in Fig. 7.1 as well as in Fig. 10.6b. That is, estimation of phase and use of the generalized maximum likelihood ratio detector are equivalent to assuming no prior knowledge of phase and integrating it away with the uniform density.

Unknown Phase and Frequency

Since we have placed all unknown parameters in the density $p_1(r)$, the maximum likelihood parameter estimates can be computed by maximizing the conditional likelihood ratio Eq. (10.104). In the case of unknown phase and frequency, that is accomplished by

$$(\hat{\omega}, \hat{\theta}) \Leftarrow \max \int_0^T r(t)a(t) \cos[\omega t + \phi(t) + \theta]. \qquad (10.109)$$

Here the frequency to be estimated is the sum of the carrier and whatever Doppler shift may be present. The phase modulation is taken to be that of the transmited signal.

For any fixed ω we have already determined that the receiver using the maximizing value of θ is just the value q computed as in the previous section, which does not in fact depend on $\hat{\theta}$. To complete the receiver it remains to determine

$$\hat{\omega} \Leftarrow \max_\omega q(\omega).$$

That could be done analytically, but it is more usual to implement the Doppler filter bank of Fig. 10.5 and take as the estimate that value corresponding to the Doppler filter with largest output at the sampling instant. Since that value corresponds to the maximized likelihood ratio, it can serve as the test statistic as well. That is, we declare signal present, with the corresponding Doppler, if one of the filter outputs crosses a threshold set by the false alarm probability. Although it is seldom of interest, the maximum likelihood phase estimate can also be computed from Eq. (10.107) once the frequency estimate has been determined.

Unknown Phase, Frequency, and Amplitude

The maximum likelhood estimate of amplitude must satisfy

$$\partial \ln(\lambda)/\partial A = 0.$$

From Eq. (10.104) this leads to

$$\hat{A} = \frac{2 \int_0^T r(t) a(t) \cos[\hat{\omega} t + \phi(t) + \hat{\theta}] \, dt}{\int_0^T a^2(t) \, dt} = \frac{2q(\hat{\omega})}{\int_0^T a^2(t) \, dt},$$

where we have used the result of the previous section to write the integral in the numerator as $q \cos(\hat{\theta} + \alpha) = q$. That is, the estimate of amplitude is obtained by scaling the largest output of the filter bank of Fig. 10.5. The likelihood ratio using the maximum likelihood estimates is easily found by substitution into Eq. (10.104):

$$\lambda = (\hat{A}^2/2N_0) \int_0^T a^2(t) \, dt. \tag{10.110}$$

This indicates that the receiver can be implemented by thresholding the estimate $\hat{A}$.

Unknown Phase, Frequency, Amplitude, and Time of Arrival

Finally, we assume that the time of arrival of the beginning of the signal is known only to within $0 \le \tau \le \tau_m$. (The preceding sections assumed $\tau = 0$ was known.) Since now the signal exists over $\tau \le t \le T + \tau$, the likelihood ratio Eq. (10.104) involves integration over that time range for arbitrary (unknown) τ. As the signal is in fact integrated over the region of length T where it exists, the maximum likelihood estimates of phase, frequency, and amplitude are just those already calculated, with the integration taken over this new range of t. The final step is computation of the value of τ which maximizes the likelihood ratio Eq. (10.110). If the amplitude function is smooth and the maximum delay uncertainty τ_m is small, the denominator is approximately constant and the maximum likelihood delay estimate is that value which maximizes the outputs q of the filter bank of Fig. 10.5. That is, the joint estimate of arrival time and Doppler is provided by the largest output of the filter bank, taken jointly over running time and filter number. The detection statistic can be taken as that output, so that finally we declare a target with the time and Doppler at which some filter output crosses a threshold.

10.12 LINEAR MINIMUM ERROR VARIANCE ESTIMATION

If an estimator has an error variance that attains the Cramér–Rao bound, it is called *efficient*. Certainly that is a good property, but it is not always attainable. As we mentioned earlier, if an unbiased estimator exists that attains the Cramér–Rao bound, then it must be a maximum likelihood estimator and it can be found by the methods of the previous sections. Going the other way, a maximum likelihood estimate is *asymptotically* efficient if the errors become arbitrarily small in the limit. The maximum likelihood errors will be small if the estimate is based on a sufficiently large number of independent data measurements or (usually) if we are estimating signal parameters in the case of large signal-to-noise ratio. In the cases we have already treated, we assumed large signal-to-noise ratio, and the estimators we have been discussing in fact attain the Cramér–Rao bounds we have been calculating in that case. On the other hand, for a finite number of measurements or small signal-to-noise ratio that may not be the case. Further, a maximum likelihood estimator may be difficult to compute or difficult to apply to the data. In such cases, another method may be adequate and appropriate.

The procedure we will discuss now is that of *linear* (unbiased) *minimum* (error) *variance estimation.* We will proceed in general for the case of a vector random variable $\mathbf{x}$ to be estimated in terms of a data vector of random variables $\mathbf{y}$. Either or both of these might or might not be interpreted as time samples of a random process.

We first insist that the estimator of $\mathbf{x}$ be a linear function of $\mathbf{y}$:

$$\hat{\mathbf{x}} = L\mathbf{y} + \mathbf{b},$$

where L and $\mathbf{b}$ are a (possibly nonsquare) matrix and a vector to be found. We require the estimator be unbiased:

$$\mathcal{E}(\hat{\mathbf{x}}) = \mathbf{m}_x = L\mathcal{E}(\mathbf{y}) + \mathbf{b} = L\mathbf{m}_y + \mathbf{b}.$$

This identifies

$$\mathbf{b} = \mathbf{m}_x - L\mathbf{m}_y.$$

Therefore the unbiased estimator can be written

$$\hat{\mathbf{x}} - \mathbf{m}_x = L(\mathbf{y} - \mathbf{m}_y).$$

The error made by the unbiased estimator is

$$\begin{aligned}\varepsilon &= \hat{\mathbf{x}} - \mathbf{x} = (\hat{\mathbf{x}} - \mathbf{m}_x) - (\mathbf{x} - \mathbf{m}_x) \\ &= L(\mathbf{y} - \mathbf{m}_y) - (\mathbf{x} - \mathbf{m}_x).\end{aligned}$$

The trace of the error covariance matrix is the quantity to be minimized by choice of the matrix L:

$$C_\varepsilon = \mathrm{Cov}(\varepsilon) = \mathcal{E}(\varepsilon\varepsilon^{\mathrm{T}}). \tag{10.111}$$

It is an important fact (the *orthogonality principle*) that the estimator is optimum in this sense if and only if it produces an error which has zero covariance with the data upon which the estimate is based. That is, we require

$$\mathscr{E}[(\mathbf{y} - \mathbf{m}_y)\boldsymbol{\varepsilon}^T] = 0. \qquad\qquad *(10.112)$$

That is

$$\mathscr{E}(\mathbf{y} - \mathbf{m}_y)(\mathbf{y} - \mathbf{m}_y)^T L^T$$
$$-\mathscr{E}(\mathbf{y} - \mathbf{m}_y)(\mathbf{x} - \mathbf{m}_x)^T = C_y L^T - C_{yx} = 0,$$

where C_y, C_{yx} are the indicated covariance matrices. This at once gives the solution for the estimator matrix:

$$L = C_{yx}^T C_y^{-1} = C_{xy} C_y^{-1}. \qquad\qquad (10.113)$$

It is assumed that there are no deterministic relations among the elements of the data vector $\mathbf{y}$, in which case the indicated matrix inverse exists. Then the final form of the estimator is

$$\hat{\mathbf{x}} = C_{xy} C_y^{-1}(\mathbf{y} - \mathbf{m}_y) + \mathbf{m}_x. \qquad\qquad *(10.114)$$

This is the most general form of the *Wiener filter*.

The orthogonality principle is of enough importance and generality to call for some further discussion. We will derive it in Sect. 11.3. It is certainly reasonable, because here we are dealing only with second-order statistics, the means and variances of the quantities involved. In that case, we might as well assume all the quantities involved are Gaussian, because we will not be able to tell the difference if they are not. Then the orthogonality principle simply says that the components of the error vector should be independent of the data. If they were not, the data could presumably tell us something about the error which would allow us to reduce it on the average, so that the estimator would not be optimum in the sense of minimizing the error variance.

We should also take note that we cannot literally minimize a matrix. That is, the problem of minimizing the error covariance matrix C_ε is not well defined, because only the real numbers are ordered, and we can only minimize real scalar functions of the variables in question. What we really do is minimize the sum of the variance of the components of the error vector:

$$\alpha = \sum_i \mathscr{E}(\varepsilon_i^2) = \operatorname{tr} \mathscr{E}(\boldsymbol{\varepsilon}\boldsymbol{\varepsilon}^T) = \operatorname{tr} C_\varepsilon,$$

where tr means the matrix trace.

A useful geometric picture can be set forth. The quantity α can be taken as the length of the vector $\boldsymbol{\varepsilon}$. More generally, we can define the dot product between two vectors of random variables $\mathbf{u}$, $\mathbf{v}$ as

$$(\mathbf{u}, \mathbf{v}) = \mathscr{E}(\mathbf{u} - \mathbf{u}_m)^T(\mathbf{v} - \mathbf{v}_m)$$
$$= \operatorname{tr}(\mathbf{u} - \mathbf{u}_m)(\mathbf{v} - \mathbf{v}_m)^T = \operatorname{tr} C_{uv}, \qquad\qquad (10.115)$$

where C_{uv} is the indicated covariance matrix.

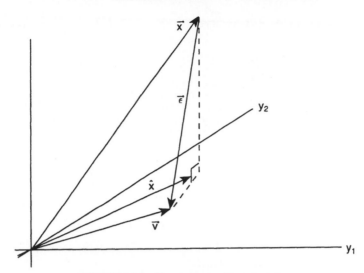

FIGURE 10.7 The orthogonality principle.

Now consider Fig. 10.7. There we show two zero-mean data random variables y_1, y_2, forming a data plane, and a zero mean vector **x** to be approximated by some vector $\hat{\mathbf{x}}$ which is a linear combination of y_1, y_2 and hence lies in the data plane. If we choose a general vector **v** in the data plane, the error will be the vector $\boldsymbol{\varepsilon} = \mathbf{v} - \mathbf{x}$ which is shown. Its length, in the sense Eq. (10.115), will be minimum if it is perpendicular to the data (y_1, y_2) plane, so that it is perpendicular to the estimator vector **v**. That is, the optimum vector $\mathbf{v} = \hat{\mathbf{x}}$ is defined as that vector **v** which is uncorrelated with the error it produces. That is just what Eq. (10.112) states, in the sense Eq. (10.115). The analogy is far reaching.

The (minimum) variance attained by the estimator Eq. (10.114) is of interest. We have

$$
\begin{aligned}
\hat{C}_\varepsilon &= \mathscr{E}(\boldsymbol{\varepsilon}\boldsymbol{\varepsilon}^T) = \mathscr{E}[(\hat{\mathbf{x}} - \mathbf{x})\boldsymbol{\varepsilon}^T] \\
&= \mathscr{E}\{[L(\mathbf{y} - \mathbf{m}_y) - (\mathbf{x} - \mathbf{m}_x)]\boldsymbol{\varepsilon}^T\} = -\mathscr{E}[(\mathbf{x} - \mathbf{m}_x)\boldsymbol{\varepsilon}^T] \\
&= -\mathscr{E}\{(\mathbf{x} - \mathbf{m}_x)[L(\mathbf{y} - \mathbf{m}_y) - (\mathbf{x} - \mathbf{m}_x)]^T\} \\
&= C_x - C_{xy}L^T = C_x - C_{xy}C_y^{-1}C_{xy}^T.
\end{aligned}
\qquad *(10.116)
$$

In this we used the orthogonality principle Eq. (10.112), which must hold true for the attained minimum error ε. The attained minimum sum of variances of the error vector components is of course the trace of the matrix Eq. (10.116).

EXAMPLE 10.12 Suppose the data are in fact a linear operation on the unknown, in zero-mean noise which is independent of the quantity being estimated:

$$\mathbf{y} = H\mathbf{x} + \mathbf{n}.$$

We assume mean $\mathbf{m}_x$ and covariances C_x, C_n, all of which are specified. The data mean is $\mathbf{m}_y = H\mathbf{m}_x$. The required covariances are

$$C_{xy} = \mathscr{E}(\mathbf{x} - \mathbf{m}_x)(\mathbf{y} - \mathbf{m}_y)^T$$
$$= \mathscr{E}(\mathbf{x} - \mathbf{m}_x)[H(\mathbf{x} - \mathbf{m}_x) + \mathbf{n}]^T = C_x H^T,$$
$$C_y = \mathscr{E}(\mathbf{y} - \mathbf{m}_y)(\mathbf{y} - \mathbf{m}_y)^T$$
$$= \mathscr{E}\{[H(\mathbf{x} - \mathbf{m}_x) + \mathbf{n}][H(\mathbf{x} - \mathbf{m}_x) + \mathbf{n}]^T\} = HC_x H^T + C_n.$$

The estimator Eq. (10.114) is

$$\hat{\mathbf{x}} = C_x H^T (HC_x H^T + C_n)^{-1}(\mathbf{y} - \mathbf{m}_y) + \mathbf{m}_x. \tag{10.117}$$

That the estimator is unbiased can be checked as

$$\mathscr{E}\hat{\mathbf{x}} = C_x H^T (HC_x H^T + C_n)^{-1}(0) + \mathbf{m}_x = \mathbf{m}_x.$$

It is not, however, conditionally unbiased. The attained minimum error variance is the trace of the matrix

$$C_\varepsilon = C_x - C_x H^T (HC_x H^T + C_n)^{-1}HC_x. \tag{10.118}$$

10.13   THE DISCRETE KALMAN FILTER

One of the most successful of all signal processing algorithms is the *Kalman* (or Kalman–Bucy) *filter*. It can be viewed as a sequential formulation of the minimum variance estimation procedures of the preceding section, and that is the view we will take here. Alternatively, it produces Bayesian estimates in a Gaussian framework, or simply weighted least-squares estimates. It is a generalization which is particularly well adapted to processing of data for which a dynamic system model exists. Since often such a model is at hand from the physics of an ongoing process, such as in prediction of satellite orbits, the filter is much used in applications.

The filter exists in versions suitable for both discrete and continuous time. For simplicity of analysis, and because it is the formulation usually needed for application, we will consider only discrete time. We will also assume a linear, but possibly time-varying, dynamic model for the data. Extensions have been developed which deal successfully with the more usual case of a nonlinear data model, and we will briefly mention them. Our development will rest on the *orthogonality principle*, which we discussed in the preceding section and which we will prove in the next chapter.

Suppose then that a discrete time linear dynamic system has state vector $\mathbf{x}(t)$ which is sampled in time to yield random vectors $\mathbf{x}_i$. The system evolves in time according to the model

$$\mathbf{x}_i = \Phi_{i,i-1}\mathbf{x}_{i-1} + G_{i-1}\mathbf{w}_{i-1}, \tag{10.119}$$

where $\Phi_{i,i-1}$ and G_{i-1} are specified matrices and $\mathbf{w}_{i-1}$ is a vector of zero-mean random variables with specified covariance $C_{w,i-1}$. The random variables $\mathbf{w}_i$, $\mathbf{w}_{j\neq i}$ are independent. Any state vector $\mathbf{x}_i$ gives rise to noisy measurements through the measurement equation

$$\mathbf{y}_i = H_i\mathbf{x}_i + \mathbf{n}_i, \tag{10.120}$$

where H_i is a known matrix and $\mathbf{n}_i$ is a vector of zero-mean random variables. The random vectors $\mathbf{n}_i$, $\mathbf{n}_{j\neq i}$ are independent, as are $\mathbf{n}_i$, $\mathbf{w}_j$.

Suppose at time $i - 1$ we have available an unbiased estimate $\hat{\mathbf{x}}_{i-1}$ of the state $\mathbf{x}_{i-1}$ with known error covariance $C_{\varepsilon,i-1}$. We seek an unbiased linear minimum variance estimator of the state at time i. The orthogonality principle requires that the sought estimate $\hat{\mathbf{x}}_i$ be the projection of the (unknown) true state vector $\mathbf{x}_i$ onto the i-dimensional space formed by the past and current measurement random variables $\mathbf{y}_1, \ldots, \mathbf{y}_i$. Let these be symbolized by the set $Y_{i-1} = \{\mathbf{y}_{j,i-1}\}$ of past measurements and the current measurement $\mathbf{y}_i$.

The computation is made easier by introducing the important vector

$$\mathbf{i}_{i,i-1} = \mathbf{y}_i - H_i\hat{\mathbf{x}}_{i,i-1}. \qquad *(10.121)$$

This is called the *innovations vector* associated with the new measurement $\mathbf{y}_i$. The vector $H_i\hat{\mathbf{x}}_{i,i-1}$ is the minimum variance unbiased estimator of the current data sample $\mathbf{y}_i$ in terms of the set Y_{i-1} of past data samples. The innovations $\mathbf{i}_{i,i-1}$ itself is the error of that estimator. To see this, use Eq. (10.119) to write Eq. (10.121) as

$$\mathbf{i}_{i,i-1} = H_i(\mathbf{x}_i - \hat{\mathbf{x}}_{i,i-1}) + \mathbf{n}_i = H_i\varepsilon_{i,i-1} + \mathbf{n}_i, \tag{10.122}$$

where $\varepsilon_{i,i-1}$ is the error in the estimate of $\mathbf{x}_i$ by $\hat{\mathbf{x}}_{i,i-1}$. This is zero mean, because the estimator is unbiased. Since this estimate is also minimum variance, the covariance between the corresponding error and the data which underlie the estimate (the set Y_{i-1}) must be zero. Also, by assumption the noise sample $\mathbf{n}_i$ has zero covariance with all previous measurements and states. Hence the covariance of the innovations $\mathbf{i}_{i,i-1}$ with the set Y_{i-1} of previous measurements is zero. Therefore, the innovations must be the error vector in the minimum variance approximation, as claimed. The argument is sketched in Fig. 10.8. The procedure is exactly that of Gram–Schmidt orthogonalization of the data vectors $\mathbf{y}_1, \ldots, \mathbf{y}_i$.

Now it is easy to form the new estimate $\hat{\mathbf{x}}_i$ of $\mathbf{x}_i$. The procedure is sketched in Fig. 10.9. The estimate $\hat{\mathbf{x}}_i$ is the projection of the true state $\mathbf{x}_i$ onto the data set Y_i, which consists of the space of the old data set Y_{i-1} and the new measurement $\mathbf{y}_i$. But projection onto that space can be done by projecting onto the space of the old data set Y_{i-1} and the innovations vector Eq. (10.121). That is easier, because the innovations is orthogonal to the space Y_{i-1}, and the sought projection is just

$$\hat{\mathbf{x}}_i = \hat{\mathbf{x}}_{i,i-1} + K_i\mathbf{i}_{i,i-1}. \tag{10.123}$$

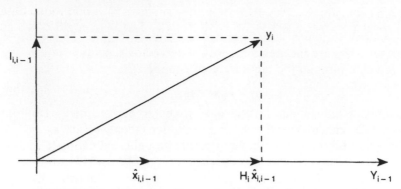

FIGURE 10.8   Formation of innovations sequence from measured data.

Using the measurement model Eq. (10.120) and the innovations definition Eq. (10.122) in this, and the assumption that the prior estimate $\hat{\mathbf{x}}_{i,i-1}$ is unbiased, this results in

$$\mathscr{E}(\mathbf{x}_i - \hat{\mathbf{x}}_i) = \mathscr{E}(\mathbf{x}_i - \hat{\mathbf{x}}_{i,i-1}) - K_i \mathscr{E}[H_i(\mathbf{x}_i - \hat{\mathbf{x}}_{i,i-1}) + \mathbf{n}_i] = 0.$$

The (matrix) multiplier K_i is the coefficient Eq. (10.113) that corresponds to approximating $\mathbf{x}_i$ by $\mathbf{i}_{i,i-1}$. (The symbol K is used here rather than L to symbolize the Kalman filter gain.) That is,

$$K_i = \text{Cov}(\mathbf{x}_i, \mathbf{i}_{i,i-1})[\text{Cov}(\mathbf{i}_{i,i-1}, \mathbf{i}_{i,i-1})]^{-1}.$$

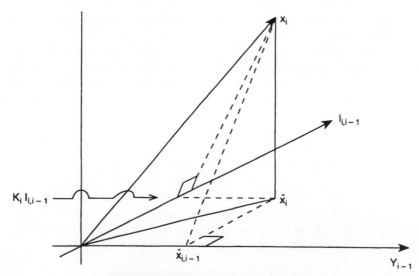

FIGURE 10.9   Formation of Kalman gain.

Since the estimator $\hat{\mathbf{x}}_{i,i-1}$ involves only the past data Y_{i-1}, and those data have zero covariance with the innovations $\mathbf{i}_{i,i-1}$, we can also write this as

$$K_i = \text{Cov}(\mathbf{x}_i - \hat{\mathbf{x}}_{i,i-1}, \mathbf{i}_{i,i-1})[\text{Cov}(\mathbf{i}_{i,i-1}, \mathbf{i}_{i,i-1})]^{-1}.$$

All quantities have zero mean. The first covariance is

$$\mathcal{E}\{(\mathbf{x}_i - \hat{\mathbf{x}}_{i,i-1})[H_i(\mathbf{x}_i - \hat{\mathbf{x}}_{i,i-1}) + \mathbf{n}_i]^T\} = P_{i,i-1}H_i^T,$$

where $P_{i,i-1}$ is the error covariance of the estimate of $\mathbf{x}_i$ based on data Y_{i-1}. Furthermore,

$$\begin{aligned}
\text{Cov}(\mathbf{i}_{i,i-1}, \mathbf{i}_{i,i-1}) \\
= \mathcal{E}\{[H_i(\mathbf{x}_i - \hat{\mathbf{x}}_{i,i-1}) + \mathbf{n}_i][H_i(\mathbf{x}_i - \hat{\mathbf{x}}_{i,i-1}) + \mathbf{n}_i]^T \\
= H_i P_{i,i-1} H_i^T + N_i.
\end{aligned} \tag{10.124}$$

Then

$$K_i = P_{i,i-1}H_i^T[H_i P_{i,i-1} H_i^T + N_i]^{-1}. \qquad *(10.125)$$

Using the gain Eq. (10.125), the estimate Eq. (10.123) is unbiased. This follows from

$$\mathcal{E}\hat{\mathbf{x}}_i = \mathcal{E}\hat{\mathbf{x}}_{i,i-1} + K_i \mathcal{E}\mathbf{i}_{i,i-1} = \mathbf{x}_i.$$

An alternative form of the gain is

$$K_i = (P_{i,i-1}^{-1} + H_i^T N_i^{-1} H_i)^{-1} H_i^T N_i^{-1}. \qquad *(10.126)$$

The error covariance for the estimate $\hat{\mathbf{x}}$ can be read from Fig. 10.9. From the right triangle with vertices $[\mathbf{x}_i, \hat{\mathbf{x}}_i, \hat{\mathbf{x}}_{i,i-1}]$ we have the squared lengths as:

$$\|\boldsymbol{\varepsilon}_i\|^2 = \|\mathbf{x}_i - \hat{\mathbf{x}}_i\|^2 = \|\mathbf{x}_i - \hat{\mathbf{x}}_{i,i-1}\|^2 - \|K_i \mathbf{i}_{i,i-1}\|^2.$$

With the interpretation of squared lengths as covariance matrices, this reads

$$P_i = P_{i,i-1} - K_i(H_i P_{i,i-1} H_i^T + N_i)K_i^T, \qquad *(10.127)$$

using Eq. (10.124) and defining P_i as the indicated error covariance. Using Eq. (10.125), this last can be written

$$P_i = (I - K_i H_i)P_{i,i-1}. \tag{10.128}$$

This last is sometimes not a good formula for calculation, however, because as the iterative process continues, the decrease in estimator error made by the addition of each new data point becomes small, and roundoff error may cause the computation to underflow and "stick" at the same value.

An alternative formula for the gain can be obtained from Eq. (10.127). Substituting Eq. (10.125) into Eq. (10.128) yields

$$P_i = P_{i,i-1} - P_{i,i-1}H_i^T(H_i^T P_{i,i-1} H_i + N_i)^{-1}H_i P_{i,i-1}.$$

Provided only that the matrices are conformable and the indicated inverses exist, it is an identity that

$$(A + BC^{\mathrm{T}})^{-1} = A^{-1} - A^{-1}B(I + C^{\mathrm{T}}A^{-1}B)^{-1}C^{\mathrm{T}}A^{-1}.$$

Applying this to the preceding equation with $A^{-1} = P_{i,i-1}$, $B = H_i^{\mathrm{T}}$, $C^{\mathrm{T}} = N_i^{-1}H_i$ results in

$$P_i^{-1} = P_{i,i-1}^{-1} + H_i^{\mathrm{T}}N_i^{-1}H_i. \qquad *(10.129)$$

This has the disadvantage that often the dimension of the state vector **x**, and therefore that of the covariance P_i, is larger than that of the observation vector y_i and therefore that of the matrix N_i.

Yet another expression for the estimator error covariance, and usually the preferred one for computation, results as follows. If we substitute Eq. (10.121) for the innovation into the estimator equation Eq. (10.123) we have

$$\hat{\mathbf{x}}_i = \hat{\mathbf{x}}_{i,i-1} + K_i(\mathbf{y}_i - H_i\hat{\mathbf{x}}_{i,i-1}). \qquad *(10.130)$$

Since the estimator is unbiased, the error covariance is

$$\begin{aligned}
P_i &= \mathscr{E}(\mathbf{x}_i - \hat{\mathbf{x}}_i)(\mathbf{x}_i - \hat{\mathbf{x}}_i)^{\mathrm{T}} \\
&= \mathscr{E}[(I - K_iH_i)(\mathbf{x}_i - \hat{\mathbf{x}}_{i,i-1}) - K_i\mathbf{n}_i] \\
&\quad \times [(I - K_iH_i)(\mathbf{x}_i - \hat{\mathbf{x}}_{i,i-1}) - K_i\mathbf{n}_i]^{\mathrm{T}} \\
&= (I - K_iH_i)P_{i,i-1}(I - K_iH_i)^{\mathrm{T}} + K_iN_iK_i^{\mathrm{T}}. \qquad *(10.131)
\end{aligned}$$

In this we used the model assumptions that the current measurement noise $\mathbf{n}_i$ has zero covariance with the past measurement noises and with the state.

To complete the recursion, we still have to bridge from the estimate $\hat{\mathbf{x}}_{i-1}$ to the estimate $\hat{\mathbf{x}}_{i,i-1}$. We have to do that so as to maintain unbiasedness. The obvious answer is the right one. Because the true state evolves as in Eq.(10.119) and we assumed the "plant noise" $\mathbf{w}_i$ is zero mean, we update the estimate as

$$\hat{\mathbf{x}}_{i,i-1} = \Phi_{i,i-1}\hat{\mathbf{x}}_{i-1}. \qquad *(10.132)$$

The error variance updates as

$$\begin{aligned}
P_{i,i-1} &= \mathscr{E}(\mathbf{x}_i - \hat{\mathbf{x}}_{i,i-1})(\mathbf{x}_i - \hat{\mathbf{x}}_{i,i-1})^{\mathrm{T}} \\
&= \mathscr{E}[\Phi_{i,i-1}(\mathbf{x}_{i-1} - \hat{\mathbf{x}}_{i-1}) + G_{i-1}\mathbf{w}_{i-1}] \\
&\quad \times [\Phi_{i,i-1}(\mathbf{x}_{i-1} - \hat{\mathbf{x}}_{i-1}) + G_{i-1}\mathbf{w}_{i-1}]^{\mathrm{T}} \\
&= \Phi_{i,i-1}P_{i-1}\Phi_{i,i-1}^{\mathrm{T}} + G_{i-1}W_{i-1}G_{i-1}^{\mathrm{T}}. \qquad *(10.133)
\end{aligned}$$

In this we took account that the state $\mathbf{x}_{i-1}$ involves only noises $\mathbf{w}_{j<i-1}$ and the plant noise sequence is white.

The Kalman filter cycle in summary is as follows:

1. Move the state estimate forward one time tick using Eq. (10.132), and compute the variance Eq. (10.133).

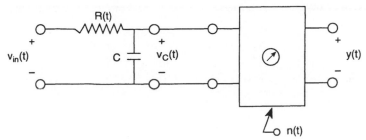

FIGURE 10.10 Time-variable circuit and noisy voltmeter.

2. Compute the gain Eq. (10.125) and apply it to the innovation as in Eq. (10.130).
3. Compute the new variance Eq. (10.131).

The procedure is initialized with some state estimate $\hat{x}_0$ and assumed variance P_0.

EXAMPLE 10.13 The circuit in Fig. 10.10 has a time-variable resistor $R(t)$ and white noise input $v_{in}(t)$ with two-sided power spectral density $N_0/2$ and one-sided bandwidth B. Time samples of the voltage $y(t) = v_{out}(t) + n(t)$ are observed, where n(t) also has bandwidth B and is zero mean with variance σ_n^2. The sampling interval is $\Delta t < 1/2B$. We want to estimate the capacitor voltage.

First discretize the system. The state variable x is what we want to estimate, the capacitor voltage. The circuit has only one energy storage element, so it is completely described by a single state variable, that is, $\mathbf{x}(t) = x(t) = v_c(t)$. We have

$$v_c(t) = \frac{1}{C} \int_{-\infty}^{t} \left\{ \frac{v_{in}(t') - v_c(t')}{R(t')} \right\} dt'.$$

That is,

$$v_c(i) = v_c(i - 1)$$

$$+ \frac{1}{C} \int_{(i-1)\Delta t}^{i\Delta t} \left\{ \frac{v_{in}(t') - v_c(t')}{R(t')} \right\} dt'$$

$$\cong v_c(i - 1) + [v_{in}(i - 1) - v_c(i - 1)] \Delta t/(R_{i-1}C).$$

We can take the model as

$$v_c(i) = (1 - \Delta t/R_{i-1}C)v_c(i - 1) + (1/R_{i-1}C)v_{in}(i - 1).$$

The measurements are given by

$$y(i) = v_c(i) + n_i.$$

That is, we have

$$\Phi_{i,i-1} = 1 - \Delta t/R_{i-1}C,$$
$$G_{i-1} = 1/R_{i-1}C,$$
$$H_i = 1,$$
$$W_i = \text{Var}[v_{\text{in}}(i)] = N_0 B,$$
$$N_i = \sigma_n^2.$$

With these, using Eqs. (10.129), (10.133), for example, the Kalman filter variance loop is described by

$$P_{i,i-1} = (1 - \Delta t/R_{i-1}C)^2 P_{i-1} + N_0 B/(R_{i-1}C)^2,$$
$$1/P_i = 1/P_{i,i-1} + 1/\sigma_n^2.$$

The filter gain is read out from the variance loop, according to Eq. (10.125):

$$K_i = 1/(1 + N_i/P_{i,i-1}).$$

The state is udpated as

$$\hat{v}_c(i, i - 1) = (1 - \Delta t/R_{i-1}C)\hat{v}_c(i - 1),$$

and the gain applied to compute

$$\hat{v}_c(i) = \hat{v}_c(i, i - 1) + K_i[y_i - \hat{v}_c(i, i - 1)].$$

Note that the filter loop can be run and the gains calculated before we ever see any data y_i. Therefore we can precompute the filter gains and store them for retrieval during operation. We can also do a study to determine, if we implemented such a filter, how well it would work in terms of decrease of the estimator variance with time as the computation proceeds. This is not possible unless the dynamical model is linear, as this one is, but the fact that the resistor is time variable is no obstruction.

Often the Kalman filter is applied to problems in which the state dynamics or the measurement equation or both are nonlinear. Then the so-called *extended Kalman filter* is used, in which everything is linearized around a nominal point at each cycle of the algorithm. That is, if the system and measurements are described by

$$\mathbf{X}_i = \mathbf{f}_{i-1}(\mathbf{X}_{i-1}) + G_{i-1}\mathbf{w}_{i-1},$$
$$\mathbf{Y}_i = \mathbf{h}_i(\mathbf{X}_i) + \mathbf{n}_i,$$

we can take the prior estimate $\hat{\mathbf{X}}_{i,i-1}$ as value around which to linearize and write (where the prime indicates differentiation)

$$\mathbf{X}_i = \hat{\mathbf{X}}_{i,i-1} + \mathbf{x}_i$$
$$\approx \mathbf{f}_{i-1}(\hat{\mathbf{X}}_{i,i-1}) + \mathbf{f}'_{i-1}(\hat{\mathbf{X}}_{i,i-1})\mathbf{x}_i + G_{i-1}\mathbf{w}_{i-1},$$
$$\mathbf{Y}_i \approx \mathbf{h}_i(\hat{\mathbf{X}}_{i,i-1}) + \mathbf{h}'_i(\hat{\mathbf{X}}_{i,i-1})\mathbf{x}_i + \mathbf{n}_i.$$

The variance equations and gain equation are then cycled once with the appropriate quantities read off from this linear model in **x**. The state update is done directly by

$$\mathbf{X}_{i,i-1} = \mathbf{f}_{i-1}(\hat{\mathbf{X}}_{i-1}),$$

and the estimate computed as

$$\hat{\mathbf{X}}_i = \hat{\mathbf{X}}_{i,i-1} + K_i[\mathbf{Y}_i - \mathbf{h}_i(\mathbf{X}_{i,i-1})].$$

Exercises

10.1 Denote the set of samples $x_n, x_{n-1}, \ldots, x_1$ by z_n. Show that the *a posteriori* probability density function of a parameter, say θ, given the set of samples may be expressed

$$p(\theta \mid z_n) = \frac{p(x_n \mid \theta, z_{n-1})p(\theta \mid z_{n-1})}{\int_{\{\theta\}} p(x_n \mid \theta, z_{n-1})p(\theta \mid z_{n-1}) \, d\theta}$$

10.2 Based on N statistically independent samples of a Gaussian process of variance σ^2 and unknown mean μ we wish to find the MAP estimator for the mean. Assume that the only *a priori* information about the mean is that it is greater than or equal to zero.
  (a) What is the estimator?
  (b) What is its probability density function?

10.3 Consider a signal $r(t) = A \sin(\omega_c t + \theta) + n(t)$ where $n(t)$ is white Gaussian noise, and θ is uniformly distributed $(0, 2\pi)$. We want a maximum likelihood estimate for the amplitude A, but since *a priori* information for the phase is known we first average over θ to obtain $p(\mathbf{r} \mid A)$. What is the equation which must be solved for the maximum likelihood estimate of A?

10.4 Using M statistically independent samples s_i, $i = 1, \ldots, M$ from a gamma distribution with n degrees of freedom so that

$$p(s_i \mid \alpha) = \frac{s_i^{n/2-1} e^{-s_i/2\alpha}}{\alpha^{n/2} 2^{n/2} \Gamma(n/2)}$$

  (a) Show that the maximum likelihood estimate of α is $\hat{\alpha} = \bar{s}/n$ where $\bar{s} = (1/M) \sum_{i=1}^{M} s_i$.
  (b) Show that this is a sufficient statistic.
  (c) Show that $\hat{\alpha}$ is an efficient estimator for α.

10.5 (a) Determine the Bayes estimator of a parameter α using the cost function

$$C(\hat{\alpha}, \alpha) = \begin{cases} 0, & |\hat{\alpha} - \alpha| < \Delta \\ 1, & \text{otherwise} \end{cases}$$

Assume that the *a posteriori* density function $p(\alpha \mid v)$, where v represents data, has definition over the range $-\infty < \alpha < \infty$. Will the estimator generally be a function of Δ?

(b)   As Δ approaches 0, what is the estimator?

(c)   Suppose $p(\alpha \mid v)$ is unimodal and symmetric about that mode. What is the Bayes estimator? Is the estimator a function of Δ?

10.6   For a signal $s(t) + n(t)$, where $n(t)$ is white Gaussian noise with power spectral density $N_0/2$, find the minimum variance of an unbiased estimate of the time of arrival of $s(t)$, where $s(t)$ is as shown in Fig. 10.11. Answer: $\sigma_{\hat{\tau}}^2 \geq \Delta(2\Delta + 3L)/(12E/N_0)$.

10.7   (a)   Repeat the preceding problem for a signal

$$s(t) = \begin{cases} A(1 + \cos \omega_0 t), & -(2m + 1)\pi/\omega_0 \leq t \leq -2m\pi/\omega_0 \\ A, & -2m\pi/\omega_0 \leq t \leq 2m\pi/\omega_0 \\ A(1 + \cos \omega_0 t), & 2m\pi/\omega_0 \leq t \leq (2m + 1)\pi/\omega_0 \end{cases}$$

to show that the minimum variance is

$$\sigma_{\hat{\tau}}^2 \geq \frac{3 + 4m}{(2E/N_0)\omega_0^2}$$

(b)   Suppose the signal is of the form $s(t) \cos(\omega_c t + \theta)$ where θ is uniformly distributed $(0, 2\pi)$. Find the minimum variance expressed in terms of A and in terms of the signal energy E.

(c)   For the signal of part (b), show that the minimum variance of an unbiased estimate of frequency (known time of arrival) is

$$\sigma_{\hat{\omega}}^2 \geq \left[\frac{2E}{N_0} \left(\frac{2}{3 + 4m} \right) \left(\frac{mT_0^2}{6\omega_0} + 1 \right) \right]^{-1}$$

where $T_0 \triangleq 4m\pi/\omega_0$.

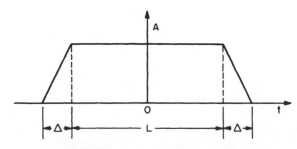

FIGURE 10.11   A trapezoidal pulse.

10.8 Repeat Exercise 10.6 for a signal

$$s(t) = \frac{A}{(2\pi)^{1/2}} e^{-t^2/2\alpha}, \qquad -\infty < t < \infty$$

to show that $\sigma_{\tau}^2 \geq \alpha/(E/N_0)$.

10.9 Consider the signal $A(t) \cos(\omega_c t + \gamma t^2 + \theta) + n(t), 0 \leq t \leq T$, where $n(t)$ is white Gaussian noise. The signal envelope and carrier frequency ω_c are known. The phase is uniformly distributed $(0, 2\pi)$. (The parameter γ might be related to the acceleration of a moving source.)

(a) Use the usual narrowband assumptions and show that the minimum variance of an unbiased estimate of γ is given by

$$\sigma_{\hat{\gamma}}^2 \geq \frac{1}{(2E/N_0)(\mu_4 - \mu_2^2)}$$

where

$$\mu_2 = \int t^2 \, | \, \tilde{a}(t) \, |^2 \, dt, \qquad \mu_4 = \int t^4 \, | \, \tilde{a}(t) \, |^2 \, dt$$

(b) What is $\sigma_{\hat{\gamma}}^2$ when $A(t)$ is a constant?

10.10 For the signal of the preceding exercise, assume the time of arrival is known, and that simultaneous estimates of frequency and acceleration parameter γ are desired. Show that

$$\sigma_{\hat{\omega}}^2 \geq [(2E/N_0)t_d^2(1 - \rho^2)]^{-1}$$
$$\sigma_{\hat{\gamma}}^2 \geq [(2E/N_0)t_4^2(1 - \rho^2)]^{-1}$$

where

$$\mu_i = \int t^i \, | \, \tilde{a}(t)|^2 \, dt, \qquad t_d^2 = \mu_2 - \mu_1^2$$

$$t_3 = \mu_3, \qquad t_4^2 = \mu_4 - \mu_2^2, \qquad \rho = \frac{t_3}{t_d t_4}$$

10.11 Consider the detection problem with hypotheses

$$H_1: \quad r(t) = B \cos(\omega t + \phi) + n(t), \qquad 0 \leq t \leq T$$
$$H_0: \quad r(t) = n(t)$$

where $n(t)$ is white Gaussian noise, B is a known constant, ϕ is uniformly distributed $(0, 2\pi)$, and ω is unknown. What is the form of the receiver if the maximum likelihood principle is used after having averaged over the phase ϕ?

10.12 Consider the signal $r(t) = A \cos \omega_1 t + B \cos(\omega_2 t + \phi) + n(t)$ where $n(t)$ is white Gaussian noise. Assume A, B, and ω_1 are known, and

ϕ is uniformly distributed $(0, 2\pi)$. How might a receiver be implemented to extract a maximum likelihood estimate of ω_2?

10.13   In the preceding problem, suppose the signal is modified to be $A \cos(\omega_1 t + \theta) + B \cos(\omega_2 t + \phi) + n(t)$ where θ and ϕ are statistically independent and uniformly distributed $(0, 2\pi)$. Assume $\int \cos(\omega_1 t + \theta) \cos(\omega_2 t + \phi)\, dt = 0$. How might a receiver be implemented to simultaneously extract maximum likelihood estimates of ω_1 and ω_2?

10.14   Consider the signal $\mathbf{r} = \mathbf{Hs} + \mathbf{n}$ where $\mathbf{r}$ and $\mathbf{n}$ are $m \times 1$ column vectors, $\mathbf{s}$ is a $p \times 1$ column vector with elements, $s_1, s_2, \ldots, s_p$, and $\mathbf{H}$ is an $m \times p$ matrix. $\mathbf{s}$ and $\mathbf{n}$ are zero mean vectors and are not necessarily Gaussian. Their covariance matrices are $\mathbf{R}_s$ and $\mathbf{R}_n$ respectively. Determine the best linear estimator of s_{p+k}. (That is, we wish to predict ahead to time $p + k$.)

10.15   Consider the signal $\mathbf{r} = \mathbf{v} + \mathbf{n}$ where each vector is a zero-mean vector sample of dimension 3×1. Denote

$$\mathbf{r} = \begin{bmatrix} r_{p-1} \\ r_{p-2} \\ r_{p-3} \end{bmatrix}, \qquad \mathbf{v} = \begin{bmatrix} v_{p-1} \\ v_{p-2} \\ v_{p-3} \end{bmatrix}, \qquad \mathbf{n} = \begin{bmatrix} n_{p-1} \\ n_{p-2} \\ n_{p-3} \end{bmatrix}$$

Consider these as being consecutive and equally spaced time samples of the processes. Assume $E\{n_{p-i}n_{p-j}\} = \rho^{|j-i|}$, and denote $E\{v_{p-i}v_{p-j}\} = R_v(j - i)$. Derive an explicit expression for the best linear estimate of v_p corresponding to the next consecutive time sample.

10.16   Consider the signal x where

$$\mathbf{x} = \begin{bmatrix} x_{p-1} \\ x_{p-2} \\ x_{p-3} \end{bmatrix}, \qquad \text{and} \qquad E\{x_{p-i}x_{p-j}\} = \rho^{|j-i|}$$

The samples represent equally spaced time samples of the signal. Show that the best linear estimator (predictor) of the next consecutive sample x_p is $\hat{x}_p = \rho x_{p-1}$.

11 *Extensions*

I n any writing work, there will always remain at the end a number of topics, not necessarily related, which should be included but which didn't fit anywhere. This chapter is devoted to those "loose ends." Some of these extend earlier material, some summarize relations scattered elsewhere, and some are just results we have found to be useful.

11.1 MATRIX RELATIONS

A useful matrix identity arises in many contexts. We have referred to it in Chapter 10. We assume everywhere in this chapter that all matrices involved are conformable, so that we can take all the indicated products. Then if A, B, and C are otherwise arbitrary matrices (or vectors), and if the indicated inverse matrices exist, it is true as an identity that

$$(A + BC)^{-1} = A^{-1} - A^{-1}B(I + CA^{-1}B)^{-1}CA^{-1}. \quad *(11.1)$$

This is often called the *matrix inversion lemma*. It is easily verified by direct multiplication of both sides by $(A + BC)$.

Equation (11.1) often arises in the special case that

$$A^{-1} = P,$$
$$B = H^{T}N^{-1}$$
$$C = H.$$

Then we have the form used in Chapter 10:

$$(P^{-1} + H^{T}N^{-1}H)^{-1} = P - PH^{T}(N + HPH^{T})^{-1}HP. \qquad *(11.2)$$

Two special cases of Eq. (11.2) are useful in considering data from a spatially distributed array of sensors, an application we will treat below. First, let the matrix H^{T} be a column vector $\mathbf{h}$, let the matrix N^{-1} be a scalar α, and again write A for P^{-1}. Then Eq. (11.2) becomes

$$(A + \alpha\mathbf{h}\mathbf{h}^{T})^{-1} = A^{-1} - A^{-1}\mathbf{h}\mathbf{h}^{T}A^{-1}/(\mathbf{h}^{T}A^{-1}\mathbf{h} + 1/\alpha). \qquad *(11.3)$$

Equation (11.3) provides a convenient way to update the inverse of an estimated covariance matrix as data arrive:

$$(1/n)C_n^{-1} = \left[\sum_{i=1}^{n} \mathbf{x}_i\mathbf{x}_i^{T}\right]^{-1} = [(n-1)C_{n-1} + \mathbf{x}_n\mathbf{x}_n^{T}]^{-1}$$
$$= C_{n-1}^{-1}/(n-1) - [C_{n-1}^{-1}/(n-1)]\mathbf{x}_n\mathbf{x}_n^{T}$$
$$\times [C_{n-1}^{-1}/(n-1)]/\{1 + \mathbf{x}_n^{T}[C_{n-1}^{-1}/(n-1)]\mathbf{x}_n\}. \qquad (11.4)$$

In another application, $A = N$ might be the covariance matrix of some noise to which is added a signal $s\mathbf{h}$ with s a zero-mean random amplitude with variance σ_s^2. Then Eq. (11.3) expresses the inverse covariance matrix of the received data $\mathbf{x} = s\mathbf{h} + \mathbf{n}$:

$$X^{-1} = (\sigma_s^2\mathbf{h}\mathbf{h}^{T} + N)^{-1}$$
$$= N^{-1} - \sigma_s^2 N^{-1}\mathbf{h}\mathbf{h}^{T}N^{-1}/(1 + \sigma_s^2\mathbf{h}^{T}N^{-1}\mathbf{h}). \qquad (11.5)$$

Yet again, we can multiply Eq. (11.2) on the right by $H^{T}N^{-1}$ to obtain

$$(P^{-1} + H^{T}N^{-1}H)^{-1}H^{T}N^{-1} = PH^{T}N^{-1} - PH^{T}(N + HPH^{T})^{-1}HPH^{T}N^{-1}$$
$$= PH^{T}[N^{-1} - N^{-1}(I + HPH^{T}N^{-1})^{-1}HPH^{T}N^{-1}].$$

Now in Eq. (11.1) we can take $A = N$, $B = I$, $C = HPH^{T}$, to obtain

$$(P^{-1} + H^{T}N^{-1}H)^{-1}H^{T}N^{-1} = PH^{T}(HPH^{T} + N)^{-1}. \qquad *(11.6)$$

Another useful special case of Eq. (11.1) can be written. In Eq. (11.1) let $C = I$. Then we have

$$A^{-1} - (A + B)^{-1} = A^{-1}B(I + A^{-1}B)^{-1}A^{-1}$$
$$= A^{-1}B(A + B)^{-1}. \qquad (11.7)$$

Partitioned Matrices

We have found occasion in the past to deal with partitioned matrices. That is, we have considered a matrix A to be of the form

$$A = \begin{bmatrix} A_{11} & A_{12} \\ A_{21} & A_{22} \end{bmatrix},$$

where the various blocks are conformable. In particular, it is often the case that $A_{12} = \mathbf{b}$, so that $A_{21} = \mathbf{c}^T$, and $A_{22} = d$, a scalar. In any event, it is easy to verify by direct multiplication and Eq. (11.1) that, in case it exists,

$$A^{-1} = \begin{bmatrix} (A_{11} - A_{12}A_{22}^{-1}A_{21})^{-1} & -(A_{11} - A_{12}A_{22}^{-1}A_{21})^{-1}A_{12}A_{22}^{-1} \\ -(A_{22} - A_{21}A_{11}^{-1}A_{12})^{-1}A_{21}A_{11}^{-1} & (A_{22} - A_{21}A_{11}^{-1}A_{12})^{-1} \end{bmatrix}.$$

*(11.8)

Furthermore, consider the matrix product

$$\begin{bmatrix} A_{11} & A_{12} \\ A_{21} & A_{22} \end{bmatrix} \begin{bmatrix} I & 0 \\ -A_{22}^{-1}A_{21} & A_{22}^{-1} \end{bmatrix} = \begin{bmatrix} A_{11} - A_{12}A_{22}^{-1}A_{21} & A_{12}A_{22}^{-1} \\ 0 & I \end{bmatrix}.$$

The determinants of these three partitioned matrices are immediately evaluated to yield

$$\det(A)\det(A_{22}^{-1}) = \det(A_{11} - A_{12}A_{22}^{-1}A_{21}),$$

which is to say

$$\det(A) = \det(A_{11} - A_{12}A_{22}^{-1}A_{21})\det(A_{22}).$$ *(11.9)

EXAMPLE 11.1 The partitioning relations above can be used to find the density of an m-dimensional Gaussian vector $\mathbf{s}$, conditioned on a second, n-dimensional, Gaussian vector $\mathbf{r}$. If we take $\mathbf{w} = [\mathbf{s}^T\ \mathbf{r}^T]^T$ as the column vector of $\mathbf{s}$ and $\mathbf{r}$ combined and assume zero means for compactness of writing, we have

$$p(\mathbf{s} \mid \mathbf{r}) = p(\mathbf{w})/p(\mathbf{r}),$$

where

$$p(\mathbf{w}) = [(2\pi)^{(m+n)/2}|C_w|^{1/2}]^{-1}\exp(-\mathbf{w}^T C_w^{-1}\mathbf{w}/2),$$
$$p(\mathbf{r}) = [(2\pi)^{n/2}|C_r|^{1/2}]^{-1}\exp(-\mathbf{r}^T C_r^{-1}\mathbf{r}/2).$$

The covariance C partitions as

$$C_w = \begin{bmatrix} C_s & C_{sr} \\ C_{sr}^T & C_r \end{bmatrix}.$$

Equation (11.8) at once yields the quantity in the exponent of the conditional density expression as

$$\mathbf{s}^{\mathrm{T}} C_{s|r}^{-1} \mathbf{s} = [\mathbf{s}^{\mathrm{T}} \, \mathbf{r}^{\mathrm{T}}] C_w^{-1} \begin{bmatrix} \mathbf{s} \\ \mathbf{r} \end{bmatrix} - \mathbf{r}^{\mathrm{T}} C_r^{-1} \mathbf{r}$$

$$= \mathbf{s}^{\mathrm{T}} (C_s - C_{sr} C_r^{-1} C_{sr}^{\mathrm{T}})^{-1} \mathbf{s}$$
$$- \mathbf{s}^{\mathrm{T}} (C_s - C_{sr} C_r^{-1} C_{sr}^{\mathrm{T}})^{-1} C_{sr} C_r^{-1} \mathbf{r} - \{\cdots\}^{\mathrm{T}} + f(\mathbf{r}),$$

where the braced quantity is the same as the second quantity on the right, and $f(\mathbf{r})$ is of no interest. Completing the square in this last expression yields

$$\mathbf{s}^{\mathrm{T}} C_{s|r}^{-1} \mathbf{s} = (\mathbf{s} - C_{sr} C_r^{-1} \mathbf{r})^{\mathrm{T}}$$
$$\times (C_s - C_{sr} C_r^{-1} C_{sr}^{\mathrm{T}})^{-1} (\mathbf{s} - C_{sr} C_r^{-1} \mathbf{r}) + g(\mathbf{r}).$$

This identifies the conditional mean and variance as

$$\mathscr{E}(\mathbf{s} \mid \mathbf{r}) = C_{sr} C_r^{-1} \mathbf{r}, \qquad \text{*(11.10)}$$

$$\mathrm{Cov}(\mathbf{s} \mid \mathbf{r}) = C_s - C_{sr} C_r^{-1} C_{sr}^{\mathrm{T}}. \qquad \text{*(11.11)}$$

Expectations

In working with vectors of observations, much simplicity follows with the use of the relation:

$$\mathrm{trace}(AB) = \mathrm{trace}(BA) = \mathrm{trace}(A^{\mathrm{T}} B^{\mathrm{T}}). \qquad (11.12)$$

This is valid for any matrices or vectors such that conformability rules are satisfied for both products AB and BA. For example, if A and B are vectors, we have

$$\mathrm{trace}(\mathbf{ab}^{\mathrm{T}}) = \sum_{i=1}^{n} a_i b_i = \mathbf{b}^{\mathrm{T}} \mathbf{a} = \mathrm{tr}(\mathbf{b}^{\mathrm{T}} \mathbf{a}).$$

Using this where appropriate, letting $\mathbf{a}$, $\mathbf{b}$, and A be constants and $\mathbf{x}$, $\mathbf{y}$ be random variables, we have

$$\mathscr{E}[\mathbf{a}^{\mathrm{T}} (\mathbf{x} - \mathbf{m}_x)(\mathbf{y} - \mathbf{m}_y)^{\mathrm{T}} \mathbf{b}] = \mathbf{a}^{\mathrm{T}} \mathscr{E}[(\mathbf{x} - \mathbf{m}_x)(\mathbf{y} - \mathbf{m}_y)^{\mathrm{T}}] \mathbf{b} = \mathbf{a}^{\mathrm{T}} C_{xy} \mathbf{b}, \qquad (11.13)$$

where C_{xy} is the covariance matrix. Also,

$$\mathscr{E}[(\mathbf{x} - \mathbf{m}_x)^{\mathrm{T}} A (\mathbf{y} - \mathbf{m}_y)] = \mathscr{E}\{\mathrm{tr}[A(\mathbf{y} - \mathbf{m}_y)(\mathbf{x} - \mathbf{m}_x)^{\mathrm{T}}]\}$$
$$= \mathrm{tr}\{A\mathscr{E}[(\mathbf{y} - \mathbf{m}_y)(\mathbf{x} - \mathbf{m}_x)^{\mathrm{T}}]\}$$
$$= \mathrm{tr}(A C_{yx}) = \mathrm{tr}(A C_{xy}^{\mathrm{T}}) = \mathrm{tr}(C_{xy} A^{\mathrm{T}}). \qquad (11.14)$$

An important relation exists for the expectation of the product of four Gaussian vector random variables. It is mainly useful in computing the statistical properties of estimators of the covariance matrix of Gaussian random variables, but it arises in other applications. For generality, we will carry out the derivation for complex vectors.

Consider then

$$\alpha = \mathscr{E}[(\mathbf{w} - \mathbf{m}_w)'(\mathbf{x} - \mathbf{m}_x)(\mathbf{y} - \mathbf{m}_y)'(\mathbf{z} - \mathbf{m}_z)], \qquad *(11.15)$$

where now the prime means the conjugate of the transpose. In Eq. (4.23) we wrote down the scalar form of this for real zero-mean variables. In the case of complex scalar variables, we are interested in the generalization of Eq. (4.23) as

$$\alpha = \mathscr{E}[(x_1 - m_1)^*(x_2 - m_2)(x_3 - m_3)^*(x_4 - m_4)].$$

By writing out real and imaginary parts and using Eq. (4.23), it results that

$$\alpha = C_{21}C_{43} + \mathscr{E}(x_1^* x_3^*)\mathscr{E}(x_2 x_4) + C_{41}C_{23}. \qquad (11.16)$$

Here the C_{ij} are the complex covariances

$$C_{ij} = \mathscr{E}(x_i - m_i)(x_j - m_j)^*.$$

In case the quantities involved are real, Eq. (11.16) is just Eq. (4.23).

Now let us write Eq. (11.15) as

$$\alpha = \sum_{i=1}^{n}\sum_{j=1}^{n} \mathscr{E}(w_i - m_{wi})^*(x_i - m_{xi})(y_j - m_{yj})^*(z_j - m_{zj}).$$

Applying Eq. (11.16), this becomes

$$\alpha = \sum_{i=1}^{n}\sum_{j=1}^{n} \{C_{xw,ii}C_{zy,jj} + \mathscr{E}[(w_i - m_{wi})^*(y_j - m_{yj})^*]$$

$$\times \mathscr{E}[(x_i - m_{xi})(z_j - m_{zj})] + C_{zw,ji}C_{xy,ij}\}.$$

Using the expression for the elements of a matrix product:

$$(AB)_{ij} = \sum_{k=1}^{n} A_{ik}B_{kj},$$

this last expression becomes

$$\alpha = \mathrm{tr}(C_{xw})\mathrm{tr}(C_{zy}) + \mathrm{tr}(C_{zw}C_{xy}) + \mathrm{tr}\{\mathscr{E}[(\mathbf{w} - \mathbf{m}_w)^*(\mathbf{y} - \mathbf{m}_y)^{\mathrm{T}*}]$$

$$\times \mathscr{E}[(\mathbf{z} - \mathbf{m}_z)(\mathbf{x} - \mathbf{m}_x)^{\mathrm{T}}]\}. \qquad (11.17)$$

In this we have explicitly shown the transpose as T.

In two common cases the final term in Eq. (11.17) vanishes. First suppose that the vectors $\mathbf{z}$, $\mathbf{x}$, say, result by time sampling the complex envelopes of two random processes $z(t)$, $x(t)$. That is,

$$(\mathbf{z})_i = \tilde{z}(t_i) = f_i + jf_q,$$
$$(\mathbf{x})_j = \tilde{x}(t_j) = g_i + jg_q,$$

where f_i, f_q are in-phase and quadrature components of the complex envelope $\tilde{z}(t)$, sampled at $t = t_i$, and similarly for g_i, g_q. Then (where $\tau = t_i - t_j$) we can use Eqs. (3.78), (3.79) to write (suppressing the means):

$$\mathscr{E}(\mathbf{z})_i(\mathbf{x})_j = [R_{fi,gi}(\tau) - R_{fq,gq}(\tau)]$$
$$+ j[R_{fi,gq}(\tau) + R_{fq,gi}(\tau)] = 0.$$

That is, in the case of sampled complex envelopes, we have

$$\alpha = \text{tr}(C_{xw})\text{tr}(C_{zy}) + \text{tr}(C_{zw}C_{xy}). \qquad *(11.18)$$

The other special case is that in which $\mathbf{z}$, $\mathbf{x}$ are vectors of Fourier coefficients of real stationary random processes. That is,

$$(\mathbf{z})_i = Z(k\,\Delta f) = Z_c + jZ_s,$$
$$(\mathbf{x})_j = X(k\,\Delta f) = X_c + jX_s.$$

Then using Eq. (3.116) we have

$$\mathscr{E}(\mathbf{z}_i)(\mathbf{x}_j) = (\mathscr{E}Z_cX_c - \mathscr{E}Z_sX_s) + j(\mathscr{E}Z_sX_c + \mathscr{E}Z_cX_s) = 0.$$

That is, we have Eq. (11.18) in this case also. As indicated in Eq. (3.116), these relations strictly hold true only in the limit of infinite observation time. In practice, the interval of time over which the Fourier coefficients are computed should be much larger than the decay time of the various covariance functions involved.

In the case of real vector random variables, Eq. (11.17) takes the more recognizable form:

$$\alpha = \text{tr}(C_{xw})\text{tr}(C_{zy}) + \text{tr}(C_{zw}C_{xy}) + \text{tr}(C_{wy}C_{zx}) \quad \text{(real).} \qquad *(11.19)$$

Various special cases of Eq. (11.17) are more usually encountered. First, suppose that $\mathbf{y} = \mathbf{x}$, $\mathbf{z} = \mathbf{w}$. Then we have

$$\mathscr{E}|(\mathbf{w} - \mathbf{m}_w)'(\mathbf{x} - \mathbf{m}_x)|^2$$
$$= |\text{tr}(C_{xw})|^2 + \text{tr}(C_{ww}C_{xx})$$
$$+ \text{tr}\{[\mathscr{E}(\mathbf{w} - \mathbf{m}_w)(\mathbf{x} - \mathbf{m}_x)^T]^*[\mathscr{E}(\mathbf{w} - \mathbf{m}_w)(\mathbf{x} - \mathbf{m}_x)^T]\}. \qquad (11.20)$$

The last term of this is real in general. In the case of real vectors, this is just

$$\mathscr{E}[(\mathbf{w} - \mathbf{m}_w)^T(\mathbf{x} - \mathbf{m}_x)]^2 = [\text{tr}(C_{xw})]^2 + \text{tr}(C_{ww}C_{xx})$$
$$+ \text{tr}(C_{wx}^2) \quad \text{(real).} \qquad (11.21)$$

As a further case of Eq. (11.21), consider that $\mathbf{x} = A\mathbf{w}$, with A being square. Then we have

$$\mathscr{E}[(\mathbf{w} - \mathbf{m}_w)^TA(\mathbf{w} - \mathbf{m}_w)]^2 = [\text{tr}(AC_{ww})]^2 + \text{tr}(AC_{ww}A^TC_{ww})$$
$$+ \text{tr}[(AC_{ww})^2] \quad \text{(real).} \qquad (11.22)$$

If A is symmetric, this last becomes

$$\mathscr{E}[(\mathbf{w} - \mathbf{m}_w)^TA(\mathbf{w} - \mathbf{m}_w)]^2$$
$$= [\text{tr}(AC_{ww})]^2 + 2\text{tr}[(AC_{ww})^2] \quad \text{(real).} \qquad *(11.23)$$

11.2 DERIVATIVE FORMULAS

In extremization problems, determining linear minimum variance filters for example, the quantity being extremized is necessarily real. In case more than one parameter is available for choice, we have to do with a scalar function of a vector variable. In addition, if we are choosing frequency domain filter coefficients or the complex envelope samples of some weighting function, the variables to be chosen are complex. In this section we write down some convenient formulas for dealing with such problems.

First we will consider a real scalar function $\phi(\mathbf{x})$ of some real parameter vector $\mathbf{x}$ or matrix A. We will define the derivative of a scalar with respect to a vector as a row vector

$$\partial\phi(\mathbf{x})/\partial\mathbf{x} = \operatorname*{row}_{i} \partial\phi(\mathbf{x})/\partial x_i. \tag{11.24}$$

Equation (11.24) extends in a natural way to a vector function $\mathbf{f}(\mathbf{x})$. Since $\mathbf{f}$ is by convention a column, we have

$$\partial\mathbf{f}(\mathbf{x})/\partial\mathbf{x} = \operatorname*{col}_{i} \partial f_i(\mathbf{x})/\partial\mathbf{x} = \operatorname*{col}_{i}\{\operatorname*{row}_{j} \partial f_i(\mathbf{x})/\partial x_j\}. \tag{11.25}$$

That is, $\partial\mathbf{f}(\mathbf{x})/\partial\mathbf{x}$ is a matrix whose ijth element is $\partial f_i(\mathbf{x})/\partial x_j$.

Finally, for a scalar function of a matrix of variables a_{ij}, we define

$$\partial\phi(A)/\partial A = \{\partial\phi(A)/\partial a_{ij}\}. \tag{11.26}$$

With these definitions, it is easy to verify the derivative formulas:

$$\partial(\mathbf{y}^T A \mathbf{x})/\partial\mathbf{x} = \mathbf{y}^T A, \tag{11.27a}$$

$$\partial(\mathbf{x}^T A \mathbf{y})/\partial\mathbf{x} = \mathbf{y}^T A^T, \tag{11.27b}$$

$$\partial[f^T(\mathbf{x}) A f(\mathbf{x})]/\partial\mathbf{x} = \mathbf{f}^T(\mathbf{x})(A + A^T)[\partial\mathbf{f}(x)/\partial\mathbf{x}], \tag{11.27c}$$

$$\partial(\mathbf{x}^T A \mathbf{x})/\partial\mathbf{x} = \mathbf{x}^T(A + A^T), \tag{11.27d}$$

$$\partial(\mathbf{x}^T A \mathbf{y})/\partial A = \mathbf{x}\mathbf{y}^T, \tag{11.27e}$$

$$\partial(\mathbf{x}^T A \mathbf{x})/\partial A = \mathbf{x}\mathbf{x}^T. \tag{11.27f}$$

These formulas are particularly useful in extremizing a function of a vector of variables.

EXAMPLE 11.2 A random data vector is modeled by $\mathbf{y} = \mathbf{h}x + \mathbf{n}$, with $\mathbf{n}$ being zero-mean white noise and x a scalar constant. We want the linear minimum variance unbiased estimator of x given $\mathbf{y}$: $\hat{x} = \mathbf{l}^T\mathbf{y}$. The unbiasedness constraint is

$$\mathscr{E}(\mathbf{l}^T\mathbf{y}) = \mathbf{l}^T[\mathscr{E}(\mathbf{h}x + \mathbf{n})] = \mathbf{l}^T\mathbf{h}x = x.$$

That is,

$$\mathbf{l}^T\mathbf{h} = 1.$$

The variance to be minimized is

$$\phi(\mathbf{l}) = \text{Var}(\mathbf{l}^T\mathbf{y}) = \mathscr{E}(\mathbf{l}^T\mathbf{n})(\mathbf{n}^T\mathbf{l}) = \mathbf{l}^T N\mathbf{l},$$

where N is the covariance of the noise. The unbiasedness constraint is introduced by a Lagrange multiplier, so we consider unconstrained minimization of

$$\alpha(\mathbf{l}) = \mathbf{l}^T N\mathbf{l} - \lambda(\mathbf{l}^T\mathbf{h} - 1).$$

Using Eqs. (11.27b), (11.27d) we have

$$\partial\alpha(\mathbf{l})/\partial\mathbf{l} = 2\mathbf{l}^T N - \lambda\mathbf{h}^T = \mathbf{0}^T,$$

so that

$$\mathbf{l}_0 = (\lambda/2)N^{-1}\mathbf{h}.$$

The multiplier results by using this solution in the constraint:

$$\mathbf{l}_0^T\mathbf{h} = 1 = (\lambda/2)\mathbf{h}^T N^{-1}\mathbf{h}.$$

With the corresponding value of λ, the filter becomes

$$\mathbf{l}_0 = N^{-1}\mathbf{h}/(\mathbf{h}^T N^{-1}\mathbf{h}).$$

Except for scale factor, this is the noise whitener and matched filter discussed in Sect. 6.6.

Sometimes we want to extremize a real function $\phi(\mathbf{z})$ of a complex vector variable $\mathbf{z} = \mathbf{x} + j\mathbf{y}$. One way to proceed is to work with real and imaginary parts and to consider the function $\phi(\mathbf{z}) = \phi(\mathbf{x}, \mathbf{y})$. In an extremization problem, we then consider the differential condition

$$d\phi = (\partial\phi/\partial\mathbf{x})\,d\mathbf{x} + (\partial\phi/\partial\mathbf{y})\,d\mathbf{y} = 0,$$

leading to the usual expressions

$$\partial\phi(\mathbf{x}, \mathbf{y})/\partial\mathbf{x} = \mathbf{0}^T,$$
$$\partial\phi(\mathbf{x}, \mathbf{y})/\partial\mathbf{y} = \mathbf{0}^T \tag{11.28}$$

We can solve these $(2n)$ equations to obtain $\mathbf{x}_0$, $\mathbf{y}_0$ and then recover the desired extremizing value as $\mathbf{z}_0 = \mathbf{x}_0 + j\mathbf{y}_0$.

It is often a tedious process to find the explicit function $\phi(\mathbf{x}, \mathbf{y})$ which is needed to use Eq. (11.28). A shortcut results by considering the differential $d\phi$ in a different notation. The technique is to consider the quantities z, z^* as being independent of one another. Clearly they are not, since z depends on (and is uniquely determined by) its conjugate z^*. Nonetheless, we can formally write

$$d\phi(\mathbf{z}, \mathbf{z}^*) = [\partial\phi(\mathbf{z}, \mathbf{z}^*)/\partial\mathbf{z}]\,d\mathbf{z} + [\partial\phi(\mathbf{z}, \mathbf{z}^*)/\partial\mathbf{z}^*]\,d\mathbf{z}^*.$$

In terms of the conjugate transpose, this is the same as

$$d\phi(\mathbf{z}, \mathbf{z}') = [\partial\phi(\mathbf{z}, \mathbf{z}')/\partial\mathbf{z}]\,d\mathbf{z} + (d\mathbf{z}')[\partial\phi(\mathbf{z}, \mathbf{z}')/\partial\mathbf{z}']^T, \tag{11.29}$$

because by Eq. (11.26) $\partial\phi(\mathbf{z}, \mathbf{z}')/\partial\mathbf{z}'$ is a column vector, since $\mathbf{z}'$ is a row vector.

The derivative formulas Eq. (11.27) in the form appropriate to complex vectors $\mathbf{w}$, $\mathbf{z}$ and a matrix A are

$$\partial(\mathbf{w}'A\mathbf{z})/\partial\mathbf{z} = \mathbf{w}'A, \tag{11.30a}$$

$$\partial(\mathbf{z}'A\mathbf{w})/\partial\mathbf{z}' = \mathbf{w}^{\mathrm{T}}A^{\mathrm{T}}, \tag{11.30b}$$

$$\partial[\mathbf{f}'(\mathbf{z})A\mathbf{f}(\mathbf{z})]/\partial\mathbf{z} = \mathbf{f}'(\mathbf{z})A[\partial\mathbf{f}(\mathbf{z})/\partial\mathbf{z}], \tag{11.30c}$$

$$\partial(\mathbf{z}'A\mathbf{z})/\partial\mathbf{z} = \mathbf{z}'A, \tag{11.30d}$$

Further, for real A:

$$\partial(\mathbf{z}'A\mathbf{w})/\partial A = (\mathbf{z}\mathbf{w}')^*, \tag{11.30e}$$

$$\partial(\mathbf{z}'A\mathbf{z})/\partial A = (\mathbf{z}\mathbf{z}')^*. \tag{11.30f}$$

Using these, we can write the differential Eq. (11.29) directly. The necessary equations for extremizing ϕ then result by setting to zero independently the partial derivatives in Eq. (11.29).

In working an extremization problem, the justification for this technique is the following. If we were to proceed using Eq. (11.28), and solve for $\mathbf{x}_0$, $\mathbf{y}_0$, we would obtain exactly the value $\mathbf{z}_0 = \mathbf{x}_0 + j\mathbf{y}_0$ that would result by setting to zero the partial derivatives in the differential Eq. (11.29). Moreover, if ϕ is real, since $d\mathbf{z}' = d\mathbf{z}^{*\mathrm{T}}$, we have

$$\partial\phi(\mathbf{z}, \mathbf{z}')/\partial\mathbf{z}' = [\partial\phi(\mathbf{z}, \mathbf{z}')/\partial\mathbf{z}]',$$

and there is actually only one equation to solve, the other being its transpose conjugate.

EXAMPLE 11.3 Suppose that, for real symmetric A (hence $A' = A$, so that ϕ is real), we are interested in the constrained extremization problem:

$$\min(\phi = \mathbf{z}'A\mathbf{z}),$$
$$\mathrm{Re}(\mathbf{a}'\mathbf{z}) = b.$$

Going one way, introducing a real Lagrange multiplier λ, we extremize

$$\alpha = (\mathbf{x} - j\mathbf{y})^{\mathrm{T}}A(\mathbf{x} + j\mathbf{y}) - \lambda\{\mathrm{Re}[(a_r - ja_i)^{\mathrm{T}}(\mathbf{x} + j\mathbf{y})] - b\}$$
$$= \mathbf{x}^{\mathrm{T}}A\mathbf{x} + \mathbf{y}^{\mathrm{T}}A\mathbf{y} - \lambda(\mathbf{a}_r^{\mathrm{T}}\mathbf{x} + \mathbf{a}_i^{\mathrm{T}}\mathbf{y} - b).$$

Using the derivative formulas Eq. (11.27), we obtain the necessary conditions for an extremum:

$$\partial\alpha/\partial\mathbf{x} = \mathbf{0}^{\mathrm{T}} = 2\mathbf{x}^{\mathrm{T}}A - \lambda\mathbf{a}_r^{\mathrm{T}},$$
$$\partial\alpha/\partial\mathbf{y} = \mathbf{0}^{\mathrm{T}} = 2\mathbf{y}^{\mathrm{T}}A - \lambda\mathbf{a}_i^{\mathrm{T}}.$$

That is,

$$\mathbf{x}_0 = (\lambda/2)A^{-1}\mathbf{a}_r,$$
$$\mathbf{y}_0 = (\lambda/2)A^{-1}\mathbf{a}_i.$$

Using these in the constraint yields

$$(\lambda/2)(\mathbf{a}_r^T A^{-1}\mathbf{a}_r + \mathbf{a}_i^T A^{-1}\mathbf{a}_i) = b,$$

from which follow $\mathbf{x}_0$, $\mathbf{y}_0$ and then

$$\mathbf{z}_0 = bA^{-1}\mathbf{a}/(\mathbf{a}'A^{-1}\mathbf{a}).$$

On the other hand, we can consider $\phi(\mathbf{z}, \mathbf{z}')$ directly. Then

$$\alpha = \mathbf{z}'A\mathbf{z} - \lambda(\mathbf{a}'\mathbf{z} + \mathbf{z}'\mathbf{a} - 2b).$$

We have at once, using Eq. (11.30d),

$$\partial\alpha/\partial\mathbf{z} = \mathbf{z}'A - \lambda\mathbf{a}' = \mathbf{0}',$$
$$\mathbf{z} = \lambda A^{-1}\mathbf{a}.$$

From the constraint,

$$\lambda(\mathbf{a}'A^{-1}\mathbf{a}) = b,$$

so that we have again

$$\mathbf{z}_0 = bA^{-1}\mathbf{a}/(\mathbf{a}'A^{-1}\mathbf{a}).$$

11.3 THE ORTHOGONALITY PRINCIPLE

Using the formulas developed in the previous section, it is easy to derive the orthogonality principle which was used in Sect. 10.12. Suppose, as there, we want to estimate a vector $\mathbf{x}$ given a data vector $\mathbf{y}$. For brevity suppose that $\mathbf{x}$ and $\mathbf{y}$ are zero mean. We require that the estimate be linear:

$$\hat{\mathbf{x}} = L\mathbf{y}.$$

The estimator matrix L is to be such that the squared error is minimum:

$$L \Leftarrow \min_L \text{ trace } \mathscr{E}(\mathbf{x} - L\mathbf{y})(\mathbf{x} - L\mathbf{y})^T.$$

That is, we seek to minimize

$$\alpha = \text{tr}(C_x - C_{xy}L^T - LC_{yx} + LC_y L^T). \tag{11.31}$$

It is tempting, and correct, to treat L as constant in this and to set the derivative with respect to L^T equal to zero:

$$-C_{xy} + LC_y = 0,$$
$$L = C_{xy}C_y^{-1}. \tag{11.32}$$

This is just Eq. (10.113). With this form of the estimator, we have

$$\mathscr{E}[\mathbf{y}(\mathbf{x} - L\mathbf{y})^T] = C_{yx} - C_y L^T = C_{yx} - C_y C_y^{-1}C_{xy}^T = 0.$$

This is the orthogonality principle, quoted without proof in Eq. (10.112).

The somewhat abrupt transition from Eq. (11.31) to Eq. (11.32) can be justified in more detail as follows. Partition the sought matrix L into rows as

$$L = \begin{bmatrix} \mathbf{l}_1^T \\ \cdots \\ \mathbf{l}_n^T \end{bmatrix}.$$

Also partition C_{xy} as

$$C_{xy} = \begin{bmatrix} \mathscr{E} x_1 \mathbf{y}^T \\ \cdots \\ \mathscr{E} x_n \mathbf{y}^T \end{bmatrix}.$$

Then the quantity Eq. (11.31) to be minimized becomes

$$\alpha = \sum_{i=1}^n [\mathbf{l}_i^T C_y \mathbf{l}_i - (\mathscr{E} x_i \mathbf{y}^T) \mathbf{l}_i - \mathbf{l}_i^T (\mathscr{E} \mathbf{y} x_i)] + tr(C_x).$$

Using Eqs. (11.27a), (11.27d), we have the necessary conditions

$$2 \mathbf{l}_i^T C_y = 2 \mathscr{E} x_i \mathbf{y}^T.$$

Collecting these together over i yields

$$L C_y = C_{xy},$$

which is just Eq. (11.32).

11.4 THE MULTIVARIATE COMPLEX GAUSSIAN DISTRIBUTION

We have often had occasion to deal with vectors of Gaussian random variables and the corresponding multidimensional Gaussian probability density. In this section, we will write down a convenient form of that density if the variables in question are the real and imaginary parts of complex numbers which obey a certain correlation property. The particular cases of interest are those in which the complex variables are either the in-phase and quadrature components of a narrowband noise process or the Fourier coefficients computed from such a process over an adequately long observation time.

Suppose then that we have four real Gaussian random variables, say x_i, y_i, x_k, y_k. We will assume zero means for brevity of notation. If we arrange these into a four-dimensional column vector $\mathbf{w}$, the joint density is

$$p(\mathbf{w}) = [(2\pi)^2 |R_w|^{1/2}]^{-1} \exp[-(\mathbf{w}^T R_w^{-1} \mathbf{w})/2], \qquad (11.33)$$

where $R_w = \mathscr{E} \mathbf{w}\mathbf{w}^T$ is the real covariance matrix and $|R_w|$ is the determinant.

Now let these four real variables be arranged into two complex variables:

$$z_i = x_i + jy_i,$$
$$z_k = x_k + jy_k.$$

Also consider the complex column vector $\mathbf{z}$ whose components are z_i, z_k and the complex covariance matrix $C_z = \mathscr{E}\mathbf{z}\mathbf{z}'$, where the prime means transpose conjugate.

Now suppose that the real vector $\mathbf{w}$ is such that its covariance matrix has the special structure

$$
R_w = \begin{bmatrix} \sigma_i^2 & 0 & \alpha_{ik} & -\beta_{ik} \\ 0 & \sigma_i^2 & \beta_{ik} & \alpha_{ik} \\ \alpha_{ik} & \beta_{ik} & \sigma_k^2 & 0 \\ -\beta_{ik} & \alpha_{ik} & 0 & \sigma_k^2 \end{bmatrix}.
\tag{11.34}
$$

By Eqs. (3.78), (3.79), (3.84) this will be the case if x_i, y_i are the in-phase and quadrature components of the complex envelope of a random process sampled at time t_i, and correspondingly for x_k, y_k. Alternatively, by Eq. (3.116) these real variables can also be the real and imaginary parts of complex Fourier coefficients, at different frequencies, computed over a sufficiently long time span for a real random process.

With the preceding definitions of the matrices and vectors involved, provided Eq. (11.34) holds true it is straightforward to calculate that

$$
\mathbf{z}'C_z^{-1}\mathbf{z} = (1/2)\mathbf{w}^T R_w^{-1}\mathbf{w},
\tag{11.35}
$$

$$
|C_z| = 2^n |R_w|^{1/2},
\tag{11.36}
$$

where currently we have the case $n = 2$. The expressions above are general for any dimension of $\mathbf{w}$, as long as Eq. (11.34) holds for the variables taken pairwise. The partitioning relations Eqs. (11.8), (11.9) are helpful in seeing these. Using Eqs. (11.35), (11.36), we can then write the density of the complex vector $\mathbf{z}$ as

$$
p(\mathbf{z}) = p(\mathbf{w}) = (\pi^n |C_z|)^{-1} \exp(-\mathbf{z}'C_z^{-1}\mathbf{z}).
\tag{*11.37}
$$

This is called the *complex Gaussian density*.

The procedure leading to Eq. (11.37) is closely related to the following way of handling computations with complex variables. There is an isomorphism between a complex number $z = x + jy$ and the 2×2 matrix

$$
Z = \begin{bmatrix} x & -y \\ y & x \end{bmatrix}.
$$

An *isomorphism* is a relation such that any algebraic expression which is correct for the complex numbers z, for example, is correct as it stands for the matrices Z. For example, for two complex numbers z_1, z_2 we can write that

$$
z_1 z_2 = (x_1 x_2 - y_1 y_2) + j(x_1 y_2 + x_2 y_1).
$$

Because the isomorphism exists, we expect that the following matrix product will be correct:

$$Z_1Z_2 = \begin{bmatrix} x_1 & -y_1 \\ y_1 & x_1 \end{bmatrix} \begin{bmatrix} x_2 & -y_2 \\ y_2 & x_2 \end{bmatrix}$$

$$= \begin{bmatrix} x_1x_2 - y_1y_2 & -x_1y_2 - x_2y_1 \\ x_1y_2 + x_2y_1 & x_1x_2 - y_1y_2 \end{bmatrix}.$$

This is obviously true.

As another example, we have

$$1/z = 1/(x + jy) = (x - jy)/(x^2 + y^2).$$

The isomorphism then dictates that

$$Z^{-1} = \begin{bmatrix} x & -y \\ y & x \end{bmatrix}^{-1} = (x^2 + y^2)^{-1} \begin{bmatrix} x & y \\ -y & x \end{bmatrix}.$$

This is also clearly true.

This isomorphic relation can be convenient in working with complex variables on a software system which is not constructed to recognize and work with complex variables. We simply replace every complex number with a 2 × 2 matrix of the preceding form and proceed to compute with real matrices. The end result of the matrix computation will have the structure indicated, from which the complex number form of the result in question can be read off.

Note, however, that the expression Eq. (11.33) is not the result of applying this isomorphism to Eq. (11.37). A direct application of the isomorphism to Eq. (11.37), for example, would result in the replacement of **z** by a 4 × 2 matrix

$$Z = \begin{bmatrix} x_1 & -y_1 \\ y_1 & x_1 \\ x_2 & -y_2 \\ y_2 & x_2 \end{bmatrix}.$$

Then the exponent of the exponential turns out to be a 2 × 2 matrix, and the matrix exponential must be used to compute the result from which the real number $p(\mathbf{z})$ would be read off.

EXAMPLE 11.4 The time samples of a complex envelope $\tilde{r}(t)$ are arranged into a complex vector $\tilde{\mathbf{r}}$. The data consist of a signal envelope and zero-mean Gaussian noise, so that we have

$$\tilde{\mathbf{r}} = H\tilde{\mathbf{s}} + \tilde{\mathbf{n}},$$

where H need not be square. We seek the maximum likelihood estimate of $\bar{s}$. The estimate results from

$$\hat{\bar{s}} \Leftarrow \max_s p(\bar{r}\,|\,\bar{s}).$$

From Eq. (11.37) this corresponds to

$$\hat{\bar{s}} \Leftarrow \min_s (\bar{r} - H\bar{s})'N^{-1}(\bar{r} - H\bar{s}),$$

where the prime means transpose conjugate and N is the complex matrix $N = \mathscr{E}\bar{n}\bar{n}'$. Using Eqs. (11.30a), (11.30c) the necessary condition is

$$(\bar{r} - H\bar{s})'N^{-1}H = 0',$$

from which

$$\hat{\bar{s}} = (H'N^{-1}H)^{-1}H'N^{-1}\bar{r}. \tag{11.38}$$

This is just the complex form of the estimator Eq. (10.130) in the case that there is no prior information, so that there we have

$$\hat{x}_{i,i-1} = 0,$$
$$P^{-1}_{i,i-1} = 0.$$

With these, the Kalman gain in the form Eq. (10.126) is just Eq. (11.38).

11.5 DETECTION

In Sect. 6.1 we considered detection of a known signal in Gaussian noise of known mean and variance. If we take the data model as

$$\mathbf{r} = H\mathbf{s} + \mathbf{n},$$

with $\mathbf{n}$ having mean zero and covariance matrix N, the Neyman–Pierson detector is given by Eq. (6.5):

$$S(\mathbf{r}) = \mathbf{s}^{T}H^{T}N^{-1}\mathbf{r} > S_t,$$

where S_t is a threshold set by the desired false alarm probability.

Suppose now that the signal $\mathbf{s}$ is not known. We can then use the generalized maximum likelihood detector. That is, the unknown signal in the above detector is replaced by its maximum likelihood estimator, Eq. (11.38). The result is the detector

$$S(\mathbf{r}) = \mathbf{r}^{T}N^{-1}H(H^{T}N^{-1}H)^{-1}H^{T}N^{-1}\mathbf{r} > S_t. \qquad *(11.39)$$

This quadratic detector is no longer the optimal Neyman–Pearson detector, but it is often satisfactory as an expedient. The corresponding formulation for complex envelope samples or Fourier coefficients could easily be written down in analogy.

Since the data vector **r** is Gaussian but not zero mean, the test statistic Eq. (11.39) has the noncentral chi-squared distribution. This was discussed in detail in Sect. 4.9. However, often the number of data values is large enough that the sum Eq. (11.39) is nearly Gaussian. In that case, we need only the mean and variance of $S(\mathbf{r})$ under the two hypotheses in order to assess performance of the detector.

We have:

$$\begin{aligned}
\mathscr{E}(S\,|\,0) &= \text{trace }\mathscr{E}[\mathbf{r}\mathbf{r}^T N^{-1} H (H^T N^{-1} H)^{-1} H^T N^{-1}\,|\,0] \\
&= \text{trace}[N N^{-1} H (H^T N^{-1} H)^{-1} H^T N^{-1}] \\
&= \text{trace}[(H^T N^{-1} H)^{-1} H^T N^{-1} H] = n,
\end{aligned}$$

where n is the dimension of **r**. Also, since

$$\mathscr{E}(\mathbf{r}\mathbf{r}^T\,|\,1) = H\mathbf{s}\mathbf{s}^T H^T + N,$$

in the same way we have

$$\mathscr{E}(S\,|\,1) = \mathbf{s}^T H^T N^{-1} H (H^T N^{-1} H)^{-1} H^T N^{-1} H\mathbf{s} + n = \mathbf{s}^T H^T N^{-1} H\mathbf{s} + n.$$

We have further that

$$S^2 = [\mathbf{r}^T N^{-1} H (H^T N^{-1} H)^{-1} H^T N^{-1} \mathbf{r}]^2.$$

Since **r** is real and Gaussian, the expected value of this can be computed from Eq. (11.23). That is,

$$\mathscr{E}(S^2) = [\text{tr}(AR)]^2 + 2\,\text{tr}(AR)^2,$$

where

$$A = N^{-1} H (H^T N^{-1} H)^{-1} H^T N^{-1}$$

and $R = \mathscr{E}\mathbf{r}\mathbf{r}^T$. Making the appropriate manipulations yields

$$\mathscr{E}(S^2\,|\,0) = n^2 + 2n,$$
$$\mathscr{E}(S^2\,|\,1) = 3(\mathbf{s}^T H^T N^{-1} H\mathbf{s})^2 + (n+2)(2\mathbf{s}^T H^T N^{-1} H\mathbf{s} + n).$$

Then

$$\text{Var}(S\,|\,0) = 2n,$$
$$\text{Var}(S\,|\,1) = 2(\mathbf{s}^T H^T N^{-1} H\mathbf{s})^2 + 4\mathbf{s}^T H^T N^{-1} H\mathbf{s} + 2n.$$

In order to use the standard Gaussian ROC curve, involving a signal-to-noise ratio (SNR), we must make the small-signal assumption that

$$\text{Var}(S\,|\,1) \cong \text{Var}(S\,|\,0) = 2n.$$

Then we can define the (power) SNR as

$$\alpha^2 = [\mathscr{E}(S\,|\,1) - \mathscr{E}(S\,|\,0)]^2/\text{Var}(S\,|\,0) = (\mathbf{s}^T H^T N^{-1} H\mathbf{s})^2/2n. \qquad *(11.40)$$

This can be used with the curve of Fig. 5.7.

11.6 GAUSSIAN SIGNAL IN GAUSSIAN NOISE

In Sect. 6.8 we treated the general discrete problem of a Gaussian signal in Gaussian noise. Here we extend the model slightly and present some alternative formulas using the material of Sect. 11.1.

Suppose that we have a real data vector

$$\mathbf{r} = H\mathbf{s} + \mathbf{n}. \tag{11.41}$$

Let $\mathbf{s}$ be Gaussian with mean $\mathbf{m}$ and covariance matrix S and let $\mathbf{n}$ be zero-mean Gaussian noise with covariance N, which is independent of the signal. To build a detector we need to threshold the likelihood ratio. Using the Gaussian density, say Eq. (11.33), in our case this results in a log likelihood ratio detector

$$\lambda = \mathbf{r}^T[N^{-1} - (HSH^T + N)^{-1}]\mathbf{r} + 2\mathbf{m}^T H^T (HSH^T + N)^{-1}\mathbf{r} > \lambda_t.$$

This is a combination of a quadratic energy detector and a matched filter, matched to the mean signal. Using Eq. (11.7), this can be written

$$\lambda = \mathbf{r}^T N^{-1} HSH^T (HSH^T + N)^{-1}\mathbf{r}$$
$$+ 2\mathbf{m}^T H^T (HSH^T + N)^{-1}\mathbf{r} > \lambda_t. \qquad *(11.42)$$

It is interesting to note in this last that the maximum *a posteriori* estimator Eq. (11.10) of the signal appears. That is, in Eq. (11.10) we have

$$C_{sr} = \mathcal{E}[(\mathbf{s} - \mathbf{m})(\mathbf{r} - H\mathbf{m})^T]$$
$$= \mathcal{E}\{(\mathbf{s} - \mathbf{m})[H(\mathbf{s} - \mathbf{m}) + \mathbf{n}]^T\} = SH^T,$$
$$C_r = \mathcal{E}[H(\mathbf{s} - \mathbf{m}) + \mathbf{n}][H(\mathbf{s} - \mathbf{m}) + \mathbf{n}]^T = HSH^T + N.$$

Then the MAP estimator Eq. (11.10) of the signal is

$$\hat{\mathbf{s}} = SH^T (HSH^T + N)^{-1}\mathbf{r},$$

so that the detector Eq. (11.42) is

$$\lambda = \mathbf{r}^T N^{-1} H\hat{\mathbf{s}} + 2\mathbf{m}^T S^{-1}\hat{\mathbf{s}} = \hat{\mathbf{s}}^T H^T N^{-1}\mathbf{r} + 2\hat{\mathbf{s}}^T S^{-1}\mathbf{m}. \tag{11.43}$$

We will again assume that the dimension of the data vector is large enough that we can consider the detection statistic λ to be Gaussian. From Eq. (11.42) we have

$$\mathcal{E}(\lambda \mid 0) = \text{tr}[HSH^T (HSH^T + N)^{-1}],$$
$$\mathcal{E}(\lambda \mid 1) = \text{tr}(N^{-1} HSH^T) + \mathbf{m}^T H^T [(HSH^T + N)^{-1} + N^{-1}]H\mathbf{m}.$$

To determine the variances, let us consider

$$\gamma = \lambda - \mathcal{E}\lambda.$$

For brevity, let

$$A = N^{-1} HSH^T (HSH^T + N)^{-1},$$
$$\mathbf{b} = (HSH^T + N)^{-1}H\mathbf{m}.$$

Then

$$\gamma_0 = \lambda - \mathscr{E}_0\lambda = \mathbf{r}^T A \mathbf{r} + 2\mathbf{b}^T\mathbf{r} - \operatorname{tr}(AN).$$

From Eq. (11.7) it is clear that A is symmetric. Then using Eq. (11.23), recalling that $\mathscr{E}(\mathbf{r} \mid 0) = 0$ and that the third central moment of a Gaussian vanishes, we obtain

$$
\begin{aligned}
\sigma_0^2 &= \operatorname{Var}(\lambda \mid 0) = \mathscr{E}(\gamma_0^2 \mid 0) \\
&= [\operatorname{tr}(AN)]^2 + 2\operatorname{tr}[(AN)^2] - 2[\operatorname{tr}(AN)]^2 + 4\mathbf{b}^T N\mathbf{b} + [\operatorname{tr}(AN)]^2 \\
&= 2\operatorname{tr}[HSH^T(HSH^T + N)^{-1}]^2 \\
&\quad + 4\mathbf{m}^T H^T(HSH^T + N)^{-1}N(HSH^T + N)^{-1}H\mathbf{m}.
\end{aligned}
$$

Further, let

$$
\begin{aligned}
\gamma_1 &= \lambda - \mathscr{E}_1\lambda = \mathbf{r}^T A \mathbf{r} + 2\mathbf{b}^T\mathbf{r} - \operatorname{tr}(N^{-1}HSH^T) \\
&\quad - \mathbf{m}^T H^T[(HSH^T + N)^{-1} + N^{-1}]H\mathbf{m} \\
&= (\mathbf{r} - H\mathbf{m})^T A(\mathbf{r} - H\mathbf{m}) + 2(AH\mathbf{m} + \mathbf{b})^T(\mathbf{r} - H\mathbf{m}) - \operatorname{tr}[A(HSH^T + N)].
\end{aligned}
$$

Then, again using Eq. (11.23) and the vanishing of the third central moment of $\mathbf{r}$, we obtain

$$
\begin{aligned}
\sigma_1^2 &= \operatorname{Var}(\lambda \mid 1) = \mathscr{E}(\gamma_1^2 \mid 1) \\
&= \{\operatorname{tr}[A(HSH^T + N)]\}^2 + 2\operatorname{tr}[A(HSH^T + N)]^2 - 2\{\operatorname{tr}[A(HSH^T + N)]\}^2 \\
&\quad + 4(AH\mathbf{m} + \mathbf{b})^T(HSH^T + N)(AH\mathbf{m} + \mathbf{b}) + \{\operatorname{tr}[A(HSH^T + N)]\}^2 \\
&= 2\operatorname{tr}(N^{-1}HSH^T)^2 + 4\mathbf{m}^T H^T N^{-1}(HSH^T + N)N^{-1}H\mathbf{m}.
\end{aligned}
$$

In the last step we used the fact that

$$AH\mathbf{m} + \mathbf{b} = N^{-1}H\mathbf{m}.$$

From the above means and variances, the performance of the detector Eq. (11.42) can be calculated, under the assumption that the dimension of $\mathbf{r}$ is large enough that the detector statistic λ is approximately Gaussian.

If we take as the small-signal case the condition

$$HSH^T + N \cong N,$$

then we have

$$\operatorname{Var}(\lambda \mid 0) \cong \operatorname{Var}(\lambda \mid 1),$$

and we can write a power signal-to-noise ratio as

$$\alpha^2 = [\mathscr{E}(\lambda \mid 1) - \mathscr{E}(\lambda \mid 0)]^2/\operatorname{Var}(\lambda \mid 1).$$

From above, and using Eq. (11.7), we have

$$
\begin{aligned}
\mathscr{E}(\lambda \mid 1) - \mathscr{E}(\lambda \mid 0) &= \operatorname{tr}\{HSH^T[N^{-1} - (HSH^T + N)^{-1}]\} \\
&\quad + \mathbf{m}^T H^T[(HSH^T + N)^{-1} + N^{-1}]H\mathbf{m} \\
&= \operatorname{tr}\{HSH^T N^{-1}HSH^T(HSH^T + N)^{-1}\} \\
&\quad + \mathbf{m}^T H^T[(HSH^T + N)^{-1} + N^{-1}]H\mathbf{m}.
\end{aligned}
$$

Again using the small-signal assumption, this is

$$\mathscr{E}(\lambda \mid 1) - \mathscr{E}(\lambda \mid 0) \cong \text{tr}(HSH^TN^{-1})^2 + 2\mathbf{m}^TH^TN^{-1}H\mathbf{m}.$$

For the special case $\mathbf{m} = \mathbf{0}$, the SNR is approximately

$$\alpha^2 \cong \tfrac{1}{2}\,\text{tr}(HSH^TN^{-1})^2. \qquad\qquad *(11.44)$$

This slightly generalizes Eq. (6.91).

Now let us consider the case that the elements in the data model Eq. (11.41) are time samples of complex envelopes, or complex Fourier coefficients, and that the underlying random processes are stationary. In the case of Fourier coefficients, we further assume that the analysis interval is long enough that coefficients relative to different frequencies are uncorrelated.

The complex Gaussian density Eq. (11.37) is the starting point for the development. The detector Eq. (11.42) remains the same as written, except that the vectors are complex and the matrices are complex covariances. In fact, all the preceding expressions for the real case remain in effect, with the prime meaning transpose conjugate rather than transpose. This is essentially because all the matrix expressions of Sect. 11.1 are true as identities for complex matrices as well as real matrices. In the next section we will elaborate on the detector Eq. (11.42) in the complex case in a problem of particular importance.

11.7 SPACE–TIME PROCESSING

In radar, sonar, and geophysical applications, among others, it is important that the instruments used have spatial directivity. That is accomplished using some type of antenna to focus the transmitted signals or the receiver directional sensitivity or both. In many current systems, rather than using a signal projector that is continuously distributed in space, such as a dish antenna, the instrument makes use of multiple distributed sources of radiation. These individually act more or less as spatially isotropic point radiators or receivers. The phases of the signals are carefully controlled to achieve constructive and destructive interference in the propagation medium to achieve spatial directivity.

In all these cases, the receiver is presented with a multiplicity of received time waveforms, one for each of the spatially distributed sensors. They are to be processed together to yield both the desired directivity (spatial filtering), and appropriate detection performance (Fig. 11.1).

We will assume a model of the signal is known, of the type shown in Fig. 11.2. A point signal $s(t)$ in space is distributed by the transmission medium across the receivers of the sensing array by known channel filters $h_{mi}(t)$. The model is also appropriate to communication with spatial diver-

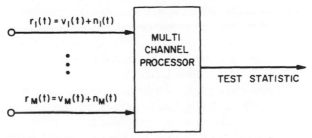

FIGURE 11.1 Configuration for space–time processing.

sity. The channel filters might represent time delay due to the different transmission path lengths, attenuation, or more complicated kinds of channel distortion. That is, we assume a known model of the type

$$r_i(t) = h_i(t)*s(t) + n_i(t), \qquad i = 1, M, \qquad (11.45)$$

where the asterisk indicates convolution. We will take the signal and noise as being stationary zero-mean Gaussian random processes. The signal is uncorrelated with every noise, and the sensor noises $n_i(t)$ are uncorrelated with one another. Neither the signal nor the noises are assumed to be white.

In geophysical applications, the signals Eq. (11.45) are usually lowpass, that is, not modulated carrier signals. In radar, on the other hand,

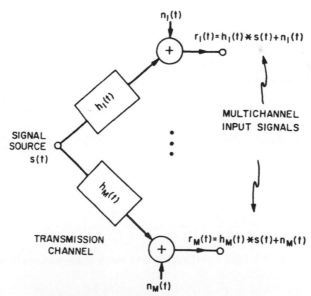

FIGURE 11.2 A model for multichannel signals.

they are always modulations on a carrier. In that case, the signals actually processed will often be the complex envelopes of the sensor signals. Equation (11.45) would then be taken as a relation among complex envelopes. In sonar, either case may be appropriate. In active sonar, there will be a carrier of some frequency and the received signals are best analyzed in terms of their complex envelopes. In the case we will treat explicitly, we will assume the actual received signals Eq. (11.45) are analyzed into complex Fourier coefficients, which are approximately uncorrelated from one frequency to another. That is appropriate for the case of passive sonar, in which a wide band of frequencies is examined, more or less one by one, for the presence of energy.

Accordingly, we will rewrite Eq. (11.45) to show explicitly the case of Fourier coefficients at various frequencies f_k. We have

$$r_i(k) = H_i(k)s(k) + n_i(k).$$

Here all quantities are complex, i indicates channel number, and k is the frequency index. To be specific, we will collect the data corresponding to each frequency into a vector

$$\mathbf{r}(k) = \operatorname*{col}_i r_i(k), \qquad i = 1, M.$$

Then at each frequency the model is

$$\mathbf{r}(k) = \mathbf{H}(k)s(k) + \mathbf{n}(k), \qquad k = 1, K.$$

In order to aggregate frequencies together in a convenient notation, define the channel matrix

$$H = \operatorname*{diag}_k \mathbf{H}(k),$$

that is, the block diagonal matrix with the vectors $\mathbf{H}(k)$ down the diagonal. Then we can take the complex data vector as

$$\mathbf{r} = \operatorname*{col}_k \mathbf{r}(k) = H\mathbf{s} + \mathbf{n}, \tag{11.46}$$

where $\mathbf{s}$ is the K-dimensional column vector of Fourier coefficients of the scalar signal $s(t)$.

If the Fourier coefficients are computed by the usual fast Fourier transform (FFT), with

$$F_k = \sum_{i=0}^{N-1} f(i\,\delta t) \exp\left(\frac{-j2\pi ki}{N}\right), \tag{11.47}$$

then, as in Eq. (3.109), the coefficients are such that, approximately,

$$\mathscr{E}\,|F_k|^2 = (N/\delta t)S_f(2\pi k/N\,\delta t),$$

where $S_f(f)$ is the power spectral density of the real process $f(t)$. As in Eq. (3.114), the coefficients at different frequencies are uncorrelated. Accordingly, we have the noise structure

$$\mathscr{E}\mathbf{n}(k)\mathbf{n}(k)' = N(k),$$

and

$$N = \mathscr{E}nn' = \underset{k}{\operatorname{diag}} N(k),$$

where we indicate a block diagonal matrix. It is the case that, for example,

$$N^{-1} = \underset{k}{\operatorname{diag}} N^{-1}(k).$$

Similarly, the signal coefficients $s(k)$ making up the vector $\mathbf{s}$ are uncorrelated, so that

$$S = \mathscr{E}\mathbf{ss}' = \underset{k}{\operatorname{diag}} \mathscr{E} |s(k)|^2 = \underset{k}{\operatorname{diag}} S(k).$$

With all these, the complex covariance structure of the model Eq. (11.46) is

$$R = \mathscr{E}(\mathbf{rr}') = HSH' + N,$$

where H and N are block diagonal, with the size of the blocks being equal to the number of sensors M, and S is diagonal of size equal to the number of frequency coefficients. It is worth mentioning that, since we assume the time signals to be real, the frequency coefficients all have conjugate symmetry in frequency, so we need to process only the coefficients for positive frequencies.

Once all these preliminary remarks have been made, we can at once write down the detector for the multidimensional problem of Fig. 11.2. It is just Eq. (11.42), where we take the mean vector $\mathbf{m}$ to be zero:

$$\lambda = \mathbf{r}'N^{-1}HSH'(HSH' + N)^{-1}\mathbf{r} > \lambda_t. \tag{11.48}$$

Since by Eq. (11.7) the matrix in this is Hermitian, the statistic λ is real, even though all the other quantities indicated are complex. The attained SNR in the small-signal case is that in Eq. (11.44).

Although in a formal sense the detection problem is now solved, it is revealing to unwind the notation in Eq. (11.48) in order to look at the structure of the processor. We have, first, that

$$\lambda = \underset{k}{\operatorname{row}}[\mathbf{r}(k)'] \underset{k}{\operatorname{diag}}\{N^{-1}(k)\mathbf{H}(k)S(k)\mathbf{H}(k)'$$

$$\times [\mathbf{H}(k)S(k)\mathbf{H}(k)' + N(k)]^{-1}\} \underset{k}{\operatorname{col}}[\mathbf{r}(k)]$$

$$= \sum_{k=1}^{K} \mathbf{r}(k)'N^{-1}(k)\mathbf{H}(k)S(k)\mathbf{H}(k)'[\mathbf{H}(k)S(k)\mathbf{H}(k)' + N(k)]^{-1}\mathbf{r}(k).$$

Continuing, recall that $S(k)$ is a scalar. Then we can use Eq. (11.6) to write

$$S(k)\mathbf{H}(k)'[\mathbf{H}(k)S(k)\mathbf{H}(k)' + N(k)]^{-1}$$
$$= S(k)\mathbf{H}(k)'N(k)^{-1}/[1 + S(k)\mathbf{H}(k)'N(k)^{-1}\mathbf{H}(k)].$$

Then the detection statistic can be written

$$\lambda = \sum_{k=1}^{K} |f(k)\mathbf{H}(k)'N(k)^{-1}\mathbf{r}(k)|^2, \qquad *(11.49)$$

where we define $f(k)$ in terms of the real quantity

$$|f(k)|^2 = S(k)/[1 + S(k)\mathbf{H}(k)'N(k)^{-1}\mathbf{H}(k)].$$

We can also put matters back into the time domain as follows. Let the frequency coefficients in the sum Eq. (11.49) be complex numbers $G(k)$. Then, were we to process the entire frequency band, we would have

$$\lambda = \sum_{k=1}^{\infty} |G(k)|^2 = \frac{N}{\delta t} \int_{-\infty}^{\infty} g^2(t)\, dt.$$

Here we have used Parseval's relation and the particular normalization that results from use of the Fourier coefficients in the form Eq. (11.47). Furthermore, define the column vectors $\mathbf{L}(k)$ by

$$\mathbf{L}(k)' = \mathbf{H}(k)'N^{-1}(k).$$

The elements $l_i(k)$ of $\mathbf{L}(k)$ are the frequency coefficients of a filter in each of the sensor channels i.

Now the detection statistic can be diagrammed in terms of the sensor signals $r_i(t)$ as shown in Fig. 11.3. The processor consists of a filter in each channel to account for the channel spatial properties and a second filter common to every channel to account for the time properties of the noise.

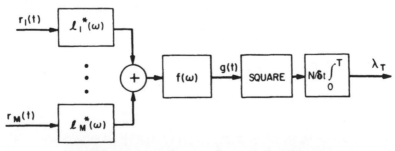

FIGURE 11.3   A realization of a space–time processor.

11.8 ESTIMATION OF THE NOISE BACKGROUND

In all of our discussions so far, when noise has been present we have assumed as part of the data model that its statistical properties were known. We have also generally assumed Gaussian noise. In applications such as radar in a cluttered background or active sonar in the presence of reverberation, other noise distributions may be more appropriate. Commonly the lognormal or Weibull distribution is used, since these have higher probability of occasional large noise "spikes" than does the Gaussian. In any event, if the noise density is known, the procedures we have discussed can go forward with more or less complexity in their realization.

On the other hand, it may be that the noise background is time variable or depends on the particular details of the situation encountered in operation. In such cases, it is necessary to estimate the noise background during operation of the system and to change the detector accordingly. There are many techniques for adaptive detector design, but to describe them in any useful detail would go beyond our aims here. We will, however, discuss the simplest, and perhaps commonest, strategy for estimating the noise background in which a detector operates as time progresses. That is the so-called *cell averaging constant false alarm rate* (CFAR) *processor*.

In Fig. 11.4a we show a notional detection system for a radar or active sonar receiver. The narrowband signal is converted down to some conve-

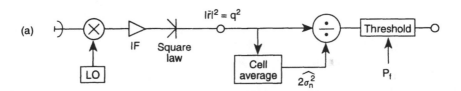

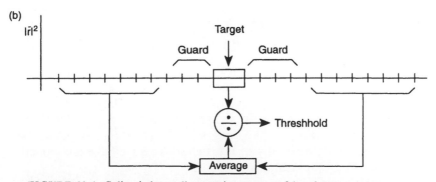

FIGURE 11.4 Split window cell averaging constant false alarm rate processor.

nient intermediate frequency, and the power in the resulting complex envelope is computed. The power waveform over each detection interval is compared with a threshold, set by the desired probability of false alarm P_f, and a target declared present if the power is above threshold. The signal is taken to be of the form

$$s(t) = A \cos(\omega_c t + \theta),$$

where θ is the uniformly distributed carrier phase. In Chapter 7 we analyzed the situation for various assumptions about the signal amplitude A. The noise is zero mean, Gaussian, and white.

If the signal has known amplitude A, in Sect. 7.2 we found that the receiver should be the quadrature receiver Eq. (7.31):

$$q^2 = (A^2/4) \left| \int_t^{t+T} \tilde{r}(t') \, dt' \right|^2$$

$$\cong (A^2 T/4) |\tilde{r}(t)|^2 > q_t^2.$$

The observation interval T in Fig. 11.4 is just the length of one range detection interval, which in Fig. 11.4b we show to be one narrow range bin, over which the received signal is approximately constant. In Eq. (7.52) we found that, for a specified P_f, the threshold should be

$$q_t^2 = 2\sigma^2 \ln(1/P_f), \tag{11.50}$$

with $\sigma^2 = N_0 T/4$ being the variances of the quadrature components of q in the case of noise only, Eq. (7.45). If we want to maintain P_f constant, from Eq. (11.50) it is clear that we must adjust the threshold in the detector to track any time variation of the detector output noise average power $2\sigma^2$.

In case the signal amplitude A is unknown, but modeled as a Rayleigh distributed random variable, in Sect. 7.5 we noted that the detector is still the quadratic matched filter, but taken with $A = 1$. The threshold is still that given by Eq. (11.50), so that again we need to track the noise variance. The noise variance enters as a multiplier on the threshold, in either case, so that the detector is conveniently implemented as

$$q^2/2\sigma^2 > q_t^2/2\sigma^2 = \ln(1/P_f).$$

This is the implementation shown in Fig. 11.4.

As a third case, we might assume the signal also to be a white Gaussian random process, with two-sided spectral density $S_0/2$. Then, as discussed at the end of Sect. 7.5, the detector is again the quadratic matched filter, with the observation time just the time of one correlation interval of the Gaussian signal. That is, in any of the signal cases we have mentioned, the scheme of Fig. 11.4 is applicable.

To be specific, we will consider a radar target detector. In Fig. 11.4b we show the various range (time) intervals in a radar received signal. In each range interval ("bin"), the detector is to make a decision whether a target is present or whether the return is due to background noise, either thermal or extraneous backscatter from terrain or sea surface ("clutter"). The strategy is to examine other range cells in the vicinity of the one of interest and to use their signals to make an estimate of the noise power $2\sigma^2$ in the cell of interest. The assumption is that there are no targets in the cells used for noise estimation and that the noise there well represents the noise in the cell of interest.

Figure 11.4b shows a particular realization of cell averaging, in which the average envelope noise power $2\sigma_n^2$ is estimated using some number L of range cells, of which $L/2$ are on either side of the range bin of interest. "Guard" intervals are placed between the target cell and the cells used for noise estimation in an attempt to keep the target return (if any) in the cell being tested from appearing also in the cells used for noise estimation. On the other hand, the length of the guard interval should be kept short, in order that the noise in the two halves of the estimation interval be nominally the same and nominally the same as that in the target cell.

Suppose then that the noise $n(t)$ in the intermediate frequency (IF) strip is zero-mean Gaussian, with envelope average power $2\sigma^2$ that we wish to estimate. Assume also that the noise is white within the IF bandwidth of the receiver and that the range sampling is at intervals long enough that the noise in the various bins is independent. The square law device in Fig. 11.4a produces the noise power, which is the squared magnitude of the complex envelope of the noise:

$$P(t) = |\bar{n}(t)|^2,$$

where

$$\bar{n}(t) = x(t) + jy(t),$$

and $x(t)$, $y(t)$ are the in-phase and quadrature components of the IF noise.

Since $2\sigma^2 = \sigma_x^2 + \sigma_y^2 = 2\sigma_n^2$ as in Eq. (3.82), we can compute $2\sigma^2$ as the sum of the maximum likelihood estimates of the variances of the quadrature components $x(t)$, $y(t)$. Since these are zero mean, as in Eq. (10.20), the maximum likelihood estimator of the variance $2\sigma^2$ is just

$$2\hat{\sigma}^2 = \frac{1}{L} \sum_{i=1}^{L} [x^2(t_i) + y^2(t_i)]$$

$$= \frac{1}{L} \sum_{i=1}^{L} P(t_i), \tag{11.51}$$

where $P(t_i)$ is the envelope signal $|\bar{r}(t_i)|^2$ in bins where we believe only noise is present. With the arrangement of Fig. 11.4a, the detector is actually

$$\lambda = q^2/2\hat{\sigma}^2 > \lambda_t. \tag{11.52}$$

For performance prediction, we are then interested in the statistics of the ratio λ, the estimated signal-to-noise ratio in the detection bin, rather than the statistics of q^2 itself.

Suppose the true noise variance is σ_n^2. Consider the case that the power in the detection bin is due to noise only:

$$q^2 = x_0^2 + y_0^2.$$

Then, as in Sect. 4.6, the density of the normalized variable q^2/σ_n^2 is chi-squared with two degrees of freedom. Furthermore, as in Sect. 4.7, the normalized variable $2L\hat{\sigma}^2/\sigma_n^2$, with $2\hat{\sigma}^2$ as in Eq. (11.51), has the chi-squared density with $2L$ degrees of freedom. Then by Sect. 4.11 the variable

$$F = q^2/2\hat{\sigma}^2 = \lambda \qquad (11.53)$$

has the Snedecor F-density. This assumes that, indeed, the power samples in the background estimate are independent of those in the signal cell power.

With the identification Eq. (11.53), we can write the false alarm probability of the detector with CFAR operating. The detector $\lambda > \lambda_t$ is the same as $F > F_t$ and the attained false alarm probability is, from Eq. (4.99),

$$P_f = P(F > F_t) = 1 - P(F < F_t) = I_U(L, 1). \qquad (11.54)$$

Here $I_U(m, n)$ is the incomplete beta function, defined in Eq. (4.95) and plotted in Fig. 4.15. The parameter U is

$$U = 1/(1 + F_t/L). \qquad (11.55)$$

The desired false alarm probability is now inserted as P_f in Eq. (11.54), and the necessary threshold is computed from Eq. (11.55). The inversion of the incomplete beta function needed to do that is discussed below.

We now want to examine the detection probability. We will consider the case of a Rayleigh fading signal, equivalent to the case of a Gaussian signal in the Gaussian noise background. Then nothing structural changes in the detection cell, since we are still examining the power of a Gaussian process (signal plus noise). However, the variance of the process increases. Instead of the noise-only variance $2\sigma_n^2$, we have the sum of this with the signal variance $2\sigma_s^2$. Therefore, the quantity with the F-density is now the ratio involving $q^2/(\sigma_s^2 + \sigma_n^2)$ and $2\hat{\sigma}^2/\sigma_n^2$. That is, we now have

$$F = [q^2/(\sigma_s^2 + \sigma_n^2)]/(2\hat{\sigma}^2/\sigma_n^2).$$

The detector is still

$$q^2/2\hat{\sigma}^2 > F_t,$$

but with Eq. (11.53) that now corresponds to

$$F = (q^2/2\hat{\sigma}^2)[\sigma_n^2/(\sigma_s^2 + \sigma_n^2)] > F_t\sigma_n^2/(\sigma_s^2 + \sigma_n^2). \qquad (11.56)$$

With Eq. (11.56), we can write the probability of threshold crossing, which is now the detection probability, as

$$P_d = P(F > F_t) = 1 - P(F < F_t) = I_U(L, 1), \quad (11.57)$$

just as in Eq. (11.54). Now, however, the index of the incomplete beta function is

$$U = [1 + F_t \sigma_n^2 / L(\sigma_s^2 + \sigma_n^2)]^{-1}$$
$$= [1 + F_t / L(1 + \alpha^2)]^{-1}, \quad (11.58)$$

where

$$\alpha^2 = \sigma_s^2 / \sigma_n^2 \quad (11.59)$$

is a signal-to-noise ratio.

In (Abramowitz and Stegun, 1965, 26.5.24) there is a convenient formula for computing the incomplete beta function which is involved in the preceding expressions. It is

$$I_U(L, K) = \sum_{i=0}^{K-1} \binom{L + K - 1}{i} (1 - U)^i U^{L+K-1-i}.$$

This is best calculated by removing the term U^{L+K-1} from under the sum and computing the binomial coefficients recursively as

$$\binom{n}{i+1} = \frac{n - i}{i + 1} \binom{n}{i}.$$

If we want to compute a ROC curve for the CFAR detector, parametrized by the SNR Eq. (11.59), we will need to compute the inverse of the beta distribution. An approximate formula for that purpose is given in (Abramowitz and Stegun, 1965, 26.2.22). If $I = I_U(L, K)$, then let

$$s = [-2 \ln(I)]^{1/2},$$

$$y = s - \frac{2.30753 + 0.27061s}{1 + 0.99229s + 0.04481s^2},$$

$$r = (y^2 - 3)/6,$$
$$2/h = 1/(2K - 1) + 1/(2L - 1),$$
$$w = (y/h)(r + h)^{1/2} - [1/(2K - 1) - 1/(2L - 1)]$$
$$\times (r + 5/6 - 2/3h),$$
$$1/U = 1 + (K/L) \exp(2w).$$

As applied to a ROC calculation, we first specify the size L of the averaging window we are willing to use, then specify a false alarm probability P_f. From Eq. (11.54) this yields an index U of the incomplete beta function, which is found using the above inversion formula. The parameter

U then determines the setting F_t of the threshold through Eq. (11.55). Now specify a SNR α^2 of interest. Then Eq. (11.58) gives the parameter U for use in the detection probability Eq. (11.57).

Unfortunately, the above inversion formula is not very accurate for small (<8, say) K or L. In our case discussed above we used $K = 1$. That is, the signal determination was based on only one range cell. Often, however, the range cell contains many more than one sampling time interval. In that case, the signal window might include a reasonably large number of sampling intervals, so that K larger than unity would be of interest. These developments are easily extended to $K > 1$. On the other hand, the approximate formula is more accurate for small values of P_f, so that the situation is often tolerable. Nonetheless, it is worth checking the inversion result by substituting into $I_U(L, K)$ in each case.

In the less useful case that the signal is modeled as having a known amplitude A, the false alarm calculations proceed exactly as above. However, now the detection statistic Eq. (11.53) in the case of signal plus noise has the noncentral F-distribution. In Sect. 4.11 we gave an expression for the distribution which, in the current notation, is

$$P(F < F_t) = P(F_t \mid 2K, 2L, \mu). \tag{11.60}$$

Here μ is the noncentrality parameter, which, for the case of one cell in the signal window, is just

$$\mu = A^2.$$

That this is A^2 and not $2A^2$ follows from Eq. (4.80) with $a(t) = A$.

Performance calculations in this case require values of the density Eq. (11.60). Although some tables exist, it is sometimes necessary to calculate needed values directly. A well-behaved iteration exists (Abramowitz and Stegun, 1965, 26.6.22 with corrections), as follows. Let x be given by

$$1/x = 1 + L/KF_t.$$

Define the values T_j iteratively as follows:

$$T_0 = 1,$$
$$T_1 = (K + L - 1 + \mu x/2)(1 - x)/x,$$
$$T_j = (1 - x)[2(K + L - j + \mu x/2)T_{j-1} + \mu(1 - x)T_{j-2}]/(2jx).$$

Then

$$P(F_t \mid 2K, 2L, \mu) = \exp\left[\frac{-\mu(1 - x)}{2}\right] x^{K+L-1} \sum_{j=0}^{L-1} T_j.$$

In addition, Robertson (1976) has carefully considered this calculation.

Exercises

11.1 Verify Eq. (11.3).

11.2 If **R** is a real symmetric matrix, show that

$$\mathbf{x}'\mathbf{R}\mathbf{x} = \mathbf{x}'\mathbf{D}\mathbf{x}$$

where **x** is a real column vector, and **D** is a triangular matrix with elements

$$d_{ij} = \begin{cases} r_{ii}, & i=j \\ 2r_{ij}, & i>j \\ 0, & i<j \end{cases}$$

11.3 Assume that **H** is a Hermitian matrix, and **y** is a complex column vector. Show that

$$\mathbf{y}'\mathbf{H}\mathbf{y} = \operatorname{Re} \mathbf{y}'\mathbf{D}\mathbf{y}$$

where the prime indicates conjugate transpose and **D** is a triangular matrix with elements

$$d_{ij} = \begin{cases} h_{ii}, & i=j \\ 2h_{ij}, & i>j \\ 0, & i<j \end{cases}$$

11.4 Invert the matrix

$$\begin{bmatrix} 1 & \rho & \rho^2 & \rho^3 \\ \rho & 1 & \rho & \rho^2 \\ \rho^2 & \rho & 1 & \rho \\ \rho^3 & \rho^2 & \rho & 1 \end{bmatrix}$$

by inverting only matrices of order two.

11.5 Invert the matrix

$$\begin{bmatrix} 1 & \rho & \rho^2 \\ \rho & 1 & \rho \\ \rho^2 & \rho & 1 \end{bmatrix}$$

by first factoring the matrix into a product of upper and lower triangular matrices.

11.6 Verify Eq. (11.35). Hint: Carry out the multiplication indicated on the right-hand side using partitioned matrices.

11.7 Consider the complex sample vector

$$\mathbf{z} = \begin{bmatrix} z_1 \\ z_2 \end{bmatrix} = \begin{bmatrix} x_1 + jy_1 \\ x_2 + jy_2 \end{bmatrix}$$

Assume that Eq. (11.34) holds. Show that

$$p(z_1) = \frac{1}{\pi \alpha_{11}} \exp\left\{ -\frac{|z_1|^2}{\alpha_{11}} \right\}$$

$$p(\mathbf{z}) = \frac{1}{\pi^2 \gamma} \exp\left\{ \frac{|z_1|^2 \alpha_{22} - 2 \operatorname{Re} z_2^* z_1 (\alpha_{12} - j\beta_{12}) + |z_2|^2 \alpha_{11}}{-\gamma} \right\}$$

where $\gamma = \alpha_{11}\alpha_{22} - (\alpha_{12}^2 + \beta_{12}^2)$.

11.8 (a) Denote the matrices of eigenvectors and eigenvalues of the symmetrized SNR $N^{-1/2}SN^{-1/2}$ by $\boldsymbol{\Phi}$ and Λ respectively. Show that the transformation $A = N^{1/2}\boldsymbol{\Phi}$ is such that $AA^T = N$, $A\Lambda A^T = \boldsymbol{\Phi}$.

   (b) Consider the detection of a zero-mean Gaussian signal vector in zero mean Gaussian noise. That is,

$$H_1: \quad \mathbf{r} = \mathbf{s} + \mathbf{n}$$
$$H_0: \quad \mathbf{r} = \mathbf{n}$$

Show that the likelihood ratio receiver Eq. (11.42) can be implemented by generating the test statistic

$$\lambda_T = \mathbf{r}'(A^{-1})^T[I - (\boldsymbol{\Phi} + I)^{-1}]A^{-1}\mathbf{r}$$

   (c) Show that the eigenvalues of $N^{-1/2}SN^{-1/2}$ are also the eigenvalues of SN^{-1} and that the eigenvector matrix of SN^{-1} is $N^{1/2}\boldsymbol{\Phi}$.

   (d) If the largest eigenvalue of $\boldsymbol{\Phi}$ is $\ll 1$, show that the test statistic is approximately

$$\lambda_T = \mathbf{r}^T N^{-1} S N^{-1} \mathbf{r}$$

   (e) Suppose the smallest eigenvalue of $\boldsymbol{\Phi}$ is $\gg 1$, show that the test statistic is approximately

$$\lambda_T = \mathbf{r}^T N^{-1} \mathbf{r}$$

11.9 Consider the space–time problem as discussed in Sect. 11.7 using the Fourier coefficients as samples. The signal and noise are assumed to be Gaussian. Consider a linear array of M equally spaced receiving elements as shown in Fig. 11.5. Assume the incident signal wavefront is known to be planar. Then the channel model is strictly one of pure delay so that at any frequency ω_k

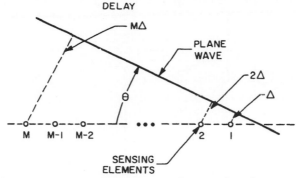

FIGURE 11.5   A linear array of M equally spaced sensors in a plane wave environment.

$$\tilde{\mathbf{H}}(k) = \begin{bmatrix} e^{-j\omega_k\Delta} \\ e^{-j2\omega_k\Delta} \\ \vdots \\ e^{-jM\omega_k\Delta} \end{bmatrix}$$

Assume that the noise in each channel has the same spectral density and is statistically independent of the noise in any other channel. Show that the optimum receiver may be implemented as shown in Fig. 11.3 with the filters $l_i^*(\omega)$ representing pure delay compensation to align the signals at each element, and that the filter $f(\omega)$ is given by

$$f(\omega) = \left\{ \frac{S_s(\omega)}{S_n(\omega)[S_n(\omega) + MS_s(\omega)]} \right\}^{1/2}$$

where $S_s(\omega)$ and $S_n(\omega)$ are the power spectral density of the signal and noise. (The compensation is the standard method employed in "phased array" antennas.)

11.10   As in the previous example, assume we have a linear array of M equally spaced receiving elements. We are interested in detecting a Gaussian signal $\mathbf{s}$ in the presence of a Gaussian interfering signal $\mathbf{m}$ and Gaussian noise $\mathbf{n}$. All are zero mean. Thus

$$H_1: \quad \mathbf{r} = \mathbf{Hs} + \mathbf{Wm} + \mathbf{n}$$
$$H_0: \quad \mathbf{r} = \mathbf{Wm} + \mathbf{n}$$

The vectors are defined by relations such as those given in Eq. (11.46). Assume that $\mathbf{H}$ and $\mathbf{W}$ correspond to plane waves (pure delay). Denote the incremental delays of each wave by Δ and δ respectively. Assume the noise $\mathbf{n}$ is of equal power at each receiving element and is statistically

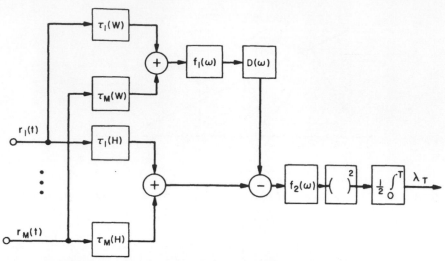

FIGURE 11.6   Signal detection in the presence of an interfering plane wave.

independent of the noise in any other element. Show that the optimum receiver can be implemented as shown in Fig. 11.6. Here the set $\tau_i(H)$ represents delay compensation to align the signal **s** at each element, and $\tau_i(W)$ represents the delay compensation to align the signal **m** at each element. (These represent "beams" steered to the signal and interfering noise respectively.) Also

$$f_1(\omega) = \frac{S_m(\omega)}{S_n(\omega) + MS_m(\omega)}$$

$$f_2(\omega) = \frac{1}{S_n(\omega)^{1/2}} \left[\frac{S_s(\omega)}{S_n(\omega) + \dfrac{S_m(\omega)S_s(\omega)|D(\omega)|^2}{S_n(\omega) + MS_m(\omega)}} \right]^{1/2}$$

$$D(\omega) = e^{j\omega(\Delta-\delta)} \frac{1 - e^{jM\omega(\Delta-\delta)}}{1 - e^{j\omega(\Delta-\delta)}}$$

It may be verified that $|D(\omega)|^2$ represents the power response of an array to a plane wave arriving at an angle θ_Δ, when the array is steered to an arrival angle θ_δ.

11.11   Design a space–time processor to detect the signal $\mathbf{r} = \mathbf{v} + \mathbf{n}$ where **v** and **n** are Gaussian vectors. Assume, as in Fig. 11.3, that there are M receiving elements. Assume that the noise at each element is of equal power, say $S_n(\omega_k)$, and is statistically independent of the noise at any other element. Furthermore, assume the same conditions for the signal

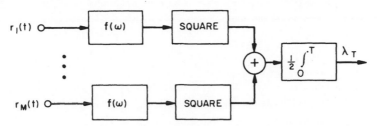

FIGURE 11.7 Receiver to detect independent signals assuming independent noise.

except for the level of the signal power, which we denote as $S_v(\omega_k)$. Show that the optimum receiver may be implemented as shown in Fig. 11.7 with

$$f(\omega) = \left\{ \frac{S_v(\omega)}{S_n(\omega)[S_v(\omega) + S_n(\omega)]} \right\}^{1/2}$$

$S_v(\omega)$ and $S_n(\omega)$ are the power spectral densities of the signal and noise, respectively. Thus the channels are incoherently combined after the quadratic detectors.

References

Abramowitz, M., and Stegun, I. A., Eds., "Handbook of Mathematical Functions with Formulas, Graphs, and Mathematical Tables." Dover, New York, 1965.

Albersheim, W. J., A closed-form approximation to Robertson's detection characteristics, *Proc. IEEE,* **69,** No. 7, 839 (July 1981).

Bendat, J. S., and Piersol, A. G., "Random Data; Analysis and Measurement Procedures," 2nd Ed. Wiley, New York, 1986.

Beyer, W. H., Ed., "Handbook of Tables for Probability and Statistics," 2nd ed., Chemical Rubber, Cleveland, Ohio, 1968.

Courant, R. and Hilbert, D., "Methods of Mathematical Physics," Vol. 1, Interscience, 1953.

Davenport, W. B., and Root, W. L. "An Introduction to the Theory of Random Signals and Noise." McGraw–Hill, New York, 1958.

DiFranco, J. V., and Rubin, W. L., "Radar Detection." Artech House, Dedham, MA, 1980.

Feller, W., "An Introduction to Probability Theory and its Applications," Vol. 2, Wiley, New York, 1966.

Gradshteyn, I. S. and Ryzhik, I. M., "Table of Integrals, Series, and Products," 4th Ed. Academic Press, New York, 1965.

Grimmett, G. R., and Stirzaker, D. R., "Probability and Random Processes," 2nd Ed. Oxford, New York, 1992.

Helstrom, C. W., "Statistical Theory of Signal Detection," 2nd Ed. Pergamon, Oxford, 1968.

Hodges, J. L., Jr., On the noncentral beta-distribution, *Ann. Math. Stat.,* **26,** 1955, 648–653.

Lehmann E. L., "Testing Statistical Hypotheses," 2nd Ed. Wadsworth, Belmont, CA, 1991.

Pachares, J., A table of bias levels useful in radar detection problems, *Trans. IRE,* **IT-4,** 38–45 (March 1958).

Proakis, J. G., "Digital Communications." McGraw-Hill, New York, 1983.

Robertson, G. H., Computation of the noncentral F distribution (CFAR) detection, *IEEE Trans. Aerospace Electron. Systems,* **AES-12,** No. 5, 568–571 (September 1976).

Robertson, G. H., Operating characteristics for a linear detector of CW signals in narrowband Gaussian noise, *Bell System Tech. J.,* **46,** No. 4, 755–774 (1967).

Sage, A. P., and Melsa, J. L., "Estimation Theory with Applications to Communications and Control," McGraw-Hill, New York, 1971.

Stuart, A., and Ord, J. K., "Kendall's Advanced Theory of Statistics," Vol. 1: "Distribution Theory," 5th Ed. Oxford, New York, 1987.

Stuart, A., and Ord, J. K., "Kendall's Advanced Theory of Statistics," Vol. 2: "Classical Inference and Relationship," 5th Ed. Oxford, New York, 1991.

Bibliography

Anderson, T. W., "An Introduction to Multivariate Statistical Analysis," 2nd Ed. Wiley, New York, 1984.

Barton, D. K., "Modern Radar System Analysis," Artech House, Norwood, MA, 1988.

Feller, W., "An Introduction to Probability Theory and Its Applications," Vol. 1, 3rd Ed. Wiley, New York, 1968.

Gagliardi, R., "Introduction to Communications Engineering." Wiley, New York, 1978.

Gelb, A., Ed., "Applied Optimal Estimation." MIT Press, Cambridge, MA, 1974.

Johnson, D. H. and Dudgeon, D. E., "Array Signal Processing," PTR Prentice-Hall, Englewood Cliffs, NJ, 1993.

Nathanson, F. E., "Radar Design Principles," 2nd Ed. McGraw-Hill, New York, 1991.

Oppenheim, A. V., and Schafer, R. W., "Discrete-Time Signal Processing." Prentice-Hall, Englewood Cliffs, NJ, 1989.

Papoulis, A., "Probability, Random Variables, and Stochastic Processes," 3rd Ed. McGraw-Hill, New York, 1991.

Picinbono, B., "Random Signals and Systems," Prentice-Hall, Englewood Cliffs, NJ, 1993.

Poor, H. V., "An Introduction to Signal Detection and Estimation," 2nd Ed., Springer, New York, 1994.

Scharf, L. L., "Statistical Signal Processing; Detection, Estimation, and Time Series Analysis." Addison-Wesley, Reading, MA, 1991.

Van Trees, H. L., "Detection, Estimation, and Modulation Theory," Part I: "Detection, Estimation, and Linear Modulation Theory." Wiley, New York, 1968.

Van Trees, H. L., "Detection, Estimation, and Modulation Theory," Part III: "Radar–Sonar Signal Processing and Gaussian Signals in Noise." Wiley, New York, 1971.

Viterbi, A. J., "Principles of Coherent Communication." McGraw-Hill, New York, 1966.

Viterbi, A. J., "Principles of Digital Communication and Coding." McGraw-Hill, New York, 1979.

Walpole, R. E. and Myers, R. M., "Probability and Statistics for Engineers and Scientists," 5th Ed., Macmillan, New York, 1993.

Historical

Gabor, D., Theory of communication; Part 1, The analysis of information, *J. Inst. Elec. Eng.*, **93** (part III), No. 93, 429–441 (November 1946). Also Parts 2, 3, pp. 442–457.

Lawson, J. L., and Uhlenbeck, G. E., "Threshold Signals." McGraw-Hill, New York, 1950.

Marcum, J. I., A statistical theory of target detection by pulsed radar, *IRE Trans. Information Theory*, **IT-6**, No. 2, 59–267 (April 1960).

Peterson, W. W., Birdsall, T. G., and Fox, W. C., The theory of signal detectability, *Trans. IRE Prof. Grp. Information Theory*, **PGIT-4**, 171–212 (September 1954).

Rice, S. O., Mathematical analysis of random noise, *Bell System Tech. J.*, **23**, No. 3, 282–332 (July 1944); **24**, No. 1, 46–156 (January 1945).

Swerling, P., Probability of detection for fluctuating targets, *IRE Trans. Information Theory*, **6**, No. 2, 269–308 (April 1960).

Woodward, P. M., "Probability and Information Theory, with Applications to Radar." Pergamon, New York, 1953.

Index

Printed and bound by CPI Group (UK) Ltd, Croydon, CR0 4YY

17/10/2024

01775563-0001